Techniques and Experiments for Organic Chemistry

SIXTH EDITION

Addison Ault
Cornell College

University Science Books
Sausalito, California

University Science Books
55D Gate Five Road
Sausalito, CA 94965
Fax: (415) 332-5393
www.uscibooks.com

Manuscript editor: Ann B. McGuire
Compositor: KP Company
Printer & Binder: Maple-Vail Book Manufacturing Group

This book is printed on acid-free paper.

Library of Congress Cataloging-in-Publication Data

Ault, Addison.
    Techniques and experiments for organic chemistry/ by Addison Ault. –
6th ed.
        p.   cm.
    Includes index.
    ISBN 978-0-935702-76-7
    1. Chemistry, Organic-Laboratory manuals.   I. Title.
    QD261.A94   1997
    547′ .0078-dc21   97-1814
                                                             CIP

Printed in the United States of America
10 9 8 7 6 5 4 3

Contents

PART I   Laboratory Operations

Preliminary Topics

v

Separation of Substances; Purification of Substances

Determination of Physical Properties

Determination of Chemical Properties: Qualitative Organic Analysis

Apparatus and Techniques for Chemical Reactions

PART II   Experiments

Isolations and Purifications

Transformations

Synthetic Sequences: Experiments that Use a Sequence of Reactions

Preface to the Sixth Edition

Like its previous five editions, this book is intended for use in the laboratory portion of an introductory course in organic chemistry. The organization of this sixth edition of *Techniques and Experiments for Organic Chemistry* is the same as that used in previous books: Part 1 describes lab techniques, and Part 2 presents experimental procedures.

This book will assist the organic lab instructor in several different ways. First, it tells the student how to perform most of the operations that are called for in introductory lab experiments. For example, it tells the student how to filter a hot solution, how to collect a solid product by suction filtration, how to recrystallize a solid, how to distill a liquid, how to perform an extraction, and how to dry a solution. Second, it provides a source of procedures or "recipes" for the preparation of a variety of organic compounds. The procedures illustrate many of the reactions that are described in the traditional lecture part of an introductory organic course, including substitutions, additions, eliminations, and "Name" reactions such as the Grignard reaction and the Wittig synthesis.

Another feature of this book that students will enjoy is its inclusion of more "interesting" compounds. Procedures are included for the preparation of thermochromic compounds, a chemiluminescent compound, a photochromic compound, aspirin, Tylenol, the alarm pheromone of the honey bee, dyes, and fragrances such as coconut aldehyde and oil of wintergreen.

Although some of the experiments in this book follow a "microscale" approach, this is not the book's primary emphasis. There is an important esthetic dimension to organic lab that risks being lost at the "microscale" level where students will miss the pleasure of watching 10 mL of a distillate collect or 5 grams of crystals form. There are many ways to reduce the cost of organic lab — such as minimizing waste—without requiring students to work with very small amounts of material.

New to this edition

The major change in this edition is the inclusion of detailed instructions for the disposal of waste. Chapter 4 on the Collection and Disposal of Waste treats this topic in a general way, and the presentation of every experiment includes specific directions for the safe disposal of all of the materials that are used or produced in that experiment.

New experiments include seven reactions of vanillin: acetylation, nitration, bromination, reduction to vanillyl alcohol, conversion to the oxime, conversion to the semicarbazide, and the aldol condensation with acetone. Vanillin is a pleasant and familiar starting material, and these transformations illustrate seven important organic reactions. Other new experiments include the isolation of ibuprofen from ibuprofen tablets, a microscale isolation of acetylsalicylic acid from aspirin, the isolation of trimyristin from nutmeg, and the nitration of phenacetin. Almost all of the experiments of the previous edition have been retained.

Acknowledgments

I appreciate the interest and encouragement of my colleagues, especially Truman, Cindy, and Jeff, and of my wife, Janet, and my children, Margaret, Warren, Tad, Peter, and Emily.

Preface to the Fifth Edition

This book is intended for use in the laboratory part of an introductory course in organic chemistry. The overall organization of this fifth edition is the same as that of the fourth edition. Part 1 contains general descriptions of the theory and practice of the most common laboratory techniques of organic chemistry; Part 1 also presents directions for a number of exercises that illustrate these techniques. Part 2 contains experiments that range from the purification of natural products to one-step transformations to multi-step syntheses. Again, as in the fourth edition, this book includes a number of variations, or alternative experiments, that call for a different starting material and that require the student to provide some of the details of the experimental procedure. By using these variations, the instructor can move away from the detailed recipe, or "cookbook," approach to laboratory work.

The first sections of Part 1 contain a discussion of laboratory safety, a description of the glassware used in the organic chemistry laboratory, advice on cleaning up, directions for writing up a laboratory notebook, and an introduction to the chemical literature. After these preliminary topics come discussions of procedures for the isolation and purification of organic substances and of techniques such as crystallization, distillation, extraction, and the chromatographic methods. These are followed by sections on physical methods for the identification and characterization of organic compounds. These methods include the determination of boiling point and melting point, the determination of properties such as density, index of refraction, and optical rotation, and the recording and interpretation of infrared and nuclear magnetic resonance spectra. Next come chemical methods of identification and characterization: qualitative tests for the elements, qualitative tests for functional groups, and procedures for the preparation of derivatives. (The physical properties of selected unknowns and the melting points of their derivatives are again listed in tables in the Appendix.) Finally, in Part 1, there are sections that describe the apparatus and techniques used in the laboratory operations of organic chemistry.

Part 2 first presents experiments that exemplify the separation and purification of substances, experiments such as the isolation of cholesterol from gallstones and of lactose from powdered milk, the recovery of (R)-(+)-limonene from grapefruit or orange peel, the isolation of the two enantiomeric forms of carvone (these enantiomers have different odors), and the resolution of *a*-phenylethylamine. Part 2 then continues with a variety of one-step transformations that illustrate the chemistry of the functional groups and concludes with a set of multi-step syntheses: the preparation of Δ^4-cholestene-3-one from cholesterol, the syntheses of tetraphenylcyclopentadienone, sulfanilamide, *p*-phenetidine, and 1-bromo-3-chloro-5-iodobenzene, and the preparation of a merocyanine dye.

The fifth edition retains all of the experiments of the fourth edition, and several new experiments have been added. These include the isolation of piperine from black pepper, the borohydride reduction of vanillin to vanillyl alcohol, the preparation of the analgesic *p*-ethoxyacetanilide from *p*-aminophenol, the synthesis of "coconut aldehyde," and a bootstrap synthesis—the preparation of two moles of *p*-phenetidine from one mole of *p*-phenetidine.

I believe that students find lab work more interesting when they understand why procedures work and can participate in the planning of experiments. I therefore emphasize explanations and provide for student participation in the planning of experiments by allowing students to choose one of several similar preparations of the same compound or to adapt a procedure for the preparation of one compound to the preparation of another, as in the variations.

PART 1

Laboratory Operations

➤ **Preliminary Topics**

➤ **Separation of Substances; Purification of Substances**

➤ **Determination of Physical Properties**

➤ **Determination of Chemical Properties; Qualitative Organic Analysis**

➤ **Apparatus and Techniques for Chemical Reactions**

Part 1 of this book describes many of the practical operations that students might perform in the organic chemistry laboratory. The first section of this part presents a discussion of laboratory safety, followed by sections that describe the glassware used in the organic laboratory, how to keep your glassware clean, how to properly dispose of waste, how to keep a laboratory notebook, how to find information in the chemical literature, and some helpful tables.

The remaining sections are presented in groups:

- methods of separation and purification

- methods for the determination of physical properties

- methods for the determination of chemical properties, and

- apparatus and techniques for carrying out organic chemical reactions.

The sections on the theory and techniques of separation and purification include discussions of filtration, crystallization, distillation, steam distillation, sublimation, chromatography, and removal of water (drying).

The sections on the determination of physical properties and the dependence of physical properties on molecular structure treat boiling point, melting point, density, index of refraction, optical rotation, molecular weight, and solubility characteristics, as well as the spectroscopic methods: infrared, ultraviolet-visible, nuclear magnetic resonance, and mass spectrometry.

The sections concerning the determination of chemical properties include qualitative tests for the elements, qualitative tests for functional groups, and reactions for the formation of derivatives.

The sections on apparatus and techniques for chemical reactions include methods for heating and cooling, for stirring, for adding reagents, for controlling gases that are evolved, for evaporating and concentrating, and for working up the reaction mixture and isolating the product.

Preliminary Topics

These first seven sections provide information and procedures that should be of use in all experiments in the organic laboratory. To avoid accidents, hazards must be recognized and certain precautions must be taken; the first section describes some of these dangers and precautions. The next section describes and illustrates some of the special items of glassware used in the organic laboratory. The third section tells how to keep your glassware clean, and the next section provides recommendations for the collection and disposal of waste. A fifth section describes the purpose of a laboratory notebook and recommends ways to keep such a notebook. The next section tells how to find certain kinds of information in the chemical literature, and the last section includes information about solutions of acids and bases, descriptions of several common reagents, and a periodic table of the elements with atomic weights.

1. Safety

Chemists must sometimes use dangerous materials, and therefore certain precautions must be regularly observed to minimize the probability and consequences of an accident. Although the number of hazardous materials called for by the procedures of this laboratory book has been held to a minimum, I have provided information about many other materials to provide a more comprehensive outline of the dangers most likely to be found in the organic laboratory, and for general information.

In my experience, the most important factors that ensure safety in the laboratory are knowledge of the hazards present and the quality of supervision provided by the laboratory instructors and assistants.

1.1 Fire

Know the location of the safety shower, the fire blanket, and the fire extinguishers. Never work alone in the laboratory.

The danger of fire is the most obvious hazard in the organic chemistry laboratory. This is an unavoidable result of the fact that most of the liquids used are relatively volatile and flammable (see Table 9.11-1) The danger and the consequences of a fire can be decreased merely by minimizing the number and size of the containers of flammable solvents that are stored in the laboratory.

The two most common reasons for a fire are

Watch your neighbor

- Boiling a flammable solvent with a flame and without a condenser, and
- Using a volatile and flammable solvent, especially during an extraction, without noticing that your neighbor's burner is on (or lighting your burner when your neighbor is using such a solvent).

Flammable liquids

If possible, then, all heating should be done with a steam bath, sand bath, hot plate, heating mantle, or electric immersion heater rather than a burner. If a burner is used for heating a flammable solvent, it is absolutely essential that a condenser be used; otherwise, the vapors that escape from the neck of the flask will flow down to the flame and ignite. When working with the most volatile and flammable solvents such as carbon disulfide, ether, petroleum ether, or pentane, you must extinguish all flames on the bench, because the vapors of these solvents can travel over the desktop and be ignited by a distant burner. Solvents should never be vaporized into the atmosphere of the laboratory but should be condensed and collected.

If the vapors from a flask do ignite, the fire can often be extinguished by

- Turning off the burner, or
- Gently placing a notebook or clipboard over the top of the vessel that contains the burning solvent.

If solvent that spilled on the desktop ignites,

- Move bottles and flasks of unspilled solvent away, if possible.

Fire extinguisher

- Use a carbon dioxide fire extinguisher on the fire. Discharge the extinguisher in this way:

1. Pull out the pin. You may need to break a wire or plastic strip.
2. Aim the cone at the base of the fire from a distance of a few feet.
3. Release carbon dioxide snow by squeezing the hand grip or turning the valve open.

Do not use a fire extinguisher with wild abandon, as a fire can easily be made worse by causing the blast from the nozzle to knock over and break bottles or flasks that contain more flammable materials.

If your clothing is burning, move under the nearest safety shower (the location of which you should know well enough that you can get there with your eyes shut) and pull the chain to turn the shower on.

If someone else's clothing has been set on fire, guide that person to the nearest safety shower and pull the chain. It is very easy for a person to panic under these conditions; the person must not be allowed to panic and run.

First aid for burns from flames, hot pieces of glass, hot iron rings, or hot plates consists only of the application of cold water. Never apply any kind of dressing. Call the doctor if the burn is more than a small blister.

Only water on burns

1.2 Explosions

Always wear eye protection in the laboratory, preferably shatterproof goggles.

An explosion is an exothermic reaction that accelerates in rate until it gets out of control and shatters its container. The severity of the blast depends on how much material is involved and how fast it all reacts.

Explosive mixtures are usually mixtures of oxidizing and reducing agents, as redox reactions are most likely to be highly exothermic. Gunpowder and many commercial explosives are mixtures of this type.

Explosive substances are (1) compounds that can undergo internal redox reactions, such as the polynitro compounds trinitrotoluene, picric acid (trinitrophenol), and nitroglycerine, or (2) compounds that can decompose to give very stable molecules. Such compounds include acetylene, nitrogen triiodide, diazonium salts, diazo compounds, peroxides, azides, and fulminates.

If you must work with a substance or mixture that is potentially explosive or is known to be explosive, it is best to work on as small a

Use a safety shield

scale as possible and behind a safety shield of shatterproof glass (hood fronts should be, but are not always, made of shatterproof glass). Just because a reaction has been run many times without incident by you or someone else does not mean that the potential danger has disappeared and that no explosion will occur the next time. New reactions should be run on a small scale, behind a safety shield, at least until their explosive potential has been estimated.

Placing a safety shield between yourself and the apparatus is always a good idea when carrying out a vacuum distillation. There is always the possibility that the flask will collapse under vacuum because of a crack or flaw in the glass. The safety shield will protect you from flying glass and the hot contents of the flask and oil bath.

Explosions are infrequent because most people know about potentially explosive systems and try to avoid them. Spectacular and tragic explosions continue to occur, however.

The three most common explosion hazards in the laboratory are

1. An exothermic reaction that gets out of control (explosion and fire).
2. Explosion of peroxide residues upon concentration of ethereal solutions to dryness, as explained in Section 1.7.
3. Explosion upon heating, drying, distillation, or shock of unstable compounds, which include diazonium salts, diazo compounds, peroxides, and polynitro compounds.

1.3 Poisoning

Know the location of a chart that indicates first aid measures for various types of poisoning.

Almost anything is harmful in large enough doses, but the chemical laboratory is an easier place than most in which to get a harmful dose of certain relatively dangerous materials.

Nothing by mouth

There is absolutely no reason to be poisoned by mouth, as nothing should be eaten or drunk in the laboratory and pipetting should be done with a suction bulb or pipetter, not by mouth. Always wash your hands after working in the laboratory—especially if you're a fingernail biter.

Certain harmful materials can be absorbed rather quickly through intact skin. These substances include dimethyl sulfate, nitrobenzene, aniline, phenol, and phenylhydrazine (see Section 1.7). Fatal doses of cyanide can be acquired through a cut in the skin.

Therefore, you should not work with cyanide salts or solutions if you have a cut on the hand.

When harmful or flammable gases are used (this group includes almost all gases except oxygen, nitrogen, helium, neon, and argon), they should be used in the hood. The harmful properties of several gases are described in Section 1.7.

The most suitable first aid treatment for poisoning depends on the nature of the poison. Consult a chart, which should be posted in the laboratory or the stockroom. Call the doctor.

1.4 Cuts

Most cuts occur during an attempt to force a thermometer or a glass tube into a stopper or an adapter. The thermometer or tube breaks, and a sharp end is driven into the palm of the hand or base of the fingers. Instead of forcing, you should lubricate the stopper or adapter with glycerine. The excess glycerine can be washed off with water. The use of thermometer adapters with ground-glass-jointed glassware has greatly reduced the frequency of this type of accident.

Don't force it!

Any time glass is forced, as in trying to unfreeze a joint, protect your hands by wrapping the glass in several thicknesses of a clean cloth towel.

First aid for a cut consists first of removing any large pieces of glass and then stopping the bleeding. For venous bleeding, the cut can be pinched together or pressure can be applied with a gauze pad or clean towel. Arterial bleeding is much more dangerous and must be controlled additionally by application of hand or thumb pressure to the appropriate pressure point. For the arms and hands, this point is where the pulse can be felt at the wrist or inside the upper arm just below the armpit.

Cuts should be treated by a medical doctor.

1.5 Spills

In general, spills on your skin or clothing should be treated by washing the affected part with plenty of water, either in the sink or under the safety shower. Section 1.7 describes how several particular spills should be treated. Refer to this section after you wash thoroughly, or while you wash, have someone look up the substance you have spilled. You should reread Section 1.7 before you work with any of the substances listed there.

Safety shower

Spills on the desk or floor should be treated as recommended in Section 1.7.

Use sodium bicarbonate

Both acids and bases should be neutralized with sodium bicarbonate. Never use strong acids or bases for this purpose.

1.6 Chemicals in the Eye

Always wear eye protection, preferably shatterproof goggles. Know the location of the eye-washing fountain.

If your eyes are protected by shatterproof goggles, you would have to be very unlucky indeed to get something in your eyes.

First aid for chemicals in the eyes is to wash the eyes thoroughly for several minutes, either by means of a special water fountain, which can direct a large but gentle flow of water into the eyes, or with the eyewash dispenser. If these cannot be found immediately, use a gentle stream from a hose attached to a water faucet or a beaker of water.

Remove contact lenses

Contact lenses must be removed to wash the eyes. Be alert to the possibility that a person with something in the eyes may not be able to see well enough to get to the eyewash fountain and may have to be led there by someone else.

1.7 A Short List of Hazardous Materials and Some of Their Properties

A discussion of the dangerous properties of many substances can be found in *Hazardous Chemicals Desk Reference,* 3rd edition, by R. J. Lewis, Sr., (New York: Van Nostrand Reinhold, 1993). If you plan to work with an unfamiliar substance, you should look it up in this book.

The purpose of the following list is to alert you to the greatest dangers presented by substances often found in the organic laboratory. The laboratory assistants and the instructor in charge should determine the dangerous properties of all substances present in the laboratory.

Abbreviations include OSHA: Occupational Safety and Health Administration; PEL: Permissible Exposure Limits (level to which industrial workers can be exposed during a normal 8-hour day, 40-hour work week without ill effects; TWA: Time Weighted Average; STEL: Short Term Exposure Limit; ppm: parts per million (molecules per million molecules of air); mg/cubic meter (milligrams per cubic meter of air).

- *Carcinogens:* Benzidine, α-naphthylamine, β-naphthylamine, certain polynuclear hydrocarbons.
- *Chemicals that can be rapidly absorbed in fatal doses through intact*

skin: Aniline, dimethyl sulfate, nitrobenzene, phenylhydrazine, phenol, 1,1,2,2-tetrachloroethane.

- *Explosives:* Polynitro compounds such as picric acid, trinitrotoluene, trinitrobenzene, 2,4-dinitrophenylhydrazine.

- *Lachrymators and vesicants:* Benzylic halides, allylic halides, α-halocarbonyl compounds (such as the phenacyl halides), dicyclohexylcarbodiimide, isocyanates.

- *Volatile and flammable solvents:* Pentane, petroleum ether, diethyl ether, and carbon disulfide present the greatest fire hazard of the common solvents (see Section 1.1). Carbon disulfide vapors can spontaneously ignite at steam bath temperatures.

This list contains some of the specific materials.

- *Acetic acid:* First aid for spills: dilute with large amounts of water.

- *Acetic anhydride:* Corrosive; will quickly blister the skin if not washed off. First aid for spills on the skin: wash off with water and finally with dilute ammonia solution.

- *Acetonitrile:* Poisonous on its own, but it can also produce hydrogen cyanide.

- *Acetyl chloride:* Reacts violently with water to produce HCl and acetic acid; see Acid chlorides.

- *Acid chlorides and other acid halides:* Acetyl chloride, benzoyl chloride, chloride, benzenesulfonyl chloride, *p*-toluenesulfonyl chloride: corrosive. First aid for spills on the skin: wash with water and finally with dilute ammonia solution.

- *Acids:* Corrosive. First aid for spills: dilute with large amount of water, then wash with sodium bicarbonate solution.

- *Aluminum chloride:* Corrosive; reacts violently with water to produce hydrogen chloride.

- *Ammonia:* First aid for inhalation: fresh air; inhalation of steam. First aid for ammonia in the eyes: irrigation with water for 15 minutes.

- *Aniline:* Fatal doses can be absorbed through intact skin.

- *Benzene:* according to the *Merck Index*, chronic toxicity involves bone marrow depression and aphasia; rarely, leukemia. OSHA PEL: TWA 1 ppm (1 ppm = 3 mg/cubic meter); STEL: 5 ppm; peak: 5 ppm.

- *Benzoyl peroxide:* Explosive; should not be heated for recrystallization or for melting-point determination; see Peroxides.

- *Bromine:* Exceedingly corrosive. Should be poured wearing gloves, face shield, and laboratory apron. Should be dispensed by means of a buret with a Teflon stopcock. First aid for spills on skin: wash instantly with water, rinse with ethanol, and rub in glycerine. First aid for inhalation: see Chlorine.

- *Carbon tetrachloride:* Poisoning by carbon tetrachloride can occur by inhalation, ingestion, or absorption through the skin. OSHA PEL: (Transitional): TWA 10 ppm (10 ppm = 63 mg/cubic meter); ceiling: 25 ppm; peak: 200 ppm for 5 minutes.

- *Chlorine:* Exceedingly corrosive. Should be handled in the hood. First aid for inhalation: fresh air; inhalation of vapors from a very dilute solution of ammonia. First aid for spills on skin: see Bromine.

CH_3Cl

- *Chloroform:* A human poison by ingestion and inhalation. OSHA PEL: (Transitional): TWA 2 ppm (2 ppm = 9.7 mg/cubic meter); ceiling: 50 ppm.

- *Chlorosulphonic acid:* Exceedingly corrosive, reacts with water with explosive violence to form sulfuric acid and hydrogen chloride. First aid for spills on skin or clothing: wash with very large amounts of water (safety shower).

- *Dichloromethane* (methylene chloride): Dichloromethane vapors are said to be narcotic in high concentrations. OSHA PEL: (Transitional): TWA 500 ppm (500 ppm = 1.7 grams/cubic meter); ceiling: 1,000 ppm; peak: 2,000 ppm for 5 minutes. Dichloromethane appears to be the least hazardous of the halogenated solvents.

- *Diethyl ether* (see also Ethers): OSHA PEL: TWA 400 ppm (400 ppm = 1.2 grams/cubic meter); STEL: 500 ppm.

$$H_3C\text{-}O\text{-}\underset{\underset{O}{\|}}{\overset{\overset{O}{\|}}{S}}\text{-}O\text{-}CH_3$$

- *Dimethyl sulfate:* Very dangerous; fatal doses can be absorbed quickly through intact skin. First aid for spills on skin: remove by washing with dilute ammonia solution; remove contaminated clothing.

DMSO

- *Dimethyl sulfoxide:* Not particularly harmful in itself, but is rapidly absorbed through intact skin and can apparently aid the absorption of other materials through the skin.

- *Ethers:* Ethyl ether, isopropyl ether, tetrahydrofuran, dioxane. In addition to being highly flammable, ethers can also absorb and react with oxygen upon storage to form dangerously explosive peroxides. Ether that has not been stored in full, airtight, amber bottles should be routinely discarded within two months. Ether of unknown vintage should be treated with the respect due to an equal amount of an exceedingly unstable high explosive; if crystals can be observed in the ether or if it appears to contain a viscous layer, the bottle should not even be touched but should be disposed of by explosives experts. Large amounts of ether should never be concentrated unless the absence of peroxides has been experimentally verified. Ethereal solutions should never be concentrated to dryness by heating with a flame. Isopropyl ether seems to be especially treacherous with respect to peroxide formation.

Test for presence of peroxides in ethers or hydrocarbons: Add 0.5–1 mL of the material to be tested to an equal volume of glacial acetic acid to which has been added about 100 mg of sodium or potassium iodide. A yellow color indicates a low concentration of peroxides, and a brown color a high concentration. A blank determination should be run.

Removal of peroxides from ethers: Stir or shake the ether with portions of a solution prepared by dissolving 60 grams of ferrous sulfate and 6 mL of concentrated sulfuric acid in 100 mL of water.

For information concerning the formation of peroxides, detection and estimation of peroxides, inhibition of peroxide formation, and removal of peroxides, see N. V. Steere, *Handbook of Laboratory Safety,* 2nd edition (Cleveland, Ohio: The Chemical Rubber Company, 1971), pp. 190–194, and H. L. Jackson et al., *J. Chem. Educ.,* **47,** A175 (1970).

- *Fuming nitric acid:* 95% nitric acid, containing oxides of nitrogen: extremely corrosive. First aid for spills: wash with large amounts of water and finally with sodium bicarbonate solution.

- *Fuming sulfuric acid:* Oleum; concentrated sulfuric acid containing dissolved sulfur trioxide: extremely corrosive. First aid for spills: wash with large amounts of water and finally with sodium bicarbonate solution.

- *Halogenated solvents:* Carbon tetrachloride, chloroform, etc. 1,1,2,2-Tetrachloroethane is the most dangerous, as it can be absorbed rapidly through the skin. Avoid breathing the vapors of these solvents.

- *Hydrazine:* Explosive; dangerous in combination with oxidizing agents.

- *Hydrides:* Lithium aluminum hydride, sodium hydride: react instantly and explosively with water, liberating and possibly igniting hydrogen. Calcium hydride is slightly less vigorous in its reaction with water. Contact with water must be scrupulously avoided.

Borohydrides react less vigorously with water but rapidly with acidic solutions to liberate hydrogen. Further information concerning the properties and procedures recommended for the safe handling of individual hydrides can be found in Reference 6.

- *Hydriodic acid:* First aid for spills: wash with large amounts of water and finally with dilute sodium bicarbonate solution.

- *Hydrobromic acid:* First aid for spills: wash with large amounts of water and finally with dilute sodium bicarbonate solution.

- *Hydrochloric acid:* First aid for spills: wash with large amounts of water and finally with dilute sodium bicarbonate solution.

- *Hydrogen bromide gas:* First aid for inhalation: fresh air; lie down and rest.

- *Hydrogen chloride gas:* First aid for inhalation: fresh air; lie down and rest.

- *Hydrogen peroxide:* Concentrated solutions can explode; powerful oxidizing agent; see Peroxides.

- *Hydrogen sulfide gas:* First aid for inhalation: fresh air; artificial respiration if necessary.

- *Lithium metal:* Reacts with water to produce hydrogen gas.

$LiAlH_4$

- *Lithium aluminum hydride:* Reacts instantly and explosively with water, liberating and possibly igniting hydrogen gas. Excess lithium aluminum hydride can be decomposed by dropwise addition of ethyl acetate. See Hydrides.

- *Methylene chloride:* see Dichloromethane.

- *Nitric acid:* Corrosive. First aid for spills: dilute with large amounts of water, then wash with sodium bicarbonate solution.

- *Nitrobenzene:* Fatal doses can be absorbed through intact skin.

- *Oleum:* See Fuming sulfuric acid.

- *Peracids:* Peracetic acid, trifluoroperacetic acid. Concentrated solutions can explode; powerful oxidizing agent; see Peroxides.

- *Peroxides:* All peroxides are potentially explosive, especially if heated. They are also powerful oxidizing agents, and mixtures with oxidizable materials are also potentially explosive. Ethers and other substances, such as tetralin, decalin, cumene, and certain other hydrocarbons, form peroxides when stored without exclusion of oxygen. Such liquids should not be distilled or concentrated without first determining that peroxides are not present; see Ethers, test for presence of peroxides.

- *Phenol:* Corrosive. Should not be handled with bare hands. Fatal doses can be absorbed through intact skin.

- *Phenylhydrazine:* Fatal doses can be absorbed through intact skin.

- *Phosphoric acid:* First aid for spills: wash with large amounts of water and finally with sodium bicarbonate solution.

- *Potassium cyanide:* Source of cyanide ion; fatal in small amounts; fatal doses can be ingested via a cut in the skin. Spills should be carefully cleaned up and disposed of immediately. Acidification of cyanide solutions will release deadly hydrogen cyanide gas.

- *Potassium hydroxide:* Caustic; corrosive solution. First aid for spills: wash with large amounts of water and finally with dilute bicarbonate solution.

- *Potassium metal:* Reacts instantly and explosively with water to form and ignite hydrogen gas. On storage without exclusion of oxygen, it forms exceedingly dangerous and explosive peroxides. If such material is possibly present, the sample should be disposed of by explosives experts. See Sodium metal.

- *Sodium amide:* Reacts violently with water. Aged samples have been reported to decompose explosively.

- *Sodium cyanide:* See Potassium cyanide.

- *Sodium hydride:* See Hydrides.

- *Sodium hydroxide:* Caustic; corrosive solution. First aid for spills: wash with large amounts of water and finally with dilute bicarbonate solution.

- *Sodium metal:* Reacts instantly and explosively with water to form and usually ignite hydrogen gas. Sodium and potassium metals are usually stored under xylene. If oxide-free metal is required, transfer a piece of the metal to a mortar containing xylene and cut off the oxide coating with a knife. To weigh the metal, remove a piece from the xylene, briefly blot it with a piece of filter paper and add it to a tared beaker of xylene. For further details and information concerning the handling of sodium and potassium metals, see Reference 5. Scrap sodium should be disposed of by adding little pieces to a large volume of methanol; scrap potassium can be disposed of similarly, using *tert*-butanol.

- *Sulfuric acid:* Corrosive. First aid for spills: dilute with large amounts of water, then wash with sodium bicarbonate solution.

- *1,1,2,2-Tetrachloroethane:* Fatal doses can be absorbed through intact skin.
- *Thionyl chloride:* Corrosive, volatile. First aid for spills: wash with large amounts of water and finally with dilute ammonia solution.

References

Sources from which you can obtain more information about laboratory safety in general, specific information about the storage and handling of dangerous substances, and hazardous properties of specific chemicals include

1. *Handbook of Laboratory Safety,* 2nd edition, N. V. Steere, editor, The Chemical Rubber Co., Cleveland, Ohio, 1971.
2. *Hazardous Chemicals Desk Reference,* 3rd edition, R. J. Lewis, Sr., New York: Van Nostrand Reinhold, 1993.
3. Wall-size chart: "Laboratory Safety," Frey Scientific Catalog Number F19818.
4. *Prudent Practices for Handling Hazardous Chemicals in Laboratories,* National Research Council, National Academy Press, Washington, D.C., 1981. This book recommends procedures for the safe handling of hazardous chemicals in the laboratory.
5. *Prudent Practices for Disposal of Chemicals from Laboratories,* National Research Council, National Academy Press, Washington, D.C., 1983. This book provides recommended practices and government regulations for managing chemicals for disposal.

Information about properties and methods of handling of many compounds can be found in

6. L. F. Fieser and M. Fieser, *Reagents for Organic Synthesis,* Wiley, New York; Volumes 1–17.

Information about the toxicity of many compounds can be found in

7. *The Merck Index,* 12th edition, S. Budavari, editor, Merck and Co., Inc., Rahway, New Jersey, 1996.
8. M. Sittig, *Handbook of Toxic and Hazardous Chemicals and Carcinogens,* 2nd edition, Noyes Publications, Park Ridge, New Jersey, 1985. As the preface puts it, "This handbook presents concise chemical, health

and safety information on nearly 600 toxic and hazardous chemicals, so that responsible decisions can be made by chemical manufacturers, safety equipment producers, toxicologists, industrial safety engineers, waste disposal operators, health care professionals, and the many others who may have contact with or interest in these chemicals due to their own or third party exposure." A useful book.

9. *Hazardous Chemicals Data Book,* G. Weiss, editor, Noyes Data Corporation, Park Ridge, New Jersey, 1980. "Instant information for decision making in emergency situations...." A compilation of information on 1,350 hazardous chemicals; raw information; no interpretation.

Information about methods of disposal of many compounds can be found in

10. *Catalog Handbook of Fine Chemicals,* Aldrich Chemical Co., Inc., Milwaukee, Wisconsin; 1994–1995.

Questions

1. Draw a plan of your laboratory that shows
   a. the location of the safety shower, the fire blanket, and the fire extinguishers.
   b. the location of the eyewash fountain.

2. Describe where the Poison Chart is located.

3. Provide a definition for each word.
   a. carcinogen
   b. lachrymator
   c. vesicant
   d. caustic
   e. corrosive

4. The OSHA standard for exposure to ethyl alcohol is 1,000 ppm.
   a. Express this standard in mg/cubic meter.
   b. How many mL of liquid ethyl alcohol, density 0.79 gm/mL, would have to be vaporized in a laboratory 10 meters by 14 meters by 5 meters to provide a concentration of vapor in the air of 1,000 ppm?

5. Both acids and bases can be neutralized by sodium bicarbonate. Explain. Illustrate your explanation with balanced equations.

2. Glassware Used in the Organic Chemistry Laboratory

Whereas reactions run in an industrial setting can involve hundreds or thousands of pounds of material, reactions in the organic laboratory are carried out on a far more modest scale. For many years normal preparation amounts were 10 to 50 grams of a substance, but more recently the trend has been to use smaller amounts of material. Thus, in the 1960s, procedures for the preparation of 1 to 10 grams of material were generally adopted (*semi micro scale*), and in the 1980s, procedures for the preparation of 0.1 to 0.5 grams of material were introduced *(micro scale)*.

The major advantages of using less material are the lower costs for obtaining the starting materials and disposing of the waste, the reduction in time for certain operations, and the reduction of the potential for a dangerous accident.

The disadvantages of experimentation on the micro scale are the difficulty in working with less than a few milliliters of a liquid, and the loss of the pleasure in seeing a distillate accumulate and watching lots of crystals form.

I therefore encourage experimentation on the semi-micro scale and suggest that you use the kind of glassware shown here for work with 1- to 10-gram amounts of material.

Flasks

In the laboratory, organic chemists usually use a flask with a round bottom and a narrow neck, or *boiling flask*, (Figure 2-1) to hold a reaction mixture that is to be heated or boiled. If the mixture is boiled, the flask is fitted with a *condenser* (Figure 2-2) so as to condense the vapors and return the liquid to the flask, as shown in Figure 31-1. Boiling flasks are also made with multiple necks; a *three-necked flask* is shown in (Figure 2-3).

Flasks with a flat bottom and a narrow neck, *Erlenmeyer flasks* (Figure 2-4) are used for reaction mixtures that will be allowed to stand for a period of time, or to hold a solution that is to be set aside for crystallization (Section 8). The narrow neck retards evaporation and can accept a cork, rubber, or glass stopper.

Beakers

Beakers (Fig 2-5) are used only as temporary containers or when a flask with a narrow neck would be inconvenient.

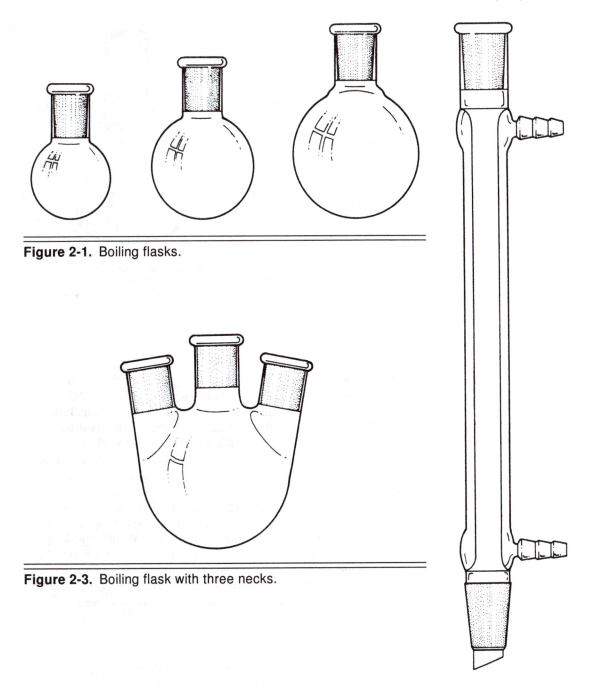

Figure 2-1. Boiling flasks.

Figure 2-3. Boiling flask with three necks.

Figure 2-2. Condenser.

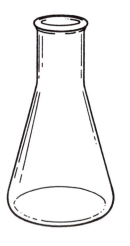

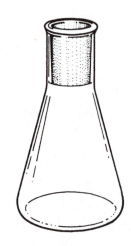

Erlenmeyer flask (plain neck)   Erlenmeyer flask (ground-glass-jointed neck)

Figure 2-4. Erlenmeyer flasks.

Adapters

Condensers and flasks can be connected together and fit with other items of equipment by means of various adapters (Figure 2-6). The *Claisen adapter* converts a single-neck flask to a two-necked flask, as shown in Figure 31-2, and the *distillation adapter* allows you to connect a condenser to a flask for a distillation, as illustrated in Figure 10.2-1.

A *vacuum adapter* is often used in a distillation to connect the receiving flask to the condenser, and the *thermometer adapter* is used to hold a thermometer in place in the distillation adapter during a distillation so as to measure the temperature of the hot vapors.

Flasks and adapters come in different sizes. Boiling flasks, for example, range in size from as small as 5 mL to 1,000 mL and larger, but flasks in the 10- to 50-mL range are the most commonly used for semi-micro scale procedures. The glass components in the figures are shown with ground glass joints by which they are connected to one another, but other types of connections are used as well.

Filtering funnels

Figure 2-5. Beaker.

Liquids and solids are often separated with a filtering funnel. When the mixture is hot, a conical *stemless funnel* (Figure 2-7) fit with a fluted filter paper is used. When the mixture is cold, you use a porcelain filtering funnel, either a *Büchner funnel* or a *Hirsch funnel,* with a *suction filtration flask* (Figure 2-8).

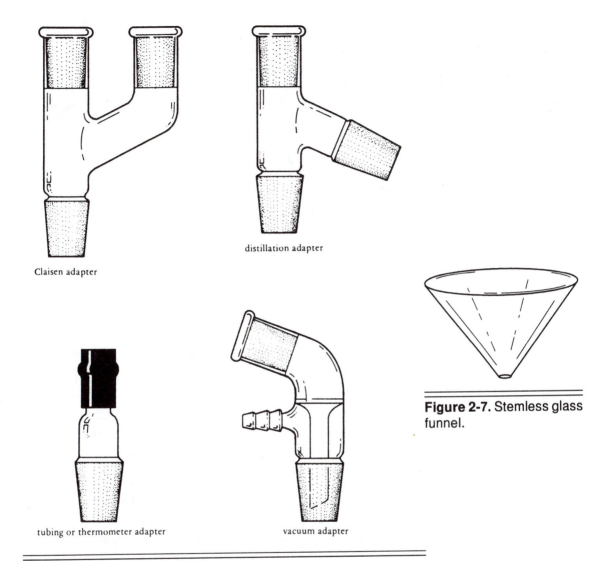

Claisen adapter

distillation adapter

tubing or thermometer adapter

vacuum adapter

Figure 2-6. Adapters.

Figure 2-7. Stemless glass funnel.

Separatory funnels

Immiscible liquids are often separated by means of a *separatory funnel* (Figure 2-9), an item of glassware that has a stopcock at the bottom for removal of the more dense liquid phase and a stoppered opening at the top for removal of the less dense liquid phase.

The separatory funnel can also be used as a dropping funnel to add a liquid to a reaction mixture, as shown in Figure 31-2.

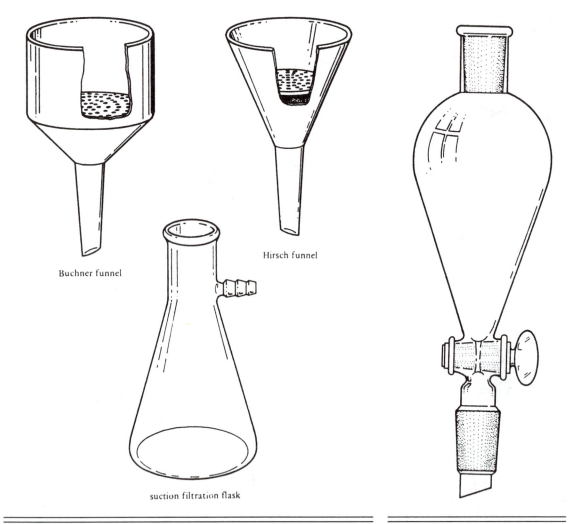

Buchner funnel

Hirsch funnel

suction filtration flask

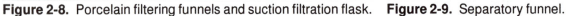

Figure 2-8. Porcelain filtering funnels and suction filtration flask.   **Figure 2-9.** Separatory funnel.

Other glassware

Other items of glassware commonly used in the organic laboratory include the *graduated cylinder*, used for measuring liquids (Figure 2-10), and the *drying tubes* (Figure 2-11), used to protect the apparatus from the moisture of the atmosphere, as illustrated in Figure E 37-1.

The style of glassware we have been describing is well suited for semi-micro scale preparations and for work on a larger scale. Micro-scale preparations, in contrast, often necessitate the use of different

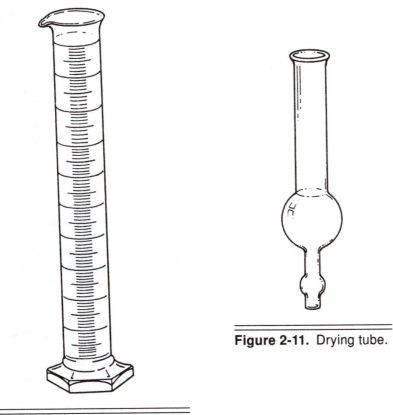

Figure 2-11. Drying tube.

Figure 2-10. Graduated cylinder.

techniques and, therefore, different apparatus. Fortunately, much microscale work can be done with quite inexpensive items, such as test tubes and Pasteur pipettes, but accurate measurement of materials necessitates the use of a balance that can read to 1 milligram for the weighing of solids, and the use of syringes or small pipettes for the transfer of liquids or solutions.

3. Cleaning Up

In chemistry, as in many things, cleaning up is a part of the job that seems to take a lot more time than it should. Some ways by which time and effort of cleaning can be minimized are described in this section.

One way to expedite cleanup is to distinguish glassware that is merely wet with a volatile solvent, glassware that can be cleaned simply by rinsing, and glassware that is too dirty to be cleaned by rinsing. Glassware that is just wet with a volatile solvent does not need to be cleaned, but simply set aside on a towel spread out on the desk top and allowed to drain and dry. Glassware that can be cleaned by rinsing should be rinsed as soon as possible either with water if it is wet from an aqueous solution, or with an organic solvent such as technical acetone if it is wet from an organic solution. Once rinsed, this glassware is wet from only a volatile solvent and need merely be set aside to drain and dry. You can save a lot of work if you avoid scrubbing with soap and water any glassware that does not need to be cleaned in this way.

Just soap, water, and acetone

Glassware that will not come clean merely by rinsing can almost always be cleaned by first adding an organic solvent such as acetone (or a solvent appropriate to the nature of the material in the flask) and allowing the flask to stand with occasional swirling to dissolve the bulk of the material. After repeating this procedure, if necessary, you can remove any stubborn residue by scrubbing with acetone and a brush, or soapy water and a brush, or both in succession. Rarely do you need to use anything stronger than soap, water, and acetone to clean glassware.

Much cleaning of glassware can be done while waiting for something to filter, cool, warm up, crystallize, distill, or react. Using odd moments to rinse, clean, and put away the glassware will save a lot of time at the end of the day. Also, the sooner something is cleaned after using, the easier it is to clean, as you should know from dishwashing experience. This is especially true of items that cannot be cleaned inside with a brush, such as pipettes or Büchner or Hirsch funnels. It should not be necessary to point out how shortsighted it is to put glassware away dirty to be cleaned before use!

3.1 Care of Ground-Glass-Jointed Glassware

Ground-glass-jointed glassware is very nice and most convenient, but it is expensive (about $5.00 per joint). All glassware can be broken; ground-glass-jointed glassware, however, can also become useless if **Avoid frozen joints** the pieces cannot be separated because the joints have become "frozen." The best way to prevent freezing is by disassembling the apparatus immediately after use, before the apparatus cools if it is hot, and by not lubricating the joints except when using strongly basic solutions.

If a lubricant has been used, the joints should be wiped as clean of grease as possible before washing so that the grease will not be

Work Smart

spread over the rest of the piece. Hydrocarbon-based greases can be removed by rinsing with dichloromethane, but silicone greases resist removal almost completely. Use dichloromethane or ether followed by soap and water. Glassware that has been used with silicone-based greases can easily be recognized because it is not wet by water; the water stands up in beads on the surface.

3.2 Separatory Funnels and Glassware with Stopcocks

Separatory funnels should be stored with the plug removed so that they cannot become frozen. Exceptions are separatory funnels with Teflon stopcocks or those whose plugs have been removed, cleaned, and regreased just before storage. When rinsing a separatory funnel with a glass stopcock plug, you should remove the plug and wipe the plug and barrel free of grease unless the apparatus has been used only briefly and the grease has not been leached out. The plug should be regreased (Teflon plugs need no lubricant) by applying a thin stripe of lubricant along the plug on the two sides without holes. After the plug is inserted into the barrel, the grease is spread out evenly by rotating the plug a few times in the barrel. Too much grease will cause the holes in the plug to become partially or completely stopped up. Other items with stopcocks, such as burets and distilling heads, should be treated similarly.

If a stopcock plug becomes frozen, pullers are available. These tools will either pull out the plug or break the piece in the attempt. While the jack screw of the puller is being tightened, the item should be wrapped in towels to contain the fragments if the piece breaks.

Frozen stopcocks or joints can sometimes be loosened by very judicious heating or tapping; these techniques are best learned by watching an expert.

3.3 Drying of Glassware

Glassware that has been rinsed with either water or acetone will dry upon storage. The outsides of all items and the insides of beakers can be dried with a towel. If you need to dry a piece of glassware such as a graduate or a flask quickly for immediate reuse, you can draw a stream of air through the inside by inserting to the bottom a length of glass tubing that is connected to the aspirator by a length of hose. Acetone or other volatile solvents will be removed from the insides of most items within a minute.

Use it wet!

A pipette can be connected to the aspirator and air drawn through the pipette by the aspirator. If the piece is wet from water, it should be rinsed with technical acetone to remove the water and then dried as just described. Obviously, something to be used with water or an aqueous solution will not normally need to be dried of water before use! It will often be fastest and most satisfactory to rinse the piece with a little of the solvent that is to be used next.

4.   Collection and Disposal of Waste

The proper collection and disposal of waste has become an important concern for all of us, and waste materials that come from the chemistry laboratory should be collected and then disposed of in a thoughtful manner.

Segregate waste!

In general, solid wastes and liquid wastes should be collected separately. Also, because it is more expensive to dispose of waste that contains covalently bound halogen, halogen-containing organic liquids and halogen-containing organic solids should be collected in containers separate from those used for nonhalogenated organic solids and liquids. Discarded or broken glass should be collected in containers reserved for glass.

Most aqueous solutions can be poured into the sink and rinsed down the drain with water. Exceptions would include aqueous solutions that contain heavy metals; these solutions should be allowed to evaporate in the hood, and the residues that remain should be disposed of in an approved manner. Small amounts of strong acids or bases can be diluted with water and then poured into the sink and rinsed down with more water. Larger amounts of acids or bases can be neutralized with sodium bicarbonate and the residue then either stored with inorganic solid waste or rinsed down the sink.

4.1 Solid Waste

Waste paper such as *filter paper* and *paper towels* should be placed in the containers reserved for waste paper.

Broken glass and *discarded Pasteur pipettes* should be placed in the container provided for waste glass.

Solid nonhalogenated organic compounds should be placed in the containers provided for solid nonhalogenated organic waste.

Solid halogen-containing organic compounds should be placed in the containers provided for solid halogenated organic waste.

Solid inorganic waste should be placed in the containers provided for solid inorganic waste..

4.2 Liquid Waste

Nonhalogenated organic solvents and *solutions that contain no halogenated compounds* should be poured into the waste containers provided for liquid organic waste.

Halogenated organic solvents and *solutions that contain halogenated compounds* should be poured into the waste containers provided for halogenated liquid organic waste.

4.3 General Instructions for Collection of Waste

After each experiment I present a section titled *Disposal* that lists suggestions for the collection and disposal of used or spilled material. This list of suggestions is presented here, along with examples of items that fall into each category.

Disposal

Sink: aqueous solutions of acids or bases; aqueous wash solutions

Solid waste: filter paper; paper towels

Nonhalogenated liquid organic waste: ether; acetone; methanol; ethanol

Halogenated liquid organic waste: dichloromethane; bromobenzene   Examples

Nonhalogenated solid organic waste: acetylsalicylic acid

Halogenated solid organic waste: *p*-bromoacetanilide

Solid inorganic waste: calcium chloride; magnesium sulfate

In some experiments, more specific suggestions are given for collection or disposal of a particular material. Local recommendations may differ from these, and your laboratory instructor may have additional suggestions.

The *Aldrich Catalog Handbook of Fine Chemicals* also suggests methods to be used for disposal of various substances.

5.   The Laboratory Notebook

The laboratory notebook serves two purposes. First, it is a place to record and keep information that should be available in the lab while you are doing an experiment. Second, it is a place to write down and preserve both the description of the experiment as it was actually done and the results of the experiment.

The plan   The information you need to have in the lab always includes a *description of the procedure* you plan to follow. When you are running a reaction, you need to know the properties of the materials with which you are working—properties such as molecular weight, melting and boiling points, density, and solubility behavior. At some time, a balanced equation must be written down, and the suitability of the molar and gram amounts called for in the procedure should be verified by reference to the balanced equation and the molecular weights. All of this information should be recorded in the lab notebook so that at some later time it will be possible to discover the source of any mistakes that may have been made.

What really happened   During the course of the experiment try to *record all relevant information about what was done and what happened.* The easiest way to do this is to state how the actual experimental conditions and results differed from the conditions and results anticipated. Sometimes, for example, it is difficult or unnecessary to duplicate exactly certain of the conditions called for (conditions such as time and temperature, if indeed they are precisely specified), but the actual conditions should be recorded in the notebook. Similarly, the behavior of the reaction may not be as anticipated, and the variations should be written down. An adequate notebook will enable another person to know exactly what you did and what happened, and your record should make it possible for your results to be verified. An adequate notebook is an essential part of good graduate school and industrial research.

Recommended format for the laboratory notebook

1. *Descriptive title and date.* This information will serve to identify the experiment that is the subject of the notebook entry. If the notebook is the record of a research project, the different attempts to carry out a particular procedure should be distinguished by a notation such as "Run One" or "Run Two" in the title. The record of the first experiment should start on page 3 or page 5 of the notebook so as to leave room for a table of contents at the beginning of the book.

2. *Balanced equation.* If you are planning to run a reaction, the equation will show at a glance exactly what you are trying to

do. It will also show the basis for the stoichiometric calculations.

3. *Molecular weights and molar, gram, and volume amounts of reagents, solvents, and products.* When an experiment is carefully planned, all of this information will be needed. Recording it in a well-organized and permanent way makes sense.

4. Relevant *physical properties of reagents, solvents, and products.* You often need to know the melting point, boiling point, density, and solubility properties of the material with which you are working. In a recrystallization, for example, you must know which solvents might freeze at the temperature of the ice bath. In a distillation, you need to know at what temperature the various components of the mixture can be expected to boil. In an extraction, you need to know which solvent should contain each substance and which solution should be the more or less dense. You must apply a certain amount of judgment as to what is relevant. For example, knowing the exact melting point or boiling point of solid sodium chloride or anhydrous magnesium sulfate should seldom be necessary. Similarly, whereas measuring liquids by volume makes sense, solids are usually measured by weight, and so the densities and volumes of solids are irrelevant.

Organize!

Items 3 and 4 can be combined, perhaps into a comprehensive table. The necessary information about the physical properties can usually be found in one of the handbooks or dictionaries (see Section 6.2).

5. A description of the *experimental procedure.* If you plan to follow a procedure from the lab manual, a reference to the page in the book where the procedure is given should be sufficient, since you will have the manual in the lab with you. If you are following a procedure from a handout or from the chemical literature, you can tape or rubber-cement a copy in your notebook; be sure to include a reference to the source of the procedure. Usually you are not allowed to bring a library book into the lab, where it might be damaged.

6. *Departures from a planned procedure.* Here, with reference to the planned procedure given in item 5, you tell what you actually did. Perhaps a slightly different amount was used than was called for. The actual time, temperature, and concentration should be recorded if they were different from those called for in the procedure or if they were unspecified. Here is where you admit that something was spilled or boiled over. Don't forget that if you have an extraordinary loss, the amounts of

materials used in the rest of the experiment must be reduced in proportion to the amount of the loss.

7. *Results, including percent yield.* The outcome of the experiment should be described. If a substance has been prepared, you should report the amount obtained, both in grams and as a percent of what could theoretically be expected on the basis of the balanced equation. For example, if you prepare acetylsalicylic acid (aspirin) from salicylic acid and acetic anhydride (Section E72), the balanced equation and the gram and molar amounts of starting materials and products are

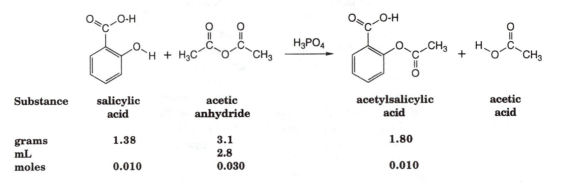

Substance	salicylic acid	acetic anhydride	acetylsalicylic acid	acetic acid
grams	1.38	3.1	1.80	
mL		2.8		
moles	0.010	0.030	0.010	

It should be clear that the *limiting reagent* in this case is salicylic acid and that the *maximum possible* or *theoretical* yield of acetylsalicylic acid is 0.010 mole, which, for a molecular weight of 180 g/mole, would correspond to 1.80 grams. If 1.20 grams was actually obtained, the actual yield would be 1.20/1.80, or 67% of theoretical.

If you have determined any physical or chemical properties, such as color, melting point, boiling point, index of refraction, density, or specific rotation, these data should be recorded along with theoretical or literature values. The sources of the literature values (lab manual, handbook, etc.) should also be recorded. If you determine the infrared spectrum of your product, you may wish to fasten the spectrum into your notebook. If you are working on a research project, you may prefer to file the spectra separately. When you do, you should make a notation in your book that a spectrum was obtained.

8. *Comments.* In case you or someone else might want to repeat the experiment, you should point out any difficulties you encountered and write any suggestions for changes in the procedure.

Because a good laboratory notebook will show all original data (such as the weight of the paper on which a substance was weighed and the weight of the paper plus the sample) and will indicate all calculations (such as the difference between the two weights just mentioned), the notebook tends to get messy. Yet the notebook will be the only record of your lab work, and you should want it to be intelligible and an example of your best efforts. One way out of this dilemma is to use the left-hand pages for recording data and calculations in a preliminary form, if desired, and the right-hand pages for the final entries. Thus, if the product of a reaction were being weighed as just described, the two weights and the subtraction could be recorded on the left-hand page, and the difference (the weight of the sample) would be recopied in the appropriate place on the right-hand page. Figure 5-1 illustrates a sample page from a laboratory notebook.

A Compromise

Theoretical yield, percent yield, and typical yield

In the example just presented, the *theoretical yield,* or maximum yield from 0.010 mole of salicylic acid that is theoretically possible assuming complete reaction and no losses during isolation and purification, is 0.010 mole or 1.80 grams of acetylsalicylic acid. The *actual yield* will always be less than the theoretical yield, and so the actual yield will be less than 100% of the theoretical yield, and this actual yield should be reported as a percent of the theoretical yield, or *percent yield.* In this example, the actual yield was 1.20 grams, or 1.2/1.80 = 0.67 = 67% of theoretical.

Sometimes a "*typical yield*" will be given for an experiment. Suppose, for example, we said that a "typical yield" of acetylsalicylic acid in this experiment is 1.30 grams or 1.30/1.80 = 0.72 = 72%. The actual yield of 1.20 grams is still 67% of theoretical. The percent yield is a function only of the theoretical yield. The "typical yield" is merely a guide for expectations; sometimes it is the yield that an experienced chemist could expect who has done the experiment several times.

Questions

1. Look up in a handbook the molecular weight, boiling point, and density of acetic anhydride.
2. Look up in a handbook the molecular weight and melting point of salicylic acid.
3. Phenacetin can be made by treating *p*-phenetidine with acetic anhydride according to the procedure described in Section E74. Assume that the experiment calls for 1.4 grams of *p*-phenetidine and 1.1 mL of acetic anhydride.

Cyclohexene   from   Cyclohexanol

In this experiment we will convert cyclohexanol to cyclohexene by heating the alcohol with 85% H_3PO_4 :

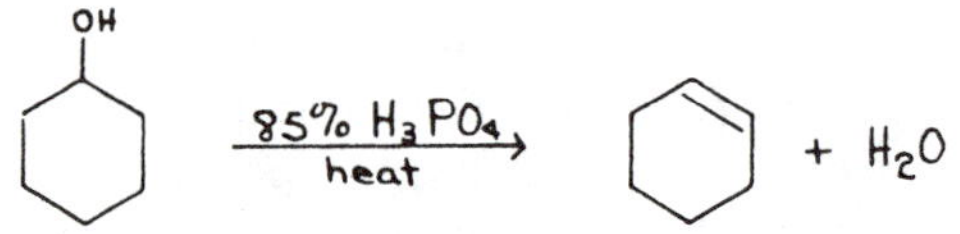

0.20 mole   0.20 mole

20.0 grams   16.4 grams

The product will be isolated by distillation.

	Amount							solubility	
Substance	moles	grams	ml	M.W.	density	m.p.	b.p.	H_2O	organics
cyclohexanol	0.20	20.0	21	100.2	0.96	25	161	no	yes
85% H_3PO_4			5	98.0		liquid	dec.	yes	no
xylene			20	106.2	mixture of isomers			no	yes
anh. $MgSO_4$		~0.5		120.4		high		yes	no
cyclohexene	0.20	16.4	20.2	82.2	0.81	-131	83	no	yes

Procedure: Ault : Techniques and Experiments for Organic
          Chemistry

Departures from procedure: washed with 20 ml. of water
          instead of saturated NaCl solution.

Yield: I obtained 5.2 grams of product boiling between
          82° and 84° C. Theoretical yield of cyclohexene: 16.4 g
          Percent yield: (5.2/16.4)(100) = 32 %

Figure 5-1. Sample page from a laboratory notebook.

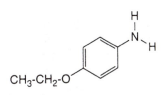

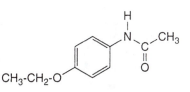

p-phenetidine
p-ethoxyacetanilide

phenacetin
p-ethoxyacetanilide

a. Write a balanced equation for the reaction.

b. Calculate the number of moles of *p*-phenetidine and of acetic anhydride that will be used.

c. Which of these substances will determine the maximum amount of product that will be produced? That is, which substance is the limiting reagent?

d. What is the maximum amount (moles and grams) of phenacetin that could be formed in the experiment?

e. If 1.6 grams of phenacetin is isolated after the reaction takes place, what is the percent yield of phenacetin?

6. The Chemical Literature

The results of chemical research are normally recorded in notes, communications, or papers in various chemical journals. If you wanted to look up the reported physical properties of a compound, methods of analysis for a compound, or methods of preparation or purification, you could search this primary literature through the index volumes of the various journals. Because there are several dozen major journals, and thousands of journals altogether, the time and effort required would be enormous.

To cope with this problem, abstracting journals have been established. Abstracting journals publish a brief summary, or abstract, of a paper and then provide detailed indexes to the abstracts. Today, *Chemical Abstracts* is the most important abstracting journal in the field of chemistry. *Chemical Abstracts* was started in 1907 by the American Chemical Society, and in 1984—to take a single year—it published abstracts of more than 443,000 papers, reports, and patents. *Chemical Abstracts* currently produces five principal indexes: Author,

General Subject, Chemical Substance, Formula, and Numerical Patent. An index entry refers you to the abstract, and if the abstract indicates that the desired information is in the original article or patent, you then look up the original by means of the reference provided with the abstract.

While using *Chemical Abstracts* is an improvement over looking things up in the indexes of individual journals, it is still a lot of work if all you want to know is the melting point of a compound or what would be a good solvent for recrystallization. For this reason, a great

Secondary sources

many specialized secondary sources have been developed. The most familiar is probably the *Handbook of Chemistry and Physics*, published by the Chemical Rubber Company, or *Lange's Handbook of Chemistry*. These and other secondary sources make certain kinds of information easy to obtain. For example, there are collections of methods of synthesis; methods of analysis; physical properties such as melting point, boiling point, solubility, vapor pressure, heat of combustion, and infrared, ultraviolet, NMR, and mass spectra; physiological properties; and hazardous properties. The key to success in quickly finding answers to specific questions about a compound is to be familiar with the secondary sources for that kind of information.

Sections 6.1 and 6.2 describe the most commonly used secondary sources for information about physical properties and methods of preparation of organic compounds, and Section 6.3 lists several collections of infrared, ultraviolet, NMR, and mass spectra. A few of the other books that organic chemists find useful are listed in Section 6.4.

The three articles by J. E. H. Hancock provide a more complete introduction to the literature of organic chemistry (Reference 1).

6.1 Secondary Sources for Physical Properties of Organic Compounds

For finding such physical properties as molecular weight, melting point, density, index of refraction, color, and solubility, the following handbooks are very convenient.

- *Handbook of Chemistry and Physics,* 75th edition, D. R. Lide, editor-in-chief; CRC Press, Inc., Boca Raton, Florida, 1994. Contains physical properties for about 12,000 organic compounds in addition to much other information.

- *Lange's Handbook of Chemistry,* 14th edition, J. A. Dean, editor, McGraw-Hill, New York, 1992. Contains physical properties for about 6,500 organic compounds, in addition to much other information. A reference to Beilstein is given for each compound. Locating the entry for a compound is much easier in the *Lange's Handbook* than in the *Handbook of Chemistry and Physics.*

* *Aldrich Catalog Handbook of Fine Chemicals,* Aldrich Chemical Co., Inc., Milwaukee, Wisconsin, 1994–1995. In addition to molecular weight and selected physical properties, this book gives methods for disposal, references to Beilstein and the *Merck Index,* and references to infrared and NMR spectra published in various "libraries" of infrared and NMR spectra that are published by the Aldrich Chemical Company (see Section 6.3).

* *Handbook of Tables for Identification of Organic Compounds,* 3rd edition, The Chemical Rubber Co., Cleveland, Ohio, 1967. Contains melting point and boiling point data, and melting points of derivatives for over 8,000 organic compounds. The organization is by functional group and by increasing melting point or boiling point within each functional group.

Three other useful secondary sources are

* *Dictionary of Organic Compounds,* J. Buckingham, executive editor, 5th edition, Volumes 1-7, Chapman and Hall, New York, 1982. This dictionary is an alphabetical listing of over 25,000 organic compounds. It contains valuable information about solvents used for recrystallization, as well as some reactions, derivatives, and literature references. Ten annual supplements have been published. Also available on CD-ROM.

* *The Merck Index,* 12th edition, S. Budavari, editor, Merck and Co., Inc., Rahway, New Jersey, 1996. This index contains information about solubility, purification, and hazardous properties as well as medicinal uses for 10,100 compounds. The *Chemical Abstracts* index name is provided for most compounds.

* *Beilsteins Handbuch Der Organischen Chemie,* 4th edition, Springer-Verlag, Berlin, 1918–present. This German language work is the most complete secondary source for information about the properties, preparation, and reactions of organic compounds. For all information, references are given to the primary literature sources. The fourth edition, in 31 volumes (Bande), was published during 1918–1938. This principal edition (Hauptwerk; H) covers the organic chemical literature from the beginning until 1909. Since 1938, there have been published a First Supplement (Erstes Erganzungswerk; E I) covering from 1910 to 1919 in an organization parallel to that

of the Hauptwerk, and a Second Supplement (Zweites Erganzungswerk; E II) similarly covering the period from 1920 to 1929. The Hauptwerk and the first two supplements list every organic compound known through 1929. A Third Supplement (Drittes Erganzungswerk; E III) covers the literature from 1930 through 1949, with many more recent references. The publication of a Fourth Supplement (Viertes Erganzungswerk) was begun in 1972. This supplement covers the chemical literature from 1950 through 1959. Starting with Volume 17, published in 1974, the Third and Fourth Supplements are being published together in common volumes that review the literature from 1930 through 1959.

The easiest way to find a compound in Beilstein is to first look it up in either *Lange's Handbook of Chemistry* or the *Aldrich Catalog Handbook of Fine Chemicals.* For example, *p*-bromoacetanilide is found in the Lange's Handbook under "Bromoacetanilide (p)"; the Beilstein reference is given as XII-642, which means that an entry for this compound is given on page 642 of Volume 12 of the Hauptwerk. In the *Aldrich Catalog Handbook* the reference given under "*p*-bromoacetanilide" is Beil. 12, 642, which provides the same information as the *Lange's Handbook.* The *Handbook of Chemistry and Physics* can also be used. Under the entry "Acetic acid, amide, *N*(4-bromophenyl)" the handbook gives the reference $B12^2$ 348, which means that an entry for *p*-bromoacetanilide can be found in Beilstein in Volume 12 of the Second Supplement on page 348. However, as the example indicates, it is sometimes difficult to find the entry in the *Handbook of Chemistry and Physics.*

Once the entry is found in either the Hauptwerk or any Supplement, the corresponding entry in any other series can be found easily through a system of double numbering of pages. In each volume of each Supplement, in addition to the ordinary page numbers, a cross-reference page number appears at the top center of each page. These numbers refer to the corresponding pages in the Hauptwerk. Thus, having found a compound in the Hauptwerk, you can find and search the corresponding pages of the same volume in each Supplement using these cross-reference page numbers. Similarly, after an entry is located in a volume of the supplement, the cross-reference page numbers can be used to search for corresponding entries in the Hauptwerk or other Supplements. Finally, direct cross-references are often given in the Second and Third Supplements. For example, at the beginning of the entry for *p*-bromoacetanilide in Volume 12, page 348, of the Second Supplement appears the notation (H 642; E I 319), which states that the corresponding entries appear (in Volume 12) on page 642 of the Hauptwerk and on page 319 of the First Supplement.

6.2 Secondary Sources for Methods of Preparation of Organic Compounds

Some of the most convenient sources of detailed directions for the preparation of specific compounds are the various laboratory manuals used in undergraduate courses in organic chemistry. Because these procedures are designed for use by students, a moderately competent organic chemist can usually make them work. The most useful book of this type is

- *Vogel's Textbook of Practical Organic Chemistry* , 5th edition, revised by B. S. Furniss, et al., Halstead Press (Wiley), New York, 1989.

There are two useful collections of specific procedures for the preparation of individual compounds. The first is a single volume, and the second is a continuing series.

- *Preparation of Organic Intermediates,* D. A. Shirley, Wiley, New York, 1951. A collection of more than 500 preparations of compounds not available commercially at the time of publication.
- *Organic Syntheses,* Wiley, New York. A continuing series of annual volumes, initiated in 1921. Each volume presents 30 to 35 detailed and tested preparations of compounds not available commercially at the time of publication. Eight Collective Volumes have been published, which include the first 69 annual volumes. In addition to a general index, the Collective Volumes contain Indexes for type of compound, type of reaction, molecular formula, apparatus, and author.

Two extensive reference works that describe general methods for the preparation of different classes of compounds are

- *Organic Reactions,* Wiley, New York. A continuing series of volumes, initiated in 1943. A typical volume presents several thorough discussions of a type of reaction or the methods of preparation of a type of compound. Exhaustive tables are included, which summarize the applications of each reaction. The information given includes reaction conditions, yields, and references to the literature.
- *Newer Methods of Preparative Organic Chemistry,* W. Foerst, editor, Academic Press, New York. A series of occasional volumes. Each volume is a collection of review articles that originally appeared in *Angewandte Chemie.* Each article discusses a particular type of reaction and includes some specific preparations and many references to the literature.

A more general reference work that contains information and references concerning the preparation of a great many compounds is the revision of the multivolume treatise originally edited by E. H. Rodd.

- *Rodd's Chemisty of Carbon Compounds,* 2nd edition, S. Coffey, editor, 4 volumes in 33 parts, including supplements, Elsevier, Amsterdam and New York, 1964–1982. A systematic presentation of the preparation and properties of organic compounds. The overall organization is by structural type.

Finally, the most comprehensive and most generally useful reference work for the preparation of organic compounds, as well as their properties and reactions, is Beilstein.

- *Beilstein's Handbuch Der Organischen Chemie,* 4th edition, Springer-Verlag, Berlin, 1918–present.

The use of Beilstein is described in Section 6.1.

6.3   Collections of Spectra

- *A Handy and Systematic Catalog of NMR Spectra,* A. Ault and M. R. Ault, University Science Books, Sausalito, California, 1980. A collection of 350 nuclear magnetic resonance spectra that includes, in addition to 290 proton NMR spectra, examples of fluorine and carbon NMR spectra.
- *The Aldrich Library of Infrared Spectra,* C. J. Pouchert, editor, Aldrich Chemical Co., Inc., Milwaukee, Wisconsin; 3rd edition (12,000 spectra), 1981.
- *The Aldrich Library of FT-IR, Spectra,* C. J. Pouchert, editor, Aldrich Chemical Co., Inc., Milwaukee, Wisconsin; 17,000 condensed- and vapor-phase FT-IR spectra, 1985.
- *High Resolution NMR Spectra Catalog,* Varian Associates, Palo Alto, California; Volume 1, 1962; Volume 2, 1963; 700 spectra.
- *The Aldrich Library of NMR Spectra,* 2nd ed, C. J. Pouchert, editor, Aldrich Chemical Co., Inc., Milwaukee, Wisconsin, 1983; two-volume set of more than 8,500 60-MHz proton NMR spectra.
- *The Aldrich Library of ^{13}C and ^{1}H FT-NMR Spectra,* 2nd ed, C. J. Pouchert and J. Behnke, editors, Aldrich Chemical Co., Inc.,

Milwaukee, Wisconsin, 1992; a three volume collection of 12,000 300-MHz proton and 75-MHz ^{13}C FT-NMR spectra arranged according to functionality.

- Sadtler *Standard Spectra,* Sadtler Research Laboratories, Inc., Philadelphia, Pennsylvania. The Sadtler standard spectra are large and continuing collections of infrared, ultraviolet, and NMR spectra. They are both in printed form and on microfilm. The largest collection, that of infrared spectra obtained with prism instruments, contains over 40,000 spectra. In addition, there are special collections available (pharmaceuticals, commonly abused drugs, etc.) and commercial special collections (agricultural chemicals, food additives, etc.).

6.4 Miscellaneous

Other useful books include

- L. F. Fieser and M. Fieser, *Reagents for Organic Synthesis,* Wiley, New York; Volumes 1 through 17. These volumes contain a wealth of interesting information about the availability, preparation, properties, and use of many substances in organic synthesis. The more recent volumes update and supplement the first volume.

- D. D. Perrin and W. L. F. Armarego, editors, *Purification of Laboratory Chemicals,* 3rd edition, Pergamon Press, New York. This book explains techniques of purification with specific methods for more than 4,000 chemicals and biochemicals.

- J. A. Riddick and W. B. Bunger, *Organic Solvents,* Volume II in Techniques of Chemistry, 3rd edition, A. Weissberger, editor, Wiley-Interscience, New York, 1970. This volume contains physical properties and purification methods for 354 solvents.

- *Hazardous Chemicals Desk Reference,* 3rd edition, by R. J. Lewis, Sr., Van Nostrand Reinhold, New York, 1993. This book provides information about 6,000 potentially hazardous compounds. The information includes physical properties and summaries of recommended exposure levels.

References

1. J. E. H. Hancock, "An Introduction to the Literature of Organic Chemistry," *J. Chem. Educ.* **45**, 193, 260, 336 (1968).

2. O. Weissbach, *The Beilstein Guide. A Manual for the Use of Beilsteins Handbuch der Organischen Chemie,* Springer-Verlag, New York, 1976.

3. *Searching the Chemical Literatue,* revised and enlarged edition, Advances in Chemistry Series #30, R. F. Gould, editor, American Chemical Society, Washington, D.C., 1961. A collection of papers on various topics concerning the chemical literature.

4. M. G. Mellon, *Chemical Publications,* 4th edition, McGraw-Hill, New York, 1965. An introduction to the nature and use of the chemical literature.

7. Tables

This chapter is a handy source of information about solutions of acids and bases, and about some common liquid and solid reagents.

Useful information

Tables 7-1 and 7-2 tell you the molar concentrations of the solutions of acids and bases that are most often found in the chemistry lab. You will find, for example, that "10% HCl" is 2.9 molar in HCl, a useful fact when your procedure calls for a certain number of moles of HCl. Table 7-1 and 7-2 also tell you how to prepare various dilute solutions of acids and bases from the concentrated forms that are commercially available.

Tables 7-3, 7-4, and 7-5 give you the mass and the volume of exactly one mole of several common acids, bases, and liquid reagents. If, for example, you need $\frac{1}{10}$ mole of something, you take $\frac{1}{10}$ of the indicated mass or volume.

Section 7.3 contains a periodic chart of the elements.

Table 7-1. Solutions of Acids

Solution	Density (grams/mL)	Concentration (moles/liter)[a]	To Make a Liter of Solution
95-98% H_2SO_4	1.84	18.1	Concentrated sulfuric acid
6 M H_2SO_4	1.34	6.0	332 mL conc. H_2SO_4 + 729 mL H_2O
3 M H_2SO_4	1.18	3.0	166 mL conc. H_2SO_4 + 875 mL H_2O
10% H_2SO_4	1.07	1.09	60 mL conc. H_2SO_4 + 957 mL H_2O
1 M H_2SO_4	1.06	1.0	55 mL conc. H_2SO_4 + 962 mL H_2O
69-71% HNO_3	1.42	15.7	Concentrated nitric acid
6 M HNO_3	1.19	6.0	382 mL conc. HNO_3 + 648 mL H_2O
3 M HNO_3	1.10	3.0	191 mL conc. HNO_3 + 831 mL H_2O
10% HNO_3	1.06	1.67	106 mL conc. HNO_3 + 905 mL H_2O
1 M HNO_3	1.03	1.0	64 mL conc. HNO_3 + 943 mL H_2O
36.6-38% HCl	1.18	12.0	Concentrated Hydrochloric acid
6 M HCl	1.10	6.0	500 mL conc. HCl + 510 mL H_2O
3 M HCl	1.05	3.0	240 mL conc. HCl + 756 mL H_2O
10% HCl	1.05	2.9	242 mL conc. HCl + 763 mL H_2O
1 M HCl	1.02	1.0	83 mL conc. HCl + 920 mL H_2O
99.7% CH_3COOH	1.05	17.5	Glacial acetic acid
6 M CH_3COOH	1.04	6.0	343 mL glacial acetic acid + 682 mL H_2O
3 M CH_3COOH	1.03	3.0	171 mL glacial acetic acid + 845 mL H_2O
10 % CH_3COOH	1.01	1.69	97 mL glacial acetic acid + 912 mL H_2O
1 M CH_3COOH	1.01	1.0	57 mL glacial acetic acid + 949 mL H_2O
48% HBr	1.50	8.9	Constant-boiling HBr
57% HI	1.70	7.6	Constant-boiling HI
85% H_3PO_4	1.70	14.7	Syrupy phosphoric acid
70% $HClO_4$	1.67	11.7	Concentrated perchloric acid

[a] Also mmole/mL.

Table 7-2. Solutions of Bases

Solution	Density (grams/mL)	Concentration (moles/liter)[a]	To Make a Liter of Solution
50% NaOH	1.53	19.1	Concentrated NaOH solution
6 M NaOH	1.21	6.0	314 mL 50% NaOH + 734 mL H_2O
3 M NaOH	1.11	3.0	157 mL 50% NaOH + 873 mL H_2O
10% NaOH	1.11	2.77	145 mL 50% NaOH + 889 mL H_2O
1 M NaOH	1.04	1.0	52 mL 50% NaOH + 963 mL H_2O
45% KOH	1.45	11.7	Concentrated KOH solution
6 M KOH	1.26	6.0	512 mL 45% KOH + 512 mL H_2O
3 M KOH	1.14	3.0	256 mL 45% KOH + 764 mL H_2O
10% KOH	1.09	1.95	167 mL 45% KOH + 850 mL H_2O
1 M KOH	1.05	1.0	85 mL 45% KOH + 927 mL H_2O
28-30% NH_3	0.90	15.0	Concentrated ammonium hydroxide
6 M NH_3	0.95	6.0	400 mL conc. NH_4OH + 590 mL H_2O
10% NH_3	0.96	5.63	375 mL conc. NH_4OH + 620 mL H_2O
3 M NH_3	0.98	3.0	200 mL conc. NH_4OH + 797 mL H_2O
1 M NH_3	0.99	1.0	67 mL conc. NH_4OH + 933 mL H_2O
10% K_2CO_3	1.09	0.79	109 g anh. K_2CO_3 + 982 mL H_2O
5% K_2CO_3	1.04	0.38	52.2 g anh. K_2CO_3 + 992 mL H_2O
10% Na_2CO_3	1.10	1.04	52.5 g anh. Na_2CO_3 + 998 mL H_2O
5% Na_2CO_3	1.05	0.50	110.3 g anh. Na_2CO_3 + 993 mL H_2O
Saturated $NaHCO_3$	1.06	1.0	85 g $NaHCO_3$ + 973 mL H_2O
5% $NaHCO_3$	1.04	0.62	52.8 g $NaHCO_3$ + 983 mL H_2O

[a] Also mmole/mL.

Table 7.3. Molar Weights and Molar Volumes of Acids

Compound	Molecular Weight	Molar Volume, mL
H_2SO_4, 98%	98.1	55
HNO_3, 70%	63.0	64
HCl, 37%	36.5	83
CH_3COOH	60.0	57
HBr, 48%	80.9	51
HI, 57%	127.9	132
H_3PO_4, 85%	98.0	68
$HClO_4$, 70%	100.5	86

Table 7-4. Molar Weights of Bases

Compound	Molecular Weight
NaOH	40.0
KOH	56.1
NH_3	17.0
K_2CO_3	138.2
Na_2CO_3	106.0
$NaHCO_3$	84.0
$NaC_2H_3O_2$	82.0
$NaC_2H_3O_2 \bullet 3H_2O$	136.1

Table 7-5. Molar Weights, Densities, and Molar Volumes of Selected Liquid Reagents

Compound	Molecular Weight	Density	Molar Volume, mL
Acetic anhydride	102	1.08	94
Ammonium hydroxide, 28%	17	0.90	67
Aniline	93	1.02	91
Bromine	60	3.12	51
Hydrazine, 64%	32	1.04	48
Phosphorus oxychloride, $POCl_3$	153	1.68	91
Phosphorus trichloride, PCl_3	137	1.58	87
Pyridine	79	0.98	81
Sodium hydroxide, 50%	40	1.53	52
Thionyl chloride, $SOCl_2$	119	1.66	72

Table 7-6. Periodic Table of the Elements

METALS — TRANSITION METALS — NONMETALS

PERIODS	IA	IIA	IIIB	IVB	VB	VIB	VIIB	VIII	VIII	VIII	IB	IIB	IIIA	IVA	VA	VIA	VIIA	O
1	1.0079 H 1																	4.00260 He 2
2	6.94 Li 3	9.01218 Be 4											10.81 B 5	12.011 C 6	14.0067 N 7	15.9994 O 8	18.9984 F 9	20.179 Ne 10
3	22.9898 Na 11	24.305 Mg 12											26.9815 Al 13	28.086 Si 14	30.9738 P 15	32.06 S 16	35.453 Cl 17	39.948 Ar 18
4	39.098 K 19	40.08 Ca 20	44.9559 Sc 21	47.90 Ti 22	50.9414 V 23	51.996 Cr 24	54.9380 Mn 25	55.847 Fe 26	58.9332 Co 27	58.71 Ni 28	63.546 Cu 29	65.38 Zn 30	69.72 Ga 31	72.59 Ge 32	74.9216 As 33	78.96 Se 34	79.904 Br 35	83.80 Kr 36
5	85.4678 Rb 37	87.62 Sr 38	88.9059 Y 39	91.22 Zr 40	92.9064 Nb 41	95.94 Mo 42	98.9062 Te 43	101.07 Ru 44	102.9055 Rh 45	106.4 Pd 46	107.868 Ag 47	112.40 Cd 48	114.82 In 49	118.69 Sn 50	121.75 Sb 51	127.60 Te 52	126.9046 I 53	131.30 Xe 54
6	132.9054 Cs 55	137.34 Ba 56	57–71 *	178.49 Hf 72	180.9479 Ta 73	183.85 W 74	186.2 Re 75	190.2 Os 76	192.22 Ir 77	195.09 Pt 78	196.9665 Au 79	200.59 Hg 80	204.37 Tl 81	207.2 Pb 82	208.9804 Bi 83	(210) Po 84	(210) At 85	(222) Rn 86
7	(223) Fr 87	(226.0254) Ra 88	89–103 †	104	105	106	107	108										

* LANTHANIDE SERIES

138.9055 La 57	140.12 Ce 58	140.9077 Pr 59	144.24 Nd 60	(145) Pm 61	150.4 Sm 62	151.96 Eu 63	157.25 Gd 64	158.9254 Tb 65	162.50 Dy 66	164.9304 Ho 67	167.26 Er 68	168.9342 Tm 69	173.04 Yb 70	174.97 Lu 71

† ACTINIDE SERIES

(227) Ac 89	232.0381 Th 90	231.0359 Pa 91	238.029 U 92	237.0482 Np 93	(242) Pu 94	(243) Am 95	(245) Cm 96	(245) Bk 97	(248) Cf 98	(253) Es 99	(254) Fm 100	(256) Md 101	(253) No 102	(257) Lr 103

Separation of Substances; Purification of Substances

A pure substance contains only one kind of molecule; an impure substance is a mixture of molecules. Thus to *purify* an impure substance is to *separate* the desired molecules from those that are not wanted; purification is separation.

The differences in the properties of different molecules are the basis for various separation, or purification, procedures, and I present the theory of each of the separtion procedures described in the next sections from the point of view of the different behavior that can be expected from different molecules under the experimental conditions.

The idea that purification procedures must be separation procedures implies that the ultimate experimental criterion for purity is that a substance is inseparable by all known separation procedures. Thus, the possibility always remains that a substance currently believed to be pure may someday be shown to be separable into components. This possibility is appreciated especially well by biochemists, who have often found that a new separation procedure shows that a substance considered pure is, in fact, a mixture. Separation of many biochemical materials is very difficult because often the molecules to be separated behave in nearly the same way in almost all separation procedures.

In practice, the question of the purity of a particular substance is usually settled indirectly. That is, the properties of a sample of the substance in question are compared with the properties of a sample judged to be pure because it could not be separated into components. If the two samples have the same properties, they may be judged to be of the same purity—both pure or, sometimes, both impure. Some of the properties by which substances can be characterized, such as boiling point, index of refraction, and infrared spectrum, and the ways by which they can be determined, are discussed in Sections 17 through 26. If the comparison shows that the samples are different, at least one sample is not pure.

Usually, however, the comparison with a reference sample is indirect; you compare your experimental values with values given in the chemical literature. Thus, you might compare your boiling point and index of refraction with the literature values and your infrared spectrum with a reference spectrum. In an indirect comparison, you would be more tolerant of differences between your experimental values and the literature values, because it is unlikely that the experimental conditions were exactly the same. Factors such as different apparatus (thermometer, refractometer, spectrometer) and different experimenters could easily account for small discrepancies; the samples could be identical.

It should now be apparent that the question "Is this sample pure?" can never be answered with an unqualified Yes. The best you can say is that it appears to be homogeneous and that its properties are the same as those of a sample believed to be pure.

8. Filtration

Filtration involves the separation of insoluble solid materials from a liquid or a solution. In this operation, the liquid or solution passes through a porous barrier (filter paper or sintered glass) and the solid is retained by the barrier. The liquid can be made to pass through the barrier by gravity alone, in which case the procedure is called a *gravity filtration*. Alternatively, the liquid can be caused to pass through by a combination of gravity and air pressure. Such an operation is called a vacuum or *suction filtration*.

8.1 Gravity Filtration

A piece of filter paper and a conical glass funnel to support the paper are all that are required for gravity filtration. To maximize the rate at which the liquid flows through the filter paper, the paper should be

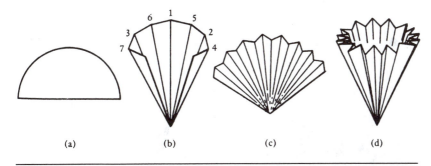

(a)   (b)   (c)   (d)

Figure 8.1-1. Folding the filter paper for gravity filtration. (a) Fold the filter paper circle (11 cm diameter) in half. (b) Crease the half to divide it into eight equal pie-shaped sections; it is easiest to make the creases in the numerical order shown. (c) Turn the piece over and pleat it into a fan by folding each pie-shaped section in half in the direction opposite to the previous creases. (d) Pull the two sides apart.

folded as indicated in Figure 8.1-1. The folded paper is then dropped into the funnel (see Figure 8.1-2). The funnel is best supported in an iron ring, as shown in the figure. The material to be filtered is poured into the filter paper cone, in portions if necessary. In this operation, gravity alone provides the force to cause the liquid to pass through the filter.

Gravity filtration is used to remove an insoluble residue during recrystallization (Section 9.4) or to remove a drying agent from the dried solution (Section 16.2).

8.2 Micro-scale Gravity Filtration

For smaller amounts, as in micro-scale work, a Pasteur dropping pipette can be used as a *micro-scale funnel* instead of the glass funnel, and a tiny piece of cotton or glass wool pushed down into the narrow part of the tip can serve as the filter, as illustrated in Figure 8.2-1.

8.3 Vacuum or Suction Filtration

In vacuum or suction filtration, a partial vacuum is created below the filter, causing the air pressure on the surface of the liquid to increase the rate of flow through the filter. A typical apparatus is illustrated in Figure 8.3-1.

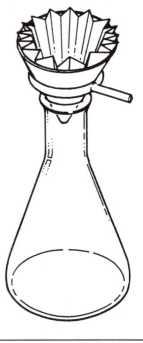

Figure 8.1-2. Arrangement of filter paper, funnel, and flask for gravity filtration.

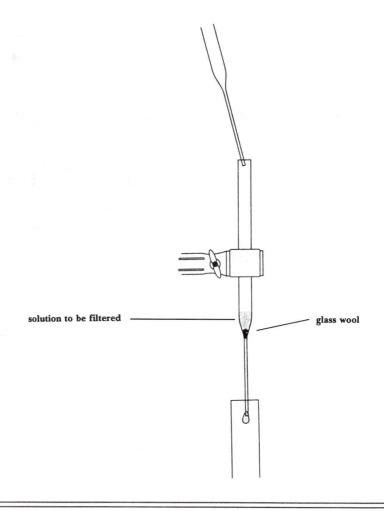

solution to be filtered ——————————————— glass wool

Figure 8.2-1. Micro-scale "filter funnel."

Since the pressure difference in a suction filtration is much greater than in a gravity filtration, the filter paper must be supported or it will be torn apart. Special funnels with a perforated solid support have been designed for use in suction filtration. Two popular models are the Büchner and Hirsch funnels, which are illustrated in Figure 2-8.

Don't make this mistake!

A circle of filter paper only just large enough to cover the holes in the bottom of the Büchner or Hirsch funnel should be used. **A common error** is to try to use a piece of filter paper so large that it must be turned up at the edges. If this is done, it is almost impossible to create a vacuum in the suction flask. Not only will the filtration take much longer, but any material that flows over the edge of the

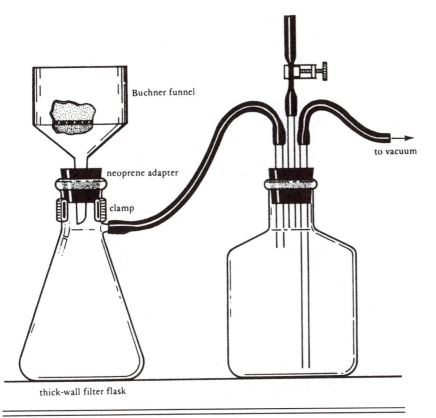

Figure 8.3-1. Apparatus for vacuum filtration.

filter paper will run down into the suction flask without being filtered.

Filtration is done by connecting the side arm of the suction flask to the vacuum source, which is almost always the water aspirator. When a water aspirator is used, the flask should be connected to the aspirator through a trap, as shown in Figure 8.3-1. The trap prevents water from the aspirator from being sucked back into the filter flask. Turn the aspirator on just a little at first so as to create a gentle vacuum, wet the filter paper with a small portion of the same solvent used in the solution being filtered while making sure that the paper is being pushed down over the holes, and pour the mixture to be filtered onto the center of the paper. Once the mixture has been added, the vacuum can be increased. When using the water aspirator, be sure to break the vacuum by disconnecting the tubing attached to the side arm of the filter flask before turning off the water.

Suction filtration is used to collect a solid product. A crude solid product can be collected by subjecting the entire reaction mixture to

Suck it down

suction filtration. A recrystallized solid product can be collected by suction filtration of the mixture of crystals and solution obtained in a recrystallization.

9. Recrystallization

Purification of a solid by recrystallization from a solvent depends on the fact that different substances are soluble to differing extents in various solvents. In the simplest case, all the unwanted materials are much more soluble than the desired compound. In this case, the sample is dissolved in just enough of the hot solvent to form a saturated solution, the solution is cooled, and the crystals, which will have separated upon cooling, are collected by suction filtration (Section 8.3). The soluble impurities remain in solution after cooling and pass through the filter paper with the solvent upon suction filtration.

If insoluble impurities are present in the sample, they are removed by filtering the hot solution by gravity (Section 8.1) before the sample is allowed to cool.

These procedures are summarized in the following flow chart.

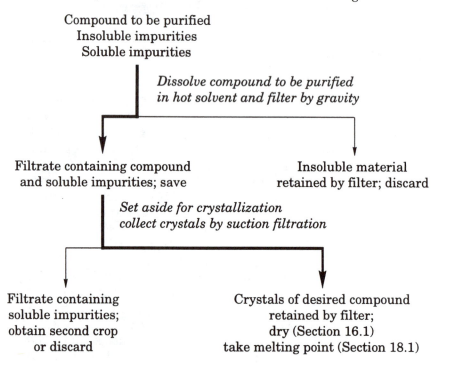

The different steps of this process are described in the next eight sections:

- **Choice of Solvent: 9.1**; usually a solvent is recommended.
- **Dissolving the Sample: 9.2**; done by warming the sample and solvent.
- **Decolorizing the Solution: 9.3**; only if hot solution is discolored.
- **Hot Filtration: 9.4**; gravity filtration; only if insoluble residue is present.
- **Cooling for Crystallization: 9.5**; the most critical step.
- **Cold Filtration: 9.6**; suction filtration; collection of the product.
- **Washing the Crystals: 9.7**; not always necessary.
- **Drying the Crystals: 9.8**; always important; not always easy.

9.1 Choice of Solvent

The choice of solvent is crucial in purification by recrystallization, but there is no easy way to know which solvent will work best. If you wish to recrystallize a known compound, the laboratory manual or the chemical literature may recommend or report the use of a certain solvent. Sometimes however, you must determine what solvent will be suitable. This problem is discussed in Section 9.11.

The good news is that a solvent will be recommended for all the recrystallizations called for in this book.

9.2 Dissolving the Sample

After a suitable solvent has been identified, a solution of the sample in the hot solvent must be prepared. Figure 9.2-1 shows a typical apparatus for this procedure.

Choose an *Erlenmeyer flask* (Figure 2-4) of such a size that it will be less than half filled with solution. Place the solid in the flask, add about 75% of the amount of solvent thought to be required, fit the flask with a reflux condenser, and bring the mixture to a boil. For solvents boiling below 95°C, use the steam bath; for higher-boiling solvents, the hot plate or sand bath is appropriate (Section 33.1). If a little more solvent is needed, add it down the condenser. Be sure to add a boiling stone (Section 10.6).

If the amount of solvent needed is not specified or has been

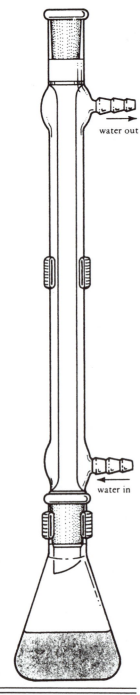

determined only roughly, place only a portion of the sample in the flask (for example, 1.0 gram in a 125-mL Erlenmeyer flask) and add measured amounts of solvent down the condenser. From the amount of solvent required to dissolve this first portion, you can calculate the amount needed for the whole sample. If the total will fill the flask much more than half full, transfer the first solution to a larger flask of appropriate size and prepare the rest of the solution in the larger flask. If no transfer is necessary, add the remainder of the sample (remove the flask from the heat and remove the condenser from the flask) followed by the rest of the solvent, using the solvent to wash the solid down from the neck of the flask as necessary.

You should keep in mind the possibilities that the compound may dissolve only slowly in the boiling solvent and that "insoluble" impurities may be present in your sample.

9.3 Decolorizing the Solution

The presence of *colored impurities* in a sample or in a solution of a colorless compound is obvious. Sometimes during recrystallization, the colored substances remain in solution upon cooling and are removed with the solvent upon suction filtration. Often the colored substances are adsorbed by the crystals as they are formed, giving an obviously impure product. Because the types of molecules that impart the color to the solution are often the types that are preferentially adsorbed by activated charcoal, or *decolorizing carbon,* the addition of a small amount of activated charcoal (1–2% by weight of the sample; or 1 mg per milliliter of solution; or enough to cover the tip of a spatula) in 50 mL of solution, followed by gravity filtration of the hot solution, can serve to remove some if not most of the color. The molecules responsible for the color will be adsorbed by the carbon and will be separated along with it during the hot, or gravity, filtration.

The carbon should be added only after the solution has cooled below the boiling point; otherwise the solution will boil over when the carbon is added. After the carbon is added, the solution can be reheated to the boiling point.

9.4 Hot Filtration

When the desired substance is in solution in the hot solvent, insoluble impurities (including dust, pieces of filter paper, glass, or cork) and decolorizing carbon, if used, can be separated by *gravity filtration.* Vacuum filtration cannot be used, because the reduced pressure in

Figure 9.2-1. Setup for dissolving a sample for recrystallization.

water out

water in

the suction flask will cause the filtrate to boil, and material in solution will be deposited over the walls of the flask.

The main problem in hot filtration is that the hot solution cools a little before it runs through the filter. This means that some crystallization can take place in the filter. You can try to avoid this undesired crystallization by warming the funnel (with steam, quickly wiping it dry with a towel; or with a flame) and pouring only a little of the solution into the filter at a time, keeping the remainder at the boiling point. If a large volume of solution must be filtered, or if the solubility of the substance decreases greatly in the range just below the boiling point, a heated funnel is a necessity. Often it is a good idea to use a small excess (10–25%) of solvent so that the solution will not become saturated until the temperature has fallen somewhat below the boiling point. A *stemless funnel* is always recommended: with a stemless funnel, you can avoid the problem of crystallization and subsequent clogging in the stem.

Although usually the filtrate should be collected directly in the Erlenmeyer flask in which the cooling for crystallization is to take place, an arrangement such as that illustrated in Figure 9.4-1 can be very useful when there is trouble with crystallization of the compound in the filter. The filtrate in the beaker can be boiled up around the funnel, keeping it hot. Some care must be used with a flammable solvent. After filtration is complete, transfer the filtrate to an Erlenmeyer flask for crystallization.

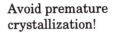

Avoid premature crystallization!

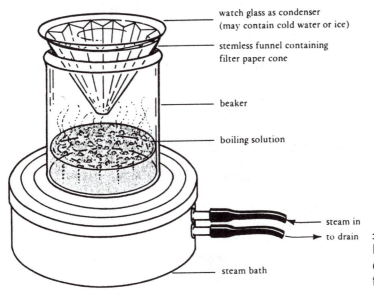

watch glass as condenser
(may contain cold water or ice)

stemless funnel containing
filter paper cone

beaker

boiling solution

steam in
to drain

steam bath

Figure 9.4-1. A way to prevent crystallization in the funnel during hot filtration.

If an excess of solvent has been used to minimize the problem of crystallization in the filter paper, some solvent can be removed at this point by distillation. This operation will also serve to bring back into solution any crystals that may have separated during filtration.

In all cases, the filtrate should be reheated, if necessary, to dissolve any crystals that may have formed and to give a clear solution. Crystals that form when the hot filtrate hits the cold flask are likely to be less pure than those that separate more slowly from solution.

9.5 Cooling for Crystallization

After the filtrate from the hot filtration has been adjusted to the desired volume and all solid has been brought back into solution, the filtrate is allowed to cool.

Don't cool too fast

The rate of cooling determines the size of the crystals. Slow cooling tends to favor fewer and larger crystals; fast cooling tends to favor more and smaller crystals. Very large crystals are to be avoided because they often occlude the solvent and its dissolved impurities. Very small crystals are undesirable because it is difficult to wash them free of the solvent and the soluble impurities, and they take longer to dry. Needles between 2 and 10 mm long are fine, as are prisms 1 to 3 mm in each dimension.

DON'T RUSH!

Usually the best compromise of speed, convenience, and quality of crystals is reached by allowing the solution to cool to room temperature on a non-heat-conducting surface such as a cork ring. Sometimes you can hurry the cooling without getting overly small crystals by swirling the flask in a beaker of water at room temperature, or even by placing the flask directly in an ice bath. The rate of cooling can be slowed greatly by supporting the flask in a beaker of water at the temperature of the solution and allowing both to cool spontaneously to room temperature. The fastest and most satisfactory procedure can be determined only by experimentation.

Because solubility decreases with decreasing temperature, it is often a good idea to finally cool the mixture from room temperature to 0°C (or to the freezing point of the solvent if it is above 0°C) in a mixture of ice and water. Cooling below 0°C is not often done, since colder baths are not so easily prepared and other problems related to the condensation of water vapor from the air in the form of water or frost require special techniques and apparatus.

When is crystallization complete?

If the crystals are collected too soon, some material will be lost that would have separated from solution on further standing. The minimum acceptable time for crystallization varies from a few minutes for some substances to days for others, and it can be determined only by experiment. To tell whether one-half hour is enough, two identical samples must be prepared. One is filtered after standing for half an hour at room temperature, and the other perhaps the next day. A comparison of the amount recovered in the two cases will tell whether anything is to be gained by letting the sample stand for more than half an hour.

9.6 Cold Filtration

When crystallization is complete, the product is collected by *suction filtration* using either a Büchner or Hirsch funnel. The size of the funnel used should be such that it will not be more than half filled with crystals. The suction flask should be large enough that the solution will not fill it above the tip of the funnel or above the side arm.

Sometimes the solvent can be decanted through the funnel until the mass of crystals remaining in the flask is just covered with solvent, and then the entire mass of crystals can be poured into the funnel in one smooth operation. More often, it is necessary to suspend the crystals in the solvent by swirling, and then to quickly pour part of the suspension into the funnel, wait until the liquid level has fallen almost to the level of the crystals in the funnel, and then repeat the swirling and pouring operation.

Pour the crystals out with the solvent

Crystals will often remain in the flask after all the solvent has been poured out. These must be scraped into the funnel, rinsed into the funnel with fresh cold solvent, if the compound is not too soluble in the cold solvent, or rinsed into the funnel with portions of the filtrate. The last process is messy, and it is far better to work hard to pour the crystals out with the liquid in the first place!

During the entire operation of getting the crystals into the funnel, it is best if the level of the liquid in the funnel does not fall below that of the crystals. Air should never be drawn through the crystals until they have been rinsed with fresh solvent, if that is necessary, as explained next. Sometimes you need to break the vacuum in the flask by disconnecting the hose at the side arm to slow the rate of filtration sufficiently.

9.7 Washing the Crystals

After the crystals have been transferred to the funnel for suction filtration and the liquid has been drawn off, some fresh solvent should be poured over the crystals to wash off the liquid that contains the soluble impurities. If this step is not done, the soluble impurities will be deposited on the crystals when the solvent evaporates. If the product is relatively soluble in the cold solvent, one washing will usually have to suffice. Two washes are generally appropriate.

Don't use too much solvent

If the crystals are relatively soluble, a minimum amount of solvent must be used, and this portion of solvent should be cooled thoroughly in an ice bath. If the crystals are not very soluble, larger amounts of solvent may be used and these portions need not be chilled. In either case, you can use some of the wash liquid to rinse any remaining crystals out of the flask in which crystallization took place.

If the crystals are not matted down tight into a solid cake, the vacuum can be released and the wash liquid poured evenly over the crystals and then drawn off by reestablishing the vacuum. If the crystals do form a solid cake, the wash liquid will have to be added to the crystallized matter in the funnel and the product mass carefully pulled apart and suspended evenly by means of a small spatula. Some care is required to suspend all of the product and not tear or dislodge the filter paper. It is best if a very slight vacuum can be maintained during this operation without drawing the wash liquid through too fast. When a procedure says "Wash thoroughly," it is probably calling for this rather tedious and delicate operation of carefully breaking up and suspending the filter cake in the wash liquid. After this step has been done, the wash liquid is drawn off by suction. An alternative procedure that is sometimes appropriate is to transfer the filter cake to a beaker, add solvent, and break up and suspend the material in the beaker. Afterwards, the product is again collected by suction filtration.

Sometimes, when a nonvolatile solvent such as acetic acid is used for recrystallization, it may be possible to wash off this nonvolatile solvent with a more volatile solvent to speed the drying of the crystals. Of course, the crystals should not be soluble in the more volatile solvent.

9.8 Drying the Crystals

After you have collected the crystals by suction filtration, remove as much solvent from the product as possible by continuing to draw air through the crystals while they remain on the filter paper in the funnel. The last traces of solvent will evaporate when the crystals are removed from the funnel and spread out to dry.

Suck them dry!

Don't make this mistake!

If the crystals collect as a solid cake, air should be drawn over them in the funnel until the solvent has almost stopped dripping. Failure to suck the filter cake as dry as possible before breaking it up and spreading it out to dry is a very common mistake. The filter cake should be a damp, friable solid, not a paste or mush, when spread out to dry.

More ways of drying solids are described in Section 16.1.

9.9 More Techniques of Crystallization

The preceding sections describe the normal techniques for recrystallization. There are, however, various problems that can be encountered during a recrystallization. This section describes some of these difficulties, including "oiling out" and failure to crystallize, and it suggests some possible solutions.

Solvent pairs

Occasionally, you will want to recrystallize a compound that is too readily soluble in some of the available solvents and not soluble enough in others. When this is necessary, a mixture of solvents, or *solvent pair*, can be useful. To test the suitability of a pair of solvents, prepare a hot solution of a sample of the compound in a small amount of the better solvent, and add slowly, while keeping the mixture hot, some of the poorer solvent. When a cloudiness is produced by slight crystallization, add a little of the better solvent, still keeping the mixture hot, until the cloudiness has been dispelled, and then allow the solution to cool. Although the solubility of a substance in a mixture of solvents usually changes gradually with the proportion of the solvents, the solubility of a substance is sometimes greatly increased or decreased by the addition of only a small amount of a better or a poorer solvent.

Solvent pairs that are most often used include toluene/hexane, acetic acid/water, and alcohol/water. Any pair of miscible liquids can be used. Although alcohol/water mixtures are often used, they seem to promote separation of the product as a liquid.

When a recrystallization is carried out using a pair of solvents, it is often a good idea to do the hot filtration before adding the poorer solvent. The addition of the poorer solvent should then be carried out as just described. **A common error** is to add more of the poorer solvent than is needed to just achieve saturation of the hot solution. In extreme cases, this error results in the precipitation of everything in solution, impurities as well as the desired compound.

Don't make this mistake!

Oiling out

Sometimes, during cooling for crystallization, the product separates not as crystals but as a liquid, an "oil." This *oiling out* shows up first as a cloudiness or opalescence, and then as visible droplets. An oil is undesirable because it is often an excellent solvent for impurities. When, or if, the oil finally freezes, the impurities that have dissolved in the oil will remain in the crystals.

If the first traces of oil can be caused to solidify by the addition of seed crystals, by vigorous stirring or swirling of the mixture, or by scratching the walls of the flask with a stirring rod, the remainder of the product will usually separate as crystals if the rate of cooling is not too great. If *oiling out* cannot be prevented—that is, if most of the product separates as an oil before it can be caused to solidify—you can hope that recrystallization of the solidified oil will give a better result, you can try a different solvent, or, probably best, you can purify the product by another method before attempting to recrystallize it.

Failure to crystallize

Occasionally, crystallization will not occur when a solution is cooled, even though it is supersaturated. The most stubborn cases are the result of impurities, or "tar," acting as a protective colloid. If the normal expedients of adding a seed crystal or scratching the flask with a stirring rod fail, crystallization can sometimes be initiated by cooling the mixture in a salt-ice bath (about -10°C) or a Dry Ice/acetone bath (about -70°C), depending on the freezing point of the solvent.

Seed crystals

For crystallization to occur in some cases, the solution must be stored in a refrigerator or freezer for long periods of time, even years.

Wet samples

A solid can often be isolated by pouring a reaction mixture into water and collecting the resulting precipitate by suction filtration. In this procedure, when the product separates as a fine powder, as it frequently does, sucking it dry using vacuum filtration is difficult. The crude, damp material can easily be more than half water. Recrystallization of such damp material from a water-miscible solvent often requires the use of more solvent than expected, because the water in the crude product will reduce its solubility. If the damp material is recrystallized from a water-immiscible solvent, an extra water phase will be present along with the hot solution. The water should be removed with a medicine dropper or pipette before hot filtration. The solution will also be saturated with water at this point.

Low-melting compounds

Recrystallization of low-melting compounds is not easy. Low melting point and high solubility in nonpolar solvents usually go together, as explained in Section 23.2. This situation gives you the choice of using very small volumes of nonpolar solvents or using solvent pairs that include water. If water is used, the product will often separate as a liquid upon cooling. Neither alternative is very attractive.

9.10 Micro-Scale Recrystallization

When the amount of solid material is only a few hundred milligrams, or the volume of the hot solution is less than about 5 mL, the usual semi-micro techniques of recrystallization give unacceptably large losses in the two filtration steps.

When working on the micro scale, you may choose an easy and inexpensive way to recrystallize that makes use of two test tubes and two Pasteur pipettes, according to the following procedure.

Choose a test tube that will be filled only about $\frac{1}{8}$ full by the solution, add the sample to the test tube, add the solvent, and warm the test tube in either the steam bath or the sand bath (Section 33.1) to dissolve the sample. The amount of material will be so small that only a brief period of heating will be needed, and the relatively large size of the test tube will provide enough surface area to condense the vapors of the solvent.

After the sample has dissolved, filter the solution into a second, somewhat smaller, test tube. For a filter, use a Pasteur pipette with a tiny piece of cotton or glass wool pushed down just to the point where the body of the pipette starts to narrow to form the long tip. Stand this filter pipette on its tip in the second test tube, as shown in Figure 9.10-1, and, using the second Pasteur pipette, transfer the hot solution from the first test tube into the top of the filter pipette, where it will run down through the cotton or glass wool into the second test tube. While the solution is draining down through the filter pipette into the second test tube, this second test tube can be warmed cautiously with either the steam bath or the sand bath to keep the solution in the filter pipette warm.

After the solution has drained down through the filter pipette, the filter pipette can be removed, and the second test tube, which contains the filtered solution, can be set aside to cool for crystallization.

After the crystals have formed, the solvent can often be removed by drawing it up into another Pasteur pipette whose tip has been pushed to the bottom of the test tube in which crystallization has taken place. If the crystals are well-formed granules, you can wash them by adding a few drops of solvent, agitating them gently, and withdrawing the solvent with another Pasteur pipette.

Alternatively, the crystals can be collected by centrifugation followed by decanting of the solvent. Washing can be done by suspending the crystals in a little cold solvent, centrifuging, and decanting. The wash liquid can be removed in the same way.

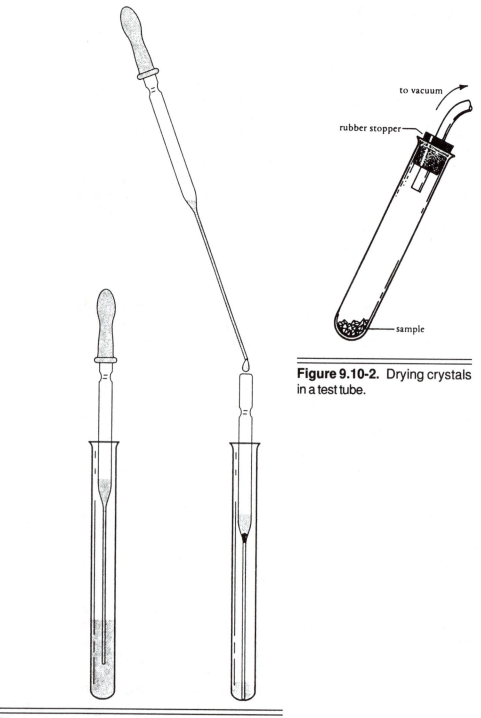

Figure 9.10-2. Drying crystals in a test tube.

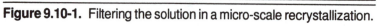

Figure 9.10-1. Filtering the solution in a micro-scale recrystallization.

The crystals can be dried in the test tube by laying the tube on its side with the bottom slightly raised or by connecting it to the vacuum as shown in Figure 9.10-2.

When the crystals are dry, they can be removed from the test tube by inverting the test tube over a piece of filter paper and tapping it with a stirring rod.

9.11 Selection of a Suitable Solvent

The procedure will almost always recommend a suitable solvent for recrystallization. Eventually, however, you will need to identify a suitable solvent yourself, or you may decide that the recommended solvent does not work well enough.

A *suitable solvent* is one that will dissolve the sample when the solution is hot but not when it is cold. If you have a large amount of material, you will want to use a solvent in which the material is fairly soluble so that the volume of solvent will not be inconveniently large. On the other hand, if you have only a small sample, a solvent in which your compound is only sparingly soluble will permit you to use a volume of solvent that is not inconveniently small. If possible, you would prefer to work with a solvent that is neither toxic nor expensive.

The best way to find a suitable solvent is to recrystallize small portions of your material in little test tubes from different solvents to see which gives the best results. If your compound is polar, alcohols and other water-soluble solvents are good possibilities. If your material is nonpolar, hexane and other nonpolar, non–water-soluble solvents can be tested. Table 9.11-1 lists some of the properties of the more common solvents.

Test the solvent

Questions

1. Suppose you are recrystallizing a sample of benzoic acid from water. The original sample is contaminated with sand and salt.

   a. Briefly describe the procedure you would use.

   b. Explain how the two impurities would be separated from the benzoic acid by your procedure.

2. Explain what effect each of the following mistakes would have on the success of a recrystallization.

   a. Too much solvent was used.

   b. Too little solvent was used.

Table 9.11-1. Properties of common solvents

Solvent	Boiling Point, °C	Density g/mL	Solubility in water	Flammability
Acetic acid	118	1.05	∞	+
Acetone	57	0.79	∞	+++
Acetonitrile	80	0.79	∞	++
Benzene	80	0.88	i	+++
Carbon disulfide	45	1.26	i	++++
Carbon tetrachloride	77	1.59	i	−
Chloroform	61	1.49	i	−
Cyclohexane	81	0.78	i	+
Dichloromethane	41	1.34	i	−
Dimethylformamide	153	0.94	∞	+
Dimethylsulfoxide	153	0.94	∞	+
Dioxane	101	1.03	∞	+++
Ethanol, 95%	78	0.81	∞	++
Ether diethyl	35	0.71	7	++++
Ethyl acetate	77	0.90	8	+++
Hexane	68	0.66	i	+++
Ligroin	60–90	0.67	i	+++
Methanol	65	0.79	∞	+
Pentane	36	0.63	i	++++
Petroleum ether	30–60	0.64	i	++++
Tetrachloroethane	146	1.60	i	−
Tetrahydrofuran	65	0.89	∞	+++
Toluene	111	0.87	i	+++
Water	100	1.00	i	−

Solubility in water: ∞ = miscible with water in all proportions; i = "insoluble" in water; numbers are grams of compound that will dissolve in 100 mL of water at room temperature.
Flammability: − = not flammable; + to ++++ = low degree to high degree of flammability.

   c. The hot solution was filtered by suction.

   d. The decolorizing carbon was not completely removed by the hot filtration.

   e. The hot solution was immediately placed in an ice bath.

   f. The crystals were washed with warm solvent.

   g. The crystals were not washed at all.

   h. There was inadequate suction during the suction filtration.

Problems

1. What is the expected percent recovery upon recrystallization of 2.00 grams of benzoic acid from 100 mL of water, assuming the solution cools to 18°C for crystallization?

Solubility of benzoic acid in water:

2.2 g/100 mL water at 75 °C

0.27g/100 mL water at 18 °C

2. What is the expected percent recovery upon recrystallization of 30 grams of benzoic acid from 100 mL of carbon tetrachloride, assuming that the solution will be cooled to 20°C for crystallization?

Solubility of benzoic acid in carbon tetrachloride:

32 g/100 mL at 60°C

5.6 g/100 mL at 20°C

3. Compare the percent recovery to be expected upon recrystallization of 2.00 grams of benzoic acid from 200 mL of water with that to be expected if only 100 mL of water is used, cooling to 18 °C in either case.

4. a. Calculate the percent recovery to be expected upon recrystallization of 2.00 grams of benzoic acid from 100 mL of water if the solution is allowed to cool to 4°C before suction filtration.

Solubility of benzoic acid in water at 4°C:

0.18 g/100 mL

   b. Do the same for the recrystallization of 30 grams of benzoic acid from carbon tetrachloride, assuming that the solution can be cooled to 0°C before suction filtration.

Solubility of benzoic acid in carbon tetrachloride at 0°C:

1.5 g/100 mL

5. a. Would you choose to recrystallize a 100-milligram sample of benzoic acid from water or from carbon tetrachloride? Present the reasons for your choice.

   b. Would you choose to recrystallize a 100-gram sample of benzoic acid from water or from carbon tetrachloride? Present the reasons for your choice.

Exercises

Any skill improves with practice. The purpose of an exercise is to provide practice. For this reason, I have included a few suggestions for practice in recrystallization.

1. Recrystallization of *endo*-5-norbornene-2,3-dicarboxylic acid. This substance can be recrystallized very nicely from water. Its tendency to remain supersaturated for a little while makes it easier to complete the hot filtration before crystallization begins. If the solution is

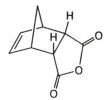

endo-5-norbornene-2,3-dicarboxylic anhydride

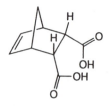

endo-5-norbornene-2,3-dicarboxylic acid

allowed to cool slowly, the crystals separate in beautiful long spars. Since the corresponding anhydride is much less expensive and is converted to the diacid on boiling with water, the instructions specify that you start with it rather than the acid. The anhydride can be prepared according to the procedure of Section E67.

a. *Semi micro-scale procedure:* Place 4.0 grams of the anhydride in a 125-mL Erlenmeyer flask and add 50 mL of water. Heat the mixture to boiling and continue to heat until the last of the melted crystals go into solution. Filter the hot solution by gravity and allow it to cool. Collect the resulting crystals by suction filtration and wash them with a little water.

b. *Micro-scale procedure:* Place 400 milligrams of the anhydride in a small test tube and add 5.0 mL of water. Heat the mixture to boiling until the last of the melted crystals go into solution. Filter the hot solution by gravity into another test tube using Pasteur pipettes as illustrated in Figure 9.10-1. Allow the solution to cool for crystallization and then remove the solvent with a second Pasteur pipette as described in Section 9.10.

2. Recrystallization of acetanilide. Water is often recommended for the recrystallization of acetanilide. Solubility information for acetanilide can be found in the *Handbook of Chemistry and Physics* (older editions).

3. Recrystallization of *m*-nitroaniline. Both water and 75% aqueous ethanol have been recommended for the recrystallization of this compound. It usually forms lovely yellow needles.

4. Recrystallization of *p*-nitroaniline. Water at 100 mL per gram has been recommended for the recrystallization of this substance.

5. Recrystallization of benzil. If benzil is dissolved in 95% ethyl alcohol at the rate of 6.5 mL per gram in a relatively large flask, and if care is taken to leave no traces of solid in any part of the flask, the solution can be cooled to room temperature without crystallization. When a minute seed crystal is added, a very beautiful phenomenon of crystal growth may be observed. If no seed crystal is added, crystallization may take a long time to occur, but sometimes the entire sample of benzil will separate as a single crystal.

10.   Distillation

In a distillation, a liquid is heated to the temperature at which it changes to a vapor. The vapor is then cooled, and thus liquefied, in another part of the apparatus. Separation of the components of a mixture by distillation takes advantage of the fact that different substances can differ in the degree to which they can be vaporized under the conditions of the experiment.

10.1 Vapor Pressure

The phenomenon of vapor pressure

If a sample of a liquid is placed in an otherwise completely empty space, some of the liquid will vaporize. As this happens, the pressure in the space above the liquid will rise and finally reach some constant value. The pressure under these conditions is due entirely to the vapor of the liquid and is called the *equilibrium vapor pressure.*

The equilibrium vapor pressure increases with temperature according to Equation 10.1-1, and as illustrated in Figure 10.1-1, where C is a constant and T is the absolute temperature:

$$P \alpha\, e^{-C/T} \quad \text{or} \quad P \alpha\, \frac{1}{e^{C/T}} \qquad (10.1\text{-}1)$$

This relationship between vapor pressure and temperature can be rewritten in a logarithmic form:

$$\ln P = \frac{-C}{T} + \text{constant} \qquad (10.1\text{-}2)$$

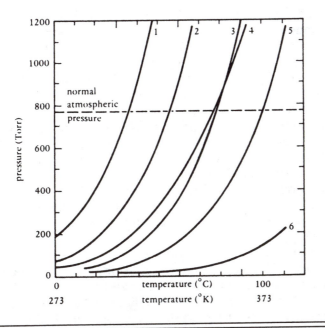

Figure 10.1-1. Graph of the equilibrium vapor pressure versus temperature for several liquids: (1) diethyl ether, (2) acetone, (3) ethyl alcohol, (4) carbon tetrachloride, (5) water, (6) bromobenzene.

or, using common logarithms:

$$2.3 \log P = -\frac{C}{T} + \text{constant}$$

$$\log P = -\frac{C}{2.3T} + \frac{\text{constant}}{2.3} \qquad (10.1\text{-}3)$$

In these last equations, "constant" is the natural log of the proportionality constant implied in Equation 10.1-1. The logarithmic form of the equation makes it apparent that a graph of $\log P$ versus $1/T$ should give a straight line of slope $-C/2.3$, as illustrated in Figure 10.1-2.

The molecular interpretation of vapor pressure

Molecular interpretation

The existence of vapor pressure is explained by molecules of liquid escaping into the empty space above the liquid. As the number of molecules in the vapor space above the liquid becomes larger, the rate of return of molecules from the vapor space to the liquid increases until the rate of return has risen to equal the constant rate of escape. This point is the *equilibrium condition,* and the corresponding concentration of molecules in the vapor space gives rise to the equilibrium vapor pressure. At higher temperatures, the greater kinetic energy of the molecules in the liquid results in a greater constant rate of escape. Equilibrium is established at higher temperatures, then, with larger numbers of molecules in the vapor phase, and at correspondingly higher pressures.

The equilibrium vapor pressure will be exerted in a closed container whether or not there are other molecules present in the gas phase (air, for example). If there are other molecules present, the equilibrium vapor pressure will not be equal to the total pressure as in the previous example. The total pressure will be the sum of the equilibrium vapor pressure of the liquid (its *partial pressure*) plus the pressure due to the other molecules (the sum of their various partial pressures).

Different substances have different vapor pressures at any given temperature, as illustrated by the examples in Figures 10.1-1 and 10.1-2. In other words, the values for the constants C and "constant" in Equations 10.1-2 and 10.1-3 are different for different molecules. In Section 17.3, boiling points will be interpreted in terms of the heat of vaporization (the amount of heat required to convert a mole of the liquid to a vapor at the normal boiling point; always positive) and the entropy of

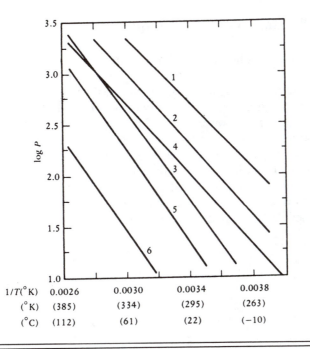

Figure 10.1-2. Graph of log (equilibrium vapor pressure) versus $1/T$, degrees K, for several liquids. The liquids are the same as those of figure 10.1-1: (1) diethyl ether, (2) acetone, (3) ethyl alcohol, (4) carbon tetrachloride, (5) water, (6) bromobenzene.

vaporization (the entropy increase that accompanies the conversion of a mole of liquid to a vapor at the normal boiling point; always positive). Because it can be shown that in Equation 10.1-3

$$C = \frac{\Delta H_{vap}}{R} \quad \text{and} \quad \text{Constant} = \frac{\Delta S_{vap}}{R}$$

where R is the ideal gas constant, then

$$\log P = -\frac{\Delta H_{vap}}{2.3RT} + \frac{\Delta S_{vap}}{2.3R}$$

A high vapor pressure at any given temperature is thus the result of a small heat of vaporization or a large entropy of vaporization, or both, and to say that a compound generally has a high vapor pressure is equivalent to saying that it has a low boiling point.

10.2 Distillation of a Pure Liquid

If a pure liquid is heated in a flask connected to a condenser that is open to the atmosphere at the other end, as in the apparatus pictured in Figure 10.2-1, its vapor pressure will rise, as explained in the preceding section. When the temperature of the liquid becomes sufficiently high, the vapor pressure of the liquid will slightly exceed that of the atmosphere, the liquid will start *boiling*, and the vapors produced will expand out of the flask and into the condenser. The condenser's function is to cool the vapors and reconvert them to liquid. In a distillation, the condenser is arranged so that the condensate does not return to the flask, in contrast to its use as a reflux condenser. As long as liquid remains in the flask, the temperature of

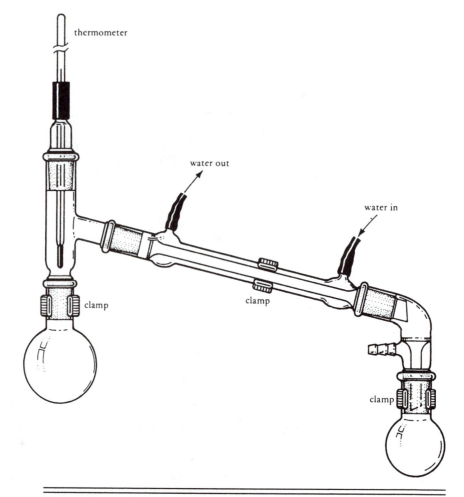

Figure 10.2-1. A simple apparatus for distillation at atmospheric pressure.

the distilling vapor will not rise. The continuing input of heat serves only to supply the required heat of vaporization, and thus to convert more liquid to vapor.

The temperature at which distillation takes place at a total pressure of 1 atmosphere is called the *normal boiling point* of the liquid. If the pressure in the apparatus, which is usually open to the atmosphere, is not equal to exactly 1 atmosphere, the temperature at which boiling will begin (the temperature at which the equilibrium vapor pressure equals external pressure) will be different from the normal boiling point. You might sometimes want to distill a substance at as low a temperature as possible, using an apparatus in which the pressure can be reduced (see Section 11).

The significant features in the *distillation of a pure liquid* are that

- the compositions of the liquid, the vapor, and the condensate (or distillate) are identical and constant during the process, and
- the temperatures of the liquid and the vapor are constant and, ideally, equal throughout the distillation.

10.3 Miscible Pairs of Liquids

Ideal behavior: Raoult's Law

When two liquids that are completely soluble in one another are mixed, the vapor pressure of each liquid at a particular temperature is diminished by the presence of the other liquid. Such mixtures can be characterized according to the contribution of each component to the total vapor pressure as a function of the composition of the mixture.

Miscible pairs of liquids are said to behave ideally if the contribution of each component to the total vapor pressure is directly proportional to its mole fraction. That is,

$$P_A = x_A P_A^0 \qquad\qquad P_B = x_B P_B^0$$
$$P_{\text{total}} = P_A + P_B = x_A P_A^0 + x_B P_B^0$$

where P_A and P_B are the vapor pressures of A and B above a solution of mole fraction x_A and x_B and where P_A^0 and P_B^0 the vapor pressures of pure A and pure B at that particular temperature. This type of behavior, often referred to as *ideal behavior* is approximated by mixtures such as benzene/toluene, n-hexane/n-heptane, and n-butyl bromide/n-butyl chloride, in which the mixture is composed of molecules of similar size and type of intermolecular interaction. A vapor pressure–composition diagram is shown for the ideal system benzene/toluene in Figure 10.3-1.

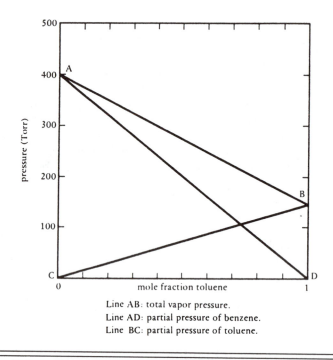

Line AB: total vapor pressure.
Line AD: partial pressure of benzene.
Line BC: partial pressure of toluene.

Figure 10.3-1. Vapor pressure–composition diagram for the ideal system benzene/toluene at 60°C.

Boiling point diagrams

Temperature versus composition

For describing the separation of a pair of miscible liquids by distillation, a boiling point diagram is very helpful. This diagram shows the *temperature* at which mixtures of various composition boil (at a given total external pressure, usually 1 atmosphere) and the *compositions* of the liquid and vapor that are in equilibrium at this temperature.

The way this information is stored in a boiling point diagram is best shown by considering how one is constructed. Suppose, for example, that you prepare mixtures of benzene and toluene of 0.1, 0.3, 0.5, 0.7, and 0.9 mole fraction benzene and heat each mixture to boiling in an apparatus open to 1 atmosphere of pressure and in which the condensed vapor is returned to the boiler. A thermometer can be used to determine the temperature at which the mixture boils, and a sample of the condensate can be removed to determine the composition of the vapor. (The composition of the liquid is known because the mixture was made up of known amounts of the components; it could also be determined experimentally.) Suppose the results are as shown here:

Boiling Point (°C)	Liquid Composition (mole fraction benzene)	Vapor Fraction (mole fraction benzene)
110.6	0.00	0
105.7	0.10	0.21
98.3	0.30	0.51
92.4	0.50	0.71
87.3	0.70	0.86
82.6	0.90	0.96
80.0	1.00	1.00

These data can be plotted as shown in Figure 10.3-2. A smooth line can then be drawn to interpolate all the other possible experimental points, and the result is shown in Figure 10.3-3. Now, from Figure 10.3-3, the composition of the vapor in equilibrium with any mixture of benzene and toluene at its boiling point (under 1 atmosphere of pressure) can be determined. For example, if a 1:4 mixture of benzene and toluene (20 mole percent benzene; 0.20 mole fraction benzene) is heated for distillation, you would expect that the mole fraction of benzene in the vapor would be about 0.39 at an initial boiling point of about 102°C. It is generally true, as in the particular case of benzene and toluene, that the vapor in equilibrium with the liquid will be richer in the more volatile component than the liquid. This seems intuitively reasonable in that the molecules of the component with the higher vapor pressure at any given temperature should tend to escape more frequently and thus be overrepresented in the vapor phase.

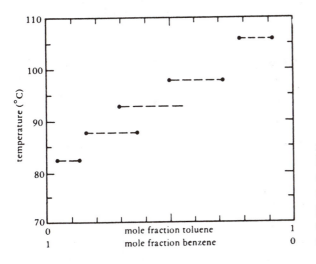

Figure 10.3-2. Construction of the boiling point diagram for the system benzene/toluene at 1 atmosphere. Each pair of points indicates the boiling point of a mixture of benzene and toluene and the composition of the liquid and the vapor that are in equilibrium.

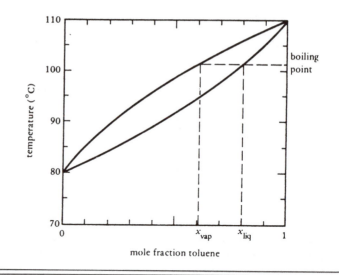

Figure 10.3-3. Boiling point diagram for the system benzene/toluene at 1 atmosphere.

Miscible liquids

As the distillation of a mixture of miscible liquids progresses, the mixture will gradually be depleted of the more volatile component. As this happens, according to the distillation diagram, *the boiling point will gradually rise,* and the distillate, though always richer than the residue in the more volatile component, will contain a continually decreasing proportion of the more volatile component. You can think of this process as being represented by the gradual movement of line AB in Figure 10.3-4 upward and to the right to, say, A′ B′. Thus, the significant features in the distillation of a miscible pair of liquids that serve to distinguish it from the distillation of a pure liquid are that

Distillation of a miscible pair of liquids

- the compositions of the liquid and vapor (or distillate) are not the same, and
- the boiling point of the liquid will gradually rise during the distillation.

Exceptions to this generalization are discussed in the section on azeotropic mixtures (Section 10.5).

10.4 Fractional Distillation

From Figure 10.3-3 and the example cited in the previous section, you can see that if you started to distill a mixture of benzene and

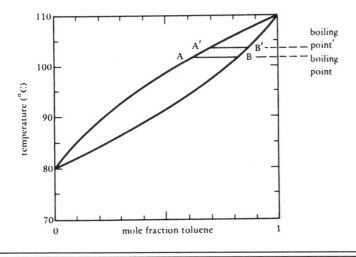

Figure 10.3-4. Boiling point diagram for the system benzene/toluene at 1 atmosphere. The horizontal lines AB and A'B' connect points on the two curves that represent the compositions of liquid and vapor that are in equilibrium at two different temperatures.

toluene that was 0.2 mole fraction in benzene, the vapor initially in equilibrium with the mixture would be about 0.4 mole fraction in benzene. If you started with a large sample, this would be the composition of the first part of the distillate. If the first part of the distillate were then redistilled, the vapor in equilibrium with it would be approximately 0.6 mole fraction in benzene and the first few drops of the distillate would have this composition. You can see that if you started with a large enough sample and repeated this process several times, collecting only the very first part of the distillate each time, you could obtain a very small sample of fairly well purified benzene. A four-step process of this type could be represented by the movement of a point from A to B...to I in Figure 10.4-1.

    The process just described would be too inefficient to be practical. If 10% of the material were distilled each time, then, starting with 1 liter, only 0.1 milliliter would be obtained at the end of the fourth stage. Also, if a finite amount were collected in each step, the final purity would be less than that indicated in Figure 10.4-1.

Fractional distillation

Fractionating columns

Fortunately, a device called a *fractionating column* has been developed that can effect a high degree of purification much more quickly and easily and with little loss or rejection of material. One type of fractionating column is shown in Figure 10.4-2.

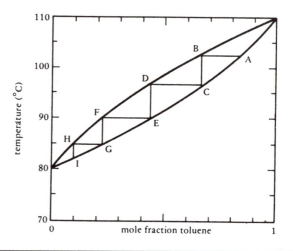

Figure 10.4-1. Boiling point diagram for the system benzene/toluene at 1 atmosphere. The series of four horizontal and four vertical lines connecting the points A and I indicate that after four successive steps of vaporization followed by condensation, a large sample of liquid whose boiling point and composition correspond to point A would yield a small sample of a liquid whose boiling point and composition would correspond to point I.

Figure 10.4-2. A fractionating column.

The significant feature of a fractionating column is that it provides for efficient exchange of heat and material between the condensate flowing down the column and the vapors flowing up. The material finally coming out of the top of the column as a vapor has been subjected to multiple condensations and evaporations on the way up, each of which has served to enrich the vapor in the more volatile component. A good column can produce a distillate in which the enrichment corresponds to between 25 and 100 steps like the four in Figure 10.4-2. Because some industrial fractionating columns are built up of units called plates, each of which theoretically provides enrichment corresponding to one step, the efficiency of enrichment of a column is expressed as *theoretical plates*, rather than as steps.

10.5 Azotropic Mixtures

Nonideal behavior: deviations from Raoult's Law

In many mixtures of pairs of miscible liquids, the vapor pressure of each component cannot be represented by Equation 10.3-1, and a plot of the total vapor pressure of the mixture versus composition at any given temperature will not give a straight line such as the line AB in

Figure 10.3-1. However, so long as the total vapor pressure of any mixture at a certain temperature is neither greater than nor less than the vapor pressure of either pure component at that temperature, mixtures of these substances will approximate the expected behavior of ideal mixtures upon distillation, and the components of the mixture can be completely separated by fractional distillation. The system water/methanol is such a system; its vapor pressure-composition diagram is given in Figure 10.5-1 and its distillation diagram in Figure 10.5-2.

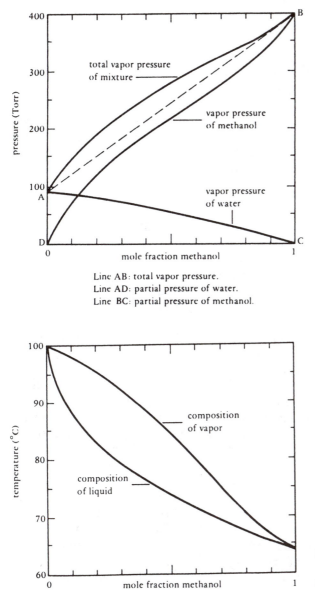

Line AB: total vapor pressure.
Line AD: partial pressure of water.
Line BC: partial pressure of methanol.

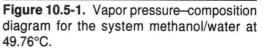

Figure 10.5-1. Vapor pressure–composition diagram for the system methanol/water at 49.76°C.

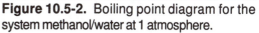

Figure 10.5-2. Boiling point diagram for the system methanol/water at 1 atmosphere.

Some mixtures, however, depart from ideal behavior to such an extent that they form mixtures that cannot be separated by distillation.

Maximum boiling mixtures

If the total vapor pressure of mixtures goes through a maximum at intermediate compositions, the boiling point of the mixture whose composition corresponds to the vapor pressure maximum will be lower than that of either pure component, and the vapor in equilibrium with this mixture will have the same composition as the liquid.

Minimum boiling mixtures

The opposite is also observed. If the total vapor pressure of mixtures goes through a minimum at intermediate composition, the boiling point of the mixture whose composition corresponds to the vapor pressure minimum will be higher than that of either pure component, and the vapor in equilibrium with this mixture will have the same composition as the liquid.

Azeotropes

Because the composition of the vapor in equilibrium with these minimum- or maximum-boiling mixtures is identical to the composition of the liquid, complete separation of the components of such a mixture by distillation is impossible. Such mixtures are called *azeotropes*.

Minimum boiling azeotrope   If a pair of liquids can form a minimum-boiling azeotrope, a perfectly efficient fractionating column will produce a distillate of the composition of the azeotrope, no matter what the initial composition of the mixture, and the residue in the boiler after complete removal of the other components as the azeotropic mixture will be pure A or pure B, depending on whether the mixture initially contained more A or more B than the azeotropic composition. Thus, although any mixture of methanol and water can produce 100% methanol by fractional distillation, the most ethanol-rich material that can be separated by fractional distillation of a dilute aqueous solution of ethanol is a product that is only 96% by weight ethanol. The reason is that 96% by weight ethanol has a boiling point of 78.1°C, a boiling point below that of pure ethanol, 78.3°C.

Maximum boiling azeotrope   If, on the other hand, a pair of liquids can form a maximum-boiling azeotrope, a perfectly efficient fractionating column can produce a distillate of either pure A or pure B, depending on whether the mixture initially contained more A or more B than the azeotropic composition. The residue, after complete removal of the amount of either A or B initially present in excess over the material of azeotropic composition, will be the pure azeotrope.

10.6 Technique of Distillation

Constant rate of heating

A *constant rate of heating* is essential for optimum performance of any distillation apparatus. Of the methods of heating described in Section 33.1, an electrically heated oil bath or an electric heating mantle gives the most constant and most easily regulated rate of heat input. If the separation is very easy and high efficiency is not needed, a steam bath can be used at lower temperatures with flammable materials (for example, to remove the volatile solvent from a high-boiling product), or a flame can be used with higher-boiling substances (for example, to complete the distillation after the solvent has been removed by heating with the steam bath).

Rate of distillation

The rate of distillation is controlled by the rate of heating. The higher the temperature of the bath above the boiling point of the mixture, the greater the rate of distillation. The lower the rate of distillation, the greater the efficiency of any distillation apparatus. With an easy separation, a collection rate for the distillate of one to three drops per second (3 to 10 mL per minute) will often be a reasonable compromise between speed and efficiency. In careful fractionations, the reflux ratio may be quite high, and the collection rate much lower than this. Because the required rate of heating for a given rate of distillation depends on how well the column is insulated against heat loss, the rate must be determined by experimentation.

Bumping; boiling stones

Even with a constant source of heat, the rate of boiling will not always be constant. Sometimes alternate periods of superheating followed by vigorous boiling will occur. More than being just an annoyance, this phenomenon, called *bumping*, will prevent equilibrium from being established within the fractionating column. Smooth boiling can usually be promoted by adding several *boiling stones*. These may be bits of unglazed porous clay plate, pieces of carborundum, or bits of a specially prepared anthracite coal. The tiny air bubbles trapped in the pores of such materials prevent superheating by providing the nuclei for bubble formation.

The use of several *boiling stones* is recommended whenever a liquid is to be boiled, in recrystallizations as well as in distillations. Should you forget to add the boiling stones, let the mixture cool a bit before you drop them in; if you do not allow a brief cooling time, you run the risk of having the mixture boil over.

Insulation

For a distillation column to operate at maximum efficiency, it must be insulated against heat loss. The most usual form of *insulation* is a high-vacuum jacket (like a Thermos bottle) that is an integral part of the column. Sometimes provision is made for electrical heating, with or without the vacuum jacket. No insulation at all is provided with the column shown in Figure E 117.1.

Wrap with foil

The simplest way of providing some insulation for these columns is to wrap them, not too smoothly, with a couple of layers of aluminum foil. Asbestos tape, glass wool, and many other materials have been used, either singly or in combination. The higher boiling a substance is, the more easily it will be condensed by the column. If the column is tall and poorly insulated, it may be impossible to put as much heat in at the boiler as can be lost in the column. In this case, it will not be possible to drive the material over.

Avoid superheating

When you are heating with a flame, it is easy to superheat the vapor, upset the equilibria within the column, and lower column efficiency. *Superheating* can be minimized by heating through an abestos board, and by wrapping the upper part of the boiling flask and the column in a layer of aluminum foil.

Use the right size flask

The flask used as the boiler in a distillation should not be more than half full at the start. If a large amount of solvent must be removed from a small amount of a high-boiling compound, it is best to remove most of the solvent in a first distillation and then to transfer the residue to a smaller flask to complete the distillation. Losses in transfer can be prevented by rinsing with a small amount of the solvent that was just removed. If you attempt to complete the distillation from a large flask, not only will the losses be greater because of the larger vapor volume and surface area, but the walls of the flask will act as a condenser and may make it impossible to drive the higher-boiling material over.

10.7 Small-Scale Distillation

When small amounts of material are distilled, the losses can be significant. Vapors will remain in the boiling flask, and the condenser
and adapters will retain liquid as well. The distillation apparatus illustrated in Figure 10.2-1 can give losses of 1 to 3 mL, depending on
the size of the flask, and if a fractionating column is used, the total
may be between 3 and 5 mL. Thus, if the volume of material to be
distilled is less than 10 mL, a smaller apparatus should be used.

Use a small boiling flask

Omit the condenser

One way to shrink the apparatus is to use the smallest size of components and to omit the condenser, as shown in Figure 10.7-1.

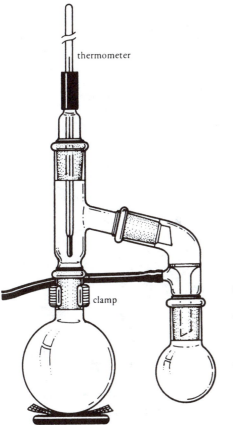

Figure10.7-1. An apparatus for distillation. The receiver and
vacuum adapter should each be held on with a spring clamp or a
rubber band. The hose should lead over the edge of the desk to
keep flammable fumes from the burner, or to the vacuum pump in
the case of a vacuum distillation. A small piece of stainless steel
sponge may be placed just below the thermometer as a packing
for better fractionation. This set-up, which does not make use of a
condenser, should not be used with low-boiling compounds.

A better but more expensive apparatus is shown in Figure 10.7-2 It is very easy to superheat the vapors in a small apparatus if the procedures described above are not used.

A very common error

Position of thermometer

Because the temperature of the vapor is usually taken to be the boiling point of the corresponding distillate, care must be taken to avoid superheating the vapor and to make certain that the entire mercury-containing bulb of the thermometer is below the bottom of the side arm leading to the condenser. **A very common error** is to position the thermometer so high that the bulb is not entirely heated to the temperature of the vapor; too low a temperature will then be recorded.

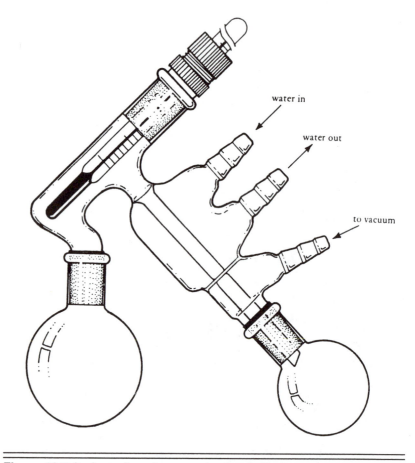

water in

water out

to vacuum

Figure 10.7-2. A small-scale apparatus for distillation.

Questions

1. Explain what effect each of the following mistakes would have on the success of a distillation.

   a. You forgot to add boiling stones.

   b. You attempted to distill a low-boiling, flammable liquid (such as ether) using a flame but without using a condenser.

   c. Your thermometer was positioned too high in the distilling adapter. That is, the top of the mercury bulb was above the bottom of the opening of the side arm of the adapter that leads to the condenser.

   d. You distilled too fast.

   e. When isolating carvone (boiling point around 230°C; Experiment 11) from either oil of spearmint or oil of caraway, you collected as product the material boiling from 110°C to 231°C.

Problems

1. a. Estimate from Figure 10.5-2 the boiling point of a water/methanol mixture 0.6 mole fraction in water.

   b. What would be the composition of the vapor in equilibrium with this mixture at the boiling point?

2. Referring to Figure 10.5-2, and assuming the use of a magic fractionating column with an infinite number of theoretical plates and zero holdup,

   a. What would be the boiling point and the composition of the initial distillate, starting with an equimolar mixture of methanol and water?

   b. What mole percent of the total mixture would distill with this composition?

   c. What would be the boiling point and composition of the remainder?

   d. Present the results of your calculations for b and c as a plot of boiling point of the distillate (vertical axis) versus mole percent distilled (horizontal axis).

Exercises

1. Distillation of methanol. The distillation of methanol provides an example of the distillation of a pure liquid.

Procedure: Using a clamp and a ring stand, support a 100-mL round-bottom boiling flask on your steam bath. Add 50 mL of methanol and one boiling stone to the flask. Fit the flask with a distilling adapter, as illustrated in Figure 10.2-1. Support a thermometer in the top joint of the adapter, and connect a condenser, with distillation adapter and receiver, to the side arm of the distillation adapter. Using two lengths of rubber tubing, provide for a gentle flow of water from the cold water tap to the lower hose connection of the condenser, and from the upper hose connection of the condenser to the drain. Adjust the height of the thermometer so that the top of the bulb of mercury is just below the bottom of the opening of the side arm.

Heat the contents of the flask with the steam bath, vigorously at first and then more gently as the contents of the flask appears to approach the boiling point. Slowly distill the contents, keeping a record of the temperature of the vapor, as indicated by the thermometer, versus the volume of the distillate.

After the distillation is complete, make a plot, in your lab notebook, of the temperature (vertical) versus the volume of distillate (horizontal).

2. Distillation of a mixture of methanol and water. Distill 100 mL of an equimolar mixture of methanol and water, using a simple fractionating column as illustrated in Figure 10.4-2. Keep a record of the temperature of the vapor at the top of the column versus the volume of the distillate collected. When the distillation is complete, plot your data as in problem 3, and compare the actual result with the ideal.

3. Isolation of (R)-(–)- or (S)-(+)-carvone from oil of spearmint or oil of caraway by distillation (Experiment 11).

11. Reduced-Pressure Distillation

A liquid will boil when its vapor pressure equals the pressure on its surface. Thus, you can make a substance boil at a temperature lower than its normal boiling point by distilling it in an apparatus in which the pressure can be reduced. Because many substances undergo noticeable decomposition at elevated temperatures, distillation under reduced pressure is desirable for these substances to minimize decomposition. With substances whose thermal stability is not known, distillation is usually carried out under reduced pressure if it appears that the normal boiling point will be greater than 125–175°C.

11.1 Estimation of the Boiling Point at Reduced Pressure

In general, the boiling point of a compound will decrease by about 20–30°C each time the external pressure is reduced by a factor of two. Thus, the boiling point of a compound in an apparatus in which the pressure is maintained at 45-50 Torr ($\frac{1}{16}$ atmosphere) would be expected to be 80-100°C lower than its normal (atmospheric pressure) boiling point.

A nomograph

The *nomograph* in Figure 11.1-1 is useful for estimating both expected reduced-pressure boiling points from normal boiling points and normal boiling points from observed reduced-pressure boiling points. The nomograph applies to liquids that are not associated in the liquid phase. The variation of boiling point with pressure for associated liquids, such as alcohols, is 10-20% less than for nonassociated liquids.

Reference 1 describes a more accurate method of calculating changes in boiling point with variations in pressure.

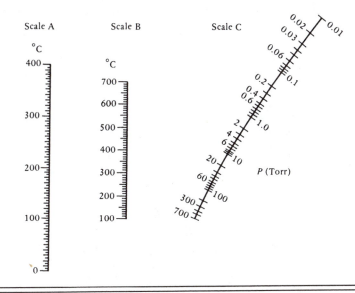

Figure 11.1-1. A nomograph for estimating the boiling point as a funtion of pressure. The nomograph relates the normal boiling point of a substance (scale B) to boiling points at reduced pressures (scales A and C). A line connecting points on two scales will intersect the third scale at some point. Thus, from values for A and C, you can estimate B. Knowing B (or having estimated B), you can estimate the boiling point A′ at a reduced pressure C′. A transparent plastic ruler works as well as anything for connecting points and noting intersections.

11.2 Apparatus

Semi-micro-scale distillation

With larger amounts of material, a distillation can be carried out at reduced pressure in a regular distillation apparatus like that shown in Figure 10.2-1 by connecting a source of vacuum to the side arm of the distillation adapter.

Micro-scale distillation

For small-scale work, arrangements like those shown in Figures 10.7-2 or 11.2-1 can be used with a source of vacuum to distill a high-boiling product after removal of the solvent.

Fractional distillation; fraction collector

If more than one component of a mixture is to be collected, a specialized receiver must be used that permits the collection of fractions of different boiling ranges without variation of the internal pressure or

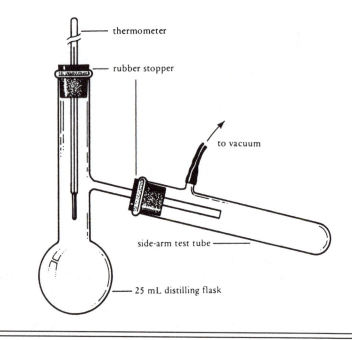

Figure 11.2-1. A simple apparatus for vacuum distillation of a small sample.

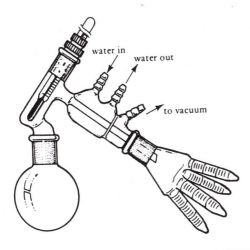

Figure 11.2-2. Small-scale distillation apparatus with a rotating fraction collector.

interruption of the distillation process. Figure 11.2-2 shows the small-scale distilling apparatus of Figure 10.7-2 fit with a *fraction collector.*

11.3 Source of Vacuum

The aspirator

The most common and convenient source of a vacuum is the *water aspirator.* The ultimate pressure attainable with a water aspirator equals the vapor pressure of water at temperature of flow. Between 5 and 30°C. the vapor pressure of water, in Torr, is numerically equal to the temperature, in degrees C, within 2.5 Torr. Thus, in the winter, you might hope to attain a pressure below 15 Torr, whereas in the summer, 25 Torr might be the ultimate pressure.

An arrangement like that illustrated in Figure 11.3-1 is suitable for reduced-pressure distillation when the desired pressure is equal to or above the pressure attainable with the water aspirator. The suction flask not only serves as a trap to prevent water from being sucked back into the apparatus in the event that the water pressure decreases momentarily, but it also buffers the system against changes in pressure that result from variations in the flow rate of the water or from changing distillation receivers. The Bunsen burner valve can serve to provide a controlled rate of leakage of air into the system in case you want to operate at a pressure above the ultimate pressure you can attain with the aspirator.

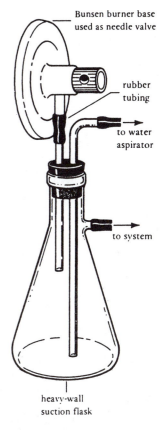

Bunsen burner base used as needle valve

rubber tubing

to water aspirator

to system

heavy-wall suction flask

Figure 11.3-1. Pressure regulation with the water aspirator.

The vacuum pump

For pressures below 10 to 25 Torr, a mechanical *vacuum pump* is generally used. A pump in average condition can give an ultimate pressure of 3 to 5 Torr, and 1 Torr if in good condition.

11.4 Pressure Measurement

For measuring pressures in excess of 5 to 10 Torr, you will find a simple *manometer* such as that illustrated in Figure 11.4-1 to be quite adequate. For determination of pressures down to 1 Torr, a tilting McLeod gauge is useful (Reference 2).

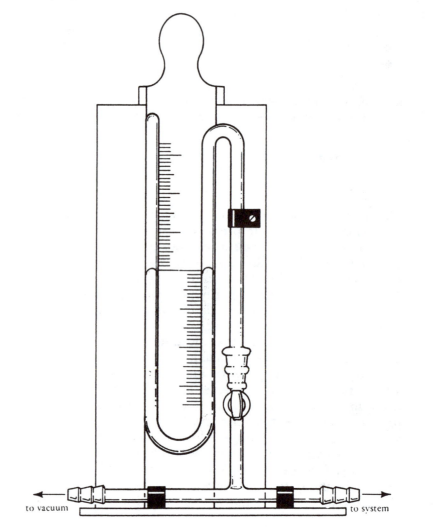

Figure 11.4-1. A simple manometer. The pressure equals the difference in height of the two columns of mercury. The difference in height can be read with the help of the movable scale.

to vacuum

to system

11.5 Technique of Distillation Under Reduced Pressure

The volume of the vapor that results from the vaporization of a given amount of substance is many times larger at a reduced pressure than at 1 atmosphere (100 times larger at 7.6 Torr; 760 times larger at 1 Torr). Thus, maintenance of smooth boiling and prevention of bumping are even more important in vacuum distillations than in distillations at atmospheric pressure; a burst of vapor produced in a bump during vacuum distillation can easily splash material over into the receiver and can upset the vapor-liquid equilibria.

Avoid bumping

Boiling stones are often used to prevent bumping in a brief distillation, but they usually lose their effectiveness fairly quickly. If, for some reason, the pressure in the apparatus rises momentarily, the boiling stones often cease to function afterward. Microporous carbon boiling chips (available from Fisher Scientific Company, Pittsburgh, PA.) are often recommended as being most suitable.

Use a magnetic stirrer

One of the most convenient methods of heating the boiling flask during distillation is to use a magnetically stirred and electrically heated oil bath (see Figure 34-1). If this method of stirring is used, bumping can usually be eliminated by placing a small magnetic stirring bar in the boiling flask and adjusting the level of the heating bath until it is somewhat higher than the level of the liquid in the flask. If the temperature of the bath is not too far above the boiling point of the liquid, and the liquid and bath are well stirred (the stirring bar in the bath turns the smaller one in the flask), boiling will usually be very smooth.

Reduced pressure boiling points

Experimental measurement of the boiling point of a substance in a reduced-pressure distillation is much less certain than in distillations at atmospheric pressure, and extrapolations of the reduced-pressure boiling points to atmospheric-pressure boiling points will contain these uncertainties. Because, for a given heating rate, the vapor velocity in a reduced-pressure distillation may be several hundred times greater (because of the greater volume) than in an atmospheric-pressure distillation, it is much easier for superheated vapors to reach the thermometer. In addition, the rapid flow of large amounts of vapor through the apparatus may result in a large pressure gradient. That is, the pressure may be considerably higher in the boiler than at the manometer. Both of these effects can be minimized by using a heating bath, and by using it at a temperature that provides a minimum rate of distillation. An additional uncertainty is introduced when an air bleed is used to promote even boiling, because the air will contribute an unknown amount to the total pressure of the system.

Other techniques of distillation, which also apply to distillation under reduced pressure, are described in Section 10.6.

Problems

1. a. Benzyl alcohol is reported to boil at 93°C at 10 Torr. At what temperature would it be expected to boil at atmospheric pressure?

    b. 1,2-Dibutoxybenzene is reported to boil at 135 to 138°C at 12 Torr. What would you expect its normal boiling point to be?

2. a. α-Phenylethylamine boils at 187°C at atmospheric pressure. At what temperature would you expect it to distill under the vacuum that can be obtained with a water aspirator (assume 20 Torr)?

    b. The normal boilng point of nitrobenzene is 211°C. At what temperature would it distill under a pressure of 15 Torr?

3. a. Diethyl ether has a normal boiling point of 34.6°C. Calculate the vapor pressure of ether at O°C.

    b. Diethyl phthalate is reported to have a normal boiling point of 296°C. How low would the pressure have to be to distill this substance at a temperature below 100°C?

Exercises

1. In the isolation of eugenol from oil of cloves (Experiment 8), the procedure can be extended to include vacuum distillation of the eugenol.

2. (R)-(−)- or (S)-(+)-Carvone can be isolated from oil of spearmint or oil of caraway by distillation under reduced pressure (Experiment 11).

3. In the preparation of dibenzylketone (Experiment 94), the final distillation of the product can be carried out much more easily under reduced pressure.

References

1. H. B. Hass, *J. Chem. Educ.* **13,** 490 (1936).

2. K. B. Wiberg, *Laboratory Technique in Organic Chemistry*, McGraw-Hill, New York, 1960.

12. Distillation of Mixtures of Two Immiscible Liquids; Steam Distillation

In a mixture of two completely immiscible liquids, each exerts its own vapor pressure independently of the other. As the temperature of such a mixture in an apparatus open to the atmosphere is raised, the vapor pressure of each substance increases until the total vapor pressure equals the pressure of the atmosphere. Because the total vapor pressure is the sum of the individual vapor pressures, the total vapor pressure must become equal to atmospheric pressure at a temperature below the boiling point of either pure substance. The mixture thus distills at a temperature below the boiling point of either pure component.

Because organic compounds are generally miscible with one another, this phenomenon is usually observed only when one of the liquids is water; in these cases, the distillation process is called *steam distillation*. Steam distillation is a useful technique for effecting certain separations and for distilling certain high-boiling compounds at a temperature no greater than 100°C.

12.1 Theory of Steam Distillation

Composition of the distillate

The composition of the vapor at the boiling point, and therefore the *composition of the distillate,* can be calculated by realizing that the ratio of the number of molecules of each type in the vapor must equal the ratio of the vapor pressures of the pure substances at that temperature:

$$\frac{P_B}{P_A} = \frac{\text{molecules of } B}{\text{molecules of } A} = \frac{\text{moles of } B}{\text{moles of } A} = \frac{n_B}{n_A} \qquad (12.1\text{-}1)$$

For example, a mixture of the immiscible liquids bromobenzene and water steam distills at 95.3°C at 760 Torr. The vapor pressure of water at this temperature is 641 Torr, and therefore the vapor pressure of bromobenzene must be 760 − 641 = 119 Torr. The vapor pressure ratio of 119/641 must equal the mole ratio of the vapor and thus of the distillate, as in Equation 12.1-2:

$$\frac{119}{641} = \frac{n_{\phi Br}}{n_{\text{water}}} = 0.186 \text{ mole } \phi Br \,/\, \text{mole water} \qquad (12.1\text{-}2)$$

Because 0.186 mole of bromobenzene at 157 grams per mole is (0.186 mole)(157 grams per mole) = 29.2 grams of bromobenzene, and 1 mole of water is 18 grams of water, 29.2 grams of bromobenzene will distill with each 18 grams of water, or 1.6 grams of bromobenzene will distill with each gram of water.

Composition of the distillate

The higher the ratio of organic compound to water the better, for less distillate will have to be collected (and hence it will take less time) to collect a given amount of the organic compound. The higher the vapor pressure of the organic compound, the more efficient the steam distillation. Solids with sufficiently high vapor pressures can also be steam distilled.

12.2 Technique of Steam Distillation

A substance can be distilled with steam by adding water to a mixture in which the substance (for example, a crude product from a reaction) is present and boiling the resulting suspension. If the desired compound has a significantly higher vapor pressure at about 100°C than any of the other components of the mixture, it can be selectively removed by this process. An apparatus suitable for this kind of operation is illustrated in Figure 12.2-1.

Test for completion of distillation

Look for oily droplets. The easiest way to determine whether a steam distillation is complete is to see that the distillate currently being produced contains no water-insoluble material. This can be done by collecting a sample of the distillate separately, or by observing either that the distillate flowing down the condenser no longer contains little oily droplets or that solid is no longer collecting in the cold condenser. (If solid tends to collect and block the condenser, the cooling water should be turned off until the solid has melted and run into the receiver.) No harm is done in allowing the steam distillation to continue somewhat past the point at which insoluble material can be observed in the newly formed distillate.

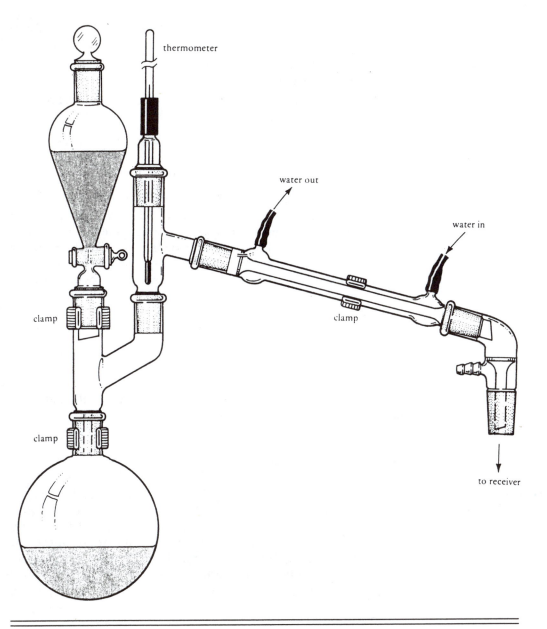

thermometer

water out

water in

clamp

clamp

to receiver

Figure 12.2-1. Apparatus for steam distillation.

Problems

1. In Experiment 9, oil of cloves is isolated from cloves by steam distillation. Assuming oil of cloves to be composed mostly of eugenol, which has a normal boiling point of 252°C,

   a. Estimate the vapor pressure of eugenol at 100°C.

   b. Calculate how many grams of water would be required to steam distill a gram of eugenol (oil of cloves).

Exercises

1. Isolation of oil of cloves from whole cloves (Experiment 9).

2. Other experiments that involve a purification by steam distillation include the oxidation of cyclohexanol to cyclohexanone (Experiment 20), the preparation of isoamyl bromide (Experiment 24), the preparation of aniline (Experiment 41), the preparation of chlorobenzene (Experiment 54), and the synthesis of l-bromo-3-chloro-5-iodobenzene (Experiment 105).

13.   Sublimation

Small samples of volatile solids can be purified by sublimation or vaporization followed by condensation.

13.1 Theory of Sublimation

Distillation versus sublimation

The processes of sublimation and distillation are similar. In the distillation of a liquid, the liquid is heated until it reaches the temperature at which its vapor pressure equals the total pressure of the system. The vapors then expand to a cooler part of the apparatus where they condense. If a solid is distilled, it first melts, then vaporizes, and then condenses and solidifies; the distillation of a solid is just like distillation of a liquid except for melting and solidification.

Sublimation

Some solids, however, have a sufficiently high vapor pressure that they will vaporize, or "distill" without melting, and such behavior is called *sublimation*. The loss of water from food that has been stored in a freezer for a long time takes place by sublimation ("freezer burn"). Similarly, wet clothes hung out to dry on a winter day first freeze and then dry by sublimation.

Very rarely does a substance have a vapor pressure of 760 Torr or greater at the normal melting point T_f, but many substances have a large enough vapor pressure near their melting point to be sublimed at the reduced pressures easily obtained in the laboratory. Table 13.1-1 lists a number of substances and their vapor pressures at the melting point.

Sublimation is a useful procedure for purification if the impurities are essentially nonvolatile and if the desired substance has a vapor pressure of at least a few Torr at its melting point.

Sublimation is most useful with small samples, as mechanical losses can be kept very low. One apparatus for sublimation is illustrated in Figure 13.1-1.

Table 13.1-1. Some substances that are easily sublimed under laboratory conditions

Compound	Melting Point $(T_f, °C)$	Vapor Pressure at T_f
Hexachloroethane	185	780 Torr
Perfluorocyclohexane	59	950 Torr
Camphor	179	370 Torr
Anthracene	218	41 Torr
Naphthalene	80	7 Torr
Benzene	5	36 Torr
p-Dibromobenzene	87	9 Torr
p-Dichlorobenzene	53	8.5 Torr
Phthalic anhydride	131	9 Torr
Benzoic acid	122	6 Torr
Carbon dioxide	−57.5	2 atm
Iodine	114	90 Torr

13.2 Technique of Sublimation

In a sublimation at atmospheric pressure, spread out the material to be purified in the bottom of the sublimator (Figure 13.1-1). Heat the bottom part cautiously without melting the solid until crystals appear on the cold inner surface of the condenser. At this point, only a little heat is needed to keep the sublimation going. The rate of sublimation is much easier to control if a sand bath or an oil bath is used for heating (Section 33.1). When only a nonvolatile residue remains, carefully remove the condenser and scrape out the product.

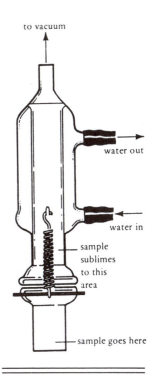

to vacuum

water out

water in

sample sublimes to this area

sample goes here

Figure 13.1-1. An apparatus for sublimation.

In sublimation at reduced pressure, the procedure is the same except that the top of the sublimator is connected to a vacuum source such as the water aspirator or the oil pump (Section 11.3).

Exercises

1. Benzoic acid has a vapor pressure of 6 Torr at the melting point and can be purified by vacuum sublimation as well as by crystallization. A mechanical vacuum pump should be used so that the process will not take too long.
2. In the procedure of Experiment 5, caffeine is purified by recrystallization from ethanol. An alternative method should be purification by sublimation, because caffeine is reported to sublime slowly at atmospheric pressure at temperarures above 120°C, to sublime quickly at 1 Torr at 160 to 165°C, and to sublime at 89°C and 15 Torr.

14. Extraction by Solvents

The fact that substances can differ in solubility in various liquids can serve as a basis for separating them. Recrystallization (Section 9) makes use of these differences, and extraction and chromatography (Section 15) do so as well.

14.1 Theory of Extraction

The separatory funnel

If a mixture of substances, say A and B, is dissolved in an organic solvent, such as diethyl ether, and the solution is mixed thoroughly with the immiscible solvent water, an essentially complete separation of A and B can be effected if one of the substances, say A, is much more soluble in water than in ether, and the other, B, is much more soluble in ether than in water. If at equilibrium most of the A molecules are in the water phase and most of the B molecules are in the ether phase, separating the water from the ether will separate the A molecules from the B molecules. A separation like this is done in a *separatory funnel,* shown in Figure 14.1-1, which allows one phase to be drawn off from the bottom, and the other to be poured out the top.

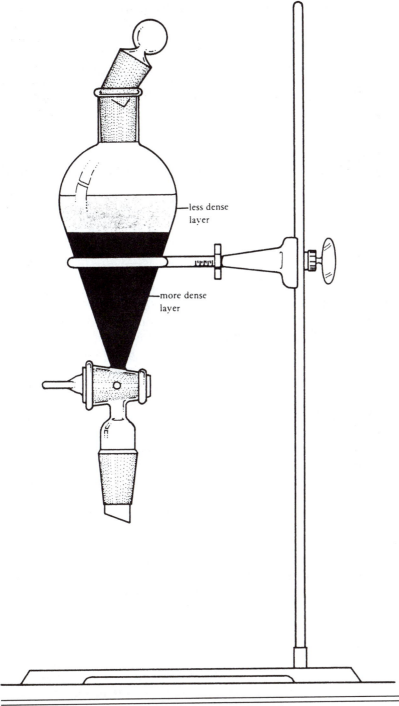

less dense layer

more dense layer

Figure 14.1-1. Extraction using a separatory funnel.

Table 14.1-1. Estimated relative solubility in oganics and water

Type of Compound	Estimated ratio of solubility in organic solvents to solubility in water
Covalent substances that contain only carbon, hydrogen, and halogen	Very much greater than 1
Covalent substances that containoxygen and/or nitrogen in addition to carbon, hydrogen and halogen	
a. 5 carbons per functional group	10:1
b. 2 carbons per functional group	1:1
c. 1 carbon atom per functional group	1:10
Salt of an organic acid	very much less than 1
Salt of an organic base; amine salt	very much less than 1
Inorganic salts	very much less than 1

Sometimes a small amount of a liquid, such as methyl alcohol, that is very soluble in water can be removed from an organic solution by shaking the organic solution with a portion of water. Shaking the mixture in the separatory funnel produces many small droplets and increases the surface between the organic and liquid phase, which facilitates movement of molecules from one phase to the other. After the mixture is allowed to stand so that the organic and aqueous phases form separate layers, most of the methyl alcohol molecules will be in the aqueous phase, and removal of this phase from the separatory funnel serves to remove these methyl alcohol molecules.

Table 14.1-1 gives a general indication of the relative solubility of different types of compounds in water and organic solvents.

14.2 Extraction of Acids and Bases

The separation of neutral organic molecules by extraction is not often successful because most organic molecules are generally much more soluble in organic solvents than in water. However, acids and bases, which can be converted to water-soluble ions, often can be completely removed from organic solutions.

Acids are extracted by base

The extraction of an acidic substance from an organic solution into water can be effected by mixing (shaking) the organic solution in the separatory funnel with an aqueous solution of base.

$$H-A \xrightarrow{\text{Base}} A:^-$$

acidic substance	**conjugate base of acidic substance**
covalent	**ionic**
soluble in organic solvents	**insoluble in organic solvents**
insoluble in water	**soluble in water**

In the extraction of an acidic material from an organic solvent with aqueous base, molecules of the acid in the organic solvent occasionally wander into the aqueous phase and are immediately converted to their ionic, conjugate base, form. The ionic form is very much more soluble in water than in the organic solvent, and so the ions never go back into the organic layer.

Bases are extracted by acid

Conversely, the extraction of a basic substance from an organic solution into water can be effected by mixing (shaking) the organic solution in the separatory funnel with an aqueous solution of acid.

$$B: \xrightarrow{\text{Acid}} H-B^+$$

basic substance	**conjugate acid of basic substance**
covalent	**ionic**
soluble in organic solvents	**insoluble in organic solvents**
insoluble in water	**soluble in water**

In the extraction of a basic material from an organic solvent with aqueous acid, molecules of the base in the organic solvent occasionally wander into the aqueous phase and are immediately converted to their ionic, conjugate acid, form. Again, the ionic form is very much more soluble in water than in the organic solvent, and so the ions never return to the organic layer.

For the extraction of an acidic substance from an organic solution into water to be successful, the aqueous solution must be basic enough to convert the acid to its conjugate base. A useful guide is that the pH of the aqueous solution used to extract an acid from an organic solvent must be at least 4 pH units on the basic side of the pK_a of the acid to be extracted. Thus, aqueous sodium carbonate solution (pH 11) can be used to extract acids stronger than $pK_a = 7$ (see Table 14.2-1).

To extract an acid

Similarly, to extract a base from an organic solvent, you must use an aqueous solution whose pH is at least 4 units on the acidic side

To extract a base

of the pK_a of the conjugate acid of the amine. According to Table 14.2-1, aqueous HCl or aqueous sulfuric acid will be needed to extract all but aliphatic amines.

Recovery of extracted organic acids and bases

Often, unwanted acids and bases are removed from organic solutions by extraction with aqueous base or acid. Sometimes, however, the acidic material or the basic material is the desired compound. If you want the acidic material, for example, you acidify the separated basic aqueous solution, thereby converting the conjugate base of the desired acid to the acidic form. If the solid is water insoluble, it can be recovered by suction filtration. Alternatively, the desired acid can extracted into an organic solvent, such as ether or dichloromethane, to yield, after evaporation of the solvent, the acidic product. This procedure is presented in detail in Section 39.

Table 14.2-1. Required pH for extraction of different acids and bases

Appoximate pH Needed to Extract an Acid from an Organic Solvent		
Compound	pK_a	pH as Basic or More Basic Than:
Mineral acids	>1	about 4
Carboxylic acids	about 5	about 9
Phenols	about 10	about 14

Approximate pH Needed to Extract a Base from an Organic Solvent		
Compound	pK_a	pH as Acidic or More Acidic Than:
Anilines	about 5	about 1
Pyridines	about 6	about 2
Aliphatic amines	about 11	about 7

Approximate pH of Aqueous Solutions, 5 to 10% by Weight	
Compound	Approximate pH
HCl; H_2SO_4	0
Acetic acid	3
$NaHCO_3$	8
Na_2CO_3; K_2CO_3	11
NaOH; KOH	14

An analogous treatment of an acidic extract with base can lead to recovery of a basic product, as explained more fully in Section 39.

14.3 Technique of Extraction

The objective of a simple extraction is to partition one or more substances between two immiscible solvents. This is usually accomplished with the use of a *separatory funnel* (Figure 14.1-1). If the separatory funnel has a glass stopcock, prepare the funnel for use by making sure that the stopcock is lightly greased and will turn without difficulty; Teflon stopcocks need not be greased.

Close the stopcock

With the separatory funnel supported in a ring (Figure 14.1-1), check to make sure that the stopcock is closed and pour in the solution to be extracted. Then add the extracting solvent (the funnel should not be filled to more than about three-fourths of its height), replace the stopper after wetting it with water (to keep the organic solvent from creeping out around the stopper), and swirl or shake the contents to mix them. With vigorous shaking, a total mixing time of 10 to 30 seconds is usually enough to establish equilibrium. After allowing the mixture to stand in the separatory funnel until the two immiscible layers have separated cleanly, remove the stopper at the top and draw off part of the lower layer through the stopcock at the bottom. Wait a little while for the remainder of the lower layer to drain down (gentle swirling of the separatory funnel can speed this up), and draw this off also. Then pour the upper layer out the top.

Mix thoroughly

Allow to stand

Let the air in!

Release the pressure

When a volatile solvent is involved in an extraction, the establishment of the equilibrium vapor pressure of the solvent will cause the pressure to rise inside the stoppered separatory funnel. The pressure is best released by turning the funnel upside down (with the stopper held in place with the palm of the hand) and cautiously opening the stopcock. When a very volatile solvent such as ether is being used, the first mixing should consist only of a slow inversion of the separatory funnel followed by release of the pressure. After alternate cautious sloshing of the contents of the separatory funnel and then release of pressure, the sound of the escaping vapors will indicate that the pressure is not being built up so fast and the periods of mixing can be longer and more vigorous. If you neglect to release the pressure inside the funnel, the stopper may be forced out. If the contents of the funnel are forced out as well, the escape of a volatile and flammable solvent can result in a dangerous fire.

Cool the solution

It is dangerous to try to extract a solution if its temperature is near or above the boiling point of the extracting solvent. Thus, if you plan to extract with pentane, ether, or dichloromethane, you may need to cool the solution. When these solvents are used, it is also a good idea to hold the separatory funnel by the ends so that the contents will not be warmed by your hands.

Beware of CO_2

If a strong acid is to be extracted with carbonate or bicarbonate solution, the carbon dioxide produced can cause a large buildup in pressure unless mixing is done very cautiously with frequent release of pressure. In cases for which much carbon dioxide production is anticipated, it is best to do the mixing in a flask or beaker and then to transfer the mixture to the separatory funnel for separation.

Choice of solvent

If an organic product is to be purified by dissolution in an organic solvent followed by extraction of the solution with two or more portions of aqueous solution, the whole process will be much faster and easier and will involve less loss if the organic solution is less dense than water. When the water layer is the more dense layer, the water layer can be drawn off through the stopcock, and the organic solution is retained in the funnel, ready for the next extraction. If, however, the organic phase is heavier than water, the organic layer will have to be drawn off through the stopcock, the aqueous layer poured out, and the organic layer returned to the separatory funnel for the next extraction. Each such transfer will take time and may involve a loss of material. Conversely, if it is necessary to extract an aqueous solution with several portions of solvent to achieve the maximum recovery of a substance, it will be more convenient to extract with a solvent heavier than water so that the solvent can simply be drawn off each time through the stopcock without removal of the water layer first.

ether or hexane

dichloromethane

The typical first choices for an organic solvent less dense than water are diethyl ether and hexane or petroleum ether (a mixture of low-boiling alkanes). A common first choice for an organic solvent more dense than water is dichloromethane.

The densities of a number of solvents are listed in Table 9.11-1.

Extraction in batches

If you have to extract a larger volume of material than will fit into your separatory funnel, the extraction can be done in batches. Small amounts of organics can be removed from large amounts of water by adding a little ether (or other solvent less dense than water) to the flask containing the water/product mixture, swirling the mixture well, and then adding it in portions to the separatory funnel, drawing off most of the water layer after each addition. The converse of this procedure can be used to wash a large amount of an organic solution with a little water, if the organic solution is heavier than water.

More suggestions

Often you can tell which layer in the separatory funnel is organic and which is aqueous by knowing the relative volumes used or relative densities of the two solvents. Sometimes, however, the transfer of material from one layer to the other or the presence of several substances of different densities can make the identification of the layers uncertain.

Smell won't tell! You cannot tell by smell which layer is which, since the vapor pressure of each component in each phase will be the same. A better way is to add a little water and see with which layer the added water combines. The best way is to withdraw a little of the lower layer and see if it is miscible with water.

Save everything! If doubt still remains, each layer should be carried on in the procedure as if it were the desired layer until it becomes obvious that one of the two cannot be the right one. It is very common to discard the product layer through error or ignorance; thus, you should always *save everything* until the product has been safely isolated.

Interfacial "crud." In most extractions, at least a trace of insoluble material collects at the interface between the two immiscible layers. Separating the layers without taking along some of this insoluble material is often impossible. This is not a crucial matter, because whatever is picked up can always be removed by filtration at the end of the extraction or at some later stage in the purification. For example, in the very common case in which the organic product of a reaction is isolated by dissolving it in an organic solvent, extracting ("washing") the solution with one or more aqueous solutions to remove certain undesired materials, drying it (Section 16.2), and removing the drying agent by gravity filtration, any insoluble impurity that may have been carried along in the organic layer will be removed along with the drying agent in the gravity filtration.

Salting out. Organic substances are less soluble in solutions of salts than in pure water. Thus, adding a salt such as sodium chloride, sodium sulfate, or potassium carbonate to the aqueous layer will permit more complete extraction of the desired organic material.

Only one layer? Sometimes the mixture in the separatory funnel does not separate into two phases but forms a single, homogeneous solution. This can happen when large amounts of methanol, ethanol, or tetrahydrofuran are present, as these liquids are good solvents for both water and organic materials.

Sometimes separation into two liquid phases can be brought about by the addition of more water and more of the organic solvent, or by adding sodium chloride or potassium carbonate to the mixture. To avoid this problem, remove most of the methanol, ethanol, or tetrahydrofuran, possibly by distillation, before doing the extraction.

Crystallization. Occasionally, the solid being purified will begin to crystallize during the extraction. This happens when the amount of organic solvent is reduced below that required to dissolve all of the solid. The loss of the organic solvent occurs because of its slight solubility in the aqueous liquids that are used to wash the organic solution. Or, this can happen because another substance, perhaps an alcohol, that helped dissolve the solid product in the organic layer has itself been removed by extraction. No matter what the cause, addition of more of the same solvent or of a better solvent should bring the solid material back into solution.

Emulsions

A common problem in extraction is failure of the immiscible solutions to separate completely and cleanly into two layers; a certain volume of the mixture at the interface sometimes consists of droplets of one solution suspended in the other, an *emulsion*. Sometimes, the separation becomes clean and complete if the separatory funnel and its contents are allowed to stand undisturbed for a few minutes. Other times, the emulsion may persist for hours or days. If the volume of the emulsion is small, it can sometimes be temporarily ignored and the extraction procedure continued in the hope that the emulsion will disappear in later extractions. But if most of the mixture is emulsified, it must either be allowed to stand (if the time is available) or be broken in some other way.

Try adding solvent. If the emulsion is caused by too small a difference in density between the two layers, the addition of solvent can produce a larger density difference. Pentane will most efficiently decrease the density of an organic solution, whereas dichloromethane can be used to increase the density of an organic solution. Adding

Change the density

water can either increase or decrease the density of the aqueous phase, depending on its composition. If the aqueous phase contains appreciable amounts of organic solvents (such as alcohols), it may have a density of less than one gram per milliliter; if, on the other hand, it is a solution of inorganic materials, its density will be greater than one gram per milliliter. Adding salt or saturated sodium chloride solution can also increase the density of the aqueous phase.

Emulsions are most often found in extractions involving basic solutions. Presumably, this is because traces of higher-molecular-weight organic acids are converted to their salts, and the resulting soap causes emulsions to form.

The tendency of an extraction with a "neutral" aqueous solution to emulsify can sometimes be overcome by adding a few drops of acetic acid, thus suppressing soap formation. Stubborn emulsions can sometimes be broken by centrifugation or suction filtration of the emulsified material. Suction filtration of an emulsion is a very messy operation and should be done only in desperation.

Change the pH

With emulsions, prevention is far better than cure. If the mixture is sloshed or shaken gently at first and then allowed to stand, the tendency toward emulsion formation can be estimated. When it appears that emulsion formation may be a problem, it is wise to mix the layers more gently for a longer time. In extreme cases, it may be desirable to carry out the mixing in a round-bottom flask by slow stirring with a magnetic stirrer. In this case, because the area of the interface will be much less than if the mixture were broken up into a suspension of tiny bubbles by vigorous shaking or stirring, it may take 30 to 60 minutes to approach equilibrium.

Prevention of emulsions

14.4 Micro-Scale Extraction

When you are working with 100- to 200-milligram amounts and solution volumes of less than 10 mL, extraction using a separatory funnel can lead to substantial losses as the solutions are transferred, and an alternative micro-scale extraction technique can be useful.

Use a stoppered test tube

In one micro-scale approach, the separatory funnel is replaced by a stoppered test tube, and solutions are transferred by use of Pasteur pipettes.

Less dense organic layer on top

If, for example, an ethereal solution is to be extracted or "washed" with a series of aqueous solution, the ethereal solution is transferred to the test tube and the first wash solution is added. After the test tube is stoppered, the tube is shaken briefly to mix the contents, and then the tube is set aside to allow the layers to separate. After the layers have separated, a Pasteur pipette is used to reach the bottom of the test tube, and the aqueous layer alone is drawn into the pipette, which is then withdrawn to leave the ethereal layer behind in the test tube. Now another aqueous solution can be added, and the process repeated.

More dense organic layer on the bottom

If, on the other hand, you want to wash a dichloromethane solution with several less-dense aqueous solutions, you add the aqueous wash to the test tube with the dichloromethane solution to be extracted, and, after mixing and allowing the layers to separate, you use a Pasteur pipette to transfer the more dense dichloromethane layer to a second test tube just like the first. The next wash is done in the second test tube, and the more dense dichloromethane layer is again removed by means of a Pasteur pipette and returned to the first tube, which you have cleaned if necessary.

Problems

1. Estimate whether or not each of the following acids could be "completely extracted" with one portion of an excess of (a) aqueous sodium bicarbonate solution, (b) aqueous sodium carbonate solution, and (c) aqueous sodium hydroxide solution.

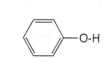

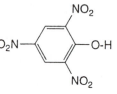

trichloroacetic acid	butyric acid	phenol	picric acid
$K_a = 0.2$	$K_a = 0.000015$	$K_a = 1.3 \times 10^{-10}$	$K_a = 0.16$
$pK_a = 0.70$	$pK_a = 4.52$	$pK_a = 9.89$	$pK_a = 0.80$

2. Using extraction procedures only, how would you separate a mixture of p-dichlorobenzene, p-chlorobenzoic acid, and p-chloroaniline?

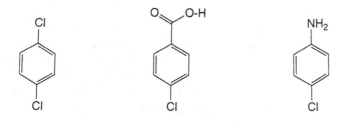

p-dichlorobenzene   **p-chlorobenzoic acid**   **p-chloroaniline**

3. Why is carbon dioxide produced when a strong acid, such as sulfuric acid or hydrochloric acid, is neutralized with sodium bicarbonate or with sodium carbonate? Illustrate your answer with balanced equations.

Exercises

1. Extraction of caffeine from tea or NoDōz (Experiment 5).
2. Extraction of eugenol from oil of cloves (Experiment 9).
3. Separation of a three-component mixture: an acid, A—H, a base, B:, and a neutral substance, N.

Procedure

Separation of the components. Dissolve about 5 grams of the mixture in 50 mL of diethyl ether and transfer the solution to a 125-mL separatory funnel. Add to the funnel 30 mL of 1 M HCl solution. Shake the mixture well to extract the basic substance as its hydrochloric acid salt into the water layer:

$$\text{B:} + H_3O^+ + Cl^- \longrightarrow H—B^+ + Cl^- + H_2O$$

Draw off the lower, aqueous layer and save it.
Add to the ether solution in the separatory funnel about 30 mL of 1 M NaOH solution. Shake the mixture well to extract the acidic substance as its sodium salt into the water layer:

$$A—H + HO^- + Na^+ \longrightarrow A:^- + Na^+ + H_2O$$

Draw off the lower, aqueous layer and save it. The ether layer should now contain only the neutral substance N.

Isolation of the neutral substance, N. Wash the ether layer by adding to the separatory funnel about 25 mL of water, shaking the mixture, allowing the layers to separate, and then drawing off and discarding the lower, aqueous, layer. Transfer the ethereal solution of the neutral compound to a small Erlenmeyer flask, dry the solution over 1 to 2 grams of anhydrous sodium sulfate for a few minutes (Section 16.2), remove the drying agent by gravity filtration (Section 8.1), and remove the ether by distillation on the steam bath (Section 37). Collect the ether in a cooled receiver, and then pour it into the container reserved for used ether. The infrared spectrum of the residue may be determined (Section 24); or, if the residue is a solid, it can be recrystallized (Section 9) and its melting point determined (Section 18.1).

Isolation of the basic substance, B. Transfer the aqueous solution of the hydrochloric acid salt of the basic substance to a clean 125-mL separatory funnel. Make the solution strongly basic by adding about 2 mL of 50% aqueous sodium hydroxide. Make sure the solution is well mixed. The basic substance will now be present as the free base:

$$H\!-\!B^+ + Cl^- + Na^+ + HO^- \longrightarrow B\!: + Na^+ + Cl^- + H_2O$$

Add 25 mL of ether to the separatory funnel. Shake the mixture well to extract the free base into the ether layer. Draw off the lower, aqueous layer (which may now be discarded) and transfer the ethereal solution of the basic substance to a small Erlenmeyer flask. Dry the solution over 1 to 2 grams of anhydrous potassium carbonate for a few minutes, remove the drying agent by gravity filtration, and remove the ether by distillation on the steam bath. Collect the ether in a cooled receiver, and then pour it into the container reserved for used ether. The infrared spectrum of the residue can be determined; or, if the residue is a solid, it can be recrystallized and its melting point determined.

Isolation of the acidic substance, A—H. Transfer the aqueous solution of the sodium salt of the acidic substance to a clean 125-mL separatory funnel. Make the solution strongly acidic by adding about 3 mL of concentrated HCl. Make sure the solution is well mixed. The acidic substance will now be present as the free acid:

$$A\!:^- + Na^+ + H_3O^+ + Cl^- \longrightarrow A\!-\!H + Na^+ + Cl^- + H_2O$$

Add 25 mL of ether to the separatory funnel. Shake the mixture well so as to extract the free acid into the ether layer. Draw off the lower, aqueous layer (which can now be discarded, and transfer the ethereal solution of the acidic substance to a small Erlenmeyer flask. Dry the solution over 1 to 2 grams of anhydrous sodium sulfate for a few minutes, remove the drying agent by gravity filtration, and remove the ether by distillation on the steam bath. Collect the ether in a cooled receiver, and then pour it into the container reserved for used ether. The infrared spectrum of the residue can be determined; or, if the residue is a solid, it can be recrystallized and its melting point determined.

Figure 14.4-1 is a flow chart that summarizes the procedures for **separation of acidic, basic, and neutral substances.**

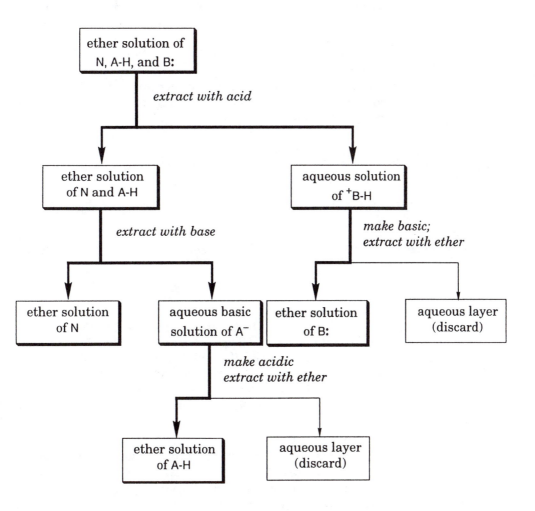

Figure 14.4-1. Flow chart for the separation of acidic, basic, and neutral substances.

4. Separation of the components of a commercial mixture of aspirin, phenacetin, and caffeine.

Some brands of headache or cold tablets, the so-called APC tablets, contain a mixture of *acetylsalicylic acid* (aspirin), *phenacetin*, and *caffeine*. It is possible to take advantage of the acid–base properties of these compounds to separate them by an extraction procedure. *Caffeine*, whose conjugate acid has a pK_a of –0.16, can just be extracted as the conjugate acid from chloroform into 4 M HCl. After the acidic extract has been neutralized, caffeine can be reextracted from the water with additional chloroform and isolated by distillation of the chloroform. *Aspirin*, having a pK_a of 3.49, can be extracted from chloroform by 0.5 M aqueous sodium bicarbonate solution. After the aqueous extract is neutralized by addition of HCl, the precipitated aspirin can be isolated by reextraction with more chloroform and recovered by distillation of the chloroform. *Phenacetin*, a substance that is neither acidic nor basic, remains in the original chloroform solution and is recovered by distillation of the chloroform after the other two substances have been removed.

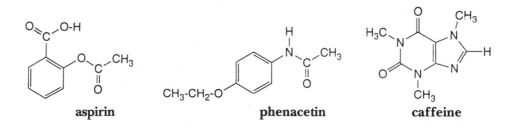

aspirin   phenacetin   caffeine

Procedure

Isolation of caffeine. Crush 3 APC tablets (Note 1) and add them to a separatory funnel that contains 25 ml of chloroform. Then add to the funnel 20 mL of 4 M HCl (80 milliequivalents of acid). Shake the funnel to thoroughly mix the contents (Note 2). Allow the funnel to stand so that the layers will separate. Draw off and save the lower, chloroform layer that contains the unextracted aspirin and phenacetin (Note 3).

Recovery of caffeine. Neutralize the aqueous acidic extract of caffeine in the separatory funnel by adding 7.0 grams of solid sodium bicarbonate (83 meq of base). Because a lot of carbon dioxide will be produced, you should add the sodium bicarbonate in portions. Swirl the contents of the funnel after each portion has appeared to react. After all the base has been added and the reaction appears to be complete, add 10 mL of chloroform, stopper the funnel, invert it, and release the pressure by opening the stopcock. Cautiously mix the contents of the funnel and frequently release the pressure by opening the stopcock of the inverted separatory funnel. Finally, when the pressure does not build up any more, shake the funnel to thoroughly mix the contents of the funnel (Note 2), allow the layers to separate (Note 4), and draw off the lower, chloroform layer into a small Erlenmeyer flask labeled "caffeine." Reextract the neutralized acid solution with a second 5-mL portion of chloroform and add this further chloroform extract to the labeled flask. Dry the combined extracts with a small portion of anhydrous magnesium sulfate (Section 16.2) and filter the mixture by gravity (Section 8.1) into a small, weighed Erlenmeyer flask containing a boiling stone (Note 5). Remove the chloroform by distillation on the steam bath (Note 6) and determine the weight of the residual solid by reweighing the flask. The recovered caffeine (Note 7) can be recrystallized from 10 mL of carbon tetrachloride, if desired. The melting point of caffeine is reported to be 238°C, with sublimation starting at 170°C, and its infrared and NMR spectra are shown in Experiment 5.

Isolation of aspirin. Place the original chloroform solution that was saved from the first extraction in a clean separatory funnel. Then add 25 mL of 0.5 M sodium bicarbonate solution (12.5 meq of base) and thoroughly mix the contents of the funnel (Note 2) so as to extract the aspirin into the basic water layer. Because carbon dioxide will be formed, the funnel must be vented occasionally to prevent too great an increase in pressure. After mixing, allow the funnel to stand undisturbed so that the layers can separate. Draw off the lower, chloroform layer into a small Erlenmeyer flask labeled "phenacetin" and add to this a small amount of anhydrous magnesium sulfate.

Recovery of aspirin. Add to the basic aqueous solution of aspirin in the separatory funnel 5 mL of 4 M HCl (20 meq of acid). Mix the contents of the funnel by swirling it gently. Carbon dioxide will be evolved, and the aspirin will separate as a solid. Recover the aspirin by extracting first with a 15-mL portion of chloroform and then with a 5-mL portion. Combine these two extracts in a small Erlenmeyer flask labeled "aspirin." Dry the aspirin extracts with a small portion of anhydrous magnesium sulfate, and then filter the mixture by gravity into a small, weighed Erlenmeyer flask containing a boiling stone (Note 5). Remove the chloroform by distillation on the steam bath (Note 6) and determine the weight of the residue. The recovered aspirin (Note 8) can be recrystallized by adding 5 mL of water, heating the mixture on the steam bath, and adding slightly more than the minimum amount of 95% ethanol required to dissolve the solid; about 1.5 mL should be added. Aspirin is reported to melt at 135°C with rapid heating. Its IR and NMR spectra are shown in Experiment 3.

Recovery of phenacetin. Remove the magnesium sulfate from the chloroform solution of phenacetin by gravity filtration, collecting the filtrate in a small, weighed Erlenmeyer flask containing a boiling stone (Note 5). Remove the chloroform by distillation on the steam bath (Note 6) and determine the weight of the residual solid. The recovered phenacetin (Note 9) can be recrystallized from a very small amount of 95% ethanol. The melting point of phenacetin is reported to be 134 to 135°C, and its IR and NMR spectra are shown in Experiment 74.

Figure 14.4-2 is a flow chart that summarizes the procedures for the **separation of aspirin, caffeine, and phenacetin.**

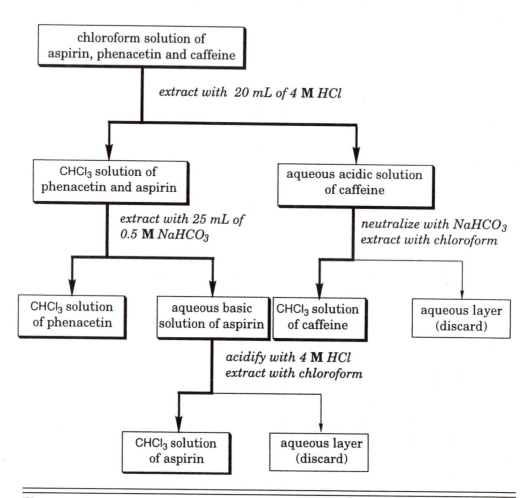

Figure 14.4-2. Flow chart of the separation of aspirin, caffeine, and phenacetin

Notes

1. Three APC tablets weigh 1.5 grams and contain 10.5 grains aspirin (680 mg), 7.5 grains phenacetin (486 mg), and 1.5 grains caffeine (97 mg).
2. One full minute of continuous shaking is sufficient.
3. If you have a second separatory funnel, this extract can be added directly to it, and two parts of the experiment can be carried out simultaneously.
4. If an emulsion is formed, it can be broken by adding several drops of glacial acetic acid to the separatory funnel.
5. Weigh the flask with the boiling stone in it.
6. Put the recovered chloroform into the container reserved for it. The last of the chloroform may have to be removed by heating the flask without the adapters and condenser attached.
7. Only about half the caffeine is recovered, and NMR analysis shows that it is contaminated with phenacetin.
8. About three-fourths of the aspirin is recovered, and it appears to be pure by NMR analysis.
9. Almost all the phenacetin is recovered, but it contains a small amount of a less soluble material, perhaps part of the binder. Analysis by NMR shows it to contain between 5 and 10% caffeine and something that contributes a single peak that overlaps the highest field component of the methyl triplet (Reference 2).

References

1. P. Haddad and M. Rasmussen, *J. Chem. Educ.* **53,** 731 (1976).
2. D. P. Hollis, *Anal. Chem.* **35,** 1682 (1963).

15. Chromatography

Chromatographic separations, like extractions, take advantage of the fact that different substances are partitioned differently between two phases. In technique, however, the two methods differ greatly, and therefore chromatographic methods are discussed separately in this section.

15.1 Theory of Column Chromatography

In column chromatography, a solid phase, the *adsorbent,* is held in a vertical tube, the mixture to be separated (*A* plus *B*) is placed on top of the column of adsorbent, and a solvent, the *eluant,* is allowed to flow down through the column. At all times, a certain fraction of each component of the mixture will be adsorbed by the solid and the remainder will be in solution. Any one molecule will spend part of the time sitting still on the adsorbent with the remainder flowing down the column with the solvent. A substance that is strongly adsorbed (say, *A*) will have a greater fraction of its molecules adsorbed at any one time, and thus any one molecule of *A* will spend more time sitting still and less time moving. In contrast, a weakly adsorbed substance (*B*) will have a smaller fraction of its molecules adsorbed at any one time, and hence any one molecule of *B* will spend less time sitting and more time moving. Thus, the more weakly a substance is adsorbed, the faster it will get to the bottom of the column and flow out with the eluant. Because the eluant is collected in small portions (fractions), the early fractions will contain *B* and the later ones will contain *A*. The name *chromatography* was given to this process because the substances to which this method of separation were first applied were colored plant pigments.

Adsorbent

Selectivity of
adsorbent

Several factors determine the efficiency of a chromatographic separation. The *adsorbent* should show a maximum of selectivity toward the substances being separated so that the differences in rate of elution will be large. For the separation of any given mixture, some adsorbents may be too strongly adsorbing (holding all components near the top of the adsorbent) or too weakly adsorbing (allowing all components to move through the adsorbent almost as fast as the eluting solvent). Table 15.1-1 lists a number of adsorbents in order of adsorptive power.

Eluting solvent

Selectivity of
solvent

The *eluting solvent* should also show a maximum of selectivity in its ability to dissolve or desorb the substances being separated. The fact that one substance is relatively soluble in a solvent can result in its being eluted faster than another substance. However, a more important property of the solvent is its ability to be itself adsorbed on the adsorbent. If the solvent is more strongly adsorbed than the substances being separated, it can take their place on the adsorbent and all the substances will flow together. If the solvent is less strongly

Table 15.1-1. Chromatographic adsorbents. The order in the table is approximate, because it depends on the substance being adsorbed and the solvent used for elution.

Most Strongly Adsorbent	
Alumina	Al_2O_3
Charcoal	C
Florisil	MgO/SiO_2 (anhydrous)
Silica gel	SiO_2
Lime	CaO
Magnesia	MgO
Magnesium carbonate	$MgCO_3$
Calcium phosphate	$Ca_3(PO_4)_2$
Calcium carbonate	$CaCO_3$
Potassium carbonate	K_2CO_3
Sodium carbonate	Na_2CO_3
Talc	MgO/SiO_2 (hydrous)
Sucrose	Carbohydrate (polyhydroxylic)
Starch	Carbohydrate (polyhydroxylic)
Least Strongly Adsorbent	

adsorbed than any of the components of the mixture, its contribution to different rates of elution will be only through its difference in solvent power toward them. If, however, it is more strongly adsorbed than some components of the mixture and less strongly than others, it will greatly speed the elution of those substances that it can replace on the column, without speeding the elution of the others.

Table 15.1-2 lists a number of common solvents in approximate order of increasing adsorbability, and hence in order of increasing eluting power. The order is only approximate because it depends on the nature of the adsorbent. Mixtures of solvents can be used, and, because increasing eluting power results mostly from preferential adsorbtion of the solvent, addition of only a little (0.5–2%, by volume) of a more strongly adsorbed solvent will result in a large increase in the eluting power. Because water is among the most strongly adsorbed solvents, the presence of a little water in a solvent can greatly increase its eluting power. For this reason, solvents to be used in column chromatography should be quite dry. (See Section 16.3 for methods of drying solvents.)

Solvents should be dry

The particular combination of adsorbent and eluting solvent that will result in the acceptable separation of a particular mixture can be determined only by trial. Alumina is the most commonly used adsorbent because it is readily available at relatively low cost and has

Table 15.1-2. Eluting solvents for chromatography

Least Eluting Power (alumina as adsorbent)

Petroleum ether (hexane; pentane)
Cyclohexane
Carbon tetrachloride
Benzene
Dichloromethane
Chloroform
Ether (anhydrous)
Ethyl acetate (anhydrous)
Acetone (anhydrous)
Ethanol
Methanol
Water
Pyridine
Organic acids

Greatest Eluting Power (alumina as adsorbent)

a wide range of adsorptive power, depending upon how much water it has adsorbed. The adsorptive power of alumina can be decreased by adding up to 15% by weight of water. If alumina has become too heavily hydrated in storage, it can be "activated" by heating it at 200°C for about three hours. A trial to determine the conditions for a chromatographic separation might be made by preparing a small column using partially hydrated alumina, adding a sample of the mixture, and starting elution with a solvent of weak eluting power (hexane, for example). Solvents of successively greater eluting power (dichloromethane, dry ether, dry acetone, methanol; see Table 15.1-2) can then be tried until one is found that will move the material down the column. Further, more sensitive, trials can be made using different solvents, or mixtures of solvents, of similar eluting power. If the trial adsorbent is too strongly adsorbing, a more completely hydrated alumina can be tested or a weaker adsorbent can be tried; if it is too weakly adsorbing, a less hydrated grade of alumina can be tested.

   If the substances in the mixture differ greatly in adsorbability, it will be much easier to separate them. Often, when this difference is present, a succession of solvents of increasing eluting power is used. One substance may be eluted easily while the other stays at the top of the column, and then the other can be eluted with a solvent of greater eluting power. Table 15.1-3 indicates an approximate order of adsorbability by functional group.

Table 15.1-3. Adsorbability of organic compounds by functional group

Least Strongly Adsorbed

Saturated hydrocarbons; alkyl halides
Unsaturated hydrocarbons; alkenyl halides
Aromatic hydrocarbons; aryl halides
Polyhalogenated hydrocarbons
Ethers
Esters
Aldehydes and ketones
Alcohols
Acids and bases (amines)

Most Strongly Adsorbed

Comparison of this table with Table 15.1-2 indicates that there is a relationship between the eluting power of a solvent and the tendency of the solvent to be adsorbed.

15.2 Technique of Column Chromatography

The column

Several types of tubes can be used to support the adsorbent. One of the most satisfactory and least expensive is illustrated in Figure 15.2-1. A chromatographic column like this, essentially a giant medicine dropper, can be made by heating, drawing down, cutting off, and fire polishing a long piece of glass tubing 10 to 25 mm in diameter. The column should be at least 50 cm long so that the column of eluting solvent above the adsorbent can be high enough to contribute a sufficient hydrostatic head to achieve an acceptably large flow rate. If the tip is not too narrow, the flow of solvent through the column can be stopped by slipping a rubber policeman over the tip. Otherwise, a piece of rubber tubing and a clamp can be used. Burets and other columns with stopcocks are sometimes used, but they have the disadvantage that the stopcock grease will be eluted along with the other substances in the system, and the grease will contaminate the material being purified. Small trial chromatograms can be run in medicine droppers.

Figure 15.2-1. A chromatographic column.

Packing the column

You can pack the column with adsorbent by partially filling the column with a liquid of low adsorbability (petroleum ether, for example) and slowly adding the dry powdered adsorbent down the top so that it settles evenly and uniformly. Alternatively, the adsorbent can be mixed with the liquid to form a thin slurry, which is then poured and rinsed into the column. The objective is to form a uniform bed of adsorbent without holes, channels, or air bubbles. The separation will be best if the bottom of the bed of adsorbent is flat. Therefore, if you are using a column that is tapered or has some other odd shape at the bottom, you should first partially fill the column with solvent, then push a plug of glass wool down to fill up the odd volume, and finally add a little sand, rinsing it down to form a flat base on which the adsorbent can come to rest. After all traces of adsorbent have been rinsed down (solvent will usually have to be drained off during this process) and the top of the bed of adsorbent has been flattened by jiggling the tube, you should drop a disk of filter paper just a little smaller than the diameter of the tube flat on the adsorbent, and then sprinkle enough sand to form a layer 2 or 3 mm deep on the paper. The purpose is to protect the top of the column of adsorbent so that it will not be disturbed when the sample or the eluting solvent is poured onto it. Figure 15.2-2 shows a column ready for use.

The ratio of the weight of adsorbent to the weight of sample may vary greatly but usually will not be much outside the range of 25 or 50 to 1. The ratio of the height of the bed of adsorbent to its diameter should normally be between 3:1 and 10:1. If the ratio is greater, the rate of flow of eluant may be too low. Thus, when larger samples are to be chromatographed, large-diameter columns must be used and not just deeper beds of adsorbent. The finer the adsorbent, the greater its surface area and hence its adsorptive capacity, but the lower the flow rate. Sometimes it is necessary to use a mixture of adsorbent plus diatomaceous earth (Celite) as a nonadsorbing diluent to attain a minimum acceptable flow rate.

Adding the sample

Once the column has been prepared, the solvent should be allowed to drain until it is just level with the top of the sand that covers the adsorbent. If the column is allowed to drain dry, the bed of adsorbent usually will crack, and its ability to separate will be greatly diminished because the solution will be able to flow through the cracks. It is a waste of time to try to use a column with cracked or channeled adsorbent.

The sample is then added. A liquid can be added directly; a solid should be dissolved in as little as possible of a solvent of low

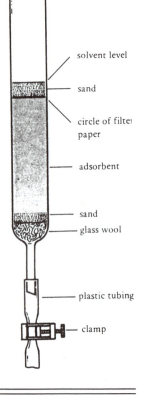

Figure 15.2-2. A simple chromatographic column ready for use.

solvent level

sand

circle of filter paper

adsorbent

sand

glass wool

plastic tubing

clamp

eluting power. After the addition, the column is allowed to drain again until the level of liquid has fallen just to the top of the adsorbent. If necessary, rinse the sample down from the walls of the column with additional small portions of solvent, draining the column each time. You will find that a medicine dropper is useful for these additions. The goal is to get the sample adsorbed in a minimum layer of adsorbent. The narrower the sample band, the better the separation, because narrow bands will overlap less.

Elution

After the sample has been added, the column is eluted using a solvent or series of solvents as recommended or as determined by trial. In a new situation, you should start with a solvent of low eluting power (petroleum ether, hexane) and work up through solvents of increasing eluting power (dichloromethane, ether, etc.) either as mixtures or as pure solvents. When the level of liquid in the column is low, the eluting solvent should be added very cautiously to avoid disturbing the top of the bed of adsorbent. During elution, the solvent should not be allowed to flow through any sort of rubber or plastic tubing, since material can be eluted from the tubing as well.

Collecting the fractions

If the substances to be separated are colored, their progress down the column can be followed visually, and the eluant that contains each component can be collected separately. If the substances are colorless, the eluant must be collected in successive fractions, and the presence of components of the original mixture must be determined by some analytical procedure. The most common method is to collect fractions of approximately equal volume in tared (previously weighed) flasks, evaporate or distill the solvent (Section 37) and reweigh to determine the weight of the residue. If the residue is a solid, it will usually crystallize and can then be easily seen. The residue can be identified by determining one or more physical properties such as the melting point or the infrared absorption spectrum. With substances that fluoresce (many aromatic compounds do), their progress down the column can be followed by illuminating the column with ultraviolet light ("black light"). Often, the presence of colorless substances in the eluant can be established by determining its ultraviolet absorption; negative fractions need not be evaporated.

The rate of flow through the column can be increased by filling the column with eluting solvent to a higher level. If the flow is still too slow, it can be hastened by applying a little air pressure to the top of

to chromatographic
column

atomizer bulb

large bottle

Figure 15.2-3. A way to apply pressure to the top of a chromatographic column.

the column with an atomizer bulb connected through a large ballast volume (Figure 15.2-3). Connecting the column to a water aspirator by means of a suction flask is not recommended because it is very easy to evaporate the solvent from the bottom of the column and thus form cracks and spoil the column.

Advantages and disadvantages of column chromatography

Column chromatography is especially suited for purifying small amounts of material. Normally, it effects separations far more completely than distillation or recrystallization can. Its disadvantages are that it is a highly empirical method, often requiring several trials to establish acceptable conditions, and that it is not suited to the purification of large quantities because most columns of reasonable size have only a small capacity.

Exercise

1. Purification of technical-grade anthracene.

   Procedure

   Column. 1.5 to 2 cm diameter.

   Adsorbent. Alumina; use enough to make a bed of adsorbent 8 to 10 cm deep. Pack the column using hexane or petroleum ether.

   Sample solution. 0.3 g of technical anthracene in 150 to 200 mL of hexane.

   Eluting solvent. Hexane. Examination of the column in ultraviolet light will show a narrow, deep-blue fluorescent zone near the top, due to carbazole. Immediately below this zone should appear a nonfluorescent yellow band due to naphthacene. Anthracene forms a broad blue-violet fluorescent zone in the lower part of the column. Elution with hexane should be continued until anthracene begins to come off the column; the eluant collected before the anthracene begins to come off should be discarded. At this point, the elution should be continued with hexane: dichloromethane 1:1 (by volume), until the yellow band reaches the bottom of the column.

15.3 Theory of Thin-layer Chromatography

The theoretical basis for the separation of substances by *thin-layer chromatography* (TLC) is exactly the same as that for column chromatography. The difference between the methods is that in thin-layer chromatography the adsorbent is used in the form of a thin layer (about 0.25 mm thick) on a supporting material, usually a sheet of glass or plastic. The sample is applied to the layer of adsorbent, near one edge, as a small spot of a solution. After the solvent has evaporated, the adsorbent-coated sheet is propped more or less vertically in a closed container, with the edge to which the spot was applied down. The solvent, which is in the bottom of the container, creeps up the layer of adsorbent, passes over the spot, and, as it continues up, effects a separation of the materials in the spot, or "develops" the chromatogram, in the same way as the eluting solvent does in column chromatography. When the solvent front has nearly reached the top of the adsorbent, the sheet is removed from the container.

Small samples

Because the amount of adsorbent involved is relatively small and the ratio of adsorbent to sample must be high, the amount of sample must be very small, usually much less than a milligram. For this reason, thin-layer chromatography (TLC) is usually used as an analytical technique rather than a preparative method, although with thicker layers (about 2 mm) and large plates with a number of spots or a stripe of sample, it can be used as a preparative method. The separated substances are recovered by scraping the adsorbent off the plate (or cutting out the spots if the supporting material can be cut) and extracting the substance from the adsorbent.

R_f values

Because the distance traveled by a substance relative to the distance traveled by the solvent front depends on the molecular structure of the substance, TLC can be used to identify substances as well as to separate them. The relationship between the distance traveled by the solvent front and the substance is usually expressed by the so-called R_f value:

$$R_f \text{ value} = \frac{\text{distance traveled by substance}}{\text{distance traveled by solvent front}}$$

The R_f values are strongly dependent upon the nature of the adsorbent and solvent. Therefore, experimental R_f values and literature values do not often agree very well. To determine whether an unknown substance is the same as a substance of known structure, the two substances must be run side by side in the same chromatogram, preferably at the same concentration.

15.4 Technique of Thin-layer Chromatography

Preparation of microscope slide plate

The two most widely used adsorbents in TLC are silica gel and alumina. Adsorbent-coated TLC plates can be obtained commercially or prepared in the laboratory. Ready-made plates are much more convenient.

Application of sample

A very small spot. The sample to be separated is generally applied as a small spot (1 to 2 mm diameter) of solution about 1 cm from the end of the plate opposite the handle. The addition may be made with a microsyringe or with a micropipet prepared by heating and drawing out a melting point capillary. As small a sample as possible should be used to minimize tailing and overlap of spots; the lower limit just retains the ability to visualize the spots in the developed chromatogram. If the sample solution is very dilute, make several small applications in the same place, allowing the solvent to evaporate between additions. Do not disturb the adsorbent when you make the spots, or the flow of the solvent will be uneven. The starting position can be indicated by making a small mark near the edge of the plate.

Development

The chamber used for development of the chromatogram can be as simple as a beaker covered with a watch glass, or a cork-stoppered bottle. The developing solvent (an acceptable solvent or mixture of solvents must be determined by trial) is poured into the container to a depth of a few millimeters. The spotted plate is then placed in the container, spotted end down; the solvent level must be below the spots, as shown in Figure 15.4-1. The solvent will then slowly rise in the adsorbent by capillary action.

To get reproducible results, the atmosphere in the development chamber must be saturated with the solvent. Saturation can be accomplished by sloshing the solvent around in the container before any plates have been added. The atmosphere in the chamber is then kept saturated by keeping the container closed

top of adsorbent

solvent front

solvent

spot

Figure 15.4-1. Development of a plate in thin-layer chromatography.

all the time except for the brief moment during which a plate is added or removed.

Visualization: iodine or UV light

When the solvent front has moved to within about 1 cm of the top end of the adsorbent (after 15 to 45 minutes), the plate should be removed from the developing chamber, the position of the solvent front marked, and the solvent allowed to evaporate. If the components of the sample are colored, they can be observed directly. If not, they can sometimes be visualized by shining ultraviolet light on the plate or by allowing the plate to stand for a few minutes in a closed container in which the atmosphere is saturated with iodine vapor. Sometimes the spots can be visualized by spraying the plate with a reagent that will react with one or more of the components of the sample.

Exercises

1. Separation of leaf pigments.

   Procedure

   Sample solution. Crush green leaves in a mortar with a few milliliters of ethanol or acetone and twice this volume of petroleum ether or hexane. Filter the mixture, saving the filtrate. Extract the leaves again, filter, combine the two filtrates and transfer them to a separatory funnel. Wash the extract with three small portions of water and then dry it (Section 16.2) over anhydrous sodium sulfate.

   Developing solvent. Hexane:acetone, 7:3 (by volume).

   Visualization. The carotenes move most rapidly, then chlorophyll a, chlorophyll b, and the xanthophylls.

2. Analysis of mixtures of aspirin, phenacetin, and caffeine.

   Procedure

   Sample solution. Use methanol; known solutions should contain 5 to 10 mg per mL. Unknown mixtures can be prepared from materials available at drugstores.

   Developing solvent. Ethyl acetate or acetone.

   Visualization. Ultraviolet light or iodine vapors.

15.5 Theory of Paper Chromatography

The process of paper chromatography is very similar to that of thin-layer chromatography except that a strip of paper replaces adsorbent and support. As with TLC, a very small amount of a dilute solution of the substance is applied as a spot near one end of a strip of filter paper, and the strip is supported in a closed container with the end containing the spot hanging down into a solvent. As the solvent rises up the paper by capillary action, it effects the separation ("develops" the chromatogram) as does the eluant in column chromatography.

Under ordinary conditions, about 18% by weight of filter paper consists of adsorbed water. This means that when the molecules of a substance are sitting still on the paper, they are probably dissolved in the adsorbed water rather than adsorbed by the paper.

As in TLC, the amounts of sample must be so small that paper chromatography is used almost entirely as an analytical method by which you can demonstrate the homogeneity of a sample or qualitatively determine the composition of a mixture through determination of R_f values.

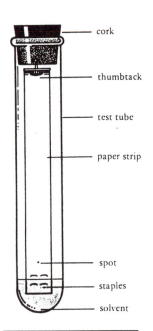

Figure 15.6-1. An arrangement for paper chromatography.

cork
thumbtack
test tube
paper strip
spot
staples
solvent

15.6 Technique of Paper Chromatography

Arrangement of paper strip

The following simple procedure can be used for the paper chromatographic analysis of a single sample spot. As a container for the development of the paper chromatogram, use a 25 mm × 125 mm or a 25 mm × 150 mm test tube. Cut a half-inch-wide strip of filter paper, and thumbtack one end of it to the bottom of the cork used to stopper the test tube. The strip should be just long enough that the lower end will dip into the half-inch layer of solvent in the bottom of the test tube. During the development of the chromatogram, the paper must not touch the walls of the test tube. A couple of staples fastened into the lower end of the strip will help it hang straight. Figure 15.6-1 illustrates the suggested arrangement.

Several samples can be run simultaneously by using a sheet of filter paper bent around to form a short tube and held that way (without the edges touching) by two staples. Development is carried out with the tube standing on end in a beaker covered with a watch glass. The grain of the paper—that is, the direction of the longer axis of the ellipse formed when a drop of water is allowed to spread on the paper—must be vertical. This arrangement is illustrated in Figure 15.6-2.

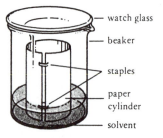

Figure 15.6-2. An arrangement for paper chromatography that allows several samples to be run at once.

watch glass
beaker
staples
paper cylinder
solvent

Application of sample

The solution of the sample of the material to be separated should be applied in as small a spot as possible. Use a micropipet, which can be made by heating and drawing out a melting point capillary or can be obtained commercially. If the solution is so dilute that a large amount must be used, several small applications should be made, allowing time for the previous one to dry before the next one is added. As with TLC, the less sample used, the smaller the spots and the better the resolution. The only limitation is the ability to see the spots of separated material. The sample spot should be made far enough from the end of the paper that it will not dip into the solvent. Mark the starting line lightly with a pencil so that you will later be able to determine the R_f values more easily.

Use a small spot

Development

The chromatogram should be developed until the solvent front has risen about 10 cm. The paper should then be removed from the developing chamber, the line of the solvent front marked with a pencil, and the solvent rinsed off or allowed to evaporate.

Visualization

If the substances are colored, the spots can be observed directly. Ultraviolet light can sometimes be used to visualize the spots, or the paper can be sprayed with a reagent that will react with some or all of the substances of the sample in such a way as to transform them to a colored derivative.

As in TLC, R_f values are fairly sensitive to the exact experimental conditions. Comparison of experimental R_f values of unknown substances with those of substances whose structures are known are best made by running a sample of the known substance alongside the unknown sample on the same strip of paper.

Exercises

1. Analysis of spinach leaves.

   Procedure

   Sample solution. Grind a few grams of fresh or frozen spinach leaves with about three volumes of acetone. Filter the mixture with suction and discard the filtrate. Grind the residue with a minimum volume of acetone and use the resulting solution.

   Developing solvent. Petroleum ether.

Visualization. Carotenes move most rapidly, followed by xanthophylls and chlorophylls; a gray area due to decomposed chlorophylls may appear after the carotenes.

2. Analysis of carrots. Analysis of carrots as in Exercise 1 shows carotene to be predominant.

3. Analysis of food coloring.

   Procedure

   Sample solution. Dilute the food coloring by a factor of five with isopropyl alcohol: water, 1:2 (by volume).

   Developing solvent. Isopropyl alcohol:water, 1:2 (by volume) (Reference 1).

4. Separation of ink pigments.

   Procedure

   Sample solution. Dilute 1 mL of Script "washable" writing fluid in 5 mL of 95% ethanol. Use 0.005 mL of solution.

   Developing solvent. 95% ethanol (Reference 2).

15.7 Theory of Vapor-Phase Chromatography

In *vapor-phase chromatography* (VPC), the stationary phase is a high boiling liquid that is present as a coating upon an inert granulated support, and the mobile phase is a gas, usually helium. As in the other chromatographic methods, separations are possible when there are differences in the way in which substances are partitioned between the stationary and mobile phases.

If the liquid phase does not preferentially dissolve molecules with certain functional groups, the order of elution is most volatile to least volatile (order of increasing boiling point). This order would always be expected for members of a homologous series and for structural isomers with the same functional group. Some liquids do appear to preferentially dissolve molecules with certain functional groups. They therefore display high selectivity toward these molecules and are especially well suited for the analysis of mixtures containing them.

Hydrocarbon greases separate molecules in order of boiling point but are not good for alcohols,

a long-chain hydrocarbon

whereas polyethylene glycol is good for polar compounds such as esters, ketones, and alcohol.

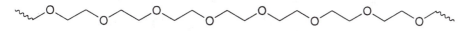

polyethylene glycol

Silicone oils are generally useful and sort the molecules of the sample according to volatility.

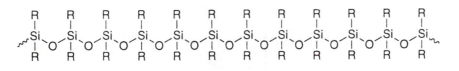

silicone oil

The apparatus required for VPC is considerably more complex and expensive than that needed for the other chromatographic methods. The essential parts, shown schematically in Figure 15.7-1, consist of a source of the carrier gas under pressure, a port through which the sample may be injected, the column (usually a 5- to 10-foot length $\frac{1}{4}$-inch-diameter metal tubing packed with the liquid-coated support), the detector, and its recorder. The detector senses the presence of material in the carrier gas, and the recorder in effect records the output of the detector as a function of time.

Thermal conductivity detector

A commonly used detector is an electrically heated wire (filament) positioned in the gas stream at the exit to the column. The wire is cooled to an equilibrium temperature by the flowing carrier gas, and at that temperature the wire has a certain resistance. When some

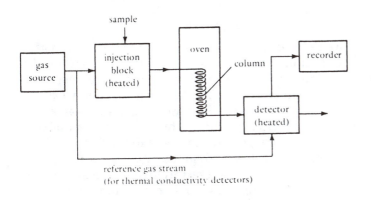

Figure 15.7-1. Schematic diagram of a vapor-phase chromatograph.

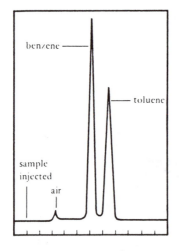

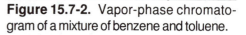

Figure 15.7-2. Vapor-phase chromatogram of a mixture of benzene and toluene.

substance other than the gas itself is present in the steam, the wire is not cooled as efficiently (helium has a very much larger thermal conductivity than any other gas but hydrogen); its temperature rises, and so does its resistance. The wire is part of a Wheatstone bridge circuit and, as the resistance of the wire varies, the resulting imbalance voltage developed in the bridge is compensated for by the recorder, which is an automatic recording null potentiometer. The position of the pen of the recorder corresponds to the position of the slide on the slide-wire of the potentiometer. The recorder output, which is a record of the pen position on the slide-wire as a function of time, is thus related to the variation of the thermal conductivity of the gas at the exit to the column. When nothing but helium is coming off the column, the pen of the recorder will draw a straight line (the baseline). As a substance comes off the column, the pen will deflect and then return to the baseline, thus drawing a "peak" (see Figure 15.7-2).

Flame-ionization detector

A more sensitive detector is the *flame ionization detector.* In a system with a flame ionization detector, a portion of the eluant from the column is fed to a hydrogen/air flame. When a substance other than the carrier gas is present in the eluant, the electrical conductivity of the hot gas of the flame will increase because of the formation of ions as the substance in the carrier gas burns. This change in conductivity is sensed as a change in voltage between a wire in the flame and the burner, and this change in voltage is followed by the recorder operating as a potentiometer, as with the hot wire detector. Nitrogen gas can be used as the carrier gas when a flame ionization detector is used, rather than the much more expensive helium gas required when using a thermal conductivity detector.

Quantitative analysis by VPC

It seems reasonable that the larger the amount of a given substance that comes off the column, the higher its concentration in the carrier gas. Thus, the area under the peak should be a measure of the amount of that substance present in the sample. If the sample is a mixture, the ratio of the areas of any two peaks should be a measure of the relative amounts of the two substances.

Determining which peak corresponds to which substance is easy if authentic samples are available. A portion of the mixture being analyzed can be spiked with one of the substances known to be present in the mixture. Analysis of this sample will show that the area of one peak has increased relative to the others, thus indicating to which component of the mixture it corresponds. The more difficult problem is to determine exactly in what way the area ratio is a measure of the relative amounts of the two substances. That is, does the area ratio correspond to the weight ratio, the mole ratio, or neither? The only way to know for sure is to analyze mixtures of known composition. When the area ratio has been determined, it generally corresponds approximately to the weight ratio, the agreement being better for similar compounds.

Qualitative analysis by VPC

Vapor-phase chromatography is often used to roughly estimate the purity of a substance by comparing the area of the main peak, which corresponds to the compound of interest, to the area of the other peaks. VPC is often used to determine whether or not a certain substance is present in a mixture. This is done by running first a sample of the mixture and then a sample of the mixture plus a little of the compound of interest. If a peak present in the first chromatogram is larger in the second, the substance added probably has the same structure as a component of the original mixture. If the added substance results in a new peak in the second chromatogram, it must not be present in the original mixture. Similarly, as implied above, it is possible to identify the substance responsible for a given peak in the chromatogram of a mixture by "spiking" portions of the mixture with small amounts of various known substances and comparing the resulting chromatograms of the spiked mixtures with the chromatogram of the original mixture. Although this method of identification is fairly reliable, especially if the result is the same when several different columns are used, it requires that you be able to guess the identity of the substance responsible for the peak and that you have a pure sample of that material available. Sometimes it is necessary to collect the substance responsible for a certain peak as it leaves the gas chromatograph, determine some physical property such as its infrared or mass spectrum, and deduce from that the identity of the substance.

Preparative VPC

So far, the applications of VPC that have been described have been analytical. As in other chromatographic methods, VPC can be used for preparative purposes by leading the carrier gas stream from the detector through a cooled tube in which the sample components of the gas stream can condense. If the tube is changed as each peak comes through, the components of the mixture can be collected separately. To purify larger amounts of material this way, you cannot just inject larger samples, because peak width increases with increasing sample size, and you will soon reach the point at which the impurity peaks overlap the peak of the desired substance.

Distillation versus VPC

Vapor-phase chromatography is one of the most widely used methods for separation or analysis of small amounts of reasonably volatile substances. Although a very good fractionating column may have an efficiency corresponding to about 100 theoretical plates, an ordinary gas chromatographic column may have an efficiency corresponding to more than 1000 theoretical plates, and columns are available with efficiencies of greater than 10,000 theoretical plates. Holdup is essentially zero, but throughput is very small.

15.8 Technique of Vapor-phase Chromatography

The gas chromatograph is ready to use when these things have been done:

1. The appropriate *column* has been installed, and the inlet port, column, and detector have been brought to their operating temperatures by their corresponding *heaters.*

2. The *carrier gas flow rate* has been set by adjustment of the *pressure regulator* and a *needle valve.*

3. The *filament current* has been set to the proper operating value.

4. The *sensitivity* has been set so that the pen of the recorder will not go off scale on the largest peak. If the pen does go off scale, the sensitivity or the sample size should be decreased; if the largest peak gives less than 50% of full-scale deflection, the sensitivity or the sample size should be increased.

5. The *balance* and *recorder zero* have been set to put the pen at the position desired for the base line.

6. The *recorder speed* has been set at the desired speed. One inch per minute is normal;. if the sample comes through very quickly, a greater recorder speed is probably desirable.

Injection of the sample

Now you must start the recorder and inject the sample. Liquid samples are injected as is; solids must be dissolved in a volatile solvent such as ether. Normally, the microsyringe used for injection will be wet with the previous sample. If so, you can rinse it by drawing in some of your sample and then squirting it out into a waste solvent container. Two or three repetitions should be adequate. The next step is to draw your sample into the syringe without any air bubbles. Sometimes, the air can be removed by dipping the needle into the sample and pumping the plunger a few times. Sometimes, you may need to draw in some liquid, hold the syringe with the needle up, and tap the barrel with your finger until the bubbles rise to the needle end. They can then be forced out by pushing the plunger in, and then the rest of the syringe can be filled. The desired amount of sample, between 0.001 and 0.005 mL (1 to 5 microliters), is taken by adjusting the end of the plunger so that the desired volume fills the barrel of the syringe. Now you are ready to inject the sample. Turn the recorder from *Standby* to *On*. Insert the needle of the syringe all the way into the rubber septum of the inlet port, slide in the plunger, and pull the needle straight back out.

Syringe is *expensive*

Follow the progress of the chromatogram by watching the recorder. A chromatogram may take less than a minute or more than half an hour; a typical time is a few minutes. When the chromatogram is complete, turn the recorder from *On* to *Standby*. If the pen goes off scale, or if the peaks are too small, change either the sensitivity setting or the sample size and try it again.

It is not unusual to decide that a chromatogram is complete when there is still material in the column. If the gas chromatograph is not used again immediately, this residue may go undetected. If, however, it is used again soon, very shallow, broad peaks can sometimes be observed under the relatively sharp peaks of the new sample. The broad peaks can be attributed to the previous sample, because the longer a substance remains in the column, the more time it has to diffuse and the shorter and wider will be its peak for a given area.

Determination of peak areas

The integrator. If the recorder has either a mechanical or an electronic integrator built into it, the area of a peak can be measured automatically. Otherwise, peak areas can be measured with a planimeter or by counting squares. If the peak is symmetrical, the product of the height of the peak times the width of the peak at half-height equals 93% of the area of the peak. If area ratios are to be used, as is usually the case, the product of peak height times width at half-height is just as useful as the area. Experimental error in measuring the width of the peak can be minimized by using a higher chart speed. If the peaks are not completely resolved but overlap somewhat, estimating their area is much more difficult. Resolution is increased by using a longer or narrower column, smaller sample size, and lower operating temperature. Two substances that cannot be resolved on one column may be resolved on another

Exercise

1. The ratio of the products of the acid-catalyzed dehydration of 2-methylcyclohexanol (Experiment 18) can be determined by VPC.

15.9 High-Pressure Liquid Chromatography

High-pressure liquid chromatography (HPLC) is a widely used variation of column chromatography. In contrast to the particles in ordinary column chromatography, Section 15.1, the particles of the solid phase used for high-pressure liquid chromatography are very finely divided, and sometimes are as small as 5 microns in diameter. Furthermore, the solid phase often consists of a glass core coated with a thin layer of adsorbent. Therefore, because the depth of adsorbent is small and the volume of liquid relative to the volume of adsorbent is also very small, equilibrium between the adsorbent and the liquid phase is established very rapidly, and columns about $\frac{1}{2}$ meter long can contain the equivalent of several thousand theoretical plates. Thus, the resolving, or separating, power of HPLC now available for samples of liquids and for solutions of solids is similar to that provided by vapor-phase chromatography, Section 15.7, for samples of gases and vapors. The finely divided nature of the solid phase used in HPLC, however, provides great resistance to the flow of the eluant. High pressures, often several thousand pounds per square inch, are required to force the eluant through the 3- to 6-mm diameter stainless steel columns, and the necessary pumps and valves are quite expensive.

15.10 Batchwise Absorption; Decolorization

Sometimes it is possible to adsorb an undesired colored compound by treating a solution with an adsorbent such as activated charcoal or alumina and then filtering the mixture by gravity to remove the adsorbent and the adsorbed material (Section 9.3). The ideal conditions are that the impurity is much more strongly adsorbed than the desired substance, and that an amount of adsorbent just sufficient to adsorb the impurity is used.

Section 9.3, page 50

Occasionally, it may be possible to decolorize a solution by filtering it through a short column of alumina, thus preferentially adsorbing the colored substance. A poorly adsorbed solvent (hydrocarbons, ether; not alcohols) must be used so that the solvent will not take the place of the colored material.

References

1. E. S. and D. Kritchevsky, *J. Chem. Educ.* **30,** 370 (1953).
2. L. F. Druding, *J. Chem. Educ.* **40,** 536 (1963).

    Sources from which more information about chromatographic methods can be obtained include:
3. E. Lederer and M. Lederer, *Chromatography*, Elsevier, Amsterdam and New York, 1967.
4. E. Heftmann, *Chromatography*, 3rd edition, Reinhold, New York, 1975.
5. R. I. W. Scott, "Contemporary Liquid Chromatography," in *Techniques of Chemistry*, Vol. XI, Wiley, New York, 1976.

16. Removal of Water; Drying

One very common separation problem is that of removing water. Because the products of many reactions are separated by procedures that use water or aqueous solutions, water usually must be removed during the purification process. Occasionally, the presence of even small amounts of water in a reaction mixture is undesirable. If so, you must remove even the water present in the reagents and solvents as a result of exposure to water vapor in the air. The following sections describe some of the methods for removal of water from solids, liquids, and gases.

16.1 Drying of Solids

Spread out the crystals

A damp solid, such as that obtained when isolating a solid by suction filtration, can often be dried simply by spreading it out on a sheet of filter paper. If the solid is only slightly damp and the crystals are not so small that the material is a damp powder or paste, it may dry within an hour to the point that the only water associated with the solid is that which must be adsorbed for it to be in equilibrium with the water vapor present in the air.

If the material is quite damp or pasty, the bulk of the water can often be removed by pressing the material between sheets of adsorbent paper (filter paper) or by spreading it out and pressing it down on a piece of unglazed porcelain plate or a block of plaster of Paris.

Increasing rate of drying

Drying oven. The overall rate of drying can be increased by increasing the rate of evaporation (by raising the temperature) and by minimizing the rate of recondensation (by decreasing the partial pressure of water in the atmosphere around the sample). To increase the rate of evaporation, the sample can be heated in a *drying oven* to a temperature somewhat below its melting point (but usually not above 110°C for organic compounds, to minimize the rate of reaction of the substance with oxygen in the air).

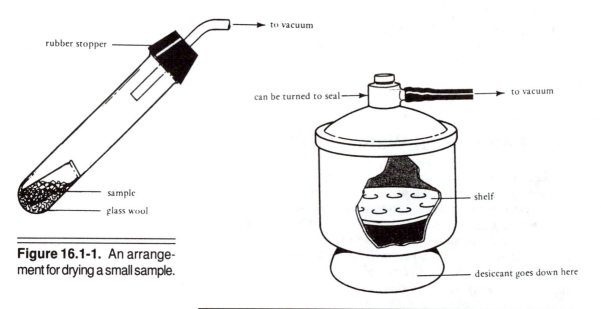

Figure 16.1-1. An arrangement for drying a small sample.

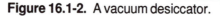

Figure 16.1-2. A vacuum desiccator.

Heat lamp. Another very convenient method of heating is to shine an infrared *heat lamp* on the sample. In all cases, the thinner the layer in which the sample is spread, the faster it will dry. The rate of recondensation of water vapor on the sample can be decreased by providing for circulation of air over the sample, or by drying the sample in a container in which the vapor pressure of water can be reduced. Figure 16.1-1 shows a simple arrangement for drying a small sample.

Vacuum desiccator. Figure 16.1-2 illustrates a *vacuum desiccator.* This device can be evacuated, and it also can contain a desiccant, a material that combines readily with water and has, in the hydrated form, a low equilibrium vapor pressure of water. The drying process inside an evacuated vacuum desiccator is evaporation of water from the sample, diffusion to the desiccant, and irreversible combination of the water with the desiccant. The absence of air in the evacuated space greatly increases the rate of diffusion of water, which is usually the slow step. Vacuum ovens are useful for drying larger quantities under vacuum at an elevated temperature.

Increasing extent of drying

The extent to which a substance can be dried depends on the partial pressure of water in the atmosphere surrounding it. In the open air, it is determined by the partial pressure of water in the air. For example, at 25°C and 60% relative humidity, the partial pressure of water will equal 0.60 times the equilibrium vapor pressure of water at this temperature, or about 14 Torr. For many substances, the amount of adsorbed or combined water that will result in an equilibrium vapor pressure of this magnitude for the substance will be quite small, and drying in air will serve to remove practically all of the water. If a substance will adsorb or combine with an objectionable amount of water at the partial pressure of water in the atmosphere, it must be dried in a closed chamber that contains a desiccant. The desiccant can be chosen to provide a sufficiently low equilibrium vapor pressure of water (see Table 16.1-1).

Drying in the air

Removal of organic solvents

Although removal of water has been the objective of the methods previously described, they can also be used for the removal of other solvents from solids. Removal of other solvents by air drying is usually complete, because the partial pressure in air of solvents other than water should be zero. Of course, the drying process can always be speeded up by increasing the temperature or using a vacuum desiccator, or both. Occasionally, a substance will hold solvent quite tenaciously and will require long periods of drying at an elevated

Table 16.1-1. Desiccants commonly used in desiccators

Substance	Hydrated Form	Equilibrium Vapor Pressure of Hydrated Form (Torr)
Aluminum oxide	1% by weight water	0.001
Concentrated sulfuric acid	95% sulfuric acid	0.001
	80% sulfuric acid	0.6
Potassium hydroxide	$KOH \cdot H_2O$	1.5
Sodium hydroxide	$NaOH \cdot H_2O$	0.7
Calcium chloride	$CaCl_2 \cdot H_2O$	0.04
Drierite	$CaSO_4 \cdot \frac{1}{2} H_2O$	0.004

temperature. High-boiling solvents, acetic acid especially, are best removed directly after collection by suction filtration; the sample is washed on the funnel with small portions of a more volatile solvent that will dissolve the high-boiling solvent but not the compound.

16.2 Drying of Solutions

Use a desiccant

When an organic substance is isolated by extraction with an organic solvent, the aqueous phase from which the substance was extracted, or the aqueous solutions with which the extracts are washed, will leave the extract saturated with water. It is usually most convenient to remove the water by treating the extract with a solid that will combine with the water.

$$\text{Solid} + n\text{H}_2\text{O} \longrightarrow \text{Solid} \cdot (\text{H}_2\text{O})_n$$

anhydrous salt   **hydrated salt**

Gravity filtration, page 44

The hydrated salt is then removed by gravity filtration. The completeness of water removal by this method depends both on how long the drying agent is left in contact with the extract and on how low the equilibrium vapor pressure of water is for the particular hydrated salt.

Drying agents

The ideal material for this purpose is a solid that is insoluble in the organic solvent, is inert toward both the solvent and the substance in solution, can be used in small amounts, and will combine quickly and completely with the dissolved water to give a solid, which is the

Table 16.2-1. Drying agents commonly used for drying solutions in organic solvents

Substance	Capacity	Speed	Intensity	Cost	Convenience	Suitability
Calcium chloride	H	M	H	L	H	a
Calcium sulfate (Drierite)	L	H+	H+	M	H	b
Magnesium sulfate	H	H	M,H	L	M	c
Molecular sieves, 4Å	H	H	H		H	c
Potassium carbonate	M	M	M		M	d
Sodium sulfate	H+	L	L	L	M	e

H, High; M, Medium; L, Low.

a. Combines with alcohols, phenols, amines, amino acids, amides, ketones, and some aldehydes and esters. Should not be used to dry solutions containing compounds of these types unless you want to remove them as well. Some calcium hydroxide may be present that will combine with acids. The hexahydrate is unstable above 30°C.
b. Generally useful. The hemihydrate is stable to at least 100°C.
c. Generally useful.
d. Combines with acids and phenols. Should not be used to dry solutions that contain acids unless you want to remove them also.
e. Generally useful. The decahydrate is unstable above 32°C.

hydrate of the original material and can be removed by gravity filtration. Table 16.2-1 lists the drying agents most commonly used in this way. The drying agents are characterized by their capacity (the amount of water that can be removed by a given weight of drying agent), speed (rate at which water combines to form the hydrate), intensity (degree of dryness ultimately achieved), suitability for use with different classes of compounds, and cost. Because the criteria are not independent of one another (for instance, the intensity depends on the degree of hydration allowed and thus upon the capacity), and because they also depend on the solvent being dried (ether solutions dry faster than benzene or ethyl acetate solutions), the ratings can be only approximate and qualitative.

Solutions in solvents in which water is relatively soluble (ether, ethyl acetate) can be dried more economically by carrying out the drying process in two steps. A relatively inexpensive, high-capacity, low-intensity drying agent is used first, and the remaining water is then removed by treatment with a high-intensity drying agent. For example, the solution can be dried first with anhydrous sodium sulfate, which will combine with 127% of its weight of water to form a decahydrate, $Na_2SO_4 \cdot 10\,H_2O$. The solution can then be further dried with anhydrous calcium sulfate, which has a very high intensity but a low capacity, because it combines with only 6% of its weight of water to form a hemihydrate, $CaSO_4 \cdot \frac{1}{2}\,H_2O$.

Procedure

The procedure for drying a solution is to add a portion of the solid drying agent and to allow the mixture to stand for 5 to 15 minutes with occasional swirling. If a second liquid phase appears (a saturated aqueous solution of the drying agent) or if the drying agent clumps together, more drying agent should be added. You can tell when an excess of anhydrous magnesium sulfate is present, because the suspension that forms when the mixture is swirled is quite cloudy and settles only slowly.

16.3 Drying of Solvents and Liquid Reagents

Drying agents

Pure organic liquids (solvents and reagents) can be dried by treatment with various materials much in the way in which solutions are dried. The drying agent can act either by forming a relatively stable hydrate or by entering into an essentially irreversible chemical reaction with the water. Table 16.3-1 indicates which drying agents are useful for drying certain compounds or classes of compounds. If drying is the only thing to be done, you can simply add the drying agent to the liquid in the bottle in which it is stored, providing for venting if hydrogen gas will be evolved. Drying may be almost complete within half an hour but sometimes takes several days.

Table 16.3-1. Suitability of various drying agents for drying pure solvents

Solvent	Drying Agent					
	P_2O_5	KOH / NaOH	BaO / CaO	K_2CO_3	$CaCl_2$	$MgSO_4$
Alcohols	−		+			
Aldehydes, ketones	−	−	−	+		
Alkanes	+				+	
Alkenes	+				+	
Alkyl halides	+				+	+
Amines	−	+	+	+	−	
Aromatic hydrocarbons	+				+	
Aryl halides	+				+	+
Ethers	−		+		+	+[a]
Nitriles	+			+		

(+) Recommended for use; (−) advised against using
[a] $MgSO_4$ can dry ether for use in the Grignard reaction (*J. Chem. Educ.* **39,** 578, 1962).

Drying by azeotropic distillation

In the special case of solvents in which water is sparingly soluble and that form minimum-boiling azeotropes with water, the solvent can be dried by distillation. The initial distillate will contain the water as the minimum-boiling azeotrope; the remainder of the distillate will be the dry solvent. Solvents that can be dried in this way include benzene, toluene, xylene, hexane, heptane, petroleum ether, carbon tetrachloride, and 1,2-dichloroethane. As long as the condensate appears cloudy, water is being removed; after about 10% of the volume of the solvent has been distilled, all the dissolved water and the water adsorbed on the walls of the flask and condenser should be gone.

When any of these solvents must be dry for use as a solvent (or reagent) in a reaction, it is often most convenient to add a 10 to 15% excess to the reaction flask and then to remove the excess by distillation. If this is done, both the dissolved water and the water adsorbed on the inside of the apparatus will be removed as a minimum-boiling azeotrope. Sometimes one or more of the reagents can be dried as well by adding it before the distillation.

The presence of water in a liquid can sometimes be detected by adding a small amount of anhydrous cobaltous chloride or anhydrous cobaltous bromide. If water is present, the blue color of the anhydrous salt gives way to the pink color of the hydrate. One form of anhydrous calcium sulfate commercially available (Drierite) comes as granules whose surface is impregnated with cobaltous chloride. A few granules of this material can be used to test for the presence of water.

Detection of water

Reference

1. D. R. Burfield, K. H. Lee, and R. H. Smithers, "Desiccant Efficiency in Solvent Drying. A Reappraisal by Application of a Novel Method for Solvent Water Assay," *J. Org. Chem.* **42,** 3060 (1977). As the authors say in their abstract, the results range from the expected to the highly surprising.

Determination of Physical Properties

The physical and chemical properties of a substance are determined by the molecular structure of the substance. Different physical states of a substance differ not in molecular structure but only in the relationships between the molecules. The different intermolecular relationships are determined by the structure of the molecules.

Thus, samples that have identical physical and chemical properties must be identical at the molecular level, and substances that differ in chemical or physical properties must differ in molecular structure. Consequently, you can tell whether two samples are samples of the same substance by comparing their physical and chemical properties.

When a pure sample is available, its physical and chemical properties serve as the standards by which another sample of that substance can be identified; in a similar way, the properties of a mixture can serve to characterize a mixture. When a sample is not pure (when it is a mixture), the progress of its purification can be followed by comparison of its properties before and after the application of separation procedures.

In the preparative experiments in this book, the identity and purity of the products of the reactions can be established by comparing certain of their physical and chemical properties with the properties of substances whose molecular structure and degree of purity are assumed to be known. The physical properties most often used in this way are the melting or boiling point and the infrared or nuclear magnetic resonance (NMR) spectrum.

Because the physical and chemical properties of a substance are thought to be determined entirely and uniquely by the molecular structure of the substance, it follows that we can infer the molecular structure of the substance from the physical and chemical properties of the substance. After all, the properties of a substance are the only experimental knowledge we will ever have of that substance, and hence any theoretical description must be inferred from these observable properties. It is impossible to give a brief explanation of how empirical evidence has in the past been translated into structural theory and how, in general, theory is developed, but in the following discussions of physical properties and their interpretation by reference to molecular structure, many examples are presented that show how conclusions about the molecular structure of a substance can be inferred from its physical properties.

The physical properties most often used in the past to characterize substances were the boiling point, melting point, density, index of refraction, and optical rotation (Sections 17 through 21). For substances of unknown structure, determinations of molecular weight and solubility characteristics were often helpful (Sections 22 and 23). More recently, spectrometric methods have become very widely used, both for the characterization of substances and for structure determination. Of these, the most common are infrared, ultraviolet-visible, nuclear magnetic resonance, and mass spectrometry (Sections 24 through 27).

17.   Boiling Point

The boiling point is one of the most often reported physical properties of a liquid, since the boiling point can usually be determined while the liquid is purified by distillation. The boiling point is often used, with the density and index of refraction (Sections 19 and 20), to establish the identity and to estimate the purity of a liquid.

17.1 Experimental Determination of Boiling Point

If sufficient material is available, the boiling point of a liquid can be determined by distillation (Section 10.6). The temperature of the vapor should be observed during the course of the distillation, and the temperature range over which most of the material distills should be taken as the boiling point. Since the boiling point is a function of the pressure, the barometric pressure or, in a distillation under reduced pressure, the pressure of the system should be recorded.

Micro-boiling-point determination

If less than a milliliter of liquid is available, as, for instance, when the sample has been isolated by vapor-phase chromatography, a *micro-boiling-point* method can be used. In this procedure, a 10- to 15-cm length of glass tubing 3 to 5 mm in inside diameter is sealed shut at one end by heating in a flame. Two or three drops of the liquid are added to this sample tube, and a length of melting point capillary, sealed about 5 mm from one end, is dropped in with the sealed end down. (This small tube is most conveniently made by melting shut a melting point capillary near the middle and cutting off all but 5 mm on one side of the seal.) The sample tube is fastened to a thermometer by a 2- to 3-mm slice of rubber tubing used as a small rubber band (see Figure 17.1-1). The thermometer is then supported in a melting-point bath (Figure 17.1-2) and heated until a very rapid, steady stream of bubbles issues from the sealed capillary. The bath is then allowed to cool slowly, and the temperature at which a bubble just fails to come out of the capillary and the liquid starts to enter the capillary is taken as the boiling point of the liquid.

Boiling point correction

If the barometric pressure is not 760 Torr, the observed boiling point can be corrected to the temperature that would be expected at 760 Torr. This correction amounts to about 0.5°C for each 10 Torr deviation of atmospheric pressure from 760 Torr; the observed boiling point will be low if the atmospheric pressure is low.

In addition to errors introduced by experimental variables such as the rate of heating, superheating, and the presence of impurities, the observed boiling point may be in error because the thermometer may not be correctly calibrated (the scale of degrees may not be in exactly the right place on the stem of the thermometer). The extent of any error of calibration can be determined by using the thermo-meter to experimentally determine the melting point of pure samples of solids of known melting point (Section 18.1).

One of the best ways to estimate (and thus take into account) thermometer error and many experimental errors is to determine the boiling point of a pure sample of a substance of known boiling point under the same experimental conditions used to

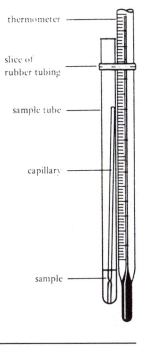

thermometer

slice of rubber tubing

sample tube

capillary

sample

Figure 17.1-1. Apparatus for micro-scale determination of the boiling point.

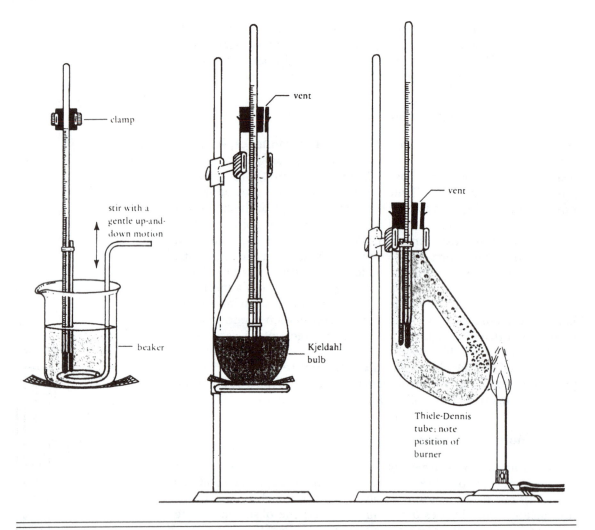

Figure 17.2-1. Heating baths for melting point or boilint point determination.

Table 17.1-1. Reference liquids for boiling point determination

Compound	Normal boiling point	$\Delta T/10$ Torr ($^\circ$C)[a]
Acetone	56.1	0.39
Water	100.0	0.37
Bromobenzene	156.2	0.53
Nitrobenzene	210.9	0.48
Quinoline	237.5	0.59
Benzophenone	305.9	0.6

[a] Variation of boiling point with pressure.

determine the unknown boiling point. Table 17.1-1 lists some liquids, and their boiling points, that have been recommended for use as standards for boiling point determinations.

17.2 Boiling Point and Molecular Structure

The variation of boiling point with the structure of covalently bonded molecules can be resolved into three contributing factors: the molecular weight, the nature of the functional group, and the degree of branching of the molecule. To see the regularity in the effect of any one factor, the other two must be held constant.

Boiling point and molecular weight

To show the effect of increasing molecular weight, Tables 17.2-1 and 17.2-2 list the boiling points of the first members of the homologous series of straight-chain alkanes and primary alkyl chlorides. These data indicate that boiling points increase with increasing molecular weight (when the nature of the functional group and degree of branching are held constant). We also see that the increase in boiling point per additional methylene group is not constant but decreases with increasing molecular weight. For this reason, a graph of boiling point versus molecular weight (or number of carbon atoms) is a curved line.

Table 17.2-1. Normal alkanes: boiling points and enthalpies and entropies of vaporization

Compound	Boiling Point °C	Boiling Point K	Molecular Weight	Increase B.P.	Increase M.W.	ΔH_{vap} (cal/mole)	ΔS_{vap} (cal/mole-deg.)
C_1	−161.49	111.66	16.042			1,955	17.51
C_2	−88.63	184.52	30.068	72.86	14	3,517	19.06
C_3	−42.07	231.08	44.094	46.56	14	4,487	19.42
C_4	−0.50	272.65	58.120	41.57	14	5,352	16.63
C_5	36.07	309.22	72.146	36.57	14	6,160	19.92
C_6	68.74	341.89	86.172	32.67	14	6,896	20.17
C_7	98.47	371.62	100.198	29.72	14	7,575	20.38
C_8	125.67	398.82	114.224	27.20	14	8,214	20.60
C_9	150.80	423.95	128.250	25.13	14	8,777	20.70
C_{10}	174.12	447.27	142.276	23.32	14	9,390	20.99

Table 17.2-2. Primary, straight-chain alkyl chlorides: boiling points and enthalpies and entropies of vaporization

	Boiling Point			Increase			
Compound	°C	K	Molecular Weight	B.P.	M.W.	ΔH_{vap} (cal/mole)	ΔS_{vap} (cal/mole-deg.)
C_1	−24.22	248.93	50.5			5,126	20.59
C_2	12.27	285.42	64.5	36.49	14	5,832	20.43
C_3	46.60	319.75	78.5	34.33	14	6,512	20.37
C_4	78.44	351.59	92.6	31.84	14	7,174	20.40
C_5	107.76	380.91	106.6	29.32	14	7,824	20.54
C_6	134.50	407.65	120.6	26.74	14	8,458	20.75
C_7	159.1	432.3	134.6	24.65	14	9,091	20.98
C_8	182.0	455.2	148.7	22.9	14	9,673	21.25
C_9	203.4	476.6	162.7	21.4	14	10,250	21.51
C_{10}	223.4	496.6	176.7	20.0	14	10,800	21.75

Table 17.2-3. Boiling points and enthalpies and entropies of vaporization of straight-chain compounds with approximately the same molecular weight but different functional groups

		Boiling Point			
Compound	Molecular Weight	°C	K	ΔH_{vap} (cal/mole)	ΔS_{vap} (cal/mole-deg.)
1-Pentene	70	30.0	303.2	6,021	19.86
1-Fluorobutane	76	32.5	305.7	6,264	20.49
Diethyl ether	74	34.3	307.5	6,355	20.67
Pentane	72	36.1	309.3	6,157	19.91
1-Chloropropane	78	45.7	318.9	6,594	20.68
Diethylamine	73	55.5	328.7	6,888	20.96
Methyl acetate	74	57.8	331.0	7,454	22.52
n-Butylamine	73	77.8	351.0	7,678	21.88
Butyraldehyde	72	74.6	347.8	7,880	22.66
2-Butanone	72	79.6	352.8	7,837	22.21
Propionitrile	69	117.4	390.6	7,937	20.32
1-Butanol	74	117.6	390.8	10,505	26.88
Propionic acid	74	139.3	412.5	7,318	17.74
Dimethylformamide	73	149.6	422.8	9,164	21.67
N-Methylacetamide	73	206	479		
Propionamide	73	213	486	14,860	30.56

Boiling point and the nature of the functional group

As an illustration of the effect of the nature of the functional group, Table 17.2-3 presents the boiling points of a number of straight-chain compounds with approximately the same molecular weight but different functional groups. Note that the more polar the functional group, the higher the boiling point. (Actually, the degree of polarity of the functional group is inferred from the boiling points of the substances.)

Boiling point and degree of branching

As an illustration of the effect of the degree of branching, Table 17.2-4 presents the boiling points and molecular structures of the eight isomers of molecular formula $C_5H_{11}Cl$. The general trend is that the more highly branched the molecule (when the functional group and the molecular weight are the same), the lower the boiling point. A comparison of the last three entries in the table illustrates the general observation that if the branches are on the same carbon atom rather than on different ones, the boiling point will be a little lower.

Table 17.2-4. Boiling points of the isomeric alkyl halides of molecular formula $C_5H_{11}Cl$

Compound	Boiling point (°C)	Compound	Boiling point (°C)
C−C−C−C−C−Cl	108	C−C(−C)−C−C−Cl	100
C−C−C(−C)−C−Cl	98	C−C−C(−Cl)−C−C	97
C−C(−Cl)−C−C−C	97	C−C(−C)−C(−C)−Cl	91
C−C−C(−C)(−C)−Cl	86	C−C(−C)(−C)−C−Cl	84

17.3 Boiling Point and the Enthalpy and Entropy of Vaporization

For processes that take place at a constant temperature, such as vaporization or freezing of a liquid, the entropy change (ΔS) is equal to the enthalpy change (ΔH) divided by the temperature at which the process occurs:

$$\Delta S = \frac{\Delta H}{T}$$

At the normal boiling point, then,

$$T_{\text{vap}} = \frac{\Delta H_{\text{vap}}}{\Delta S_{\text{vap}}}$$

Thus, the high boiling points are the result of large heats of vaporization and small entropies of vaporization. It remains to be seen how the factors of molecular weight, nature of the functional group, and degree of branching contribute to these quantities.

Enthalpy of vaporization and molecular structure

ΔH_{vap}. The *heat of vaporization* (ΔH_{vap}) is interpreted as a measure of the amount of energy required to separate the molecules against an attractive intermolecular force in the change, at constant temperature, from the liquid state to the vapor state. The stronger the intermolecular forces, the higher the heat of vaporization. Four different kinds of intermolecular forces can be distinguished:

Intermolecular forces

- van der Waals forces, proportional to the square of the polarizability,
- dipole–dipole forces, proportional to the fourth power of the dipole moments,
- induced dipolar forces, proportional to the polarizability and the square of the dipole moment, and
- hydrogen-bonding forces, usually due to the presence of O—H or N—H groups in the molecule.

Effect of molecular weight. The increase in boiling point with increasing *molecular weight* (when the nature of the functional group and the degree of branching remaining constant) is due to larger van der Waals forces between the heavier molecules. The higher the molecular weight (the more CH_2 units), the larger the total intermolecular attraction by van der Waals forces. This results in an increase in heat of vaporization with increasing molecular weight. Data for the first members of the series of normal alkanes and alkyl chlorides are presented in Tables 17.2-1 and 17.2-2.

Effect of branching. The decrease in boiling point with increasing branching (when the nature of the functional group and the molecular weight remain constant; see Table 17.2-4) can also be interpreted by looking at the influence of structure on the magnitude of the van der Waals forces. A more highly branched and thus more compact molecule will have less surface area and therefore a smaller total intermolecular attraction because of van der Waals forces.

Effect of functional group. When *functional groups* are present that cause the molecule to have a dipole moment, dipolar and induced dipolar forces can contribute to the intermolecular attraction in addition to the van der Waals forces. The higher boiling points of ethers, aldehydes, ketones, esters, nitriles, and nitro compounds (and, to a certain extent, alkyl halides) can be explained by the presence of these additional forces. The presence of O—H or N—H (or, to a smaller extent, S—H) groups in a molecule adds a further particularly effective, localized type of dipole–dipole attraction (the "hydrogen bond"), which is due almost uniquely to these functional groups. The relatively high boiling points (for a given molecular weight and degree of branching) of alcohols, thiols, primary and secondary amines, phenols, carboxylic acids, and unsubstituted and monosubstituted amides result from this additional, relatively large force. Tables 17.2-3 and 17.3-1 present data that illustrate the relative effectiveness of different functional groups in contributing to a high boiling point.

The boiling point increase that results from the introduction of a single halogen atom is due partly to the creation of a dipole moment (about the same for —F, —Cl, and —Br, and slightly less for —I) and partly to the introduction of a polarizable atom (the polarizability is least for fluorine and increases through chlorine to bromine to iodine; polarizability generally increases with atomic number or number of electrons). The effect of an iodine atom occurs mostly through its polarizability, and the effect of fluorine is almost entirely due to any dipole moment resulting from its presence. The relatively low boiling points of fluorocarbons are reasonably accounted for in terms of the low polarizability of fluorine.

Polarizability

Entropy of vaporization and molecular structure

$\Delta S_{vap.}$ The *entropy of vaporization* (ΔS_{vap}) is interpreted as a measure of the increase in disorder that results when a collection of molecules is changed from a relatively confined state to a relatively

Table 17.3-1. Boiling points of substituted butanes

Compound[a]	Boiling Point °C	M.W.	Intermolecular Forces	ΔH_{vap} (cal/mole)	ΔS_{vap} (cal/mole-deg.)
R-H	0	58	van der Waals only	5,342	19.63
R-F	33	76	v.d.W. and dipolar	6,264	20.49
ROCH$_3$	70	88	"		
R-Cl	78	93	"	7,174	20.40
R-Br	102	137	"	7,613	20.32
R-CHO	103	86	"	8,550	22.7
RCOOCH$_3$	127	116	"		
R-COCl	128	121	"		
R-I	131	184	"	7,983	19.77
R-CN	141	83	"	8,669	20.82
RNO$_2$	152	103	"	10,000	23.5
RNH$_2$	78	73	v.d.W., D., and H bonds	7,678	21.88
R-SH	99	90	"	7,700	20.72
ROH	118	74	"	10,505	26.88
R-COOH	187	02	"	11,891	25.89

[a]R = *n*-Butyl

free state (or from a small volume to a large volume, or from a state with widely spaced energy levels to a state with closely spaced energy levels). In general, the entropy of vaporization for members of a homologous series increases only slowly with increasing molecular weight (Tables 17.2-1 and 17.2-2) and is independent of the degree of branching. Thus, the effect of molecular weight and degree of branching upon boiling point is almost entirely through the heat of vaporization and not through the entropy of vaporization. In fact, for most substances of moderate boiling point, the entropy of vaporization is equal to 20 to 22 calories/mole-degree (*Trouton's rule*).

Alcohols and water. There are, however, some interesting exceptions. Alcohols (and water) generally have an unusually large entropy of vaporization. This is interpreted by saying that the increase in disorder is unusually great when an alcohol or water is vaporized, and the reason for this is that the liquid state is relatively ordered or structured compared with other liquids. This extra degree of structure, presumably due to the formation of chains or networks of hydrogen bonds, is lost upon vaporization. Thus, from the point of view of the entropy change upon vaporization, we would expect alcohols and water to have unusually low boiling points. This is not the case, however; alcohols (and water) have relatively high boiling points compared to other compounds

Hydrogen bonding

of the same molecular weight and degree of branching. The reason is that additional energy is required to break up the hydrogen bonds, which results in an unusually high heat of vaporization for alcohols and water. In fact, the greater heat of vaporization more than compensates for the increased entropy of vaporization.

Acetic acid. Acetic acid has an unusually low heat of vaporization, less than that for pentane. On the basis of heats of vaporization, acetic acid might be expected to boil below pentane. However, acetic acid also has an unusually low entropy of vaporization—about 14.5 cal/mole-degree. If acetic acid had a "normal" entropy of vaporization, it would boil at –3°C, rather than at 118°C. The low entropy of vaporization is interpreted by saying that acetic acid must retain some order or structure in the vapor phase. Vapor density measurements support this interpretation by indicating an average molecular weight considerably higher than 60, and both phenomena are interpreted in terms of partial dimerization through hydrogen bond formation in the vapor phase. Because not all the molecules are separated upon vaporization, the low heat of vaporization can thus be explained as well.

Boiling points might be expected to vary much more widely than they do, except that differences in heat of vaporization and entropy of vaporization tend to balance each other. An increase in the intermolecular force would be expected to increase the order of the liquid and thus lead to a larger entropy of vaporization, as well as to increase the heat of vaporization. Table 17.3-2 lists some compounds that have about the same boiling points (and therefore the same ratio of heat of vaporization to entropy of vaporization) but that show a considerable variation in these quantities.

Dimerization of acetic acid

acetic acid

pentane

Table 17.3-2. Enthalpies and entropies of vaporization of selected compounds boiling near 120°C

Compound	Molecular Weight	Boiling Point		(cal/mole)	(cal/mole-deg.)
		°C	K		
Acetic acid	60	118.5	391.7	5,662	14.46
Butyronitrile	71	117.4	390.6	7,937	20.32
n-Butanol	74	117.6	390.8	10,505	26.88
2-Hexanone	100	127.4	400.6	8,243	20.38
Ethyl butyrate	114	118.9	392.1	8,673	22.12
n-Octane	114	125.8	398.8	8,214	20.60
n-Butyl iodide	184	130.5	403.7	7,983	19.77

Problems

1. A liquid was observed to distill between 206 and 207.5°C when the atmospheric pressure was 743 Torr. Calculate the expected normal boiling point of the liquid. What factors contribute to uncertainty in this value?

2. A liquid was observed to distill between 121.5 and 122.5°C under the conditions described in Problem 1. Under the same conditions, toluene distilled between 109 and 109.5°C. Calculate the expected normal boiling point of the liquid. What factors contribute to the uncertainty of this value?

3. The boiling points (in °C) of a series of analogous chlorine and fluorine compounds are presented in the following table. Rationalize the different trends shown by the different halogens.

	CH_4	CH_3X	CH_2X_2	CHX_3	CX_4
Fluorine	-161	-78	-51	-82	-129
Chlorine	-161	-24	40	62	76

Which trend would you expect the analogous bromine compounds to follow?

4. The following table presents values for the enthalpy and entropy of vaporization for several aliphatic carboxylic acids. How do you interpret these data?

Compound	ΔH_{vap} (cal/mole)	ΔS_{vap} (cal/mole-degree)	Boiling point °C
Formic acid	5,318	14.26	101
Acetic acid	5,558	14.19	118
Propionic acid	9,998	24.14	141
n-Butyric acid	10,780	24.70	163
i-Butyric acid	10,630	24.92	153
Valeric acid	11,890	25.90	183

5. Rationalize the differences in boiling points of members of each group of compounds.

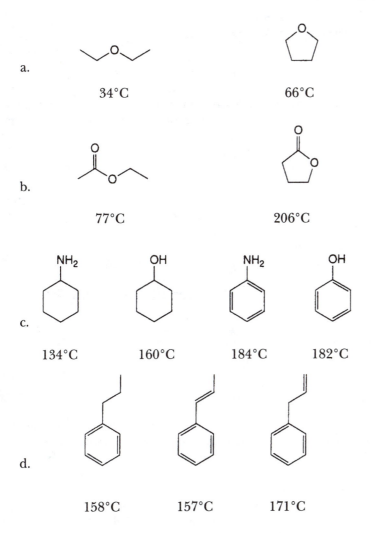

a.

34°C   66°C

b.

77°C   206°C

c.

134°C   160°C   184°C   182°C

d.

158°C   157°C   171°C

7. Assign structures to the following substances on the basis of their boiling points.

a. Isomers of molecular formula C_3H_9N. *Boiling points:* 49°, 35°, 35°, and 3 °C.

b. Isomers of molecular formula $C_4H_{10}O$. *Boiling points:* 118°, 108°, 100°, 100°, 83°, 39°, 34°, and 31°C.

c. Methyl ethers of molecular formula $C_5H_{12}O$. *Boiling points:* 70°, 61°, 61°, and 55°C.

18.   Melting Point

The normal melting point of a solid is defined as the temperature at which the solid and liquid are in equilibrium at a total pressure of 1 atmosphere (the vapor pressure of the solid is usually very much less than 1 atmosphere at the melting point). In contrast to the volume change that accompanies the vaporization of a liquid, the change in volume that takes place upon the melting of a solid is very small. Hence, the melting point of a solid, unlike the boiling point of a liquid, is practically independent of any ordinary pressure change. Because the melting point of a solid can be easily and accurately determined with small amounts of material, it is the physical property that has most often been used for the identification and characterization of solids.

18.1 Experimental Determination of the Melting Point

There are several methods by which melting points can be determined, and the choice of method depends mainly on how much material is available.

Melting points from cooling curves

Cooling curve. If large amounts of the solid are available (a gram or so), the most accurate method for the determination of the melting point is to heat the sample until it is melted (probably by means of an oil bath, Section 33.1) and then allow it to cool slowly for crystallization, keeping track of the temperature of the sample as a function of time by means of an immersed thermometer or thermistor. At first, the temperature will be observed to fall as the liquid loses heat to the surroundings. When crystallization begins, however, the heat evolved during this process (ΔH_f, the heat of fusion) will maintain the temperature at a constant value until crystallization is complete. At this point, the temperature will again fall as the solid loses heat to the surroundings. If the material is pure, the temperature of the sample remains constant during the entire process of solidification; this temperature is the melting point. Figure 18.1-1 illustrates this expected cooling curve for a pure substance. It is not unusual for the temperature to fall a little below the melting point before crystallization begins. When this happens, the sample is said to have supercooled. The heat evolved as crystallization takes place then warms the sample to the melting point (see Figure 18.1-2).

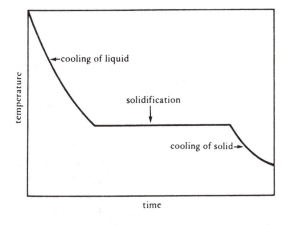

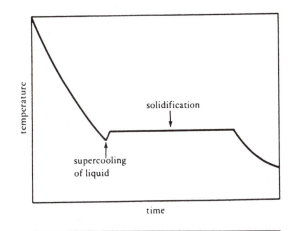

Figure 18.1-1. Expected cooling curve for a pure substance.

Figure 18.1-2. Expected cooling curve, with supercooling, for a pure substance.

This procedure can also be used for determining the freezing point of substances that are liquids at room temperature. The liquid is cooled in a cold or refrigerated bath. During the cooling time, the temperature of the bath should be only a little below the expected freezing point. (Several trials can be made with the same sample, and the temperature of the bath can be adjusted according to the results of the previous trial.)

This procedure is the one that should be used for calibrating a thermometer or checking the calibration of a thermometer by means of solids of known melting point. In this case, any disagreement between the reading of the thermometer and the true melting point, which is assumed to be known, is attributed to an error in the calibration of the thermometer. Table 18.1-1 lists a number of substances of known melting point that have been recommended for use in this method of calibration.

Table 18.1-1. Reference substances for calibration of thermometers by melting point determination

Compound	Melting Point (° C)
Ice	0.0
m-Dinitrobenzene	89.7
Benzoic acid	121.7
Salicylic acid	160.4
3,5-Dinitrobenzoic acid	205
Sym-di-*p*-tolyl urea	268

Capillary melting points

Needs only a small sample

A much smaller sample is required for the method that is most often used for the determination of the melting point of a solid. A few crystals of the compound are placed in a thin-walled capillary tube that is 10 to 15 cm long and about 1 mm in inside diameter and has been sealed at one end. The capillary, which contains the sample, and the thermometer are then suspended in an oil or air bath that can be heated slowly and evenly. The temperature range over which the sample is observed to melt is taken as the melting point. Obviously, if you are to know the temperature at which the crystals are melting, the thermometer and sample must be at the same temperature while the sample melts. This requires that the rate of heating of the bath be very low as the melting point is approached (about 1 degree per minute). Otherwise, the temperature of the mercury in the thermometer bulb and the temperature of the crystals in the capillary will not be the same as the temperature of the bath liquid, and probably not equal to each other. The transfer of heat energy by conduction takes place rather slowly.

Trial run. If the approximate temperature at which the sample will melt is not known, a preliminary melting point determination should be made in which the temperature of the bath is raised quickly. Then the more accurate determination should be carried out, with a low rate of heating near the melting point. A preliminary melting point can be determined within 10 minutes, which is the time it would take to raise the temperature of the bath 10 degrees at 1 degree per minute. It should be obvious that if you cannot estimate the melting point within about 20 degrees, then two melting point determinations (one fast, one slow) will take less time than one determination that is slow over a wide range.

Usually, the melting point capillary can be filled by pressing the open end into a small heap of the crystals of the substance, turning the capillary open end up, and vibrating it by drawing a file across the side to rattle the crystals down into the bottom. If filing does not work, drop the tube, open end up, down a length of glass tubing about 1 cm in diameter (or a long condenser) onto a hard surface such as a porcelain sink, stone desk top, or the iron base of a ring stand. The solid should be tightly packed to a depth of 2 to 3 mm.

When an oil bath is used, the capillary can be fastened to the thermometer by means of a small slice of rubber tubing used as a rubber band, as shown in Figure 18.1-3.

If a compound begins to decompose near the melting point, the capillary with the sample should be placed in the bath after the temperature has been raised to within 5 or 10 degrees of the expected melting point, to minimize the length of time the sample is heated.

A variety of oil baths can be used in a melting point determination, as well as in a boiling point determination. The simplest use a

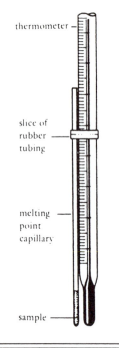

thermometer —

slice of
rubber
tubing —

melting
point
capillary —

sample —

Figure 18.1-3. Arrangement of sample and thermometer for a melting point determination.

burner flame and depend on convection for mixing; the more elaborate use an electric immersion heater and are stirred. The latter are more accurate and convenient. It is easy to heat at a low and steady rate with an electric heater, but almost impossible with a flame. Two simple heating baths are illustrated in Figure 17.2-1. It is dangerous to exceed 200°C with a bath oil such as mineral oil. Other liquids such as silicone oils, which have higher flash points, can be used at higher temperatures.

There is also a type of melting point apparatus in which the sample and thermometer are both supported in an electrically heated metal block and the sample in the capillary can be observed through a magnifying glass. Usually you can heat the block rapidly when the temperature is well below the melting point and slowly as the melting point, is approached.

As in the determination of the boiling point, the observed melting point must be corrected for any error in the calibration of the thermometer.

Thermometer correction

Capillary melting points are properly compared with one another, but occasionally they are considerably different from melting points determined from cooling curves.

Micro-hot-stage melting points

A quick and easy method to determine the melting point of a solid is to heat a few crystals of the sample between a pair of microscope cover glasses on an electrically heated metal block while observing the crystals through a magnifying glass. This method has the advantage of requiring as little as a single crystal, permitting good temperature control, and being very convenient. However, complete thermal equilibrium between the sample, block, and thermometer is not possible, because the thermometer is inside the block and the sample is on the surface, exposed to the cooler atmosphere. For this reason, observed block melting points often appear to be higher than capillary melting points; the higher the melting point, the greater the difference. However, a melting point quickly determined on a block can serve as an approximate melting point for the determination of a capillary melting point.

Needs only a few crystals

It is a melting range. Although there should be a single temperature at which a pure solid and a liquid are in equilibrium, most samples appear to melt over a small temperature range. This happens because, with capillary or block melting points, the temperature of the bath or block rises a little during the time the sample takes to melt. The presence of impurities in the sample can also cause the sample to melt over a range of temperatures, as explained in Section 18.2. Thus, the "melting point" will usually be reported as two temperatures between which the sample was observed to melt, a *melting range*.

18.2 The Melting Point as a Criterion of Purity

A dilute solution of a liquid begins to freeze at a temperature somewhat lower than the freezing point of the pure liquid. The lowering of the freezing point (assuming that the material that separates out is the pure solvent) is given by

$$\Delta T = \frac{RT_f^2}{\Delta H_f} x_B$$

where ΔT is the difference between the freezing point of the dilute solution and the pure solvent, T_f is the freezing point of the pure solvent, ΔH_f is the heat of fusion of the pure solvent, and x_B is the mole fraction of the solute (impurity) (1). Thus, the presence of an impurity makes itself known by a reduction of the freezing point of the sample. As the pure solvent crystallizes from solution, the concentration of the impurity must increase (x_B increases), and the freezing point of the solution must fall. The result is that the cooling curve for such a solution will appear as in Figure 18.2-1; freezing not only starts at a lower temperature, but as Figure 18.2-1 shows, it also becomes complete only over a range of temperatures.

Thus, a "sharp" melting point (actually, a melting *range* of less than 1°C) is often taken as evidence that the sample is fairly pure, and a wide melting range is evidence that it is not pure.

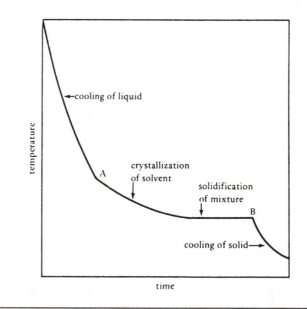

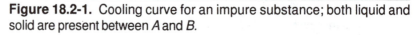

Figure 18.2-1. Cooling curve for an impure substance; both liquid and solid are present between *A* and *B*.

18.3 The Melting Point as a Means of Identification and Characterization

If two samples have different melting points, their molecules must differ either in structure or in configuration; they must be either structural isomers or diastereomers. If the melting points of two samples are the same, the structures of their molecules must be the same, although they might have enantiomeric configurations. These statements apply only to pure substances and do not take into account the fact that some substances can exist in different crystalline forms that have different melting points.

18.4 Mixture Melting Points

Mixtures of different substances generally melt over a range of temperatures, and melting is usually complete at a temperature that is below the melting point of at least one of the components. Thus, the nonidentity of two substances of the same melting point can often be established by determining that the melting point of a mixture of the two is *depressed*. If each individual sample melts "sharply" (and at the same temperature, of course) and if an intimate mixture of the two, made by rubbing approximately equal amounts together, melts over a wide range, the two substances are not the same.

Melting point depression

Usually, however, you want to establish the identity rather than the nonidentity of two samples, so it is unfortunate that the converse is not always true; the absence of a depression of the melting point or of a wide melting range of the mixture is not certain evidence that the two substances are identical in molecular structure and configuration.

18.5 Melting Point and Molecular Structure

Systematic variations of melting point with changes in structure are not as obvious or predictable as are the variations of boiling point.

Alternation of melting point. Although melting points do generally increase with increasing molecular weight, the first members of homologous series often have melting points that are considerably different from what would be expected on the basis of the behavior of the higher homologs. In some homologous series of straight-chain aliphatic compounds, melting points alternate: the melting point of successive members of the series is higher or lower than that of the previous member, depending on whether the number of carbon atoms is even or odd. Sometimes, as with the normal alkanes, the melting points of successive members of the series always increase, but by a larger or smaller amount, depending on whether the number of carbons is even or odd.

Alternation of melting points

Polarity. As with boiling points, compounds with polar functional groups generally have higher melting points than compounds with nonpolar functional groups, but unlike with boiling points, highly branched or cyclic molecules (relatively symmetrical molecules) tend to have higher melting points than their straight-chain isomers. The combined effects of branching or the presence of rings, then, reduce the range of temperature over which the liquid can exist at a vapor pressure of less than 760 Torr. In extreme cases, a liquid range does not exist at a vapor pressure of less than 760 Torr; at atmospheric pressure, the substance will sublime without melting. Hexachloroethane and perfluorocyclohexane behave in this way.

Melting point and the enthalpy and entropy of fusion

Like the boiling point of a liquid, the absolute temperature at which a substance melts equals the heat of fusion, ΔH_f, in calories per mole, divided by the entropy of fusion ΔS_f, in calories per mole-degree:

$$T_f = \frac{\Delta H_f}{\Delta S_f} \qquad (18.5\text{-}1)$$

That this should be so can be seen most quickly by realizing that at the melting point the liquid and solid phases are in equilibrium, which means that at this temperature no change in free energy will accompany the conversion of liquid to solid (or solid to liquid), i.e., $\Delta G_f = 0$. Because

$$\Delta G_f = \Delta H_f - T_f \Delta S_f = 0$$

then

$$\Delta H_f = T_f \Delta S_f$$

and Equation 18.5-1 follows from this. In other words, given definite values for ΔH_f and ΔS_f, there is only one temperature at which ΔH_f can equal $T \Delta S_f$ (that ΔG_f can equal 0; that the two phases can be in equilibrium). This temperature is called the melting point, T_f.

The enthalpy and entropy of fusion and molecular structure

The increase in boiling point with molecular weight, and the influence of the nature of the functional group and degree of branching on the boiling point, are interpreted as the effect of structure on the heat of vaporization. To a first approximation, the entropy of vaporiza-

Table 18.5-1. Melting points, heats of fusion, and entropies of fusion of selected compounds

Compound	Melting point °C	K	ΔH_f (cal/mole)	ΔS_f (cal/mole-deg.)
n-Butyl alcohol	-87.3	185.9	2,130	11.48
t-Butyl alcohol	25.2	298.4	1,620	5.44
Diethyl ether	-116.2	157.0	1,649	10.50
o-Dichlorobenzene	-16.7	256.5	3,080	12.02
m-Dichlorobenzene	-26.2	247.0	3,010	12.19
p-Dichlorobenzene	53	326	4,340	13.31
o-Dibromobenzene	1.8	275.0	2,030	7.38
m-Dibromobenzene	-6.9	266.3	3,150	11.84
p-Dibromobenzene	87	360	4,780	13.27
o-Xylene	-25.2	248.0	3,250	13.11
m-Xylene	-47.9	225.3	2,765	12.27
p-Xylene	13.3	286.4	4,090	14.28
Cyclohexane	6.6	279.7	673	2.28
Methylcyclohexane	-126.6	146.6	1,613	11.01
Cyclohexanol	23.5	296.7	406	1.37
Phenol	41	314	2,771	8.82
n-Octane	-56.8	216.4	4,957	22.91
2-Methylheptane	-109.0	164.1	2,451	14.94
2,3,3-Trimethylpentane	-100.7	172.5	370	2.12
2,2,3,3-Tetramethylbutane	100.7	373.9	1,802	4.82

The other isomeric octanes (values for four isomers are not reported) have melting points between -126 and -91°C, heats of fusion between 3,070 and 1,700 cal/mole, and entropies of fusion between 17.8 and 10.69 cal/mole-degree.

tion has a constant value of 20 to 22 calories per mole-degree. With melting points, however, both the heat of fusion and the entropy of fusion can vary widely, even for apparently closely related compounds. Table 18.5-1 presents the melting points, heats of fusion, and entropies of fusion of several sets of related compounds.

ΔH effects. The higher melting point of n-butyl alcohol compared with diethyl ether is a consequence of the larger heat of fusion of the more polar alcohol. Similarly, the higher melting points of the p-disubstituted benzenes compared with the *ortho* and *meta* isomers is due to the larger heat of fusion of the *para* isomer, because the differences in the entropy of fusion among the isomers are either small or in the wrong direction to account for the facts.

Enthalpy effects

Entropy effects

ΔS *effects.* The unusually high melting point of cyclohexane compared with methylcyclohexane is primarily the result of an unusually low entropy of fusion for cyclohexane. Similarly, the relatively high melting point of *tert*-butyl alcohol compared with its isomers is the result of a relatively low entropy of fusion. The entropy of fusion is primarily the result of increased freedom of rotation of the molecule as a whole and of parts of the molecule with respect to other parts, in the liquid phase compared with the solid phase. The increase in vibrational and translational entropy upon melting is relatively small; the large increase in translational entropy occurs upon vaporization, and there is considerable vibrational motion possible in the crystal. A small entropy of fusion, then, implies either that the molecule has relatively little rotational freedom in either the solid or liquid state (cyclic, polycyclic molecules), or that it can gain certain degrees of rotational freedom, with their associated increase in entropy, in the solid state without melting (highly branched, highly symmetrical molecules). Cyclohexane is known to have a transition at 185.9 K with an associated entropy change of 8.63 cal/mole-degree, and similar transitions have been observed for other highly symmetrical molecules.

The data for the isomeric octanes show how complex and variable the relationships can be between structure and melting point, heat of fusion, and entropy of fusion.

Problems

1. a. Calculate the freezing point depression, ΔT, for a sample of *tert*-butyl alcohol that contains 0.01 mole fraction of an impurity.

    b. Do the same for a similar solution of cyclohexanol.

2. a. A sample of a substance of known molecular structure starts to freeze at 57.00°C and is half frozen at 56.80°C. What would be the melting point of the pure substance?

    b. The sample in part a is remelted, and sufficient solute of known molecular weight is added to give a solution containing 1 mole percent of this solute (you must assume that the solvent is pure). The sample now starts to freeze at 56.90°C. What mole fraction of impurity was present in the original sample?

References

1. G. M. Barrow, *Physical Chemistry* 5th edition, McGraw-Hill, New York, 1988, p. 291.

2. A. I. Vogel, *A Textbook of Practical Organic Chemistry,* 3rd edition, Wiley, New York, 1957, p. 21.

3. E. L. Skau, J. C. Arthur, and H. Wakeham, *Technique of Organic Chemistry*, Vol. I, Part I, 3rd edition, A. Weissberger, editor, Interscience, New York, 1959, p. 287.

19. Density; Specific Gravity

Before the advent of the spectroscopic methods, density was one of the most important physical properties by which liquids were characterized. Because the density is often reported and tabulated, especially in the older literature, and is fairly easy to determine, it is still often used to characterize and identify liquids.

19.1 Experimental Determination of the Density

$$\text{Density} = \frac{\text{mass}}{\text{volume}}$$

The *density* is defined as the *mass per unit volume,* and the units usually used by chemists are grams/mL. The determination of density, then, requires the determination of both the mass and the volume of a sample of the substance. Because density is a function of temperature, the temperature at which the density is determined should also be recorded. For most organic liquids, the density decreases by about 0.001 g/mL per degree increase in temperature.

Temperature dependence

The density of a sample is determined by weighing a container first when it is empty and then when it is filled with the liquid whose density is to be determined. The difference in the weights gives the mass of the sample, and the volume of the sample is the volume of the container. Dividing the mass by the volume gives the density.

Specific gravity bottle

If a large amount of material is available, a volumetric flask can be used as the container. The density of a 1- or 2-mL sample can be determined by using a "specific gravity" bottle like that shown in Figure 19.1-1. The stopper for such a bottle has a capillary that allows it to be filled without trapping any air bubbles inside the bottle.

Micro-scale density determination

You can determine the density of only a drop or two of material by using a 0.05 mL syringe as the container. The syringe is weighed

Figure 19.1-1. A specific gravity bottle. As the stopper is put in, the air ecapes through the capillary. "Full" means filled to the top of the capillary.

empty and then reweighed after drawing in a portion of the sample. Because the amount of material used will be very small, the weighings should be done to 0.1 mg. Also, the sample should be positioned in the syringe so that it lies entirely within the range of the graduations so that its volume can be estimated by subtracting two scale readings.

Specific gravity: dimensionless

Specific gravity, which is related to density and sometimes confused with density, is defined as the *ratio* of the weight of a certain volume of liquid to the weight of an equal volume of water, or the ratio of the density of a liquid to the density of water. Since the density of water is almost exactly 1, the specific gravity of a liquid is numerically almost equal to the density. At 4°C, when the density of water is exactly 1.000 g/mL, the numerical values of the density and specific gravity are exactly equal. The specific gravity is a dimensionless number because it is a ratio of quantities with the same units.

Dimensionless number

19.2 Density and Molecular Structure

The density of a substance is determined mainly by the atomic weight of its constituent atoms: high atomic weight results in high density.

Most organic liquids have densities between 0.8 and 1.1 at room temperature. The structural possibilities for liquids that are more or less dense than this are quite limited. Aliphatic acyclic hydrocarbons, saturated and unsaturated, and aliphatic acyclic ethers and amines generally have densities less than 0.8. Aromatic hydrocarbons usually have densities between 0.86 and 0.9. Iodides and bromides have densities greater than 1.1, and alkyl chlorides have densities less than 1.

Alkanes: least dense

Alkyl halides: most dense

Compounds of density greater than 1.2 usually have at least one iodine or bromine atom, or two or more chlorine atoms. Compounds with two or more functional groups, especially compounds with relatively large intermolecular forces (relatively high boiling for their molecular weight), generally have densities greater than 1.

As the ratio of $-CH_2-$ units per functional group increases, the densities tend toward a value between 0.8 and 0.9.

Problem

1. How would densities determined on the moon differ from those determined on Earth?

20. Index of Refraction

Along with the boiling point and density, the index of refraction is one of the physical properties most often used to identify and characterize liquids.

20.1 Experimental Determination of the Index of Refraction

Index of refraction

The index of refraction of a substance is the ratio of the speed of light in a vacuum to the speed of light in the substance. It is often measured by making use of the fact that the critical angle of refraction of light passing from one medium to another is a function of the refractive indexes of the two media. A *refractometer* measures the critical angle of refraction of light passing from a liquid of unknown refractive index to a glass prism of known refractive index. The index of refraction is dimensionless because it is a ratio of quantities with the same units.

Wavelength dependence and temperature dependence

Because the index of refraction is a function of both the wavelength of light and the temperature, these values must both be specified along with the measured refraction. The D line of sodium, $\lambda = 5,890$ nm, is the wavelength for which the index of refraction is often reported, and at the temperature T, the index of refraction determined at this wavelength would be reported as n_D^T. The index of refraction of most organic liquids decreases between 0.00035 and 0.00055 per degree increase in temperature.

Abbe refractometer

With the *Abbe refractometer*, two drops of sample are required in the space between the prisms, and the instrument is adjusted until the field seen through the eyepiece appears as shown in Figure 20.1-1. The intersection of the cross-hairs should be on the border between the light and dark sections of the field, and the compensator should be set to sharpen and achromatize the border between the light and dark sections until the difference is as sharp and as near black and white as possible. When the cross-hairs and border are lined up as shown in Figure 20.1-1, the index of refraction is read from the scale.

Figure 20.1-1. View through the eyepiece of the Abbe refractometer. When the refractometer is set to be read, the intersection of the cross-hairs falls on the border between the light and dark fields; the border should be as sharp and free from color as possible.

The value is reliable to +0.0002, provided that the instrument is properly calibrated. The calibration can be checked by determining the index of refraction of water, which is 1.3330 at 20°C and decreases 0.0001 unit per degree increase in temperature between 20 and 30°C. With the compensator, the index of refraction determined with ordinary (white) light is very close to the value that would be obtained using the light from a sodium lamp.

If you cannot see a sharp boundary, insufficient sample was used, or the sample evaporated before the adjustment was completed. In making determinations on very volatile liquids, you can introduce the sample with a fine dropper through a little channel that leads to the space between the prisms, without having to open the prisms.

When you are done, use a soft tissue and a little acetone or ethanol to remove the sample.

The thermometer indicates the temperature of the sample space. It can be maintained at a particular value by circulating water from a constant-temperature bath through the prism housings, using the hose connections provided.

20.2 Index of Refraction and Molecular Structure

The index of refraction of a substance is determined mainly by the polarizability of its constituent atoms and functional groups; the presence of atoms of higher atomic number and of conjugated unsaturation results in a higher index of refraction. Alkyl iodides and aromatic compounds usually have an index of refraction greater than 1.5000; most other compounds give a value lower than this.

As with the boiling point and density, the main use of experimental values for the index of refraction is for comparison with values reported in the literature for compounds whose structure is known. It is possible, however, to calculate the expected index of refraction for a substance from its molecular structure through the following relationship:

$$\text{Molecular refraction} = M\, n_{\text{D}}^{T} = \text{molecular weight} \times n_{\text{D}}^{T}$$

Molecular refraction   Solving for n_{D}^{T} shows that the index of refraction equals the *molecular refraction* divided by the molecular weight. Contributions of individual structural features (atoms or bonds) to the molecular refraction can be found in Reference 1.

The fact that the molecular refraction can be estimated from the structural formula of a compound makes it possible to use the index of refraction as a criterion of identity even though literature values may not be available.

Reference

1. A. I. Vogel et al., *J. Chem. Soc.* **1952** (514).

21. Optical Activity

An additional physical property of use in the identification and characterization of optically active solids or liquids is the *optical rotation.*

21.1 Experimental Determination of Optical Rotation

Optical Rotation

The *optical rotation* is measured by preparing a solution of the substance and then determining, by the use of a polarimeter, the direction and degree to which the plane of polarization of a beam of plane-polarized light is rotated upon passage through the solution.

The polarimeter

    The polarimeter. The *polarimeter* consists essentially of a polarizing prism (polarizer), which transmits from the monochromatic light source a beam of plane-polarized light, a trough to hold the sample tube, and a second polarizing prism (analyzer) that can be rotated so as to exactly compensate for the rotation that the solution induces in the polarized light beam. The correct position of the analyzer is determined by looking through the central eyepiece at the light transmitted by the instrument. Figure 21.1-1 indicates the appearance of the field for possible positions of the analyzer with respect to the plane of polarization of the beam.

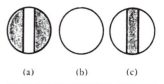

(a)   (b)   (c)

Figure 21.1-1. View through the eyepiece of the polarimeter. The analyzer should be set so that the intensity of all parts of the field is the same (b). When the analyzer is displaced to one side or the other, the field will appear as in (a) or (c).

    The observed rotation, α. The angular displacement of the analyzer is read in degrees from the graduated circle with the aid of an auxiliary eyepiece and vernier, and the difference between the angular position of the analyzer with the sample in and out of the trough is the *observed rotation, α.*

    The sample. The *sample* is prepared by making up a solution of known concentration by dissolving a weighed amount of the substance in a volumetric flask. A concentration of 0.5 to 2 grams per 100 mL is usually recommended. Solvents most commonly used are water, methyl and ethyl alcohols, and chloroform. A dilute solution is most desirable, but if the rotatory power of the substance is low, a higher concentration may be necessary. If the solution is not free from dust, it should be filtered. The volume of solution required varies with the length and diameter of the sample tube. Twenty-five milliliters should be enough in any case, and tubes are available that require 1 mL or less. With a liquid, the rotation can be determined using the neat liquid.

The polarimeter tube. The sample tube, or *polarimeter tube,* is usually a 1- or 2-decimeter-long glass tube with a fitting for a brass screw cap at each end. It is filled by placing one end glass against one end of the tube (glass-to-glass) and then screwing on the cap, using a rubber washer between the end glass and the cap. With the tube standing vertically on this cap, the solution is added through the other end (possibly by means of a dropper or pipette) until the rounded meniscus stands above the end of the tube. The other end glass is then slid across the end of the tube, without leaving an air bubble in the tube; if air is trapp′d in the tube, remove the glass, add more solution if necessary, and try sliding the glass on again. The other brass cap is then screwed on, using a rubber washer as with the first cap. Screw the ends on firmly but not too tight, as strain in the glass of the end caps may cause them to rotate the beam slightly. Some polarimeter tubes are constructed so that they can be filled through a port in the middle with both ends capped. After placing the tube in the trough of the instrument, close the cover.

The image can be brought into focus, if necessary, by moving the eyepiece in or out. Distortion of the image may be due to an air bubble in the light path or to the presence of density gradients in the sample, which are caused by variations either in concentration or in temperature.

Variables that affect optical rotation

The magnitude of the observed rotation, α, depends on several variables that must therefore be recorded along with the observed rotation.

- *Concentration.* It is reasonable that the greater the concentration of molecules in the path of the beam, the greater should be the rotation. In many cases, however, the increase in rotation is not exactly proportional to the concentration. This is explained by postulating that solute–solute interactions, which are more prevalent at higher concentrations, affect the rotation differently from solute–solvent interactions, which are predominant at low concentrations.

- *Length of sample tube.* The rotation of the beam will be proportional to the number of molecules in the path; for a given concentration, the number of molecules in the path will be proportional to the length of the path. The path length is conventionally expressed in decimeters.

- *Solvent.* The observed rotation of the same substance at the same concentration in different solvents can be very different, even of opposite sign!

- *Temperature.* A change in concentration accompanies the expansion and contraction that a solution undergoes with change in temperature. The most popular explanations of the rather large changes in the observed rotation that occur with changes in temperature are (1) the change with temperature of the relative populations of conformational isomers, which have different rotatory powers, and (2) the change with temperature of the solvation of the optically active molecules.

- *Wavelength of light.* The magnitude (and also the sign, in many cases) of the rotation depends on the wavelength of light used. Since the D line of sodium was used early, most rotations are reported at this wavelength, 5,890 nm. Generally, the magnitude of rotation is larger at shorter wavelengths, and for this reason and the fact that the human eye (especially the dark-adapted eye) is more sensitive to green light, the green line of the mercury arc (5,461 Å) has been recommended for use.

Specific rotation, $[\alpha]$. The optical rotation is usually reported as the *specific rotation,* $[\alpha]$:

for solutions:

$$[\alpha]_D^T = \frac{100\alpha_D^T}{lc}$$

for pure liquids:

$$[\alpha]_D^T = \frac{\alpha_D^T}{ld}$$

where

$[\alpha]_D^T$ = specific rotation, at the D line of sodium at the temperature T
α_D^T =   observed rotation: positive if clockwise, negative if counter-clockwise at the D line of sodium at the temperature T
l = length of sample tube, in decimeters
c = concentration of solution, in grams per 100 mL
d = density of the pure liquid at the temperature T, in g/mL

The *specific rotation* $[\alpha]$ is a measure of the rotary power per unit path length per unit weight concentration. In taking your experimental data, you must report the solvent and the concentration as well, because the variation of rotation with concentration is not necessarily linear and not usually independent of the nature of the solvent. Also, when comparing an experimental value of the specific rotation with a literature value, you must compare solvent and concentration as well as temperature and wavelength.

21.2 Optical Activity and Molecular Structure

Molecules must be chiral

A compound that is optically active must be composed of molecules for which no conformation that can be attained by the molecules is the same as the mirror image of that conformation; all conformations must be *chiral;* all molecules must be *handed;* all conformations must belong to point groups $\mathbf{C}_n$ or $\mathbf{D}_n$. Thus, if the molecules of a substance possess, in any conformation, a symmetry element whose corresponding symmetry operation involves reflection (a plane of symmetry, a center of symmetry, or a higher, even improper, axis of symmetry), the substance cannot exist in chiral forms, and no samples can be optically active.

Chiral center

Molecular chirality

The presence of a *chiral center* in the molecules of a substance permits the existence of chiral molecules and optically active samples. The most common chiral center is a carbon atom to which are bonded four different atoms or groups of atoms (a chiral carbon). In some rare cases, the expected optical activity will be too small to be observed, even though the molecules of the sample are known to be chiral. In other cases, the presence of a second chiral center similar to the first may give at least one conformation of the molecule a plane or center of symmetry (a meso form),

Meso form

Racemic mixture

and samples of the meso isomer will be optically inactive. Of course, if the sample is a racemic mixture (an equimolar mixture of enantiomers), no optical activity will be observed.

The presence of a chiral center in the molecules of a substance is not required for molecular chirality, as illustrated by the existence of certain optically active allenes and biphenyls. Excellent discussions of the details of the relationships between molecular structure, molecular symmetry, and optical activity are presented in References 1 through 3.

If a sample is optically active, then it is not a racemic mixture and it must be composed of chiral molecules. The absence of optical activity indicates that the sample is a racemic mixture or that its molecules possess a plane of symmetry, a center of symmetry, or a higher, even improper, axis of rotation in at least one conformation.

Optical rotatory dispersion

The sign and magnitude of the rotation are a function of both the configuration of the molecule and the wavelength of the light, and a graph of the specific rotation of a compound versus wavelength is called its *optical rotatory dispersion curve* (ORD curve). Because enantiomers have equal but opposite optical rotations at all wavelengths, their ORD curves are exactly similar in shape but opposite in sign. A more useful observation is that similar or homologous compounds of the same relative configuration have similar ORD curves of the same sign. This makes it possible in certain cases to establish the relative configurations of optically active molecules without interconversion. If, in addition, the absolute configuration of one has been established, the absolute configuration of all is thereby established. A more complete discussion of the interpretation of ORD curves can be found in Reference 1.

Problems

1. The observed rotation at 25°C of a sample of a-phenylethylamine (Section 21.1) in a 10-cm tube, using a sodium lamp, is -35.6°. The density, d^{25}, of the amine is reported to be 0.953 g/mL. Calculate the specific rotation, $[\alpha]_D^{25}$.

2. The observed rotation α of a solution of 232 mg of cholesterol (Section 21.1) in 10 mL of chloroform in a 1-dm tube was -0.73°. Calculate the specific rotation $[\alpha]_D^{20}$.

3. How can a clockwise rotation of 10° be distinguished from a counterclockwise rotation of 350°?

4. For each molecular structure, determine (1) the total number of stereoisomers possible, and (2) which of the stereoisomers should be optically active.

   a. 1-bromohexane
   b. 2-bromohexane
   c. 1,2-dibromohexane
   d. 2,3-dibromohexane
   e. 1,2-dibromocyclohexane
   f. 1,3-dibromocyclohexane
   g. 1,4-dibromocyclohexane
   h. chloroallene ($CH_2=C=CHCl$)
   i. 1,1-dichloroallene
   j. 1,3-dichloroallene
   k. 3-hexene
   l. 2-chloro-3-hexene
   m. 2,5-dichloro-3-hexene
   n. 2,5-dichloro-3-hexyne

References

1. K. Mislow, *Introduction to Stereochemistry*, W. A. Benjamin, New York, 1965.

2. H. H. Jaffe and M. Orchin, *Symmetry in Chemistry*, Wiley, New York, 1965, Chapters 1 and 2.

3. M. Orchin and H. H. Jaffe, *J. Chem. Educ.* **47,** 246 (1970).

4. B. Testa, *Principles of Organic Stereochemistry*, Marcel Dekker, New York, 1979.

5. H. Kagan, *Organic Stereochemistry*, Halsted Press (John Wiley and Sons), London (New York), 1979.

22. Molecular Weight

One of the bits of information most useful in establishing the molecular structure of a new substance is the molecular weight. If the molecular weight is known with confidence to 1 atomic mass unit, all but a small fraction of the total structural possibilities can be eliminated by this fact alone.

If the problem is simply to establish the identity of a substance with a known compound, the determination of the molecular weight of the unknown is much less useful, because it provides only a single data point for comparison. The infrared spectrum, for example (Section 24), would provide many more points for comparison and would usually be easier to determine than the molecular weight. Of course, the molecular weight of any substance of known structure is also automatically known and thus can be compared with that of an unknown substance, which is not true of the infrared spectrum and other physical and chemical properties.

22.1 Molecular Weight Determination by Mass Spectrometry

Mass spectrometry is currently the most commonly used method for the determination of molecular weight. It has the great advantages of being applicable to most substances that are at least slightly volatile (a vapor pressure of 10^{-2} Torr at a temperature of 100°C), of requiring a very small amount of sample, and of routinely giving the molecular weight to 1 atomic mass unit. The disadvantages of the mass spectrometric method are that mass spectrometers are expensive and difficult to maintain in top operating condition. Section 27.2 describes how the mass spectrum of a substance can be determined and how it can provide information about the molecular structure, as well as the molecular weight, of a substance.

22.2 Molecular Weight Determination by Other Methods

Freezing point depression; boiling point elevation; osmotic pressure

Other methods of molecular weight determination include measurement of *freezing point depression, boiling point elevation,* or *osmotic pressure.* In these methods, the concentration of a solution of the unknown substance is determined. Knowing the weight of the solute added, you can then calculate its molecular weight.

Equivalent weight of acids and bases

The *equivalent weight* of acidic or basic substances can be determined by titration. From the equivalents of titrant consumed, the equivalents of sample are known. Knowing the weight of the sample used, you can calculate the weight per equivalent.

Methods for the determination of molecular weights by freezing point depression, using camphor and *tert*-butyl alcohol as solvents, are given in References 1 and 2.

References

1. E. J. Cowles and M. T. Pike, *J. Chem. Educ.* **40,** 422 (1963).
2. M. J. Bigelow, *J. Chem. Educ.* **45,** 108 (1968).

23. Solubility

The solubility of a substance is significant for two reasons. First, before you work with a compound in the laboratory, you want to know whether the substance will be soluble in aqueous solutions such as water, aqueous acid, and aqueous base, or in organic solvents such as ether. Second, if you are working with a substance of unknown molecular structure, its solubility characteristics are important indicators of various structural features.

Solubility is a matter of *degree*

The concept of solubility is a little subtle in that *solubility is a matter of degree*. That is, many substances are at least slightly soluble, and the choice of where to draw the line below which substances will be called "insoluble" and above which substances will be called "soluble" is arbitrary. Often, the line is drawn at 30 mg of substance per mL of solvent (3 g per 100 mL of solvent). Be aware that solubility will vary from solvent to solvent, there will be borderline cases, and you must be alert for the use of different definitions of solubility.

23.1 Solubility of Liquids in Liquids

Mixing is normal. Although the mutual insolubility of "oil" and water is probably the most familiar example of the solubility behavior of two liquids, it is much more satisfactory to approach the phenomenon of solubility from the point of view that mutual solubility, or *mixing*, is the *expected* or *normal behavior* for two liquids, and that *insolubility* is *exceptional* or *unusual* behavior.

   The two liquids benzene and toluene are mutually soluble in all proportions (miscible) for the same reason that red and white marbles will mix if you shake them together: the free energy of the mixed state is less than the free energy of the unmixed state because of the increase in entropy associated with mixing. The mixed state has the same potential energy as the unmixed state, but it is more probable than the unmixed state. If the intermolecular forces between unlike molecules and like molecules are the same, the heat of mixing, ΔH_{mix} will be zero. Because the entropy of mixing, ΔS_{mix} is always positive, the free-energy change upon mixing will, in this case, be negative.

$$\Delta G_{mix} = \Delta H_{mix} - T\Delta S_{mix} \qquad (23.1\text{-}1)$$

In the special case when ΔH_{mix} equals zero, this equation becomes

$$\Delta G_{mix} = T\Delta S_{mix} \qquad (23.1\text{-}2)$$

Entropy makes it happen   *Mixing is "entropy-driven".* The phenomenon of mixing is one of a very few in which the free-energy change can be considered to be primarily a function of the change in entropy rather than primarily a function of the change in enthalpy (energy). Thus, the expected or normal miscibility of liquids is observed when the heat of mixing, ΔH_{mix} is negative (mixing is exothermic), zero, or only slightly positive (mixing is only slightly endothermic).

Insolubility

If ΔH_{mix} is large and positive (highly endothermic), it can more than compensate for $T\Delta S$ and will cause ΔG_{mix} to be positive (Equation 23.1-1). If this is the case, the unmixed state will have a lower free energy than the mixed state, and mixing will not occur. The enthalpy of mixing, ΔH_{mix}, can be positive if the forces between like molecules are larger than the forces between unlike molecules, because it will then take more energy (work) to separate the like molecules (of both solvent and solute) to the larger average intermolecular distance of the mixed state than will be regained upon moving the unlike molecules together in the mixed state:

$$\Delta H_{mix} = E_{separation} + E_{association} \qquad (23.1\text{-}3)$$

where $E_{separation}$ is the energy required to separate like molecules (always positive), and $E_{association}$ is the energy released upon moving unlike molecules together (always negative). The larger the intermolecular attractions, the greater the magnitude of each term of this equation.

Water "squeezes out" oil. Thus, the nonmiscibility of water with most organic liquids is the result of the large intermolecular attraction between water molecules (hydrogen bonding), which must be lost and remain uncompensated for in the mixed state when the organic liquid cannot form hydrogen bonds with water. It is in this way that the immiscibility of "oil" (or hexane, benzene, carbon tetrachloride, etc.) and water can be explained.

Water welcomes hydrogen bonding solutes. On the other hand, methanol, ethanol, acetone, and acetic acid are miscible with water because the intermolecular attractions between the molecules of water and solvent are comparable in magnitude with the intermolecular attractions between water and water, and solvent and solvent. This results in a relatively small heat of mixing and therefore, by Equation 23.1-1, a negative free-energy change upon mixing.

"Like dissolves like." The generalization that *"like dissolves like"* makes sense because substances whose molecules have similar intermolecular forces (type and magnitude) tend to mix.

Partial Solubility

Complete immiscibility, however, is an extreme behavior actually approached by few pairs of liquids; mercury and water appear to approach this extreme very closely. It is more usual for pairs of liquids that are not completely miscible to be slightly soluble in one another.

For example, 100 mL of water will dissolve 7.5 g of diethyl ether, and 100 mL of ether will dissolve 1.3 g of water at 25°C; 100 mL of water will dissolve 8.5 g of n-butyl alcohol, and 100 g of n-butyl alcohol will dissolve 25 g of water at 20°C.

Entropy effects. Partial solubility can be understood by realizing that the increase of entropy due to mixing is largest for the first material that dissolves. That is, ΔS_{mix} for a material going from mole fraction solvent = 1.0 to mole fraction solvent = 0.9 (or x_{solute} = 0.0 to x_{solute} = 0.1) is greater than ΔS_{mix} for going from $x_{solvent}$ = 0.9 to $x_{solvent}$ = 0.8 (or x_{solute} = 0.1 to x_{solute} = 0.2). Because ΔH_{mix} might be expected to be approximately the same in both cases, ΔG_{mix} (which equals $\Delta H_{mix} - T\Delta S$) can be negative for the formation of a dilute solution (when the average ΔS_{mix} per increase in concentration of solution is larger) and positive for the formation of a concentrated solution (when the average ΔS_{mix} per increase in concentration of solute is smaller). The larger and more positive the enthalpy of mixing ΔH_{mix}, the sooner ΔH_{mix} will equal $T\Delta S_{mix}$; the larger and more positive the enthalpy of mixing, the sooner the solution should be

Table 23.1-1. Solubility in water of straight-chain organic compounds with different functional groups; branching of the alkyl group increases the solubility

Type of Compound	Number of Carbon Atoms						
	1	2	3	4	5	6	7
Alcohol	· – – – ———————
Aldehyde	· · · · · · · · · · · · · · · · – – – – ———————
Alkane	————————————————————————
Alkene	————————————————————————
Alkyl halide	————————————————————————
Amide	· – – – – ———————
Amine	· · · · · · · · · · · · · · · · · · – – – – ———————
Carboxylic acid	· · · · · · · · · · · · · · · · · · · – – – – – – ———
Methyl ester	· · · · · · · · · · · · · · · · · · – – – – ———————
Methyl ether	· · · · · · · · · · · · · · · – – – – ———————
Methyl ketone	· – – – – – ———
Nitrile	· · · · · · · · · · · · · · · · · · – – – ———————
Thiol	· · · · · · · ————————————————————

Solubility is greater than 5 grams per 100 mL of water: · · · · · · · · · · · ·

Solubility is between 1 and 5 grams per 100 mL of water: – – – – – –

Solubility is less than 1 gram per 100 mL of water: ——————

expected to become saturated with respect to the solute, the point at which ΔG_{mix} is equal to zero. Table 23.1-1 presents a qualitative summary of the solubility in water of organic compounds containing different functional groups. Note that branching of the alkyl group increases the solubility.

Enthalpy effects. Because both the heat of vaporization of a liquid (and thus the boiling point of a liquid) and the heat of mixing are functions of intermolecular forces, there is an understandable correspondence between the solubility of certain organic compounds in water and their boiling points. For the four isomeric butyl alcohols, for example, the order of decreasing boiling point is the order of increasing solubility in water. In Section 17.2, lower boiling point was interpreted in terms of smaller intermolecular forces. Similarly, ΔH_{mix} would be expected to be more negative (or less positive) when less energy is required to separate the molecules from their average distance in the unmixed state to their average distance in the mixed state. Table 23.1-2 gives the boiling points and solubilities of the isomeric butyl alcohols and several other compounds.

R-group effect. In general the lower homologs corresponding to a given functional group are more soluble in water than are the higher homologs (see Table 23.1-1). This can be understood by realizing that the larger the hydrocarbon part of the molecule, compared with the functional group that can hydrogen bond to water, the smaller the net intermolecular attraction between water molecules and the molecules of the organic liquid. This will result, according to Equation 23.1-3, in a more positive heat of mixing. In other words, the larger the alkyl group, the more the molecule will behave like the water-insoluble hydrocarbons.

Table 23.1-2. Boiling point and solubility in water of isomeric compounds

Compound	Boiling Point (°C)	Solubility in Water
n-Butyl alcohol	118	8.3[a]
Isobutyl alcohol	108	9.6[a]
Sec-butyl alcohol	100	13.0[a]
Tert-butyl alcohol	83	Miscible in all proportions
Diethyl ether	35	7.5[a]
Methyl *n*-butyl ether	70	1.00[b]
Methyl isobutyl ether	58	1.24[b]
Methyl *sec*-butyl ether	59	1.79[b]
Methyl *tert*-butyl ether	54	5.89[b]

[a] Grams per 100 mL water at 20°C. [b] Weight percent of ether in saturated aqueous solution at 20°C.

Solubility increases with increasing temperature

Temperature effect. The general trend of greater solubility at a higher temperature can be explained by Equation 23.1-1. Greater solubility means that the same amount of material can be dissolved in a smaller volume of solvent to form a more concentrated solution. Although the formation of a more concentrated solution involves a smaller ΔS_{mix} than does the formation of a dilute solution, the term $T\Delta S_{mix}$ can maintain the same value despite a decrease in ΔS_{mix} if T is greater.

"Salting out." The lower solubility of most organic compounds in aqueous salt solutions than in water makes sense because ΔH_{mix} for a salt solution should be more positive than for water; mixing with salt solution requires more energy to separate the charged ions. Thus, relatively soluble organic substances can often be *salted out of* aqueous solution by saturating the solution with NaCl or Na_2SO_4.

Entropy wins at high temperatures

Because ΔS_{mix} must always be positive, Equation 23.1-1 predicts that anything ought to be miscible with anything else if the temperature is high enough. This prediction will certainly be true when all the substances are above their critical temperatures because they will then behave as gases. As vapors, mercury and water are completely miscible.

23.2 Solubility of Solids in Liquids

The process of dissolution of a solid in a solvent can be considered to be the sum of two consecutive processes: the melting of the solid (to give the supercooled liquid) followed by the mixing of the liquid with the solvent. Thus, the free-energy change for the dissolving of a solid can be considered to be the sum of the free-energy changes for these two processes:

$$\Delta G_{dis} = \Delta G_f + \Delta G_{mix} \qquad (23.2\text{-}1)$$

Because

$$\Delta G_f = \Delta H_f - T\Delta S_f \quad \text{and} \quad \Delta G_{mix} = \Delta H_{mix} - T\Delta S_{mix}$$

it follows that

$$\Delta G_{dis} = \Delta G_f + \Delta H_{mix} - T\Delta S_{mix} \qquad (23.2\text{-}2)$$

or

$$\Delta G_{dis} = \Delta H_f - T\Delta S_f + \Delta H_{mix} - T\Delta S_{mix} \qquad (23.2\text{-}3)$$

Solids are more soluble at higher temperature

From equation 23.2-2, you can see why *solids are generally more soluble at higher temperature,* which is why purification by recrystallization is possible. At a higher temperature, less solvent will be needed to dissolve a given amount of material to give a saturated solution ($\Delta G_{dis} = 0$) because, although ΔS_{mix} will be smaller because a more concentrated solution is being formed, a larger value for T will compensate partially through the $T\Delta S_{mix}$ term, just as in the mixing of liquids discussed in the previous section, and through the $T\Delta S_f$ term, because both entropy terms should be positive.

Solubilities of isomeric solids

Finally, equation 23.2-1 shows why isomers, which might be expected to have the same tendency to mix in a solvent, can have different solubilities. For example, the *cis/trans* isomers, maleic and fumaric acid, might be expected to mix equally well with water, but the *cis* isomer, maleic acid, is 100 times more soluble than the *trans* isomer at room temperature.

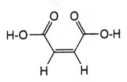

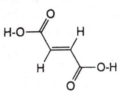

maleic acid; m.p. 143°C
60 g/100 mL H$_2$O

fumaric acid; m.p. 286°C
0.6 g/100 mL H$_2$O

Similarly, the three position isomers of dinitrobenzene might be expected to mix equally well with benzene, but their solubilities range over a factor of 10 at 50°C. (Table 23.2-1). These differences in solubility can be understood by observing that the less soluble isomer has the higher melting point. Below the melting point, ΔG_f will be positive, and the higher the melting point, the greater the magnitude of this term. Thus, the compensating $T\Delta S_{mix}$ term must remain larger for the higher-melting isomer, and saturation must occur at higher dilutions.

23.3 Classification of Compounds by Solubility; Relationships Between Solubility and Molecular Structure

In the following sections, structural features of molecules that lead to solubility under various conditions in different solvents will be described. This information allows us to predict the solubility

Table 23.2-1. Solubilities and melting points of some isomeric compounds

Compound	Melting point (°C)	Solubility in		
		Water	Alcohol	Benzene
Maleic acid (*cis*)	143	60[a]	51[a]	
Fumaric acid (*trans*)	286	0.6[a]	5[a]	
o-Dinitrobenzene	116			17.5[b]
m-Dinitrobenzene	90			37.6[b]
p-Dinitrobenzene	170			3.1[b]

[a] Grams per 100 mL solvent at 20°C. [b] Grams per 100 mL solvent at 50°C.

behavior of a compound of known structure and to exclude certain possibilities for the structure of an unknown substance on the basis of its solubility under different conditions. Again, a compound will be classified as "soluble" if it is soluble to the extent of 30 or more milligrams per milliliter of solvent (3 or more grams per 100 mL of solvent).

Solubility in water

Covalent substances are soluble in water at room temperature only if the molecules have a functional group that can form hydrogen bonds with water—a functional group, in other words, that contains a nitrogen, oxygen, or sulfur atom. The higher homologs are decreasingly soluble in water, and with monofunctional compounds they become "insoluble" at about five carbon atoms (see Table 23.1-1). If two or more functional groups are present, especially if they include the amino or hydroxy group, compounds with more than five or six carbon atoms may be water soluble.

Ionic substances mix with water

Ionic substances (salts) are often very soluble in water. Although these compounds are solids and may have very large heats of fusion, solvation of the ions is often sufficiently exothermic that they are soluble in water.

If an organic liquid is soluble in water, it is most likely to be a relatively low molecular weight mono- or difunctional compound. Higher molecular weight and polyfunctional compounds are much more likely to be solids.

If an organic solid is soluble in water, it is most likely to be either a salt (the alkali metal salt of an acid or the hydrogen halide or sulfate salt of an amine), a polyfunctional compound that happens to be a solid (polyhydric phenols, sugars), or an amino acid (which exists as an inner salt).

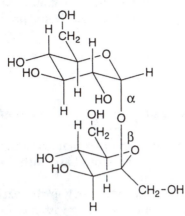

sucrose (table sugar); very soluble in water

Acidification of the aqueous solution of the salt of an organic acid will convert it to the free acid, which will separate from solution if it is not itself soluble in water or aqueous acid; similarly, making the solution of the salt of an amine basic will liberate the free base, which will separate from solution if it is not soluble in water or aqueous base. Salts of weak acids will hydrolyze in aqueous solution to form hydroxide ion, and salts of amines will hydrolyze to produce hydronium ion; the decrease or increase in acidity of the water in which the salt is dissolved can be detected by means of pH paper or a pH meter.

Solubility in diethyl ether and dichloromethane

Almost all organic liquids are soluble in ether. A great many organic solids are soluble in ether. Those that are not are compounds with large heats of fusion and thus include salts and high-molecular-weight, high-melting covalent compounds. Dichloromethane is an excellent solvent for nonpolar organic liquids and solids but not as good a solvent as diethyl ether for more polar organics. Unlike ether, dichloromethane will not burn.

Covalent substances mix with ether

Solubility in 5% aqueous sodium hydroxide

Because the sodium salts of acids are generally very soluble in water, water-insoluble organic compounds that can be converted to their conjugate bases by 5% aqueous sodium hydroxide will dissolve in this solvent. Such compounds include carboxylic acids, sulfonic acids, sulfinic acids; phenols; sulfonamides of primary amines, imides; some β-diketones and β-keto esters; mercaptans and thiophenols; some

Acids dissolve in base

primary and secondary nitro compounds; and oximes. Acidification of the basic solution of such compounds should precipitate the sample.

a carboxylic acid   **a sulfonic acid**

It should be obvious that substances that are soluble in water would be expected to be soluble in aqueous base, with the exception of the salts of water-insoluble amines.

Some acid halides and some easily hydrolyzed esters will react to give soluble products upon standing with 5% aqueous sodium hydroxide.

Solubility in 5% aqueous sodium bicarbonate

Strong acids dissolve in weak base

Of the substances listed as soluble in 5% aqueous sodium hydroxide, only carboxylic acids, sulfonic acids, sulfinic acids, and certain phenols with multiple electronegative substituents, such as 2,4,6-tribromophenol or 2,4-dinitrophenol, are sufficiently acidic to be converted to their conjugate bases by 5% aqueous sodium bicarbonate solution and thus to dissolve in this solvent. Dissolution of acids in sodium bicarbonate solution will result in the evolution of carbon dioxide gas.

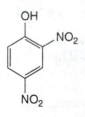

2,4-dinitrophenol

Solubility in 5% aqueous hydrochloric acid

Bases dissolve in acid

Five percent aqueous hydrochloric acid will convert many amines to their water-soluble hydrochloride salts and thus cause them to dissolve. Most aliphatic amines (primary, secondary, and tertiary) will be converted to their conjugate acids under these conditions, as will most aromatic amines in which no more than one aromatic ring is attached directly to the nitrogen. Diphenylamine, carbazole, 2,4,6-tribromoaniline, and the nitroanilines are not sufficiently basic to be converted to their conjugate acids under these conditions.

N,N-disubstituted amides and N-benzylacetamide can be converted to their conjugate acids under these conditions but not N-unsubstituted or most N-monosubstituted amides.

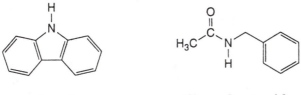

carbazole   **N-benzylacetamide**

The hydrochloride salts of some basic substances are not soluble in 5% hydrochloric acid, although the salt may be soluble in water. Thus, if a solid remains after treatment with 5% hydrochloric acid, it should be separated from the supernatant liquid by suction filtration or centrifugation and its solubility in water determined. If doubt still remains as to whether the solid is the salt or the original compound, its melting point or infrared spectrum can be compared with that of the original material. Basification of the supernatant liquid should give a precipitate if the hydrochloride salt was partially dissolved.

It should be obvious that substances that are soluble in water would be expected to be soluble in aqueous acid, with the exception of the salts of water-insoluble acids.

Solubility in concentrated sulfuric acid

Compounds containing an oxygen or nitrogen atom will dissolve in or react with cold concentrated sulfuric acid. Alkenes will react to form either a soluble alkyl hydrogen sulfate or an insoluble polymer, or will oxidize, usually with the evolution of heat. Polyalkylbenzenes will undergo sulfonation to give a product that is soluble in the mixture.

Even weak bases dissolve in conc. sulfuric acid

All saturated compounds containing only carbon, hydrogen, or halogen will be insoluble in concentrated sulfuric acid, as will those aromatic hydrocarbons or their halogenated derivatives that cannot be easily sulfonated.

Solubility in 85% phosphoric acid

Certain oxygen-containing compounds such as alcohols, aldehydes, methyl and cyclic ketones, and esters are soluble in 85% phosphoric acid, provided that they contain fewer than nine carbon atoms. Ethers are somewhat less soluble: di-*n*-propyl ether is soluble, whereas di-*n*-butyl ether and anisole are not.

23.4 Techniques for Determination of Solubility

A sample size of 30 mg of a solid or one or two drops of a liquid to 1 mL of solvent may be satisfactory in most cases. The solid should be powdered. All solubility determinations should be made at room temperature, without heating.

In determining solubility in water or ether, add the solvent in portions to the sample in a small test tube (8-mm diameter by 50-mm length). Mix the contents of the tube well after each addition of solvent. Adding the solvent in portions will allow you to distinguish between substances that are freely soluble, barely soluble, and relatively insoluble. If the substance is insoluble in water, the resulting suspension can be used to determine solubility in hydrochloric acid or sodium hydroxide: add a drop or two of concentrated hydrochloric acid or 50% sodium hydroxide solution.

In determining solubility in 5% sodium hydroxide, 5% sodium bicarbonate (only for substances found to be soluble in 5% sodium hydroxide), or 5% hydrochloric acid (only for compounds that contain nitrogen), add the sample in portions to the solvent, mixing well after each addition. If the first portion does not dissolve completely, the fact that the material is of limited solubility is established, and the remainder need not be added. The solution should be neutralized after the removal of any insoluble residue.

When determining solubility in concentrated sulfuric acid, add the sample to the acid. Color changes, evolution of heat or a gas, or other evidence of reaction can be as significant as mere dissolving.

Solubility in 85% phosphoric acid should be determined only if the compound is soluble in concentrated sulfuric acid.

Problems

1. Perfluoroheptane (C_7F_{16}) dissolves in heptane only to the extent of 0.27 mole per mole of heptane. Account for the immiscibility of these two substances.

2. Account for the miscibility with water of polyethyleneglycol, a high-boiling substance whose formula is

$$HO—(CH_2CH_2O—)_n CH_2CH_2OH$$

3. Account for the fact that, although diethyl ether boils lower than the isomeric butyl alcohols, it is less soluble in water (see Table 23.2).

4. Evaluate the relative effectiveness of nitrogen, oxygen, and sulfur atoms in hydrogen bonding to water (see Table 23.1).

5. Rationalize the trends in the solubilities of the following isomeric compounds:

Compound	Boiling point (°C)	Melting point (°C)	Solubility in water (grams per 100 mL)
Valeric acid	187		3.4
Isovaleric acid	177		4.2
Trimethylacetic acid		36	2.2
n-Propyl acetate	101		1.9
Isopropyl acetate	88		3.2

6. The solubility of caffeine in several solvents is given; rationalize the high solubility in chloroform.

Solvent	Solubility (grams per 100 g solvent; 18°C)
Chloroform	11.2
Carbon tetrachloride	0.09
Ether	0.12
Ethyl acetate	0.73
Benzene	0.91

7. Explain the fact that some organic compounds, for instance, many proteins, are more soluble in salt solutions than in pure water; that is, they can be "salted in" to aqueous solution.

References

Sources from which more information about the relationships between solubility and molecular structure can be obtained include

1. R. L. Shriner, R. C. Fuson, D. Y. Curtin, and T. C. Morrill, *The Systematic Identification of Organic Compounds*, 6th edition, Wiley, New York, 1980, p. 90.

2. N. D. Cheronis and J. B. Entrikin, *Identification of Organic Compounds*, Interscience, New York, 1963, p. 77.

3. A. I. Vogel, *A Textbook of Practical Organic Chemistry*, 3rd edition, Wiley, New York, 1957, p. 1045.

24.   Infrared Absorption Spectrometry

Absorption spectrometric methods are relatively recent developments. Ultraviolet(UV)-visible spectrometry (Section 25) has been known the longest. It was followed by infrared (IR) spectrometry and, most recently, by nuclear magnetic resonance (NMR) spectrometry (Section 26). Because equipment is widely available for the determination of IR, UV, and NMR spectra, these methods are now commonly and routinely used for the analysis and characterization of pure substances and mixtures.

Absorption spectrometry

Absorption spectrometry is the determination of the degree to which electromagnetic radiation (light; light energy) is absorbed by a substance over a range of wavelengths. The record of energy absorption versus wavelength is the absorption spectrum. The usual wavelength range for IR spectrometry is from 2 microns to 16 microns, although instruments are available that scan up to 200 microns. Table 24-1 indicates the relationship between the IR region and other regions of the electromagnetic spectrum, including the UV-visible region and the NMR region.

Table 24-1. The electromagnetic spectrum: energy, frequency, and wavelength.

Energy, calories/mole of quanta	10^{12}	10^{10}	10^{8}	10^{6}	10^{4}	10^{2}	1	10^{-2}	10^{-4}	10^{-6}		
Frequency, Hz	10^{22}	10^{20}	10^{18}	10^{16}	10^{14}	10^{12}	10^{10}	10^{8}	10^{6}	10^{4}		
Wavelength, microns (μ)	10^{-8}	10^{-6}	10^{-4}	10^{-2}	1	10^{2}	10^{4}	10^{6}	10^{8}	10^{10}		
Wavelength, nanometers (nm)		10^{-4}	10^{-2}	1	10^{2}	10^{4}	10^{6}	10^{8}	10^{10}	10^{12}		

Type of Radiation

Gamma rays
X-rays
Ultraviolet
Visible
Infrared
NMR

24.1 Wavelength, Frequency, and Energy of Electromagnetic Radiation

Electromagnetic radiation

Electromagnetic radiation (light) can be specified by wavelength, frequency, or energy. The product of wavelength (λ; length) and frequency (v; time^{-1}) equals the speed of light (c; length per time):

Light

$$\lambda v = c \qquad (24.1\text{-}1)$$

According to Equation 24.1-1, wavelength and frequency are inversely proportional to one another, that is, short wavelengths correspond to high frequencies, and vice versa.

Light as energy

Light energy is specified by the size of the *quantum,* or least increment of energy. The size of the quantum is directly proportional to the frequency of the light, as shown by Equation 24.1-2:

$$E = hv \qquad (24.1\text{-}2)$$

where E is the energy of the quantum, v is the frequency, and h is the proportionality constant, Planck's constant. Thus, large quanta are associated with high frequency and, by inverse proportionality, with short wavelength.

The relationships between wavelength, frequency, and energy are also illustrated in Table 24-1.

24.2 Units of Light Absorption

Transmittance

The degree to which light is absorbed by a sample at a particular wavelength can be expressed in several ways. The first is as *transmittance, T,* which is defined in Equation 24.2-1 as the ratio of the intensity of light that passes through the sample to the intensity incident upon the sample (that is, the ratio that expresses the fraction of light that gets through the sample):

$$\text{Transmittance} = T = \frac{I_{\text{transmitted}}}{I_{\text{incident}}} = \frac{I_{\text{t}}}{I_{\text{i}}} \qquad (24.2\text{-}1)$$

Percent transmittance

The second way is the *percent transmittance, %T*, which is defined as the transmittance × 100, Equation 24.2-2:

$$\text{Percent transmittance} = \%T = T \times 100 \qquad (24.2\text{-}2)$$

Absorbance

The third way is the *absorbance, A*, which is defined as the logarithm of the reciprocal of the transmittance, Equation 24.2-3:

$$\text{Absorbance} = A = \log\frac{1}{T} = \log\frac{I_i}{I_t} \qquad (24.2\text{-}3)$$

The relationships between these units are illustrated in Table 24.2-1.

Extinction coefficient

It is too bad to have to use "complicated" units such as absorbance, but only absorbance is directly proportional to sample concentration and sample thickness (path length). The relationship between absorbance and concentration is given by Equation 24.2-4:

$$A = \varepsilon c l \qquad (24.2\text{-}4)$$

where A is the absorbance, c is the concentration in moles/liter, and l is the path length, in cm; ε, the extinction coefficient, is the proportionality constant.

Table 24.2-1. Relationships between transmittance, percent transmittance, and absorbance units

%T	T	1/T	Absorbance
100	1	1	0
75	3/4	1.33	0.125
50	1/2	2	0.301
25	1/4	4	0.602
10	1/10	10	1.000
5	1/20	20	1.301
2	1/50	50	1.699
1	1/100	100	2.000
0	0	infinite	infinite

The practical consequences of this relationship between concentration and path length will be brought out in the discussions of sample preparation for infrared and ultraviolet spectroscopy.

Most IR and UV spectrometers are calibrated in both percent transmittance and absorbance units.

24.3 Infrared Light Absorption and Molecular Structure

Mechanism of energy absorption

In all forms of absorption spectrometry, an interpretation of a spectrum is a description of the ways by which light energy is absorbed by the molecules of the sample. In the infrared range, energy absorption is explained by an increase in the amplitude of various molecular vibrations. The frequency of the light absorbed equals the frequency of the molecular vibration, and thus the higher-frequency vibrations will absorb the larger quanta, or the shorter-wavelength light.

Bonds are springs; atoms are masses

To interpret infrared spectra, bonds between atoms in molecules can be thought of as springs, and the atoms as masses at the ends of the springs. According to this model, then, parts of the molecule can stretch and bend with respect to the rest of the molecule. The frequency of stretching and bending can be related to the atomic masses and the force constants of the springs by Equation 24.3-1:

$$v \propto \sqrt{\frac{k}{\mu}} \qquad (24.3\text{-}1)$$

Here, v is the frequency of stretching or bending, k is the force constant (a measure of the stiffness of the spring), and μ is the reduced mass. The reduced mass is defined by Equation 24.3-2:

Force constant

$$\mu = \frac{M_1 M_2}{M_1 + M_2} \qquad (24.3\text{-}2) \quad \text{Reduced mass}$$

where M_1 and M_2 are the total masses at the two ends of the spring. If one part of the molecule is much lighter than the remainder ($M_1 \ll M_2$), then μ approximately equals the mass of the light part, M_1.

When this is true, Equation 24.3-1 reduces to Equation 24.3-3:

$$v \propto \sqrt{\frac{k}{M_1}} \qquad\qquad (24.3\text{-}3)$$

and the frequency of bending or stretching of a small part of the molecule with respect to the remainder is proportional to the square root of the stiffness of the spring and inversely proportional to the square root of the mass of the small part. The fact that the absorption of energy by the increase in amplitude of the stretching vibration of C—H, C—F, C—Cl, C—Br, and C—I bonds occurs at lower and lower frequencies (longer wavelength; lower energy) can thus be interpreted by saying that the force constant k appears to remain about the same, while M_1 increases. Similarly, the fact that the stretching frequency of a multiple bond is higher than that of the corresponding single bond can be interpreted by saying that the force constant is larger for the multiple bond than for the single bond. It is harder to stretch a double or triple bond than a single bond.

24.4 Interpretation of Infrared Spectra

Establishment of identity of samples

The simplest use of infrared spectra is to determine whether two samples are identical or not. If the samples are the same, their IR spectra, obtained under the same conditions, must be the same. If the samples are different, their spectra will be different.

Enantiomers: spectra are the same; diastereomers: spectra are different

If the two samples are both pure substances very similar in structure, the differences in the spectra may be so small that it is not easy to see them; it may even be beyond the power of the instrument to detect them. The absorption peaks of the spectrum of an impure sample should be less intense than those of a pure sample, assuming equal amounts of sample, and the spectrum will contain additional peaks.

The infrared spectra of enantiomers should be identical, and those of diastereomers should be different. Figures 1 and 2 in Experiment 11 show the IR spectra of a pair of enantiomers, the R and S isomers of carvone; the spectra are identical within experimental uncertainty.

Figures 2 and 4 in Experiment 14 illustrate the IR spectra of a pair of diastereomers, the *trans* and *cis* isomers of 1,2-dibenzoylethylene. The spectra are obviously different.

Use of IR spectra to monitor a purification

Comparison of IR spectra is a convenient and relatively sensitive way to establish the identity of a substance with a sample of known structure and purity. Comparison of the IR spectra of different fractions obtained in a fractional distillation or of material before and after recrystallization is a good way to determine and follow the progress of a purification. Most IR spectra are obtained for use in these ways.

Use of IR spectra for the determination of molecular structure

A more sophisticated level of interpretation of an IR spectrum is to use it to establish the structure of an unknown material. Comparison of IR spectra of substances of known structure has led to the establishment of a great many correlations between wavelength (or frequency) of IR absorption and features of molecular structure. The presence or absence of absorption at certain wavelengths thus indicates the presence or absence of certain structural features. Table 24.4-1 presents a number of these correlations.

Detection of functional groups

Certain structural features can be established fairly easily. For example, if a substance contains only C, H, and O, the oxygen can be present only as C=O, O—H, or C—O—C (or a combination of these, such as the ester or carboxylic acid group). The presence or absence of absorption in the carbonyl region (~5.8–6.0 microns; ~1730–1670 cm^{-1}) or O—H region (~2.7–3.0 microns; ~3,700–3,300 cm^{-1}) can serve to eliminate or establish some of these possibilities. The nature of a functional group involving nitrogen can be inferred in a compound containing only C, H, and N in a similar way. If the compound contains both O and N and/or other atoms other than C and H, the problem is more difficult but can be approached in the same way.

Establishing whether a compound is primarily aromatic or primarily aliphatic is also fairly easy. The IR spectrum of an aromatic compound (without a long aliphatic side chain) generally has only weak absorption in the stretching region (3.3 microns; 3,000 cm^{-1}), sharp bands in the 6–7-micron region (1,660–1,430 cm^{-1}), and strong absorption at wavelengths greater than 12 microns (less than

Alipatic or aromatic?

Table 24.4-1. Infrared absorption–structure correlations

	Range (microns)	Intensity	Range (cm⁻¹)
C—H stretching vibrations			
Alkane	3.38–3.51	m–s	2962–2853
Alkene	3.23–3.32	m	3095–3010
Alkyne	3.03	s	3300
Aromatic	3.30	v	3030
Aldehyde	3.45–3.55	w	2900–2820
and	3.60–3.70	v	2775–2700
C—H bending vibrations			
Alkane	6.74–7.33	v	1485–1365
Alkene			
monosubstituted (vinyl)	7.04–7.09	s	1420–1410
	7.69–7.75	w–s	1300–1290
	10.05–10.15	s	995–985
and	10.93–11.05	s	915–905
disubstituted, *cis*	14.5	s	690
disubstituted, *trans*	7.64–7.72	m	1310–1295
and	10.31–10.42	s	970–960
disubstituted, *gem*	7.04–7.09	s	1420–1410
and	11.17–11.30	s	895–885
trisubstituted	11.90–12.66	s	840–790
Aromatic			
5 adjacent H atoms	13.3	v,s	750
and	14.3	v,s	700
4 adjacent H atoms	13.3	v,s	750
3 adjacent H atoms	12.8	v,m	780
2 adjacent H atoms	12.0	v,m	830
1 isolated H atom	11.3	v,w	880
N—H stretching vibrations			
Amine not hydrogen bonded	2.86–3.03	m	3500–3300
Amide	2.86–3.2	m	3500–3140
O—H stretching vibrations			
Alcohols and phenols			
not hydrogen bonded	2.74–2.79	v,sh	3650–3590
hydrogen bonded	2.80–3.13	v,b	3750–3200
Carboxylic acids			
hydrogen bonded	3.70–4.00	w	2700–2500
C—O stretching vibrations			
Esters			
formates	8.33–8.48	s	1200–1180
acetates	8.00–8.13	s	1250–1230
propionates, etc.	8.33–8.70	s	1200–1150
benzoates; phthalates	7.63–8.00	s	1310–1250
and	8.69–9.09	s	1150–1100

Table 24.4-1. *continued*

	Range (microns)	Intensity	Range (cm⁻¹)
Carbon–halogen stretching vibrations			
C—F	7.1–10.00	s	1400–1000
C—Cl	12.5–16.6	s	800–600
C—Br	16.6–20.0	s	600–500
C—I	approx. 20	s	approx. 500
C=C stretching vibrations			
Isolated alkene	5.99–6.08	v	1669–1645
Conjugated alkene			
C=C conjugated	6.25	m–s	1600
C=O conjugated	6.07–6.17	m–s	1647–1621
phenyl conjugated	6.15	m–s	1625
Aromatic	6.25	v	1600
	6.33	m	1580
	6.67	v	1500
and	6.90	m	450
C=O stretching vibrations			
Aldehydes			
saturated aliphatic	5.75–5.81	s	1740–1720
α,β-unsaturated aliphatic	5.87–5.95	s	1705–1680
Ketones			
saturated acyclic	5.80–5.87	s	1725–1705
saturated 6-ring and larger	5.80–5.87	s	1725–1705
saturated 5-membered ring	5.71–5.75	s	1750–1740
α,β-unsaturated acyclic	5.94–6.01	s	1685–1665
aryl alkyl	5.88–5.95	s	1700–1680
diaryl	5.99–6.02	s	1670–1660
Carboxylic acids			
saturated aliphatic	5.80–5.88	s	725–1700
aromatic	5.88–5.95	s	1700–1680
Carboxylic acid anhydrides			
saturated acyclic	5.41–5.56	s	1850–1800
and	5.59–5.75	s	1790–1740
Acyl halides			
chlorides	5.57	s	1795
bromides	5.53	s	1810
Esters and lactones (cyclic esters)			
saturated acyclic	571–5.76	s	1750–1735
saturated 6-ring and larger	5.71–5.76	s	1750–1735
α,β-unsaturated and aryl	5.78–5.82	s	1730–1717
vinyl esters	5.56–5.65	s	1800–1770
Amides and lactams (cyclic amides)	5.88–6.14	s	1700–1630
Triple-bond stretching vibrations			
C≡N	4.42–4.51	m	2260–2215
C≡C	4.42–4.76	v,m	2260–2100

Abbreviations: w = weak absorption; m = medium absorption; s = strong absorption;
v = variable intensity of absorption; sh = sharp absorption; b = broad absorption.

840 cm⁻¹), and it gives the general impression of having sharp and symmetrical absorption bands. The IR spectrum of an aliphatic compound generally has relatively strong absorption in the C—H region (3.4 microns; 2,950 cm⁻¹) and little or no absorption at wavelengths greater than 12 microns (less than 840 cm⁻¹) and appears to have broad and not particularly symmetrical bands. Compare, for example, the spectra of methyl heptanoate and methyl benzoate, Figures 24.4-1 and 24.4-2. Cyclic and conformationally rigid compounds generally give spectra that contain a relatively small number of sharp, symmetrical bands; the spectrum of cyclohexanone, Figure E16-1 in Experiment 16, is an example.

methyl heptanoate
aliphatic

methyl benzoate
aromatic

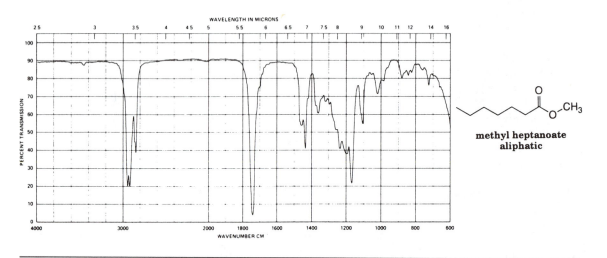

Figure 24.4-1. IR spectrum of the aliphatic compound methyl heptanoate; thin film.

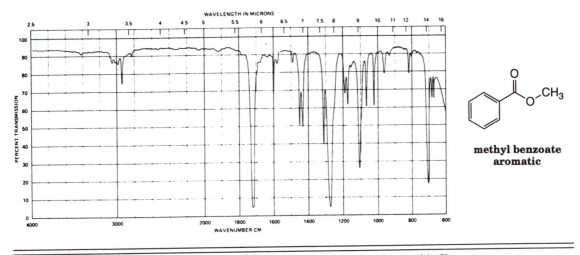

Figure 24.4-2. IR spectrum of the aromatic compound methyl benzoate; thin film.

24.5 Sample Preparation

For infrared absorption spectrometry, the sample whose absorption is to be determined must be placed in the beam of infrared radiation. The container (cell) or support for the sample must be transparent to infrared radiation and must therefore be made of one of a small number of materials, not including glass. The most commonly used material is sodium chloride, which is transparent between 2 and 16 microns (5,000 and 600 cm^{-1}). Certain other metal halide salts are transparent over other wavelength ranges.

IR cell

Liquid samples: the liquid film method

The spectrum of a pure liquid is most easily determined as a *liquid film* between a pair of sodium chloride plates. One salt plate is positioned in the holder, a drop of the liquid is placed in the center, and the second plate is put on top. The top of the holder is then pressed or screwed on gently and the holder placed in the sample (near) beam of the instrument. Nothing need be placed in the reference (far) beam, since the only absorbing material in the sample beam is the liquid whose absorption spectrum is being determined. If the strongest peak in the spectrum absorbs more than about 98% of the light, tighten the top of the holder to make the film thinner and

Salt plates

rerun the spectrum. The spectra shown in Figures 24.4-1 and 24.4-2, for example, were obtained from thin films of liquid.

The IR spectrum of a pure liquid can also be determined in solution, as with solid samples.

Solid samples: the solution method

The IR spectra of solids are most often determined in solution. Table 24.5-1 suggests appropriate concentrations for an average sample; the concentration should be adjusted, if necessary, so that the strongest peak will absorb between 90% and 98% of the light. As Equation 24.2-4 indicates, the absorbance of a sample is proportional to both its concentration and its thickness. Table 24.5-1 is based on the assumption that the sample thickness (the path length of the infrared cell) will be 0.2 mm. If a cell of 0.4 mm path length will be used instead, the amount of sample required will be only half that specified by the table. I assume that 0.5 mL of solution will be prepared in either case.

The ideal solvent would be transparent over the entire infrared range of wavelength, but no liquids attain this ideal. For this reason, the spectrum of the solvent must be subtracted from that of the sample in solution in this solvent. This subtraction is done in a double-beam instrument by placing a second cell that contains pure solvent in the reference beam, or in a single-beam instrument by recording the spectrum of the pure solvent as background and then subtracting this background spectrum.

Solvents most often used for infrared spectrometry are carbon tetrachloride, and chloroform, whose infrared spectra are shown in Figures 24.5-1 and 24.5-2. Water is never used, except with special cells with water-resistant windows.

In wavelength regions where the solvent absorbs strongly (more than about 95% of the light), the absorption recorded by the spectrophotometer is meaningless. This happens because, as the solvent absorption increases, the net absorption due to the sample becomes a smaller fraction of the total absorption, and the difference in intensity of the two beams finally falls below the limit detectable by the instrument. You can verify this if you are working with a double-beam instrument by placing your hand in the sample or reference beam when the instrument is scanning a range of strong solvent absorption (for carbon tetrachloride, between 12.3 and 13.7 microns; 810 and 730 cm^{-1}; Figure 24.5-1) and seeing that the spectrum is unaffected. Except for this short range, sample absorption can be determined at all wavelengths between 2 and 16 microns (5,000 and 600 cm^{-1}).

Table 24.5-1. Milligrams of sample required for 0.50 mL of solution

Molecular weight	100	200	300	400	500
Mg of sample	9	18	27	36	45

A cell path of 0.2 mm is assumed. For a different cell path length, adjust the amount of sample to keep the number of molecules in the beam the same as for the examples in the table.

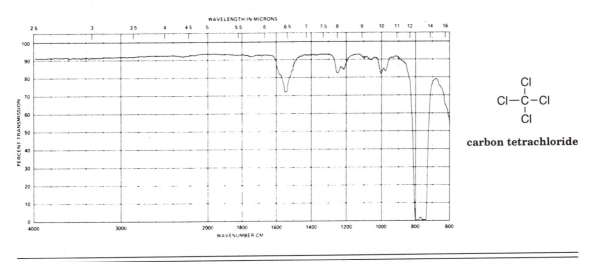

Figure 24.5-1. IR spectrum of carbon tetrachloride; 0.05-mm path length.

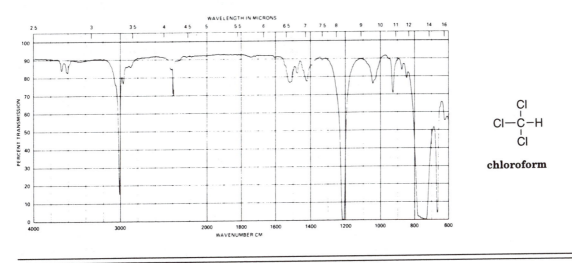

Figure 24.5-2. IR spectrum of chloroform; 0.05-mm path length.

carbon tetrachloride   chloroform

In a double-beam instrument, if the recorder pen runs either upscale or down scale in a range of intense solvent absorption, the balance control is not adjusted correctly, and the peaks of the rest of the spectrum will be distorted.

Solid samples: the KBr method

The infrared spectrum of a solid can also be determined in the solid state. One way is to grind 1 to 2 mg of sample with about 100 to 400 mg of anhydrous potassium bromide in a clean mortar and to press the resulting mixture into a translucent wafer, using a die. The wafer is then mounted in the sample beam, and, in a double-beam instrument, an attenuator is placed in the reference beam. The function of the attenuator is to compensate for the loss of sample beam intensity due to scattering. The screen of the attenuator is set so that the recorder pen indicates 100% transmission at the wavelength at which the sample is most transparent. The appropriate wavelength can be found by a trial scan of the spectrum or by a shortcut appropriate to the individual instrument. Use of the attenuator gives a more normal-looking spectrum, because it in effect moves the spectrum to the part of the chart where a given amount of absorbance corresponds to the largest movement of the pen; compare, for example, the distance of pen travel from 0.0 to 0.1 and from 1.0 to 1.1 absorbance units.

O—H absorption. Absorption in the O—H region in a spectrum obtained by the KBr pellet method must be interpreted with great caution, because it is hard to make sure that no water is in the KBr or gets into the sample during preparation.

Solid samples: the mull method

Mineral oil
Hexachlorobutadiene

A second method of determining the IR spectrum of a solid in the solid state is to grind (mull) 2 to 5 mg with a drop of *mineral oil* (paraffin oil) or *hexachlorobutadiene*. The spectrum of the mull is then determined as a liquid film. Because the recorded spectrum will be that of the compound plus the mulling agent, the spectrum of the mulling agent must be mentally subtracted from the recorded spectrum. The spectra of mineral oil and hexachlorobutadiene are presented in Figures 24.5-3 and 24.5-4.

mineral oil

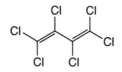

hexachlorobutadiene

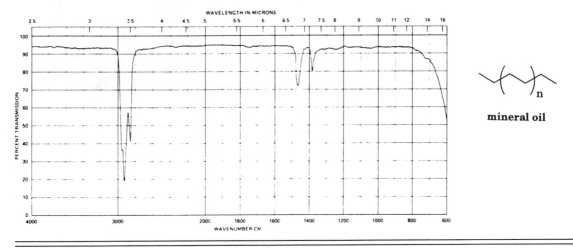

Figure 24.5-3. IR spectrum of mineral oil; thin film.

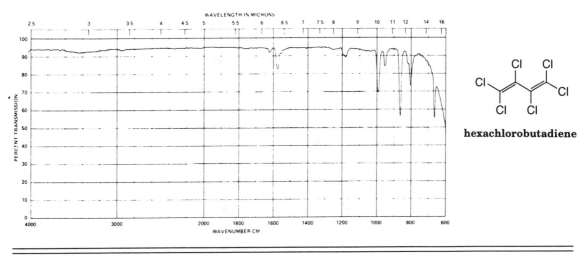

Figure 24.5-4. IR spectrum of hexachlorobutadiene; thin film.

Recall from Section 24.4 that the spectra of identical materials should be the same if the spectra are obtained under the same conditions. The reason for this qualification is that for some compounds, the appearance of the spectrum depends on the nature of the sample. For example, the appearance of the part of the spectrum of an alcohol depends on the concentration of the solution. The reason is that in dilute solution, a smaller fraction of the molecules are hydrogen bonded than in a concentrated solution, and these two molecular species—hydrogen bonded and not hydrogen bonded—absorb at different wavelengths in the O—H region. Also, the appearance of the spectrum of a sample obtained by the KBr technique often depends on the exact details of its preparation, presumably because of differences in particle size and distribution. The spectra of solids obtained in the solid state sometimes differ from spectra obtained in solution, and spectra of the same compound in different solvents will almost always look different because of different regions of strong solvent absorption.

Reference 11 presents an excellent introduction to the practice of infrared spectrometry, especially for sample preparation.

Problems

1. Saturated aliphatic esters (1) absorb at about 5.71–5.76 microns. α, β-Unsaturated esters (2) absorb at longer wavelength, and vinyl esters (3) absorb at shorter wavelength. Explain.

(1)   **(2)**   **(3)**

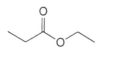

5.71–5.76 μ

2. Interpret the trend in the wavelength of the carbonyl absorption of these compounds:

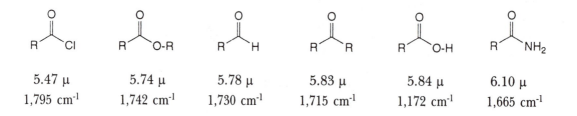

R–CCl	R–C–O-R	R–C–H	R–C–R	R–C–O-H	R–C–NH$_2$
5.47 μ	5.74 μ	5.78 μ	5.83 μ	5.84 μ	6.10 μ
1,795 cm^{-1}	1,742 cm^{-1}	1,730 cm^{-1}	1,715 cm^{-1}	1,172 cm^{-1}	1,665 cm^{-1}

3. Figures 24.5-5 and 24.5-6 show the infrared spectra of androstenedione and testosterone (introduction to Experiment 88). Which is the spectrum of androstenedione and which is the spectrum of testosterone?

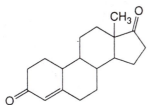

androstenedione

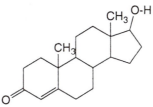

testosterone

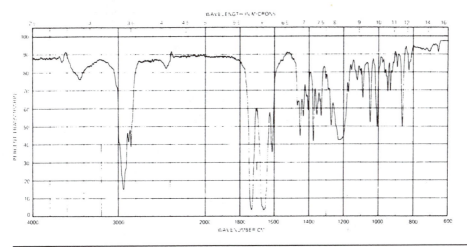

Figure 24.5-5. IR spectrum of either androstenedione or testosterone.

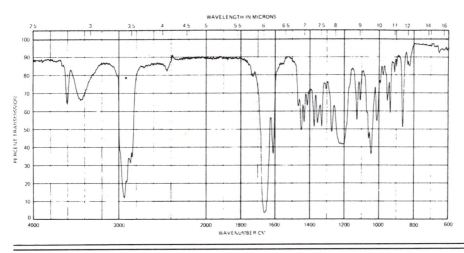

Figure 24.5-6. IR spectrum of either androstenedione or testosterone.

4. Figures 24.5-7 and 24.5-8 show the infrared spectra of pregnenolone acetate and progesterone (introduction to Experiment 88). Which is the spectrum of pregnenolone acetate and which is the spectrum of progesterone?

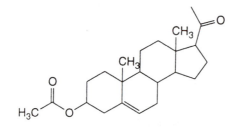

pregnenolone acetate

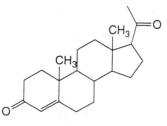

progesterone

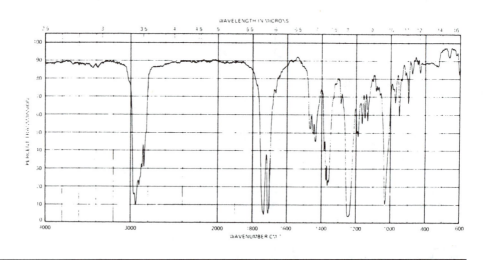

Figure 24.5-7. IR spectrum of either pregnenolone acetate or progesterone.

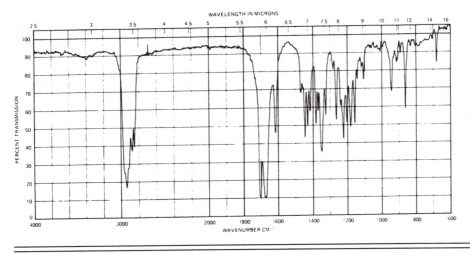

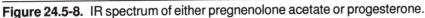

Figure 24.5-8. IR spectrum of either pregnenolone acetate or progesterone.

References

A brief introduction to infrared absorption spectrometry is

1. P. A. Jones, *Chemistry* **38,** 5 (1965).

Introductions that include spectra–structure correlations are

2. *Identification of Organic Compounds,* 4th edition, Wiley, New York, 1981, pp. 95–180.
3. J. A. Dyer, *Applications of Absorption Spectroscopy of Organic Compounds,* Prentice-Hall, Englewood Cliffs, N.J., 1965, pp. 22–57.
4. D. J. Pasto and C. R. Johnson, *Organic Structure Determination,* Prentice-Hall, Englewood Cliffs, N.J., 1969, pp. 109–158.
5. A. D. Cross and R. A. Jones, *Introduction to Practical Infrared Spectroscopy,* 3rd edition, Plenum, New York, 1969.
6. R. T. Conley, *Infrared Spectroscopy,* 2nd edition, Allyn and Bacon, Boston, 1972.

A fine presentation of spectra–structure correlations is made in

7. L. J. Bellamy, *The Infrared Spectra of Complex Molecules,* 3rd edition, Halsted, New York, 1975.

Indexes of published infrared spectra include

8. H. M. Hershenson, *Infrared Absorption Spectra: Index for 1945–1957,* Academic Press, New York, 1959.

9. H. M. Hershenson, *Infrared Absorption Spectra: Index for 1958–1962,* Academic Press, New York, 1964.

A catalog of about 8,000 infrared spectra is

10. C. J. Pouchert, *The Aldrich Library of Infrared Spectra,* Aldrich Chemical Co., Inc., Milwaukee, Wisconsin, 1970 (2nd edition, 1975; 3rd edition, 1981).

Finally, an extensive discussion of methods of sample preparation is presented in

11. J. S. Swinehart, *Organic Chemistry,* Appleton Century-Crofts, 1969, Appendix III.

25. Ultraviolet-Visible Absorption Spectrometry

Ultraviolet-visible spectrometry is the determination of the degree to which a substance absorbs ultraviolet and visible light; the output of a UV-visible spectrometer is a record of energy absorption versus wavelength, the UV-visible spectrum. Usually the wavelength ranges are from 200 nanometers to 360 nm (near ultraviolet range) and from 360 nm to 800 nm (visible range; see Table 24.1). Absorption at shorter wavelengths is more difficult to measure because air absorbs in this region, and absorption at longer wavelengths involves a different mechanism of energy absorption (IR spectrometry, Section 24.3). Because absorption of light energy in these wavelength ranges is the result of the promotion of electrons in lower energy levels to higher energy levels, UV-visible spectrometry is often called electronic absorption spectrometry.

Mechanism of UV energy absorption

Table 25.1-1. Conjugated dienes: correlations between structure and wavelength of absorption maximum

	Nanometers
Base values	
Acyclic conjugated dienes and conjugated dienes contained in two nonfused 6-membered-ring systems	217
Conjugated dienes contained in two fused 6-membered-ring systems (heteroannular dienes)	214
Conjugated dienes contained in a single ring (homoannular dienes)	253
Increments	
C—C extending conjugation	+30
Each alkyl substituent including ring residues	+5
Each ring, 6-membered or less, to which the diene double bonds are exocyclic	+5
Each -Cl or -Br substituent	+17
Alkoxy or acyloxy substituent	0
Solvent corrections	
No solvent corrections are necessary	

From A. Ault, *Problems in Organic Structure Determination*, McGraw-Hill, New York, 1967, p. 16. Reprinted with permission.

25.1 Ultraviolet-Visible Light Absorption and Molecular Structure

Ultraviolet-visible spectrometry, the oldest of the spectrometric methods, found its greatest use in the determination of the structure of molecules that contain conjugated systems of double bonds. The presence of conjugated unsaturation always produced one or more intense absorption maxima at wavelengths greater than 200 nm. Useful correlations were deduced for the positions and intensities of these peaks as a function of structure for conjugated dienes (Table 25.1-1) and α,β-unsaturated ketones (Table 25.1-2).

Table 25.1-2. α,β-Unsaturated aldehydes and ketones: correlations between structure and wavelength of absorption maximum

	Nanometers
Base values	
α,β-Unsaturated ketones	215
α,β-Unsaturated aldehydes	210
Cyclopentenones	205
Increments	
C=C extending conjugation	+30
α-Alkyl substituent including ring residue	+10
β-Alkyl substituents including ring residues	+12
γ,δ, or further alkyl substituents including ring residues	+18
Each ring, 6-membered or less, to which a C=C is exocyclic	5
C=C contained in a 5-membered ring (except cyclopentenones)	+5
Enolic α or β-OH	+35
(Alkoxy and acyloxy substituents are treated as alkyl groups.)	
Solvent corrections	
Water	-8
Methanol and ethanol	0
Chloroform	+1
Dioxane	-5
Ether	+7
Hexane	+11

From A. Ault, *Problems in Organic Structure Determination*, McGraw-Hill, New York, 1967, p. 15. Reprinted with permission.

25.2 Interpretation of Ultraviolet-Visible Spectra

The two features of an ultraviolet-visible spectrum that can give information concerning molecular structure are

Features of the UV spectrum

- the wavelengths at which the maximum light is absorbed, and
- the intensity of absorption at that wavelength.

The way in which the degree of absorption is expressed and measured experimentally is explained in Section 24.2. Certain correlations of these two parameters with molecular structure are summarized in the two preceding tables.

Since the development of infrared and nuclear magnetic resonance spectrometry, ultraviolet spectrometry has become less important for the determination of structure of organic molecules. For this purpose, the newer methods are simply more convenient and more reliable. For example, the establishment of the presence of a carbonyl group is much easier and more certain through infrared spectrometry. The principal application of UV spectrometry to determination of molecular structure lies at present in the study of more complex molecules containing conjugated systems.

Ultraviolet spectrometry is, however, still widely used as a quantitative method of analysis to determine concentrations. This is because many compounds absorb strongly even at low concentration, and because UV spectrometers can determine the degree of absorption of a sample with great accuracy.

Quantitative analysis

25.3 Color and Molecular Structure

Color is the result of a selective reflection or transmission of light in the visible range of the electromagnetic spectrum. A white or colorless substance reflects or transmits all incident light equally well. For any substance to appear colored, then, it must selectively absorb light of certain wavelengths in the visible part of the spectrum. For most organic compounds, the absorption of visible light requires the presence of a π electronic system that contains four or more conjugated double bonds. Thus, colored compounds are most often found to be aromatic rather than aliphatic substances.

A substance that absorbs light near the blue end of the visible spectrum appears yellow. As the wavelength of visible absorption moves toward the red end of the spectrum, the observed color becomes orange, red, and finally purple or blue. Colored organic compounds that are not yellow or orange are very unusual.

The most common types of colored compounds are nitro-substituted phenol or aniline derivatives. Less common types are quinone and compounds containing the azo (—N=N—) system. Many dyes and colorings combine these two structural features, as illustrated in Experiments 62 through 65.

25.4 Sample Preparation

Usually, a suitable sample for UV-visible spectrometry will be a solution of the substance in a solvent transparent to UV and visible radiation. Thus, choices of sample concentration and solvent must be made, in addition to a choice of cell type.

Concentration

As stated in Section 24.2, absorbance is proportional to the concentration of the sample (c, in moles per liter) and the sample thickness, or path length (l, in cm):

$$A = \varepsilon c l$$

Extinction coefficient — The value of the extinction coefficient, ε, at the absorption maximum can help determine the molecular structure of the sample. Therefore, you will want to determine the spectrum at a concentration that will give an absorbance of about 1 or a little less, because the value of ε can be estimated most accurately under these conditions. Because the standard sample cell has a path length of 1.00 cm, the concentration, c, in moles per liter, will equal $1/\varepsilon$ when $A = 1$; and for A not to exceed 1, c should be equal to or less than $1/\varepsilon$. Table 25.4-1 summarizes the concentration, in moles per liter, required for an absorbance of 1 for absorption bands of various extinction coefficients. Table 25.4-2 presents some data that may be useful in preparing samples of various concentrations.

Table 25.4-1. Concentration c required for an absorbance of 1 for absorption bands of various extinction coefficients ε

	ε	c
Very weak band	10	0.1 M
Weak band	100	0.01 M
Medium band	1000	0.001 M
Strong band	10,000	0.0001 M

There are two important practical consequences of the fact that values for ε for different functional groups can differ by several powers of 10. First, observing that a sample of 0.0001 *M* concentration appears to be completely transparent (*A* = 0) does not necessarily mean that absorption maxima with ε = 100 or less are absent. This occurs because, at a concentration this low, the maxima would show an absorbance of only 0.01 or less, which could easily be overlooked. Similarly, a strong absorption band can completely obscure a weak one. The second consequence is that for the accurate determination of extinction coefficients that differ by much more than a power of 10, spectra obtained at different concentrations will be needed.

Solvent

The ideal solvent for UV-visible spectrometry is completely transparent over the entire spectral range. In contrast to the case with IR spectrometry, many solvents do approach the ideal fairly closely. These include cyclohexane, ethanol, methanol, and water. To compensate for light absorption by the solvent and the cell (and any absorption by impurities in the solvent), UV-visible spectrometers are generally run in the double-beam mode wherein the absorption of pure solvent in a second cell is automatically subtracted from the absorption of the sample solution. The spectrum recorded is the difference spectrum, and it represents the net absorption due to the solute.

Because small concentrations of many compounds result in very intense absorption, a small amount of an impurity introduced during careless sample preparation may dominate the spectrum and completely obscure the spectrum of the compound you are interested in.

Beware of impurities

Table 25.4-2. Milligrams of sample required for 5 mL of solution as a function of molecular weight and desired concentrations[a]

	Molecular Weight				
Concentration	100	200	300	400	500
0.1 *M*	50	100	150	200	250
0.01 *M*	5	10	15	20	25
0.001 *M*	0.5	1	1.5	2	2.5
0.0001 *M*	0.05	0.1	0.15	0.2	0.25

[a] Prepare the most dilute solutions by dilution of a more concentrated solution.

Cell type

The standard UV-visible cell is a rectangular cell that has a 1-cm path length and approximately 3-mL volume. Circular cells and cells of different path lengths are also available.

Pyrex cells

For visible spectra, Pyrex glass cells are appropriate. For spectra in the UV range, the much more expensive quartz cells will be needed, because Pyrex absorbs strongly at wavelengths shorter than 320 nm.

Problems

1. Four derivatives of cholesterol, A, B, C, and D, have the following spectral properties:

   A IR: strong absorption near 5.95 microns

   UV: λ_{max} = 230 nm; ε = 10,700

   B IR: strong absorption near 5.95 microns

   UV: λ_{max} = 241 nm; ε = 16,600

   C IR: no absorption near 5.95 microns

   UV: λ = 234 nm; ε = 20,000

   D IR: no absorption near 5.95 microns

   UV: λ = 315 nm; ε = 19,800

Which structure corresponds to which derivative?

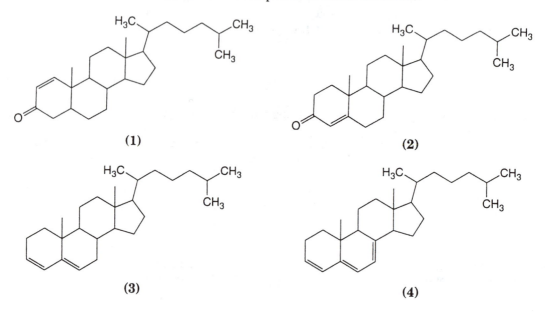

(1) (2)

(3) (4)

2. Absolute ethanol (anhydrous ethanol) can be prepared by adding benzene to 95% ethanol and distilling. The first distillate is a ternary azeotrope of benzene, water, and ethanol; the next is a binary azeotrope of benzene and ethanol. Benzene has a maximum extinction coefficient of 200 at 256 nm. What is the minimum concentration of benzene (mg/liter) that could be detected by UV spectroscopy, assuming that an absorbance as little as 0.02 could be determined?

3. LSD has a maximum extinction coefficient of less than 1000. What are the chances of detecting 10 micrograms of LSD by means of UV spectrometry?

References

1. R. M. Silverstein, G. C. Bassler, and T. C. Morrill, *Spectrometric Identification of Organic Compounds*, 4th edition, Wiley, New York, 1981, pp. 305–331.

2. J. R. Dyer, *Applications of Absorption Spectroscopy of Organic Compounds*, Prentice-Hall, Englewood Cliffs, N.J., 1965, pp. 4–21.

26. Nuclear Magnetic Resonance Spectrometry

Mechanism of energy absorption

Nuclear magnetic resonance spectrometry (NMR) measures the absorption of "light" energy in the radio frequency portion of the electromagnetic spectrum. The mechanism of energy absorption is the *reorientation of magnetic nuclei* with respect to a large external magnetic field, H_{ext}, in which the sample is placed. Unlike a compass needle in the Earth's magnetic field, which can have a great many different orientations, a nuclear magnet can have only a few. The proton, for example, can have only two orientations with respect to the field: one "with" the field (of lower energy) and one "against" the field (of higher energy). Proton NMR spectrometry is based on the fact that the energy of the quantum of absorbed radio frequency (rf) radiation goes into changing the orientation of the nuclear magnet, the proton, from being "with" the magnetic field to being "against" the field.

Because almost every organic compound contains hydrogen, proton NMR is of special interest to organic chemists, and we will

Nuclear magnets

Two orientations

Proton NMR

introduce NMR as it applies to protons. Because the letter H is used by physicists to refer to the strength of a magnetic field, and the letter H is used by chemists to refer to the element hydrogen, a chance for confusion is present here.

26.1 Shielding; Chemical Shift

Because the difference in energy of the proton in the two states "with" and "against" depends on the size of the magnetic field that the proton experiences (H_{nuc}) and because the energy for transitions between the two states ($h\nu$) is supplied by the rf field, the following relationship holds

$$h\nu \propto H_{nuc}$$

where $h\nu$ is, according to Planck's relationship, the energy of the quanta of frequency ν.

Resonance condition

Protons experiencing different magnetic fields H_{nuc} will therefore absorb rf energy at different frequencies. For an NMR spectrometer operating at a frequency of 60 Megahertz (60 MHz), H_{nuc} must equal 14,092 gauss for energy to be absorbed by a proton, that is, for *resonance* to occur. Higher frequencies and higher fields go together, and for an NMR spectrometer operating at a frequency of 300 MHz, H_{nuc} must be five times greater, or 70,460 gauss, for resonance for protons.

Resonance

Shielding

The field at the nucleus, H_{nuc}, is not exactly equal to the external field, H_{ext}, however, because electrons near the nucleus react to the external field in such a way that the field at the nucleus is somewhat less than that produced by the magnet in whose field the sample is placed. That is, H_{nuc}, is less than H_{ext} by $H_{shielding}$, the degree to which the field at the nucleus is diminished by the reaction of the electrons near the nucleus:

$$H_{nuc} = H_{ext} - H_{shielding}$$

Chemical shift

The usefulness of NMR spectrometry to the organic chemist stems from the fact that protons in different structural environments experience a different $H_{shielding}$. Thus, if the sample in the magnetic field H_{ext} is irradiated with a constant rf field of exactly 60 MHz and H_{ext} is slowly increased, energy will be absorbed by the sample each time H_{nuc} (which equals $H_{ext} - H_{shielding}$) becomes equal to 14,092 gauss. The graph of energy absorption versus H_{ext} is the NMR spectrum. The NMR spectrum of benzene shows only a single instance of energy absorption (a single resonance), as illustrated in Figure 26.1-1, whereas the NMR spectrum of *p*-xylene shows two resonances, one for the methyl protons and one for the ring protons, as shown in Figure 26.1-2.

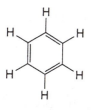

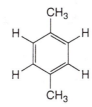

benzene
6 equivalent ring protons
one resonance

p-xylene
4 ring protons and 6 methyl protons
two resonances

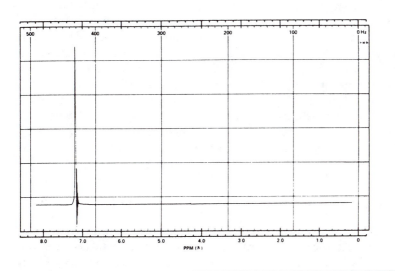

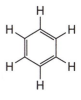

benzene

Figure 26.1-1. NMR spectrum of benzene; neat.

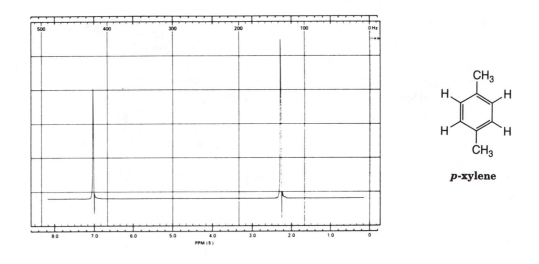

Figure 26.1-2. NMR spectrum of *p*-xylene; neat.

Sorting the protons of a molecule

Proton NMR, then, is a method for determining the *number of different types of protons* in a molecule. We can expect that sets of protons that are structurally distinct will have resonances at different chemical shifts. We can also expect that diastereotopic protons will have resonances at different chemical shifts and that enantiotopic protons will have identical chemical shifts except in the presence of a chiral solvent or other chiral additive. As we saw in the proton NMR spectrum of benzene, all 6 ring protons had the same chemical shift, and in the proton NMR spectrum of *p*-xylene, we saw that the 4 equivalent ring protons had a resonance at one chemical shift and that the 6 equivalent methyl protons had a resonance at another chemical shift. Proton NMR told us that the 6 protons of benzene were equivalent, and that the 10 protons of *p*-xylene were of two different types. We will see that the integral can tell us the relative numbers of each type of proton and that splitting can tell us about the relationships between the protons of each type.

Structurally distinct protons
diastereotopic protons
enantiotopic protons

Scale and units of chemical shift; the δ Scale

Shielding is reported as chemical shift

Chemical shifts are usually reported as the amount by which the H_{ext} required for the resonance of the sample substance varies from the H_{ext} required for the resonance of a reference substance, usually tetramethylsilane (TMS). A unit of convenient size is one part per million (ppm) of H_{ext}, and the zero of the scale is usually taken as the position of the resonance of the 12 equivalent protons of TMS. This scale, called the δ *scale*, is shown in Table 26.1-1.

$$\begin{array}{c} CH_3 \\ | \\ H_3C-Si-CH_3 \\ | \\ CH_3 \end{array}$$

tetramethylsilane; TMS
12 equivalent protons

Shielding is proportional to H_{ext}

Because shielding is proportional to H_{ext}, differences in shielding will also be proportional to H_{ext}. Thus, even though shielding will be greater in instruments with higher fields, the fractional degree of shielding will be constant in different instruments, and shielding reported in fractional units such as ppm will be the same for different instruments.

For comparison with coupling constants (see Section 26.2), chemical shift differences must be expressed in frequency units; 1 ppm in frequency units with a 60-MHz instrument will be 60 Hz, while in a 300-MHz instrument, 1 ppm in frequency units will be 300 Hz. The relationships between the various units and scales are shown in Table 26.1-1.

26.2 Splitting

Coupling

Although the presence of resonances at different chemical shifts gives information about the numbers of different functional groups in

Table 26.1-1. Units of chemical shift

Increasing H_{ext}:	→
Increasing Frequency:	←
Increasing Shielding:	→

11	10	9	8	7	6	5	4	3	2	1	0	−1	δ
660	600	540	480	420	360	300	240	180	120	60	0	−60	Hz[a]
2970	2700	2430	2160	1890	1620	1350	1080	810	540	270	0	−270	Hz[b]
3300	3000	2700	2400	2100	1800	1500	1200	900	600	300	0	−300	Hz[c]

[a] 60-MHz instrument. [b] 270-MHz instrument. [c] 300-MHz instrument.

which protons are present, the features of the NMR spectrum that provide the most detailed information about molecular structure are those that are the result of the effect upon one another of protons in different groups at different chemical shifts, or *coupling*.

Singlets and multiplets

Whereas the NMR spectrum of methyl iodide (Figure 26.2-1) consists of a single peak, or *singlet,* the NMR spectrum of ethyl iodide (Figure 26.2-2) shows two sets of peaks. The resonances of the two different kinds of protons in ethyl iodide, the methyl protons and the methylene protons, are not each a single peak, but each is a set of peaks, or *multiplet*. The methylene protons have caused the methyl resonance to appear as a triplet of approximate relative intensity 1:2:1, and the methyl protons have caused the resonance of the methylene protons to appear as a quartet of approximate relative intensity 1:3:3:1.

methyl iodide   ethyl iodide

In general, the methods of quantum mechanics are needed to calculate the expected multiplicity of a resonance. In many cases, however, a much simpler analysis is adequate.

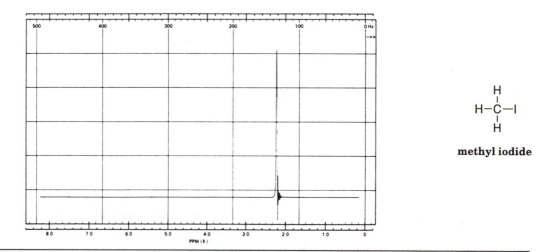

methyl iodide

Figure 26.2-1. NMR spectrum of methyl iodide; neat.

Interpretation of splitting of the CH$_3$ resonance of CH$_3$CH$_2$I

A "1:2:1 triplet." Consider first the appearance of the methyl resonance of ethyl iodide as a *1:2:1 triplet* due to the presence of the two adjacent methylene protons. Because of the small energy difference for a proton in either of the two states, a particular proton in a magnetic field of 14,092 gauss at room temperature has an almost equal probability of being in either the lower or the upper spin state—of being oriented, that is, either "with" or "against" the external magnetic field. Therefore, for the purpose of this analysis, you can estimate that of the ethyl iodide molecules, one-fourth will have both the methylene protons oriented "with" H_{ext} (↑↑), one-fourth will have both the methylene protons "against" H_{ext} (↓↓), and one-fourth will be in each of the two states possible, where one proton is "with" and one "against" H_{ext} ($\frac{1}{4}$ as ↑↓ and $\frac{1}{4}$ as ↓↑).

The effect of having both of the two adjacent methylene protons "with" the external field is to bring the methyl protons of those molecules into resonance at a slightly lower value of H_{ext}, because the two protons "with" the field will augment H_{ext} for the neighboring methyl group. This effect of the adjacent protons can be expressed by Equation 26.2-1, where $H_{coupling}$ in this case is positive:

$$H_{nuc} = H_{ext} - H_{shielding} + H_{coupling} \qquad (26.2\text{-}1)$$

Thus, the methyl resonance for one-fourth of the molecules will occur at a slightly lower field (smaller H_{ext}) than would be expected in the absence of coupling.

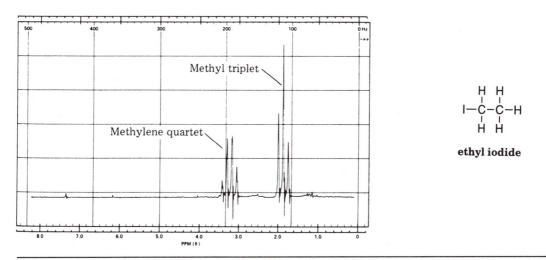

Figure 26.2-2. NMR spectrum of ethyl iodide; neat.

The methyl resonance for half of the molecules will occur at the same chemical shift as would be expected in the absence of coupling. These are the molecules in which the two methylene protons are oriented in opposite directions, canceling one another, so that in this case, in Equation 26.2-1, $H_{coupling}$ has a value of zero.

In the remaining fourth of the molecules, the adjacent methylene protons will have their spins oriented so that both oppose H_{ext}, diminishing the effect of H_{ext} on the methyl groups of these molecules. These molecules will thus have their methyl resonances at a slightly higher H_{ext} than would be expected in the absence of coupling; in this case, in Equation 26.2-1, $H_{coupling}$ is negative. In ethyl iodide, as in most ethyl groups, the magnitude of this effect, the coupling constant J, is about 7 Hz, which is $\frac{7}{60}$ of 1 ppm in a 60-MHz machine, or $\frac{7}{270}$ of 1 ppm in a 270-MHz machine.

Splitting is due to neighbors. It is important to notice in this analysis that the multiplicity of the methyl resonance is interpreted as an effect of the two *neighboring* protons. In this case, the protons of the methyl group have no apparent effect upon the multiplicity of their own resonance.

Interpretation of the splitting of the CH$_2$ resonance of CH$_3$CH$_2$I

A "1:3:3:1 quartet." The multiplicity of the resonance of the methylene group of ethyl iodide can be explained in a similar way. One-eighth of the molecules would be expected to have the protons of the adjacent methyl group all "with" H_{ext} (↑↑↑), three-eighths would be expected to have two "with" and one "against" (↑↑↓), three-eighths would be expected to have one "with" and two "against" (↑↓↓), and one-eighth would be expected to have all three "against" (↓↓↓). Thus, the molecules are divided into four groups of relative population 1:3:3:1, in which the methylene protons are expected to go into resonance at slightly different values of H_{ext}, because of the effect of the adjacent methyl protons. Again, the multiplicity of the resonance of a group of protons is explained by the effect of the adjacent protons; the multiplicity of a resonance does not depend upon the number of protons in the group corresponding to that resonance.

The N+1 Rule

In general, when a simple analysis such as this applies, the number of peaks in a multiplet is one more than the number of adjacent protons; the *"N + 1 Rule"*: a single adjacent proton will split the resonance of a group of neighboring protons into a 1:1 doublet; two adjacent protons will split the resonance of a group of neighboring protons into a 1:2:1 triplet, etc., as outlined in Table 26.2-1. The relative intensities of the peaks of a multiplet follow the coefficients of the binomial expansion

$$1:1 \qquad 1:2:1 \qquad 1:3:3:1 \qquad 1:4:6:4:1 \qquad 1:5:10:10:5:1 \qquad \text{etc.}$$

Magnetic equivalence. The N+1 Rule analysis is appropriate only if two conditions are met:

- **First,** the average coupling constant, J, between each proton in one group and each proton in the other must be the same, and
- **second,** the coupling constant, J, must be small relative to the chemical shift difference between the two groups, $\Delta\delta$.

Magnetic equivalence

That is, for the N+1 Rule to apply, *magnetic equivalence* must prevail.

Figures 26.2-3 through 26.2-8 present the NMR spectra of some compounds that can be readily interpreted by this simple method of analysis. Figure 26.2-3 shows the spectrum of 1,1,2-trichloroethane, in which the methylene protons appear as an approximately 1:1 doublet and the single proton as an approximately 1:2:1 triplet. In Figure 26.2-4, the two equivalent end protons of 1,1,2,3,3-pentachloropropane appear as an approximately 1:1 doublet, and the middle proton appears as an approximately 1:2:1 triplet.

Table 26.2-1. The N+1 rule

Number of Neighbors	Number of Peaks in Multiplet	Name of Multiplet	Relative Intensities within multiplet
0	1	singlet	1
1	2	doublet	1 : 1
2	3	triplet	1 : 2 : 1
3	4	quartet	1 : 3 : 3 : 1
4	5	quintet	1 : 4 : 6 : 4 : 1
5	6	sextet	1 : 5 : 10 : 10 : 5 : 1
6	7	septet	1 : 6 : 15 : 20 : 15 : 6 : 1
⋮	⋮	⋮	⋮
N	$N+1$		

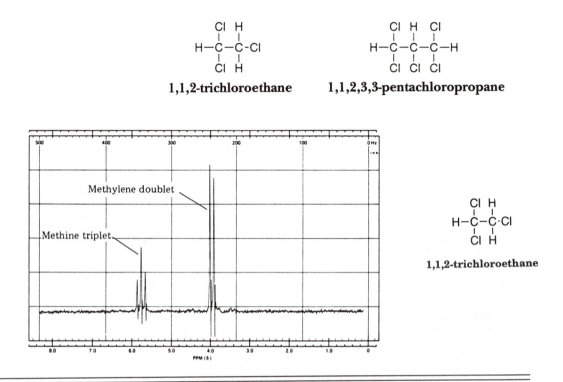

1,1,2-trichloroethane   1,1,2,3,3-pentachloropropane

Figure 26.2-3. NMR spectrum of 1,1,2-trichloroethane; CCl₄ solution.

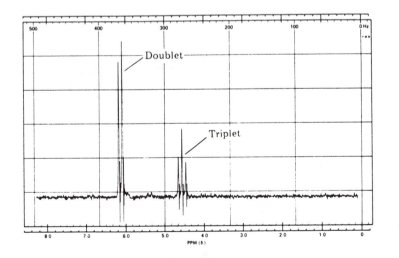

Figure 26.2-4. NMR spectrum of 1,1,2,3,3-pentachloropropane; CCl₄ solution.

Figure 26.2-5 shows the NMR spectrum of 1,1-dichloroethane, in which the methyl group appears as an approximately 1:1 doublet and the single proton as an approximately 1:3:3:1 quartet.

1,1-dichloroethane   **ethyl chloride**

Figure 26.2-6 shows the NMR spectrum of ethyl chloride, which illustrates again the characteristic resonance of the ethyl group: an

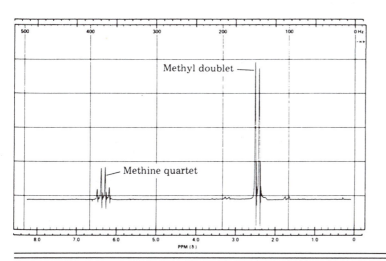

1,1-dichloroethane

Figure 26.2-5. NMR spectrum of 1,1-dichloroethane; neat.

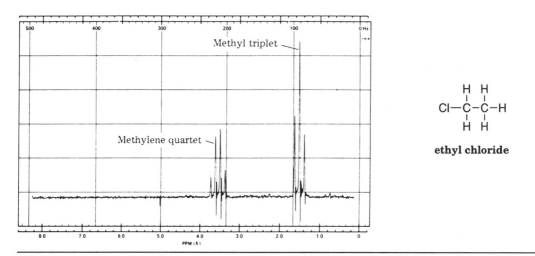

ethyl chloride

Figure 26.2-6. NMR spectrum of ethyl chloride; CCl$_4$ solution.

218   LABORATORY OPERATIONS

The "ethyl resonance"

"Pointing"

approximately 1:3:3:1 quartet for the methylene protons and an approximately 1:2:1 triplet for the methyl protons.

Notice that the actual relative intensities of the peaks within a multiplet differ slightly from the predicted 1:1, 1:2:1, etc., relative intensities that are predicted by the N+1 Rule. The distortion is such that the stronger peaks are on the side nearer the resonance of the nuclei to which they are coupled. In this sense, the peaks of a multiplet can be said to "point" toward the resonance of the nuclei with which they are coupled. Thus, in Figure 26.2-6, the methylene quartet is said to point toward the methyl resonance, and the methyl triplet is said to point toward the methylene quartet. In more complex spectra, mutual "pointing" indicates which resonances are related to one another by coupling.

By now you should expect that the NMR spectra of 1,2-dichloroethane and 1,1,1-trichloroethane should each consist of a single peak. They do, and their spectra are presented in Figures 26.2-7 and 26.2-8.

1,2-dichloroethane

1,1,1-trichloroethane

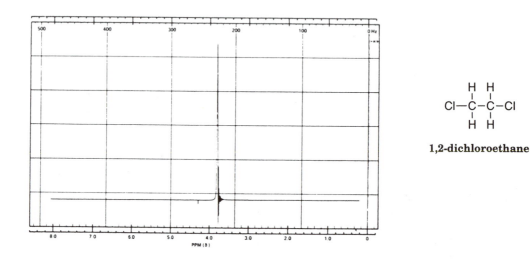

1,2-dichloroethane

Figure 26.2-7. NMR spectrum of 1,2-dichloroethane; neat.

Figure 26.2-6. NMR spectrum of ethyl chloride; CCl₄ solution.

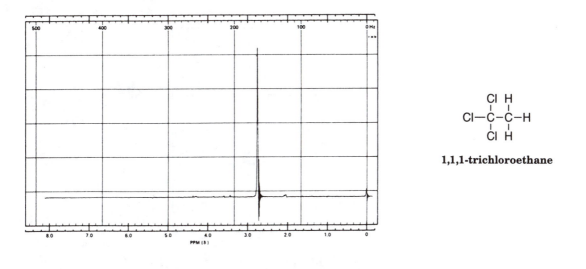

Figure 26.2-8. NMR spectrum of 1,1,1-trichloroethane; CCl₄ solution.

You should also notice from this series of spectra that protons bonded to a carbon bearing a chlorine atom are deshielded; two chlorine atoms appear to deshield more than one. The dependence of shielding upon molecular structure is discussed in more detail in Section 26.4.

26.3 The Integral

The third important feature of the NMR spectrum is the area under the absorption peaks, or the *integral*. Whereas *chemical shift* data provide information about the number and type of functional groups in which sets of protons occur, and the *splitting patterns* indicate the structural and geometrical relationships between protons, it is the total area under the peaks of a multiplet that corresponds to any set of protons, or **integral,** that provides information about the number of protons in that set. Because the degree to which energy is absorbed by a set of protons is independent of their structural environment, the integral is proportional to the number of protons.

The area under the peaks of an NMR spectrum, or **integral**, is determined electronically by use of the NMR spectrometer in a separate operation after obtaining the spectrum. Figure 26.3-1 shows the NMR spectrum of ethyl iodide and its integral. The vertical displacement of the second (uppermost) trace is proportional to the area

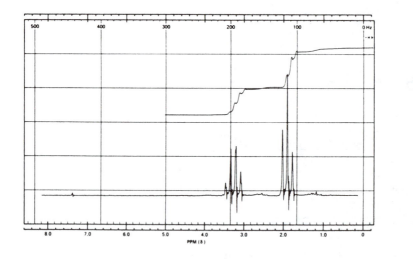

ethyl iodide

Figure 26.3-1. NMR spectrum and integral of ethyl iodide; neat.

under the corresponding peaks. The total area of the methylene quartet is seen to be two-thirds that of the methyl triplet. You can also estimate in the spectra of Figures 26.2-3 through 26.3-1 that the relative area of each multiplet corresponds to the relative number of protons involved.

26.4 Nuclear Magnetic Resonance and Molecular Structure

Chemical shift and molecular structure

The amount of energy absorbed by a proton is independent of its structural environment, and therefore the integral has the same value for every proton. In contrast, the shielding or chemical shift (shielding relative to a standard) does depend on the structural environment of a proton. Table 26.4-1 presents some of the correlations between *chemical shift and molecular structure.*

From the table, you can see two trends:

- Methyl protons are more shielded than analogous methylene protons, which, in turn, are more shielded than methine protons.

- Electron-withdrawing groups tend to deshield adjacent protons.

Table 26.4-1. Correlation between chemical shifts and molecular structure

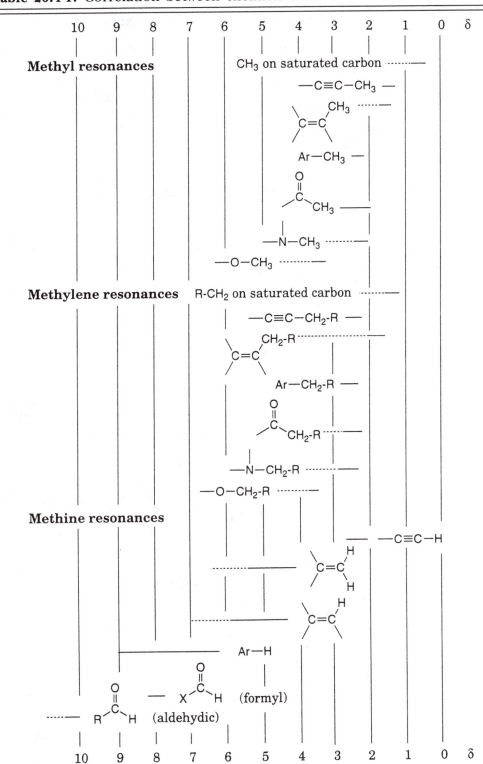

Coupling constants and molecular structure

Coupling constants depend in a very sensitive way on the structural and geometrical relationships between nuclei. Examples of the ways in which coupling depends on relationships between nuclei are presented in Table 26.4-2. A study of the table suggests that, in general, coupling does not extend over more than three bonds but that sp^2 hybridized atoms are better conductors of coupling than are sp^3 hybridized atoms.

3 bonds Yes;
4 bonds No

26.5 Interpretation of NMR Spectra

If the NMR spectrum is that of a pure substance of unknown molecular structure, the goal of interpretation is to deduce the structure of the molecules of the sample. In contrast to IR spectrometry, it is sometimes possible to deduce the structural formula of a substance from its molecular formula and NMR spectrum alone, without the use of reference spectra. The isomers of molecular formula $C_2H_4Cl_2$, for example, can be distinguished by their NMR spectra alone (Figures 26.2-5 and 26.2-7).

These are the steps of such an interpretation:

- From the integral, determine the relative number of protons responsible for each multiplet. If the molecular formula is known, the absolute number of protons can be calculated.
- Determine the chemical shift of the protons of each multiplet. The chemical shift is the difference between the shielding of the protons of the multiplet and the protons of tetramethylsilane (TMS). When the spectrum is run, the instrument is normally adjusted so that the resonance of TMS falls on the zero of the δ scale.
- Determine the coupling constant between nuclei in different sets at different chemical shifts. In the examples illustrated by the figures up to this point, the spacing between peaks of a multiplet equals the coupling constant.
- The final and most difficult step is to think of a molecular structure that accounts for the information determined in steps 1, 2, and 3. Correlation tables such as Tables 26.4-1 and 26.4-2 help out here.

Distorted N+1 Rule spectra

Unfortunately, few spectra will be as easy to interpret as the examples presented. One reason is that in the examples, the chemical shift

Table 26.4-2. Correlations between coupling constants and molecular structure

Relationship	J_{ab}(Hz)	
	range	*typical values*
(acyclic)	0–30	12–15
H_a—C—C—H_b (acyclic)	6–8	7
H_a—C—C—C—H_b	0–1	0
	0–3	0–2
	5–11	10
	11–19	17
H_a—C—C—H_b (with C=O)	1–3	2–3
H_a—C—C≡C—H_b	2–3	
aromatic *ortho*	6–10	9
aromatic *meta*	2–3	3
aromatic *para*	0–1	0

difference was always large compared with the coupling constant. Figure 26.5-1 shows the NMR spectrum of 1,2,3-trichlorobenzene, which, according to the simple analysis, should show a 1:1 doublet for the B protons and a 1:2:1 triplet for the A proton. The NMR spectrum actually consists of seven lines. We could, however, recognize this as a *distorted N+1 Rule spectrum.*

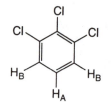

1,2,3-trichlorobenzene
J **small relative to** Δδ

Magnetic nonequivalence

Another reason for difficulty in interpreting spectra is that in the examples described so far, the coupling constants between each proton in one set and every proton in the other set were exactly equal

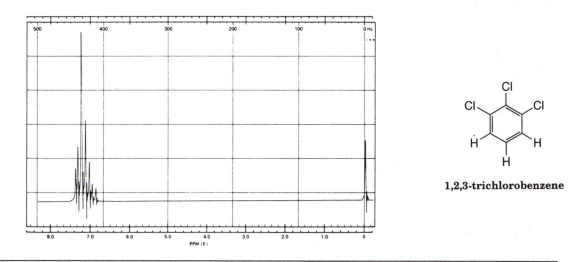

1,2,3-trichlorobenzene

Figure 26.5-1. NMR spectrum of 1,2,3-trichlorobenzene; CCl_4 solution.

(the interset coupling constants were equal; the two sets of chemical shift equivalent nuclei were *magnetically equivalent*). Figure 26.5-2 shows the NMR spectrum of 1-bromo-4-chlorobenzene. Here, the interset coupling constants are not equal: $J_{AB(ortho)}$ is not equal to $J_{AB(para)}$

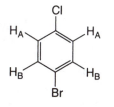

1-chloro-4-bromobenzene

$$J_{AB(ortho)} \neq J_{AB(para)}$$

Similarly, as shown in Figures 26.5-3 and 26.5-4, the spectra of *o*-dichlorobenzene and 1-bromo-2-chloroethane are complex because of the inequality of the interset coupling constants. The analysis of the NMR spectra of compounds such as these in which the protons experience *magnetic nonequivalence* is discussed in References 1 and 2.

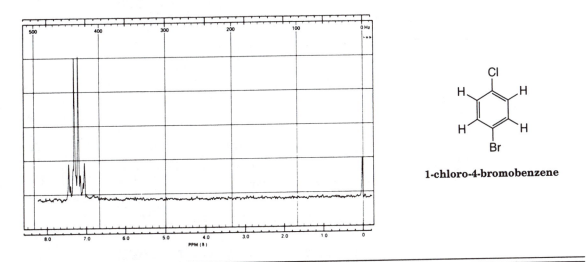

1-chloro-4-bromobenzene

Figure 26.5-2. NMR spectrum of 1-chloro-4-bromobenzene; CCl_4 solution.

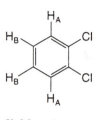

o-dichlorobenzene
$J_{AB(ortho)} \neq J_{AB(meta)}$

1-bromo-2-chloroethane
J_{AB} **are not all equal**

References 3 and 4 consider in more detail the concepts presented in this section and discuss the interpretation of the proton NMR spectra of a number of additional compounds of known structure. Reference 5 is a collection of more than 300 proton NMR spectra of simple compounds that illustrate in a systematic way the various features of NMR spectra.

26.6 Sample Preparation

Solvent

The ideal solvent for proton NMR would be a liquid in which most organic compounds would dissolve easily and that did not itself contain any protons. Carbon tetrachloride has been used as an NMR

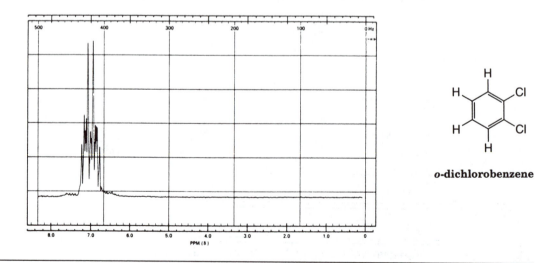

o-dichlorobenzene

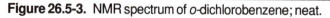

Figure 26.5-3. NMR spectrum of *o*-dichlorobenzene; neat.

solvent, but the fact that many organic solids are only slightly soluble in carbon tetrachloride and the fact that carbon tetrachloride is relatively toxic has led chemists to discontinue its use.

Deuterochloroform

At this time, most routine nuclear magnetic resonance spectra are obtained from a dilute solution of the sample in deuterochloroform. Many organic compounds are quite soluble in chloroform, and replacement of the normal hydrogen atom in chloroform by a deuterium atom eliminates the normal proton resonance of chloroform. Deuterochloroform can be obtained 99.96% pure with respect to contamination with the proton-containing version.

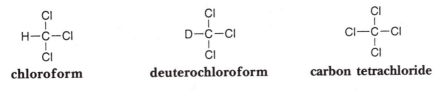

chloroform   **deuterochloroform**   **carbon tetrachloride**

Other deuterated solvents, such as hexadeuteroacetone, are available but expensive. Water-soluble samples can be run in deuterated water (deuterium oxide; D_2O).

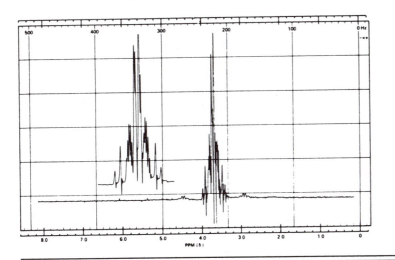

1-bromo-2-chloroethane

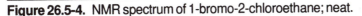

Figure 26.5-4. NMR spectrum of 1-bromo-2-chloroethane; neat.

hexadeuteroacetone

D—O—D

deuterium oxide

Concentration

Sample *concentration* should be about 10 to 30% by weight or volume. A total volume of 0.5 mL is sufficient.

NMR tube

The spectrum is determined with the sample in an *NMR tube,* a thin-walled glass tube 5 mm in diameter and 180 to 200 mm long, sealed at the bottom. A tight-fitting plastic cap closes the top of the tube.

The reference standard, tetramethylsilane

TMS   The **reference standard**, TMS, must be added if it is not already present (1% or less) in the NMR solvent as purchased. Because TMS boils at 27°C, it is best added by a micro-syringe that has been stored in the freezer. Only 5 to 10 microliters is needed. Recap the NMR tube immediately after adding the TMS; invert the tube several times to mix. Before attempting to insert the sample tube into the probe of the instrument, wipe the outside clean and insert the tube into the

Spinner   white Teflon turbine ("spinner"), which the tube to be spun in the probe. Position the tube in the spinner with the depth gauge, and then place the tube and spinner in the probe.

Problems

1. Which isomer will give exactly one peak in its NMR spectrum?
    a. 1,1-Dichloroethane or 1,2-dichloroethane?
    b. Cyclobutane or methylcyclopropane?
    c. CH_3-O-CH_3 or CH_3CH_2-O-H?
    d. *Ortho, meta,* or *para*-dichlorobenzene?
    e. Isomers of molecular formula C_5H_{12}?
    f. Isomers of molecular formula C_4H_9Br?
2. Predict the appearance of the NMR spectra of the following compounds:

a.

 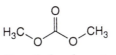

acetone   dimethylcarbonate   methyl acetate

b.

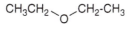

diethyl ether

c.

2-butanone   **ethyl acetate**   **methyl propionate**

3. Figure 26.6-1 shows the NMR spectrum of a compound of molecular formula $C_6H_{12}O_2$. What molecular structure is most consistent with the spectrum?

4. Figure 26.6-2 shows the NMR spectrum of a compound of molecular formula $C_7H_{14}O$. What molecular structure is most consistent with the spectrum?

5. Figure 26.6-3 shows the NMR spectrum of a compound of molecular formula C_5H_8. What molecular structure is most consistent with the spectrum?

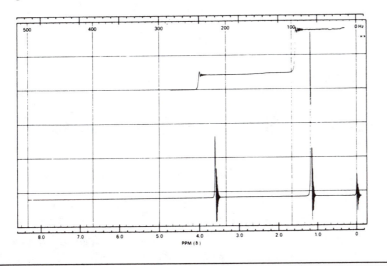

Figure 26.6-1. NMR spectrum of a compound of molecular formula $C_6H_{12}O_2$; CCl_4 solution.

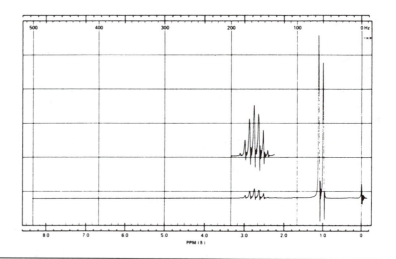

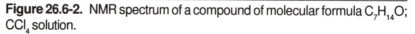

Figure 26.6-2. NMR spectrum of a compound of molecular formula $C_7H_{14}O$; CCl_4 solution.

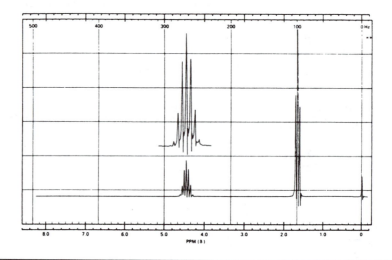

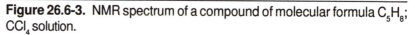

Figure 26.6-3. NMR spectrum of a compound of molecular formula C_5H_8; CCl_4 solution.

References

1. Ault, A., "*Classification of Spin Systems in NMR Spectroscopy,*" *J. Chem. Educ.* **47,** 812 (1970).

2. Ault, A., "*Tests for Chemical Shift Equivalence and Magnetic Equivalence in NMR Spectroscopy,*" *J. Chem. Educ.* **51,** 729 (1974).

Introductory texts

3. A. Ault and G. O. Dudek, *An Introduction to Proton Nuclear Magnetic Resonance Spectroscopy*, Holden-Day, San Francisco, 1976.
4. R. M. Silverstein, G. C. Bassler, and T. C. Morrill, *Spectrometric Identification of Organic Compounds,* 4th edition, Wiley, New York, 1981, pp. 181–303.

Collections of spectra

5. A. Ault and M. R. Ault, *A Handy and Systematic Catalog of NMR Spectra*, University Science Books, Mill Valley, California, 1980.
6. *High Resolution NMR Spectra Catalog*, Varian Associates, Palo Alto, California; Vol. 1, 1962; Vol. 2, 1963.
7. *The Aldrich Library of NMR Spectra*, 2nd ed, C. J. Pouchert, editor, Aldrich Chemical Co., Inc., Milwaukee, Wisconsin, 1983; two-volume set of more than 8,500 60-MHz proton NMR spectra.
8. *The Aldrich Library of ^{13}C and ^{1}H FT-NMR Spectra*, 2nd ed, C. J. Pouchert and J. Behnke, editors, Aldrich Chemical Co., Inc., Milwaukee, Wisconsin, 1992; a three-volume collection of 12,000 300-MHz proton and 75-MHz ^{13}C FT-NMR spectra arranged according to functionality.

27. Mass Spectrometry

Mass spectrometers are not generally available in the undergraduate organic laboratory, but mass spectrometry is of sufficient importance that we present a brief introduction to the theory of mass spectrometry and the interpretation of mass spectra.

27.1 Theory of Mass Spectrometry

Mass spectrum

In mass spectrometry, a vaporized sample of a substance is bombarded with a beam of electrons, and the relative abundance of the resulting positively charged molecular fragments is determined. The mass-to-charge ratio of each fragment is measured, but because almost every

Mass-to-charge ratio

fragment will bear a single positive charge, the mass-to-charge ratio will be numerically equal to the mass of the fragment. The record of relative abundance versus mass-to-charge ratio that is produced from the substance by the mass spectrometer is called the *mass spectrum* of the substance.

Ionization

The production of positively charged ionic fragments of various masses occurs in this way. An electron from the electron beam knocks an electron from a molecule to give the parent ion, or molecular ion, M^+,which must now have an unpaired electron.

$$M + e^- \longrightarrow M^{\bullet+} + 2e^- \tag{27.1-1}$$

The molecular ion may fall apart very quickly (within 10^{-8} to 10^{-10} second) to give one charged and one uncharged fragment

$$M^{\bullet+} \longrightarrow A^+ + B^{\bullet} \quad \text{or} \quad A^{\bullet} + B^+ \tag{27.1-2}$$

$$M^{\bullet+} \longrightarrow C^+ + D^{\bullet} \quad \text{or} \quad C^{\bullet} + D^+ \tag{27.1-3}$$

and so forth.

Fragmentation

The charged fragmentation products may themselves undergo further *fragmentation* to give smaller charged and neutral pieces. The relative abundance of positively charged fragments of various mass-to-charge ratios is characteristic of the molecules of a substance and can serve to identify the substance.

Acceleration

The relative abundance of the positively charged ions of various mass-to-charge ratios (m/e) is determined by *accelerating* the ions with an electrostatic field. The kinetic energy acquired by each ion in this process is equal to eV, where e is the charge on the particle and V is the potential difference through which the ion has fallen

$$\text{Kinetic energy} = eV = \frac{1}{2}mv^2 \tag{27.1-4}$$

where m is the mass of the ion, v is its velocity, and "kinetic energy" is the kinetic energy gained upon acceleration in the electrostatic field. Solving for v^2 gives

$$v^2 = \frac{2eV}{m} \qquad (27.1\text{-}5)$$

Sorting by mass

The accelerated ions then move into a magnetic field whose direction is perpendicular to their path. The action of the magnetic field is to force the ions to follow a circular path; the force of the magnetic field is a centripetal force. Because the magnitude of the force on an ion equals Hev (where H is the strength of the magnetic field, e is the charge of the ion, and v is its velocity) and the centripetal acceleration is given by v^2/r (where v is the tangential velocity and r is the radius of the path), we can write, according to Newton's second law, the equation

$$Hev = m\frac{v^2}{r} \qquad (27.1\text{-}6)$$

Solving for v and squaring gives

$$v^2 = \frac{H^2 e^2 r^2}{m^2} \qquad (27.1\text{-}7)$$

Substituting into this equation the value of v^2 that results from the acceleration by the electrostatic field (Equation 27.1-5) gives

$$\frac{2eV}{m} = \frac{H^2 e^2 r^2}{m^2}$$

and solving for m/e gives

$$\frac{m}{e} = \frac{H^2 r^2}{2V}$$

Detection

Thus, for constant electrostatic and magnetic fields V and H, fragments of different m/e will be separated because they will follow paths of different radius in the magnetic field. A detector that can determine the relative numbers of ions that have followed paths of different radius can thus provide the desired record of relative abundance of fragments of different m/e. An alternative often used is to position a *detector* so that it can continuously record the intensity of the beam of particles that follows a path of a given, fixed, radius and vary either the magnetic field H or the electrostatic field V in such a way that ions of increasing m/e will fall in succession upon the detector. A record of the output of the detector versus H^2 or $1/V$ is equivalent to a record of relative abundance versus m/e. Reference 1 describes how a mass spectrometer is constructed so that it will perform these functions.

27.2 Interpretation of Mass Spectra

Three types of information can be obtained from a mass spectrum:

- the molecular weight of a substance,
- the molecular formula of a substance, and
- the molecular structure of a substance.

Determination of molecular weight

Parent peak. You might expect that the mass that corresponds to the largest m/e of the mass spectrum would be the molecular weight of the substance. This is essentially true but with two important qualifications. The first is that if the initially formed molecular ions undergo rapid fragmentation, there may not be enough of these ions left to accelerate and detect; the peak for the molecular ion, the *parent peak* P, may not be strong enough to be seen. This is not often the case, but the possibility must be kept in mind.

The second qualification is best introduced by an example. In the mass spectrum of carbon monoxide, a peak at $m/e = 29$ appears with about 1.12% of the intensity of the parent peak P, which corresponds to the molecular ion at $m/e = 28$. The reason for this is that carbon from natural sources contains an amount of ^{13}C equal to 1.08% of the amount of naturally occurring ^{12}C, and that oxygen from natural sources contains an amount of ^{17}O equal to 0.04% of the amount of ^{16}O. The peak at $m/e = 29$ is due to the presence of $^{13}C^{16}O$ and $^{12}C^{17}O$ isotopes of carbon monoxide, which are 1.08% + 0.04% = 1.12% as abundant as $^{12}C^{16}O$. Thus, the molecular weight of a substance can be determined from its mass spectrum when the molecular ion is not too unstable, by identifying the peak of highest m/e (the parent ion P), *not counting* peaks due to molecules that contain small amounts of heavier isotopic atoms (P+1, P+2, etc.). The molecular weight obtained from a mass spectrum is the exact molecular weight, not an approximate weight, as with molecular weights determined from freezing point depression or titration, for example.

Determination of molecular formula

The fact that substances from natural sources are composed of a certain calculable fraction of molecules that contain atoms of heavier isotopes often makes it possible to eliminate all but a few of the molecular formulas that correspond to a given molecular weight. This is done by comparing the predicted intensities of the P+1 and P+2 peaks that can be calculated for different molecular formulas with the experimental P+1 and P+2 intensities.

For example, as we have seen, from carbon monoxide we would expect a P+1 peak at $m/e = 29$ that is 1.12% as intense as the parent peak at $m/e = 28$. Also, because ^{18}O is 0.20% as abundant as ^{16}O, a P+2 peak due to $^{12}C^{18}O$ that is 0.2% as intense as P would be expected. (The amounts of $^{12}C^{18}O$ and $^{14}C^{16}O$ are negligible.) On the other hand, molecular nitrogen, which also has a nominal molecular weight of 28, would be expected to have a P+1 peak at $m/e = 29$ that would be 0.76% as intense as P, because ^{15}N occurs 0.38% as abundantly as ^{14}N (0.38% $\times$ 2 atoms of N = 0.76%). Also, the P+2 peak at $m/e = 30$ would be expected to be 0.0038 $\times$ 0.0038 = 0.001444% as intense as P. If the experimental uncertainty is sufficiently small, the intensities of P+1 and P+2 relative to P will allow the assignment of CO or N_2 as the molecular formula of a sample of a gas whose parent peak appears at $m/e = 28$.

Tables have been prepared (Reference 1) of calculated intensities of P+1 and P+2 relative to P for many combinations of C, H, O, and N for integral values of P. These tables help you greatly reduce the number of possible molecular formulas for a particular molecular weight, providing, however, that accurate intensities of P+1 and P+2 can be obtained from the mass spectrum. If the molecular ion, M·, decomposes too quickly, the intensities of P+1 and P+2 may be too low to be measured with the necessary accuracy. The variations of this kind of analysis that must be made if the substance contains elements other than C, H, O, and N are described in Reference 1.

Determination of molecular structure

Because the relative intensities of the various positive ions produced upon electron bombardment in the mass spectrometer (fragmentation patterns) are different for substances of different molecular structure, it must be possible, at least in principle, to deduce the molecular structure of a substance from its mass spectral fragmentation

Fragmentation patterns

pattern. Comparison of the mass spectra of many substances containing different functional groups has led to correlations of many features of the fragmentation patterns with the presence of various functional groups in the molecule. Conversely, then, it is possible to infer the presence of certain functional groups or structural features from the fragmentation patterns. The subject of the correlations between molecular structure and fragmentation patterns is very large and can only be hinted at in this brief discussion. Reference 1 is an excellent introduction.

27.3 High-Resolution Mass Spectrometry

So far, I have assumed that mass spectrometers can distinguish between ions that differ in mass by one atomic mass unit (unit-resolution mass spectrometry). Instruments are available, at a high price, that can resolve masses to approximately 0.001 atomic mass unit (high-resolution mass spectrometry). With this kind of resolution, it would be possible to distinguish between $^{12}C^{16}O$ (molecular weight = 27.9949) and $^{14}N^{14}N$ (molecular weight = 28.0062) without reference to the relative intensities of P+1 and P+2. In fact, the molecular formula of every fragment in the mass spectrum can be determined, if its mass can be measured to about 0.001 atomic mass unit.

Problems

1. For each case, explain why the parent peak will always be either odd or even. If the parent peak can be either odd or even, explain how this can be so.

   a. The compound contains carbon and hydrogen.

   b. The compound contains carbon, hydrogen, and oxygen.

   c. The compound contains carbon, hydrogen, and fluorine.

   d. The compound contains carbon, hydrogen, and nitrogen.

2. Bromine is made up of almost equal parts of ^{79}Br and ^{81}Br.

   a. Taking the parent peak of CH_3Br to be 94 (or 12 + 3 + 79), estimate the intensity relative to the parent peak of P+1, P+2, etc.

   b. Taking the parent peak of CH_2Br_2 to be 172 (or 12 + 2 + 79 + 79), estimate the intensity relative to the parent peak of P+1, P+2, etc.

   c. Do the same for $CHBr_3$ and CBr_4.

   d. In what simple way might the number of bromine atoms per molecule in a compound containing bromine be determined from its mass spectrum?

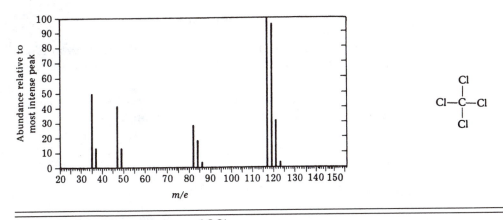

Figure 27.3-1. Mass spectrum of CCl$_4$.

3. Chlorine is made up of ^{35}Cl and ^{37}Cl in a ratio of about 3 to 1. Interpret the mass spectrum of carbon tetrachloride, shown in Figure 27.3-1. The parent peak of carbon tetrachloride fragments very rapidly.

4. The mass spectrum shown in Figure 27.3-2 is that of either *p*-methoxybenzoic acid or methyl *p*-hydroxybenzoate. Which substance can account for the spectrum? Explain.

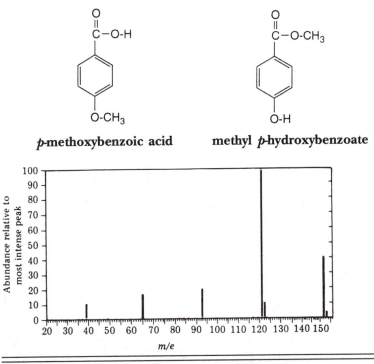

p-methoxybenzoic acid   methyl *p*-hydroxybenzoate

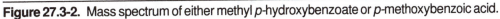

Figure 27.3-2. Mass spectrum of either methyl *p*-hydroxybenzoate or *p*-methoxybenzoic acid.

References

1. R. M. Silverstein, G. C. Bassler, and T. C. Morrill, *Spectrometric Identification of Organic Compounds,* 4th edition, Wiley, New York, 1981, pp. 3–93.

2. D. J. Pasto and C. R. Johnson, *Organic Structure Determination,* Prentice-Hall, Englewood Cliffs, N. J., 1969, pp. 243–294.

Determination of Chemical Properties: Qualitative Organic Analysis

The chemical properties of a substance, like its physical properties, can be used both for characterization and for the determination of molecular structure.

Tests for the Functional Groups

Usually the presence or absence of a functional group is deduced either by infrared spectrometry or by chemical methods. When infrared spectrometry, Section 24, is available, chemical methods will be less often used, and the information that can be obtained from the procedures of Section 29 will complement the information that can be obtained from a routine infrared spectrum.

Preparation of Derivatives

You can often confirm the identity of a sample of a substance of unknown structure and a sample of a substance of known structure by comparing the melting points of their solid derivatives, which can be prepared by the procedures of Section 30. If the melting points are the same, the compounds are probably the same; if they are different, the compounds must be different. If a mixture of the two derivatives has the same melting point as the individuals, the derivatives and therefore the original compounds must be identical.

Qualitative Organic Analysis

Part of an organic laboratory program is often devoted in *qualitative organic analysis.* During this portion, you will receive a sample of a substance whose molecular structure is not known to you, an unknown, and you will be asked to identify the compound. While the best strategy will be different for each substance, there are several experiments that can be done routinely with any *unknown.*

Liquids

For a liquid, you usually determine the boiling point, either by distillation or by the micro method (Section 17), the density (Section 19), the index of refraction (Section 20), and the solubility behavior (Section 23).

Solids

For solids, you normally determine the melting point (Section 18) and the solubility behavior (Section 23).

Spectrometry

Other physical properties that can be determined routinely include the infrared spectrum (Section 24) and the proton NMR spectrum (Section 26).

Qualitative Tests for the Elements

Fusion with sodium followed by qualitative tests for nitrogen (as cyanide), sulfur (as sulfide), and the halogens (Section 28.3), often provide important information, especially when the tests are positive.

Because time is always in short supply, you should choose your qualitative tests for functional groups with care so that the information you obtain will not merely duplicate that already in hand. It is often a good idea, however, to confirm tentative conclusions that have been based on other evidence.

If you cannot compare IR or NMR spectra with those of an authentic sample, you will probably have to prepare one or more solid derivatives (Section 30) to confirm your identification. Preparing derivatives is also a very good way to develop chemical judgment and laboratory technique.

28.   Qualitative Tests for the Elements

A common first step in the identification of an unknown substance is to determine what elements (other than carbon, hydrogen, and possibly oxygen) are present in the sample. This section describes qualitative tests for the detection of the presence of metals, nitrogen, sulfur, and the halogens.

28.1 Ignition Test; Test for Metals

Procedure

Place 50-100 mg of the substance on a porcelain crucible cover and heat the cover with a flame. Heat gently at first, and then finally heat the cover to dull redness. If a metal was present in the sample, a nonvolatile residue will remain. Add a drop of water to the residue: the resulting solution should be basic; it should turn red litmus blue.

While the purpose of this test is to determine whether a solid is a metal salt or other metal-containing substance, other information can be obtained as well. If the substance burns with a fairly sooty flame, a high ration of carbon to hydrogen is indicated (an aromatic rather than an aliphatic compound is indicated). If the substance burns with a blue rather than a yellow flame, a high percent of oxy-

Aliphatic versus aromatic

gen is indicated. Certain types of compounds can explode under these conditions (see Section 1.2).

28.2 Beilstein Test; Test for Halogens (Except Fluorine)

Procedure

Form a small coil at one end of a 20-cm length of stiff copper wire by making a few turns around a glass rod 2 or 3 mm in diameter. Heat the coil in the oxidizing part of the flame of a Bunsen burner until it does not color the flame any longer. Allow the wire to cool (by waving it in the air), and then dip the coil into the material to be tested and heat it again in the same part of the flame. If the substance contains chlorine, bromine, or iodine, a transient green or blue-green color will be imparted to the flame; the color is caused by the presence of copper that has been made volatile by combination with the halogen of the sample.

The Beilstein test is very sensitive. It should be confirmed by the sodium fusion test (Section 28.3) and the tests of Sections 29.11 and 29.13 to make sure that the effect was not due to the presence of a halogen-containing impurity—or to verify that the substance is not one that gives a positive test even though it contains no halogen. (Unfortunately, the test is not specific because some organic nitrogen compounds that do not contain halogen also give the effect.)

28.3 Sodium Fusion Test; Test for Nitrogen, Sulfur, and the Halogens

In the sodium fusion test, a substance is heated with sodium under conditions that ensure the conversion of any nitrogen to cyanide ion, any sulfur to sulfide ion, and any halogens to the corresponding halide ions. The presence or absence of these ions in the solution that results from the fusion (the *fusion solution*) indicates the presence or absence of these elements in the sample.

Sodium fusion

Procedure

Support a test tube 8 mm in diameter by 50 mm long by its lip in a hole in an asbestos board or sheet that is in turn supported on an iron ring. Add to the test tube about 40 mg of sodium (Note 1), and heat the test tube with a burner until the sodium has melted and the grey vapors rise about 2 cm in the tube. Adjust the rate of heating by adjusting the height of the burner flame or the height of the iron ring, so that this condition can be maintained. Then add, in two or three portions, 50 mg (or 2 or 3 small drops) of the compound over about a 30-second interval (Note 2). After the addition is complete, turn up the burner and heat the tube as strongly as possible until the lower 1 or 2 cm is red to orange-red hot; heat the tube at this rate for about 2 minutes (Note 3). Now, holding the asbestos sheet or board with a gloved hand or tongs, lower the red-hot tube into a small beaker (obviously no more than 50 mm high) that contains about 10 mL of distilled water (Note 4). The tube will shatter, and any residual sodium will react with the water. Boil the remains of the test tube with the water and filter by gravity. The clear, almost colorless filtrate (fusion solution) will be used in the tests that follow (Note 5).

Caution

Notes

1. For the precautions to be taken in handling sodium, see Section 1.7. A piece the size of half a medium pea, or a 4-mm cube, will be satisfactory.

2. Safety glasses must be worn; nitro compounds, azo compounds, and polyhalogen compounds such as carbon tetrachloride or chloroform may cause a slight explosion. Try to drop the sample straight down, without letting it hit the side of the test tube, where it may simply evaporate.

3. It is not possible to heat too strongly. Some burners are too feeble; make sure yours can give a strong, blue, nonluminous flame.

4. Safety glasses must be worn; hold the asbestos board at arm's length; don't aim the open top of the tube at anyone.

5. If the filtrate is more than just slightly yellow, it will usually be a waste of time to continue. Better do the procedure again and heat more strongly.

Tests for the elements: test for nitrogen; test for cyanide ion (test A)

Procedure

Add 2 or 3 mL of the fusion solution to a test tube that contains 100-200 mg of powdered ferrous sulfate crystals. Heat the mixture cautiously and with shaking until it boils (Note 1). Then, without cooling, add just enough dilute sulfuric acid to dissolve the gelatinous hydroxides of iron. The appearance of a blue precipitate of ferric ferrocyanide, $Fe_4[Fe(CN)_6]_3$ (Prussian Blue), indicates that cyanide ion is present and, therefore, that the original compound contained nitrogen (Note 2). If no cyanide ion is present, the solution should be a pale yellow color (Note 3).

Notes

1. It is often recommended that 2 drops of a 30% aqueous solution of potassium fluoride be added before boiling. If sulfide ion is present, a precipitate of black ferrous sulfide may appear; it will dissolve upon acidification. Ferric ions are produced by air oxidation during boiling.

2. If a precipitate is not immediately apparent, allow the mixture to stand for 15 minutes and then filter it. After washing the filter paper with a little water to remove all the colored solution, any Prussian Blue present should be visible on the paper. If you still have doubt as to whether a precipitate of Prussian Blue was formed, you should carry out another sodium fusion and repeat the test.

3. A positive test for cyanide is good evidence for the presence of nitrogen in the original sample, but the presence of nitrogen should never be ruled out solely on the basis of a negative test here for cyanide ion. You should make sure that your technique is adequate by carrying out the sodium fusion and the test for nitrogen on a sample known to contain nitrogen. Nitrogen in a high oxidation state, for example as a nitro group, is especially difficult to detect.

Test for sulfur; test for sulfide ion (test B)

Procedure

Acidify about 2 mL of the fusion solution with dilute acetic acid and then add a few drops of approximately 1% lead acetate solution. A black precipitate of lead sulfide indicates that sulfide ion is present and, therefore, that sulfur was present in the original sample.

Alternatively, 2 or 3 drops of a freshly prepared 0.1% solution of sodium nitroprusside (Note 1) may be added to 2 mL of the fusion solution. A purple coloration indicates the presence of sulfide ion.

Note

1. This may be prepared by dissolving a tiny crystal of sodium nitroprusside ($Na_2[Fe(CN_5)NO]$) in 1 or 2 mL of water.

Test for halogen; test for halide ion (test C)

Procedure

Acidify 2 mL of the fusion solution with dilute nitric acid (Note 1) and add a few drops of 5% aqueous silver nitrate solution. A heavy precipitate indicates the presence of chloride, bromide, or iodide ion, or any combination of the three.

If only a single halide should be present, you may be able to distinguish among the possibilities because silver chloride is white (slightly purplish because of liberation of free silver by light) and easily soluble in dilute ammonium hydroxide, silver bromide is slightly yellow and only slightly soluble in ammonium hydroxide, and silver iodide is definitely yellow and insoluble in ammonium hydroxide. To determine the solubility of the precipitate in dilute ammonium hydroxide, remove the supernatant liquid by centrifugation and decantation, add 2 mL of dilute ammonium hydroxide solution, and stir the mixture to see if the solid will dissolve.

AgCl white
AgBr slightly yellow
AgI: yellow

Note

1. If either cyanide or sulfide is present (Test A and B above), the acidified solution must be boiled gently for a few minutes to expel hydrogen cyanide and hydrogen sulfide.

Test for iodine; test for iodide ion (test D)

Procedure

Acidify 1 or 2 mL of the fusion solution with dilute sulfuric acid (Note 1). After cooling, add 1 mL of carbon tetrachloride; then add dropwise and with good mixing a solution of chlorine water (Note 2). Iodide ion, if present, will be oxidized to free iodine under these conditions and will be extracted into the carbon tetrachloride to give a purple color (Note 3).

I_2 purple

Notes

1. If cyanide or sulfide is present (Tests A and B above), acidify the sample of fusion solution with dilute nitric acid, boil the mixture until its volume has been reduced by one-half to expel hydrogen cyanide and hydrogen sulfide, and dilute the resulting solution with an equal volume of distilled water.

2. A stabilized solution of sodium hypochlorite such as Clorox may be used; make sure that the test solution remains acidic to litmus by adding acid as necessary. If this type of commercial bleach solution is used, a blank should be run since often a yellow-brown color is produced just from the reagent. Chlorine water may be prepared by acidifying 10% aqueous sodium hypochlorite solution with one-fifth of its volume of dilute hydrochloric acid.

Br_2 reddish-brown

3. The absence of iodide ion but the presence of bromide ion will result in the appearance of a brown color in the carbon tetrachloride layer, due to oxidation of the bromide ion to form bromine followed by extraction of the bromine into the carbon tetrachloride. If both iodide and bromide are present, the initial purple color will give way to a reddish-brown color.

Test for bromine; test for bromide ion (test E)

In the absence of iodide ion, the condition of Test D will result in a brown color being produced in the carbon tetrachloride layer as a result of the oxidation of bromide to bromine. If iodide is present, a purple color will appear first. This will change to reddish-brown with the continued addition of chlorine water.

Test for chlorine; test for chloride ion (test F)

If the presence of halide ion was indicated by Test C and no color was produced in the carbon tetrachloride layer in Test D, the halide ion present must be chloride.

If either iodide or bromide ion has been shown to be present by means of Test D, apply the following procedure.

Procedure

Acidify 1 or 2 mL of the fusion solution with glacial acetic acid, add 0.5 gram of lead dioxide, and boil the mixture gently until all iodine and bromine has been liberated and boiled off. When the mixture is allowed to stand so the lead dioxide settles to the bottom, the solution will be colorless when this point has been reached. Dilute the mixture with an equal volume of water, remove the excess lead dioxide either by filtration or centrifugation and decantation, and test for the presence of chloride ion in the filtrate by adding a little dilute nitric acid followed by a few drops of 5% aqueous silver nitrate solution. The formation of a white precipitate confirms the presence of chloride ion.

Disposal

Sink: aqueous wash liquids.
Heavy metal solid waste: all solid residues from all tests.
Heavy metal liquid waste: all liquid residues from all tests.
Halogenated liquid organic waste: carbon tetrachloride.

29. Qualitative Characterization Tests: Tests for the Functional Groups

Before the advent of spectrometric methods such as infrared, ultraviolet, nuclear magnetic resonance, and mass spectrometry (Sections 24 through 27), the determination of chemical properties was of paramount importance for the identification, characterization, and determination of structure of pure substances. Many reagents or reaction conditions were found to give moderately characteristic and specific results for substances that contain certain functional groups. These reagents or reaction conditions therefore serve to distinguish between substances with different functional groups and to indicate the presence or absence of certain functional groups in substances of unknown structure.

Sections 29.1 through 29.14 describe a number of tests that can distinguish between certain functional groups or can indicate the presence or absence of certain functional groups. The comparison of the results of these tests on samples of substances of known and unknown structure will, in many cases, allow a tentative conclusion to be drawn about the nature of the functional group of the substance of unknown structure.

Different substances with the same functional group will, however, give slightly different results. The question is, then, how different can the results be before you must conclude that the

Functional group tests

substances have different functional groups? There will, obviously, be borderline cases. Also, certain individual substances can react atypically, to give either a "false positive" test (the unknown behaves as if a certain functional group were present, but it really isn't) or a "false negative" text (the unknown behaves as if the functional group were not there, but it really is). That is, most tests are not completely specific or characteristic. Finally, some tests are quite sensitive, and a small amount of a substance as an impurity might be sufficient to give a result that you could interpret as a "positive" test.

The uncertainty in the interpretation of these tests can be decreased by testing the behavior of compounds of known structure under the same conditions. The substances of known structure should be either substances that are said to give typical or normal results, or, better, substances that are very similar in structure (especially if more than one functional group is present) to the suspected structure of the unknown substance.

Observations not conclusions

When recording the results of one of these characterization tests, you must be careful to record *your observations* (a yellow–orange crystalline precipitate was formed within 30 seconds after adding two drops of the liquid of unknown structure; the mixture was not heated; no further changes occurred upon standing; similar behavior but slower than with two drops of acetone) and *not your tentative conclusion* (positive test for a ketone). It is impossible to reinterpret a conclusion; only data—observations—can be reinterpreted. If your conclusions are inconsistent, there is no way to reinterpret them, and you have wasted your time; data can be continually reinterpreted until the conclusions are consistent.

There are many hundreds of qualitative tests that can be used to characterize unknown substances and to distinguish between different functional groups. The procedures presented in these sections are only a very small sample of such tests. They have been chosen because they involve some of the most familiar functional groups, and because the information they give complements the information that can be obtained from infrared and nuclear magnetic resonance spectra. Table 29-1 is a summary and index for the 14 tests presented.

The references for each test are given at the end of this section. These references provide more information concerning the scope and limitations of certain procedures, variations in procedures, and exceptional behavior of certain compounds or types of compounds. You should consult these references to minimize the chances of being misled by the results of the tests.

Table 29-1. Summary of classification tests

Test	Section	Application
Ammonia	29.1	Ammonium salts; primary amides; nitriles
Baeyer's test: see Potassium permanganate test		
Benzenesulfonyl chloride (Hinsberg's test)	29.2	Amines (distinguishes among primary, secondary, or tertiary)
Bromine in carbon tetrachloride	29.3	Alkenes; alkynes
Chromic anhydride	29.4	Alcohols (distinguishes primary and secondary from tertiary)
2,4-Dinitrophenylhydrazine	29.5	Aldehydes and ketones
Ferric chloride	29.6	Phenols and enols
Ferric hydroxamate test	29.7	Acid halides and anhydrides; esters; amides; nitriles
Hinsberg's test: see Benzenesulfonyl chloride		
Hydrochloric acid/zinc chloride test (Lucas's test)	29.8	Alcohols (distinguishes among primary, secondary, or tertiary)
Iodoform test	29.9	Methyl ketones; secondary methyl carbinols
Lucas's test: see Hydrochloric acid/zinc chloride test		
Potassium permanganate test (Baeyer's test)	29.10	Alkenes; alkynes
Silver nitrate	29.11	Halides
Sodium hydroxide test	29.12	Esters; lactones; anhydrides
Sodium iodide	29.13	Halides
Tollens' test	29.14	Aldehydes

29.1 Detection of Ammonia from Ammonium Salts, Primary Amides, and Nitriles

Ammonium salts, primary amides, and nitriles that are relatively easily hydrolyzed will liberate ammonia under the conditions of the following procedure.

Materials Required

- 20% aqueous sodium hydroxide solution
- 10% aqueous copper sulfate solution

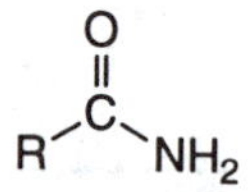

$$\underset{\text{primary amide}}{R-\overset{\overset{\displaystyle O}{\|}}{C}-NH_2}$$

$$R-C\equiv N$$
nitrile

Procedure

Place 50 mg of the substance in a small test tube and add 2 mL of 20% sodium hydroxide solution. Fasten a small circle of filter paper tightly over the mouth of the test tube and wet it with a couple of drops of the copper sulfate solution. Heat the contents of the test tube to boiling for about 1 minute, and look to see if the filter paper has turned blue. The blue color is due to the formation of the copper ammonia complex ion from ammonia either liberated from the ammonium salt or formed by hydrolysis of the primary amide or nitrile. (Ref. 1, p. 164, Ref. 2, pp. 109, 123; Ref. 3, pp. 404, 410, 798, 805.)

29.2 Benzenesulfonyl Chloride (Hinsberg's Test)

This test is useful for distinguishing among primary, secondary, and tertiary amines.

Materials Required

- 10% aqueous sodium hydroxide solution
- Benzenesulfonyl chloride

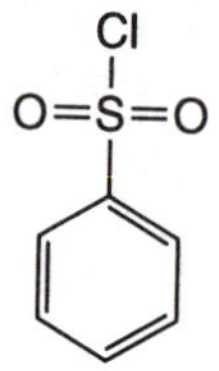

benzenesulfonyl chloride

Procedure

Add 5 mL of 10% sodium hydroxide to 0.2 mL of the liquid amine or 0.2 gram of the solid amine in a test tube, and then add 0.4 mL benzenesulfonyl chloride. Stopper the test tube and occasionally shake it vigorously over a period of 5 to 10 minutes, cooling the tube in water if it heats up a lot. By this time, all the benzenesulfonyl chloride should have reacted, either with the amine to form the sulfonamide or with the basic solution to form the water-soluble sodium benzenesulfonate; complete reaction is indicated by the disappearance of the distinctive odor of benzenesulfonyl chloride. Now make sure that the mixture is basic (if it is not, add 10% sodium hydroxide to make it strongly basic), and observe whether or not any insoluble material is present. Remove any insoluble material, by filtration if it is a solid and by decanting if it is a liquid.

Test the solubility of the insoluble material that was removed, both in water and in dilute hydrochloric acid.

Acidify the filtrate (or the basic solution that was separated by decantation, or the clear reaction mixture) and attempt to promote crystallization by scratching and by cooling. Benzenesulfonic acid is soluble in water.

Primary amines normally give a sulfonamide that is soluble in the basic reaction mixture; no insoluble material should be present at the end of the first part of the test. Acidification of the clear reaction mixture should then result in the precipitation or crystallization of the water-insoluble sulfonamide of the amine. However, the sodium salts of the sulfonamides of some primary amines—for example, cyclohexylamine through cyclodecylamine and certain high-molecular-weight amines—are not very soluble in 10% sodium hydroxide solution. If the unknown is one of these compounds, a precipitate will be present at the end of the first part of the test, but it will be found to be soluble in water.

Primary amines

Secondary amines normally give a sulfonamide that is insoluble in the basic reaction mixture; a solid residue will be present at the end of the first part of the test. This residue will be insoluble in both water and dilute hydrochloric acid. The filtrate should give no precipitate upon acidification.

Secondary amines

Tertiary amines normally give no reaction with benzenesulfonyl chloride; the amine itself should be present as the unchanged liquid or solid at the end of the first part of the test. The liquid or solid residue will be soluble in dilute hydrochloric acid.

Tertiary amines

It should be apparent that water-soluble tertiary amines and secondary amines with a carboxyl group on another part of the molecule will not show the typical behavior previously outlined. [Ref. 1, p. 119,

Ref. 3, p. 650, Ref. 4, p. 230. See also Fanta and Wang, *J. Chem. Educ.* **41,** 280 (1964). For an alternative method, see Ritter, *J. Chem. Educ.* **29,** 506 (1952).]

29.3 Bromine in Carbon Tetrachloride

This test is useful for indicating the presence of many olefinic or acetylenic functional groups. It should be used in conjunction with a similar test with aqueous potassium permanganate solution (Section 29.10).

Materials Required

- Bromine in carbon tetrachloride: a solution of 2 mL in 100 mL carbon tetrachloride
- Carbon tetrachloride

olefin

$$—C≡C—$$

acetylene

Procedure

Dissolve 0.2 mL of a liquid or 0.1 gram of a solid in 2 mL of carbon tetrachloride, and add the bromine in carbon tetrachloride solution dropwise and with shaking. The presence of an olefinic or acetylenic linkage in the sample will cause more than 2 or 3 drops of the bromine solution to be required before the characteristic orange-brown color of bromine will persist for 1 minute. The presence of electronwithdrawing groups such as phenyl or substituted phenyl, carboxyl, or halogen will reduce the rate of addition and in some cases completely inhibit the reaction. The presence of phenols, enols, amines, aldehydes, or ketones can result in the consumption (as indicated by loss of color) of large amounts of bromine, accompanied by *the evolution of hydrogen bromide*, which can be detected by exhaling cautiously over the top of the test tube and noting the fog that is produced in the moist air. The test should be carried out in diffuse light, since light can catalyze a free radical substitution reaction that will consume bromine and produce hydrogen bromide. (Ref. 1, p. 121; Ref. 2, p. 111, Ref. 3, p. 1058; Ref. 4, p. 190.)

29.4 Chromic Anhydride

This test is used to distinguish primary and secondary alcohols from tertiary alcohols.

Materials Required

- Reagent-grade acetone
- Chromic anhydride reagent: prepared by slowly pouring while stirring a suspension of 25 grams chromic anhydride (CrO_3) in 25 mL concentrated sulfuric acid into 75 mL of water and allowing the deep orange-red solution to cool to room temperature.

Procedure

Dissolve 1 drop of a liquid or 15–30 mg of a solid in 1 mL of reagent-grade acetone and then add 1 drop of the chromic anhydride reagent. Primary and secondary alcohols produce an opaque blue-green suspension within 2 seconds; tertiary alcohols give no visible reaction and the solution remains orange. Aldehydes give a positive test, but ketones, alkenes, acetylenes, amines, and ethers give negative tests for 2 seconds. Enols may give a positive test, and phenols cause the solution to turn a much darker color than the blue-green produced by primary and secondary alcohols. [See Bordwell and Wellman, *J. Chem. Educ.* **39**, 308 (1962). Ref. 1, p. 125; Ref. 2, p. 121; Ref. 4, p. 149. For a way to distinguish among primary, secondary, and tertiary alcohols according to ease of oxidation by potassium permanganate in acetic acid, see Ritter, *J. Chem. Educ.* **30**, 395 (1953).]

29.5 2,4-Dinitrophenylhydrazine

This test is useful for identifying aldehydes and ketones.

Materials Required

- 2,4-Dinitrophenylhydrazine reagent: prepared by dissolving 3 grams of 2,4-dinitrophenylhydrazine in 15 mL concentrated sulfuric acid and adding this solution, while stirring well to a mixture of 20 mL of water and 70 mL of 95% ethanol. After thorough mixing, the reagent should be filtered.

- 95% ethanol

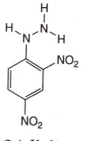

2,4-dinitro-phenylhydrazine

Procedure

Add a solution of 1 or 2 drops of the liquid, or 25-50 mg of the solid, in 2 mL of 95% ethanol to 3 mL of the 2,4-dinitrophenylhydrazine reagent. Mix the solution well and allow the mixture to stand for 15 minutes.

Most aldehydes and ketones give a solid 2,4-dinitro-phenyl-hydrazone under these conditions. The derivative is yellow if the carbonyl group is not conjugated with a double bond or an aromatic ring. The 2,4-dinitrophenylhydrazones of most conjugated aldehydes and ketones are orange-red or red. The fact that the product is nonyellow should be interpreted with caution, as the color might be due to an impurity. The most obvious impurity is 2,4-dinitrophenylhydrazine itself, which is orange-red. (Ref. 1, p. 126; Ref. 2, p. 136; Ref. 3, p. 1060; Ref. 4, p. 1620.)

29.6 Ferric Chloride Solution

This test is useful for the recognition of phenols and enols.

Materials Required

- Procedure a: 2.5% aqueous ferric chloride solution; Ethanol
- Procedure b: Ferric chloride in chloroform; prepared by dissolving 1 gram ferric chloride in 100 mL chloroform; Pyridine

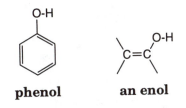

phenol   **an enol**

Procedure A

Dissolve 1 drop or 30-50 mg of the compound in 1 or 2 mL of water or a mixture of ethanol and water, and add up to 3 drops of the aqueous ferric chloride solution. Note color changes or formation of a precipitate; the color may not be permanent. Most phenols produce red, blue, purple, or green colorations; enols usually produce tan, red, or red-violet colorations. Some phenols, however, do not give color.

Procedure B

Dissolve (or suspend) 1 drop of a liquid or 30 mg of a solid in 1 mL chloroform, and add 2 drops of the solution of ferric chloride in chloroform; then add 1 drop of pyridine and note color changes. This test appears to be more sensitive than the test described in Procedure A, and some phenols that do not give a color under the conditions of Procedure A will do so under these conditions. [Ref. 1, p. 127; Ref. 2, p. 147; Ref. 4, p. 348. Proc. A: Wesp and Brode, *J. Am. Chem. Soc.* **56,** 1037 (1934). Proc. B: Soloway and Wilen, *Anal. Chem.* **24,** 979 (1952).]

29.7 Ferric Hydroxamate Test

This test is useful for the identification of acid halides and anhydrides, esters, amides, and nitriles.

Test for acid halides and anhydrides (test A)

Materials Required

- 1.0 M hydroxylamine hydrochloride in 95% ethanol
- 6 M hydrochloric acid
- 2 M hydrochloric acid

- 10% aqueous ferric chloride solution

$$H-\overset{\overset{\displaystyle H}{|}}{\underset{\underset{\displaystyle H}{|}}{\overset{+}{N}}}-O\text{-}H \quad Cl^-$$

**hydroxylamine
hydrochloride**

Procedure

Add 30-40 mg of the solid, or 1 drop of the liquid, to 0.5 mL of the hydroxylamine hydrochloride solution. Add 2 drops of the 6 M hydrochloric acid solution, warm the mixture slightly for 2 minutes, and then heat it to boiling for a few seconds. After cooling the solution, add 1 drop of the 10% ferric chloride solution.

A reddish-blue or bluish-red color will result if the original substance was an acid halide or anhydride. If the color is more red than blue, adjust the pH to 2 or 3 by dropwise addition of 2 M hydrochloric acid.

Tests for esters of carboxylic acids (test B)

Materials Required

- 1.0 M hydroxylamine hydrochloride in 95% ethanol
- 2 M potassium hydroxide in methanol
- 2 M hydrochloric acid
- 10% aqueous ferric chloride solution

**a lactone;
a cyclic ester**

Procedure

Add 30-40 mg of the solid, or 1 drop of the liquid, to 0.5 mL of the hydroxylamine hydrochloride solution. Now add the 2 M potassium hydroxide solution dropwise until the mixture is basic to litmus, and then add 4 more drops. Heat the mixture just to boiling, cool it to room temperature, and add dropwise (with good mixing) 2 M hydrochloric acid until the pH is approximately 3. Add 1 drop of the ferric chloride solution and note the color.

Esters of carboxylic acids, including lactones and polyesters, give definite magenta colors of varying degrees of intensity. Acid chlorides and anhydrides, and trihalo compounds such as benzotrichloride and chloroform give a positive test. Formic acid is the only acid that produces a color (red). Carbonic acid esters, sulfonic acid esters, urethanes, chloroformates, and esters of inorganic esters all give a negative result.

Test for amides and nitriles (test C)

Materials Required

- 1 M hydroxylamine hydrochloride in propylene glycol
- 1 M potassium hydroxide in propylene glycol propylene glycol
- 5% aqueous ferric chloride

**propylene
glycol**

Procedure

Add to 2 mL of the solution of hydroxylamine hydrochloride in propylene glycol a solution of 1 drop of the liquid, or 30-40 mg of the solid, dissolved in a minimum amount of propylene glycol. To this, add 1 mL of the potassium hydroxide solution, and then boil the mixture for 2 minutes. Cool the solution to room temperature, and then add 0.5-1.0 mL of the ferric chloride solution. A red-to-violet color is a positive test; yellow colors should be interpreted as negative tests; brown colors or precipitates are indeterminate.

You should realize that esters, anhydrides, and acid halides will be converted to hydroxamic acids under these conditions as well as under the milder conditions of Test A and B. Before a positive result in this test can be interpreted as indicating the presence of an amide or nitrile functionality, the possibility that the substance might be an ester, anhydride, or acid halide must be ruled out. [Ref. 1, p. 135; Ref. 2, pp. 119, 140, 144; Ref. 3, p. 1062. See also Davidson, *J. Chem. Educ.* **17, 81** (1940).]

29.8 Hydrochloric Acid/Zinc Chloride Test (Lucas's Test)

Useful for distinguishing among lower-molecular-weight primary, secondary, and tertiary alcohols. Since the test requires that the alcohol initially be in solution, it is limited in its application to monohydroxylic alcohols more soluble than *n*-hexanol and to certain polyfunctional molecules.

Materials Required

- Lucas's reagent: prepared by dissolving 136 grams (1 mole) of zinc chloride in 89 mL (105 grams; 1 mole) concentrated hydrochloric acid with cooling in ice.

Procedure

Add 2 mL of the reagent to 0.2 mL of the alcohol in a test tube; swirl to dissolve. Note the time required to form the insoluble alkyl chloride, which appears as a layer or emulsion.

The substitution of hydroxyl by chloride apparently takes place by an S_N1 mechanism, since tertiary alcohols, as well as allylic alcohols and benzyl alcohol, react very quickly. Secondary alcohols give indications of the formation of a second phase within 2 or 3 minutes, but primary alcohols react very much more slowly to form the insoluble alkyl chloride.

Tertiary alcohols will give a similar but slower reaction with concentrated hydrochloric acid alone. [Ref. 1, p. 131; Ref. 2, p. 121; Ref. 3, p. 261, Ref. 4, p. 150. See also Lucas, *J. Am. Chem. Soc.* **52,** 802 (1930).]

29.9 Iodoform Test

This test is useful for the identification of methyl ketones and secondary methyl carbinols.

Materials Required

- Iodine/potassium iodide reagent: prepared by adding 200 g potassium iodide and 100 g iodine to 800 mL distilled water, stirring until solution is complete
- Dioxane
- 10% aqueous sodium hydroxide solution

iodoform

Procedure

Dissolve 4 drops of a liquid or 100 mg of a solid in 5 mL of dioxane (use 1 mL of water in place of the dioxane if the compound is soluble in water to this extent). Add 1 mL of the sodium hydroxide solution and then the iodine/potassium iodide solution with shaking until the definite dark color of iodine persists. If less than 2 mL of the iodine/potassium iodide solution was consumed, place the test tube in a beaker of water at 60°C. If the dark color of iodine now disappears, continue to add the iodine/potassium iodide solution until the dark color that represents an excess of iodine is not discharged by heating at 60°C for 2 minutes. Now add 10% sodium hydroxide solution dropwise until the dark iodine color is gone, remove the tube from the heating bath, add 15 mL of water, and allow the mixture to cool to room temperature. Collect by suction filtration any solid formed and determine its melting point. Iodoform melts at 119–121°C. If the iodoform is reddish, dissolve it in 3 or 4 mL of dioxane, add 1 mL of 10% sodium hydroxide solution, and shake the mixture until the reddish color gives way to the lemon-yellow color of iodoform. Slowly dilute the mixture with water and collect the precipitated iodoform by suction filtration.

a methyl ketone

Methyl ketones (and acetaldehyde) and methyl carbinols including ethanol (compounds that can be oxidized to methyl ketones by the reagent) give iodoform under these conditions; acetic acid does not. Compounds that can react with the reagent to generate one of these functional groups will also give iodoform; conversely, it is possible that the functionality that might be expected to result in the formation of iodoform can be removed by hydrolysis before iodoform formation is complete (see Fuson and Bull). [Ref. 1, p. 137; Ref. 2, p. 112; Ref. 3, p. 1068; Ref. 4, p. 167. See also Fuson and Bull, *Chem. Revs.* **15,** 275 (1934).]

29.10 Aqueous Potassium Permanganate Solution (Baeyer's Test)

This test is useful for indicating the presence of most olefinic or acetylenic functional groups.

As a test for unsaturation, it should be used in conjunction with a similar test with a solution of bromine in carbon tetrachloride (Section 29.3).

Materials Required

- 2% aqueous potassium permanganate solution
- Reagent-grade acetone (free of alcohol)

an aldehyde	**a formate ester**	**a mercaptan**

Procedure

Dissolve 0.2 mL of a liquid or 0.1 gram of a solid in 2 mL of water or acetone. Add the potassium permanganate solution dropwise while shaking the mixture well. If more than 1 drop of the permanganate solution is consumed, as shown by the loss of the characteristic purple color, the presence of an olefin or acetylene or other functional group that can be oxidized by permanganate under these conditions is indicated. Such other groups include phenols and aryl amines, most aldehydes (but not benzaldehyde or formaldehyde) and formate esters, primary and secondary alcohols (see, however, Swinehart), mercaptans and sulfides, and thiophenols. Aryl-substituted alkenes are oxidized by permanganate under these conditions.

Certain carefully purified alkenes are not oxidized under the conditions of this test (acetone solvent), but can be oxidized if ethanol is used instead; ethanol does not react with potassium permanganate within 5 minutes at room temperature.

In this test, you must take care not to be misled by the limited reaction of impurities. [Ref. 1, p. 149; Ref. 2, p. 112; Ref. 3, p. 1058; Ref. 4, p. 192. See also Swinehart, *J. Chem. Educ.* **41,** 392 (1964).]

29.11 Alcoholic Silver Nitrate Solution

This test is useful for classifying compounds known to contain chlorine, bromine, or iodine. It should be used in conjunction with a similar test using sodium iodide in acetone (Section 29.13).

The rate of reaction of the halogen in this test gives an indication of its structural environment. Reference 1 (after Section 29.14) includes a discussion of the relationship between structure and reactivity for organic halides. Table 29.11-1 summarizes the results that can be expected for the most common functional groups.

Table 29.11-1. Alcoholic silver nitrate test

Water-soluble compounds that give an immediate precipitate at room temperature with *aqueous* silver nitrate:
  Salts of amines and halogen acids
  Low-molecular-weight acid halides (their water solubility results in their rapid hydrolysis to give halide ion)
Water-insoluble compounds that give an immediate precipitate at room temperature with *alcoholic* silver nitrate:
  Acid halides
  Tertiary alkyl halides
  Allylic halides
  Alkyl iodides
  α-Haloethers
  1,2-Dibromides
Water-insoluble compounds that react slowly or not at all at room temperature, but give a precipitate readily at higher temperatures with *alcoholic* silver nitrate:
  Primary and secondary alkyl chlorides
  Geminal dibromides
  Activated aryl halides
Water-insoluble compounds that do not give a precipitate at higher temperatures with *alcoholic* silver nitrate:
  Unactivated aryl halides
  Vinyl halides
  Chloroform, carbon tetrachloride

Materials Required

- 2% silver nitrate in 95% ethanol
- 5% aqueous nitric acid

Procedure

Add 1 drop of the liquid (or 30–40 mg of the solid) to 2 mL of the silver nitrate solution at room temperature. If no precipitate is formed upon standing at room temperature for 5 minutes, heat the solution to boiling for 30 seconds and note whether or not a precipitate is formed under these conditions. If a precipitate is formed, note the color and determine whether or not it dissolves upon addition of 2 drops of the nitric acid solution. Silver chloride is white, silver bromide is pale yellow, and silver iodide is yellow; the silver halides are insoluble in dilute nitric acid, but silver salts of organic acids will dissolve. (Ref. 1, p. 152; Ref. 2, p. 122; Ref. 3, p. 1059; Ref. 4, p. 202.)

29.12 Sodium Hydroxide Test

This test is useful for identifying esters, lactones, and anhydrides.

Materials Required

- 0.1 *M* sodium hydroxide in 95% ethanol
- Phenolphthalein indicator solution, in ethanol
- 95% ethanol

Procedure

Dissolve 0.1 gram of the substance in 3 mL ethanol. Add 3 drops of the phenolphthalein indicator solution and then add, dropwise, sufficient 0.1 *M* sodium hydroxide solution to turn the solution pink. Warm the mixture in a beaker of water 40°C. Disappearance of the pink color indicates consumption of base.

Under these conditions, esters, lactones, and anhydrides are hydrolyzed, resulting in the consumption of base; additionally, some alkyl halides, amides, and nitriles will undergo solvolysis or hydrolysis and will likewise result in the use of base. To determine that the base was not consumed by an impurity, repeat the cycle of adding base to turn the indicator pink and heating at 40°C to decolorize the solution several times. (Ref. 1, p. 164.)

29.13 Sodium Iodide in Acetone

This test is useful for classifying compounds known to contain bromine or chlorine. It should be used in conjunction with a similar test using a solution of silver nitrate in 95% ethanol (Section 29.11).

Materials Required

- Acetone, reagent grade
- Sodium iodide in acetone reagent: prepared by dissolving 15 g of sodium iodide in 100 mL reagent-grade acetone. The solution is colorless at first, but develops a pale lemon-yellow color on standing. When the color becomes definitely red-brown, the solution should be discarded.

Procedure

Add 2 drops of the liquid, or a solution of 100 mg of the solid dissolved in a minimum amount of acetone, to 1 mL of the sodium iodide in acetone solution. After mixing, allow the solution to stand at room temperature for 3 minutes. Note whether or not a precipitate has formed (sodium chloride or sodium bromide) and whether or not the solution has become reddish-brown (liberation of iodine). If no change has occurred, heat the solution in a beaker of water at 50°C for 6 minutes and note any relevant changes.

This test depends upon two facts: (1) sodium chloride and sodium bromide, which can be formed in an S_N2-type displacement of the halide by iodide, are insoluble in acetone, and (2) iodine can be formed by the reaction of iodide with 1,2-dichloro and 1,2-dibromo compounds.

Table 29.13-1 summarizes the results that can be expected for some of the more common functional groups. (Ref. 1, pp. 169, 152; Ref. 3, p. 1059; Ref. 4, p. 204.)

29.14 Tollens' Reagent: Silver-Ammonia Complex Ion

This test is useful for distinguishing aldehydes from ketones and other carbonyl compounds.

Table 29.13-1. Sodium iodide in acetone test

Formation of a precipitate within 3 minutes at 25°C:
    Primary alkyl bromides
    Benzylic and allylic chlorides and bromides
    α-Halo ketones, esters, amides, and nitriles
    Acid chlorides and bromides
Formation of a precipitate within 6 minutes at 50°C:
    Primary and secondary alkyl chlorides
    Secondary and tertiary alkyl bromides
    Cyclopentyl chloride
    Benzal chloride, benzotrichloride, bromoform, 1,1,2,2-
        tetrabromoethane
Unreactive under the conditions specified:
    Tertiary alkyl chlorides
    Cyclohexyl chloride and bromide
    Vinyl halides; aryl halides
    Chloroform; carbon tetrachloride, trichloroacetic acid

Materials Required

- 5% aqueous silver nitrate solution
- 10% aqueous sodium hydroxide solution
- 2 M aqueous ammonium hydroxide solution

Procedure

Clean a test tube thoroughly, preferably by boiling in it a 10% solution of sodium hydroxide and then discarding the sodium hydroxide solution and rinsing the test tube with distilled water. To the clean tube, add 2 mL of the silver nitrate solution and 1 drop of the sodium hydroxide solution. Add the ammonium hydroxide solution dropwise and while shaking well until the dark precipitated silver oxide just dissolves.

Add 1 drop of the liquid or 30–50 mg of the solid to be tested, shake the tube to mix, and allow it to stand at room temperature for 20 minutes. If nothing happens, heat the tube in a beaker of water at 35°C for 5 minutes.

$$Ag[NH_3]_2^+ \rightarrow Ag \downarrow$$

Aldehydes and other substances that can be oxidized by the silver ammonia complex ion will reduce the silver ion to metallic silver, which will precipitate as a "mirror" on the test tube if it is sufficiently clean, or as a black colloidal suspension if the test tube is not clean.

The solution should not be allowed to stand after the test is completed but should be discarded down the drain immediately, as the highly explosive silver fulminate may be formed on standing. Rinse the test tube with dilute nitric acid.

Substances that can be oxidized by means of the silver-ammonia complex ion include most aldehydes, "reducing sugars," hydroxylamines, acyloins, aminophenols, and polyhydroxyphenols. (Ref. 1, p. 173; Ref. 2, p. 138; Ref. 3, pp. 330, 1060; Ref. 4, p. 170.)

Disposal

Sink: sodium hydroxide solutions; potassium hydroxide solutions; aqueous ammonium hydroxide; aqueous hydrochloric acid; aqueous nitric acid; acetone; ethanol

Non-halogenated liquid organic waste: 2,4-dinitro-phenylhydrazine solution; pyridine; propylene glycol; dioxane; phenolphthalein solution.

Halogenated liquid organic waste: benzenesulfonyl chloride; bromine in carbon tetrachloride; carbon tetrachloride; chloroform; ferric chloride in chloroform; hydroxylamine hydrochloride solutions; zinc chloride solutions; iodine/potassium iodide solutions; sodium iodide solutions

Heavy metal liquid waste: chromic anhydride solution; ferric chloride solutions; potassium permanganate solution; silver nitrate solution

Heavy metal solid waste: all precipitates that contain silver or iron.

References

1. R. L. Shriner, R. C. Fuson, and D. Y. Curtin, *The Systematic Identification of Organic Compounds*, 5th edition, Wiley, New York, 1965.

2. N. D. Cheronis and J. B. Entrikin, *Identification of Organic Compounds*, Interscience, New York, 1963.

3. A I. Vogel, *A Textbook of Practical Organic Chemistry*, 3rd edition, Wiley, New York, 1957.

4. R. L. Shriner, R. C. Fuson, D. Y. Curtin, and T. C. Morrill, *The Systematic Identification of Organic Compounds*, 6th edition, Wiley, New York, 1980.

30. Characterization Through Formation of Derivatives

Derivatives

Sections 30.1 through 30.31 describe a number of procedures whereby substances containing certain functional groups may be converted into reasonably well-behaved crystalline products (derivatives) through reaction at the functional group. As in the qualitative characterization tests of Section 29, the fact that the substance of unknown structure gives a solid product is consistent with the presence in its molecules of the suspected functional group, but the formation (or nonformation) of a solid product is by no means proof of the presence (or absence) of the suspected functional group. The reasons are the same as those discussed for the qualitative characterization tests: the derivative-forming reactions are not completely specific or characteristic (some substances may react atypically), and a small amount of an impurity, possibly water, may be sufficient to result in a solid product.

If a solid product is obtained, however, it can be purified—usually by recrystallization until successive recrystallizations do not raise the melting point significantly—and the melting point determined. Comparing the melting point of the derivative with the melting points recorded in the chemical literature for the corresponding derivatives of substances of known structure can often serve to eliminate all but one or two possibilities for the structure of the substance of unknown structure. A hypothetical case will show how this is done and will show some of the problems and pitfalls that must be considered and avoided.

Suppose the substance of unknown structure is a liquid that distilled between 96.5 and 98°C at 740 Torr. On the basis of various chemical and physical properties, you deduced that the liquid must be an alcohol. When you treated a sample with 3,5-dinitrobenzoyl chloride according to the procedure of Section 30.1, you obtained a solid that melted after two recrystallizations at 74.5-75.5°C. You presumed it to be the 3,5-dinitrobenzoate ester of the unknown alcohol.

At this point, alcohols of higher and lower boiling points have already been tentatively eliminated, because their boiling points are believed to be well outside the limits of the experimental error of the observed boiling range of the unknown. Shriner, Fuson, and Curtin (Ref. 1, end of this section) offer the following additional information:

Shriner, Fuson, and Curtin

Compound	Boiling Point	Melting Point of Derivative		
		α-Naphthylurethan	Phenylurethan	3,5-Dinitrobenzoate
3-Butene-2-ol	96	—	—	
Allyl alcohol	97	109	70	48
n-Propyl alcohol	97	80	51	74
sec-Butyl alcohol	99	97	65	75
tert-Amyl alcohol	102	71	42	117

From these data, it appears that *tert*-amyl alcohol and allyl alcohol can also be eliminated, because the melting points reported for their 3,5-dinitrobenzoate esters are well outside the range of experimental error of the melting point observed for the 3,5-dinitrobenzoate of the unknown. If you knew the melting point of the 3,5-dinitrobenzoate of 3-butene-2-ol, you might be able to eliminate this as well. Now you can see that it might have been smarter to try to make the α-naphthylurethan of the unknown, because it might serve to distinguish among all possibilities.

allyl alcohol

3-butene-2-ol

Looking in Cheronis and Entrikin (Ref. 2) to try to find melting points for the derivatives of 3-butene-2-ol, you locate the following information:

Cheronis and Entrikin

Compound	Boiling Point	Melting Point of Derivative		
		α-Naphthylurethan	Phenylurethan	3,5-Dinitrobenzoate
3-Butene-2-ol	—	—	—	—
Allyl alcohol	97.1	108	—	49
n-Propyl alcohol	97.15	105	—	74
sec-Butyl alcohol	99.5	97	—	76
tert-Amyl alcohol	102.35	75	—	116

You did not find what you wanted, but you did discover a remarkable disagreement between the value reported for the melting point of the α-naphthylurethan of *n*-propyl alcohol by Shriner, Fuson, and Curtin and that reported by Cheronis and Entrikin. The disagreement between the other values reported for boiling points and melting points is typical and is an indication of the uncertainty of these values.

Vogel

Looking in Vogel (Ref. 3) to find melting points for derivatives of 3-butene-2-ol and to get another opinion on the melting point of the α-naphthylurethan of *n*-propyl alcohol, you find the following information:

		Melting Point of Derivative			
Compound	Boiling Point	α-Naphthyl-urethan	Phenyl-urethan	3,5-Dinitro-benzoate	Hydrogen 3-nitrophthalate
3-Butene-2-ol	—	—	—	—	—
Allyl alcohol	97	109	70	50	124
n-Propyl alcohol	97	80	57	75	145
sec-Butyl alcohol	99.5	98	64	76	131
tert-Amyl alcohol	102	72	42	118	—

You did not find out anything here, either, about the melting points of derivatives of 3-butene-2-ol, but you did discover that Vogel and Shriner, Fuson, and Curtin agree on the melting point of the α-naphthylurethan of *n*-propyl alcohol. However, this indicates no more than that these authors are reporting a value from the same source in the original literature, and that Cheronis and Entrikin are reporting a value from a different source (assuming no typographical or transcription errors). You also see that Vogel and Shriner, Fuson, and Curtin report significantly different values for the melting point of the phenylurethan of *n*-propyl alcohol.

CRC Handbook of Tables

One more easy place to look for melting points of derivatives is the Chemical Rubber Company's *Handbook of Tables for Organic Compound Identification*, 3rd edition (Ref. 4), which contains the following information:

Compound	Boiling Point	Melting Point of Derivative			
		α-Naphthyl-urethan	Phenyl-urethan	3,5-Dinitro benzoate	Hydrogen 3-nitrophthalate
3-Butene-2-ol	94–95	—	—	—	43–44
Allyl alcohol	97.1	108	70	49–50	124
n-Propyl alcohol	97.1	80;76	57	74	145.5
sec-Butyl alcohol	99.5	97	64.5	76	131
tert-Amyl alcohol	102.3	72	42	116	—

From this research, you see that if the compound is an alcohol and if the solid formed from the unknown by treatment of a sample by the procedure of Section 30.1 is the corresponding 3,5-dinitrobenzoate ester, the boiling point of the unknown liquid and the melting point of the derivative serve to rule out all possibilities except n-propyl alcohol, sec-butyl alcohol, 3-butene-2-ol, and other alcohols of which we are not aware that boil near 96-98°C. You also see that the derivative that should have been made was not the 3,5-dinitrobenzoate but the hydrogen 3-nitrophthalate (by the procedure of Section 30.2), which would have allowed the elimination of at least two of the remaining possibilities. The moral of your search is that a few minutes of library research can save hours of laboratory work.

A minute in the lib is worth an hour in the lab

The next step would be to prepare the hydrogen 3-nitrophthalate of the unknown to reduce the number of possibilities for its structure. The α-naphthylurethan should not be prepared for two reasons: (1) on the basis of the present information, it would not allow elimination of 3-butene-2-ol, and (2) there is uncertainty about the melting point of this derivative of n-propanol.

Of course, the number of possibilities may be reduced by consideration of other physical and chemical properties. For example, the following values for density and index of refraction are reported in the Handbook of Tables for Organic Compound Identification (Ref. 4)

Compound	d_4^{20}	n_D^{20}
3-Butene-2-ol	—	—
n-Propyl alcohol	0.80359	1.38499
sec-Butyl alcohol	0.80692	1.39495^{25}

It should be possible to distinguish between *n*-propyl alcohol and *sec*-butyl alcohol on the basis of the index of refraction, but of course 3-butene-2-ol cannot be ruled out by either the density or the index of refraction.

Assume now that you have prepared the hydrogen 3-nitrophthalate and found that it has a melting point of 144-145°C. All possibilities except *n*-propyl alcohol and alcohols of which we are not aware that boil near 96–98°C have now been eliminated. Of course, you would have to be very unlucky indeed to have an unknown sample that had the same boiling point as *n*-propyl alcohol and whose derivatives melted at the same temperature as those of *n*-propyl alcohol, but yet was not *n*-propyl alcohol.

The identity of the liquid of unknown structure and *n*-propyl alcohol could be established in several ways if a sample of *n*-propyl alcohol were available. For example, the "identity" of the infrared or mass spectra would be completely acceptable as evidence of the identity of the two substances. The comparison of spectra could also be made by way of published spectral data for *n*-propyl alcohol. Either the hydrogen 3-nitrophthalate or 3,5-dinitrobenzoate of *n*-propyl alcohol could be prepared, and the identity of the melting points and nondepression of the mixed melting point (or identity of the infrared spectra of the two derivatives) could establish the identity of the unknown liquid as *n*-propyl alcohol.

Finally, of course, it must be admitted that the NMR spectrum would immediately establish whether or not the unknown liquid was *n*-propyl alcohol, or, for that matter, any of the other four alcohols that were considered. The NMR spectrum would also reveal whether or not the unknown substance had a structure different from that of any of the five possibilities considered.

Table 30-1 is a summary and index for the 31 procedures for the formation of derivatives that are presented in the following sections. The references at the end of each procedure give more information concerning the scope and limitations of the procedures, possible procedural variations, and exceptional behavior of certain compounds or types of compounds; references that are given in abbreviated form are listed in full at the end of this section.

The book by Shriner, Fuson, and Curtin and the Cheronis and Entrikin book give many references to the original literature in which the procedures for preparations of derivatives are described. If you think that you know the identity of the substance for which you are attempting to prepare a derivative, you will often save time by looking up in these references the exact procedure used with that compound. You may find that the substance is best converted to the

Table 30-1. Procedures for the preparation of derivatives[a]

Type of Compound	Type of Derivative	Section	Appendix
Alcohol	Benzoate, p-nitrobenzoate, and 3,5-dinitrobenzoate esters	30.1	A.1
Alcohol	Hydrogen 3-nitrophthalate ester	30.2	A.1
Alcohol	Plenylurethan; α-naphthylurethan	30.3	A.1
Aldehyde	Methone derivative	30.4	A.2
Aldehyde	2,4-Dinitrophenylhydrazone	30.5	A.2
Aldehyde	Semicarbazone	30.6	A.2
Aldehyde	Oxime	30.7	A.2
Amide, unsubstituted	Hydrolysis	30.8	A.3
Amide, unsubstituted	9-Acylamidoxanthene	30.9	A.3
Amide, N-substituted	Hydrolysis	30.10	A.3
Amine, primary and secondary	N-Substituted acetamide	30.11	A.4
Amine, primary and secondary	N-Substituted benzamide	30.12	A.4
Amine, primary and secondary	N-Substituted p-toluenesulfonamide	30.13	A.4
Amine, primary and secondary	Phenylthiourea; α-naphthylthiourea	30.14	A.4
Amine, tertiary	Picrate	30.15	A.5
Amine, tertiary	Methiodide; p-toluenesulfonate	30.16	A.5
Carboxylic acid	Unsubstituted amide	30.17	A.6
Carboxylic acid	Anilide; p-toluidide; p-bromoanilide	30.18	A.6
Carboxylic acid	Phenacyl ester; p-substituted phenacyl esters	30.19	A.6
Carboxylic acid	p-Nitrobenzyl ester	30.20	A.6
Carboxylic acid anhydride	Anilide; p-toluidide; p-bromoanilide	30.18	A.6
Carboxylic acid ester	N-Benzylamide	30.21	A.7
Carboxylic acid ester	3,5-Dinitrobenzoate	30.22	A.7
Carboxylic acid ester	Hydrolysis	30.23	A.7
Carboxylic acid halide	Anilide; p-toluidide; p-bromoanilide	30.18	A.6
Ether, aromatic	Bromination product	30.24	A.8
Ether, aromatic	Picrate	30.15	A.8
Halide, aliphatic	S-Alkylthiuronium picrate	30.25	A.9
Halide, aromatic	o-Aroylbenzoic acid	30.26	A.10
Halide, aromatic	Oxidation	30.27	A.10
Halide	Anilide; p-toluidide; a-naphthalide	30.28	A.9
Hydrocarbon, aromatic	o-Aroylbenzoic acid	30.26	A.11
Hydrocarbon, aromatic	Oxidation	30.27	A.11
Hydrocarbon, aromatic	2,4,7-Trinitrofluorenone adduct	30.29	A.11
Hydrocarbon, aromatic	Picrate	30.15	A.11
Ketone	2,4-Dinitrophenylhydrazone	30.5	A.12
Ketone	Semicarbazone	30.6	A.12
Ketone	Oxime	30.7	A.12
Nitrile	Hydrolysis	30.8	A.13
Phenol	α-Naphthylurethan	30.3	A.14
Phenol	Bromination product	30.30	A.14
Phenol	Aryloxyacetic acid	30.31	A.14

[a] The section numbers refer to the section in which the procedure for preparation will be found. The appendix numbers refer to the appendix in which the melting points of the derivatives can be found.

derivative by a modification of the general procedure, or that the derivative is formed in poor yield (by an experienced chemist, which might mean that you should consider another derivative), or that the product is best isolated in a certain way or recrystallized from a certain solvent, or that it has some interesting or characteristic property by which it may be recognized. This kind of information can sometimes save hours of lab time.

These two very useful books also give literature references to many more procedures for preparation of derivatives, both for substances with the functional groups for which procedures are given in this book and for compounds with other types of functional groups.

Questions

1. Consider the alcohols 3-butene-2-ol, allyl alcohol, *n*-propyl alcohol, *sec*-butyl alcohol, and *tert*-butyl alcohol. Which of these alcohols could be distinguished by the chromic anhydride test, Section 29.4?

2. Which of the alcohols of Problem 1 could be distinguished by Lucas's Test, Section 29.8?

30.1 Benzoates, *p*-Nitrobenzoates, and 3,5-Dinitrobenzoates of Alcohols

acid chloride   **ester**

Procedure

Treat a solution of 0.5 gram of the alcohol in 3 mL of pyridine with about 2 grams of the appropriate acid chloride (benzoyl chloride, *p*-nitrobenzoyl chloride, or 3,5-dinitrobenzoyl chloride) with cooling in an ice bath. Heat the resulting mixture on the steam bath with exclusion of moisture for 10 minutes if the alcohol is primary or secondary, and for 30 minutes if it is tertiary (Note 1). Pour the mixture while stirring into 10 mL of ice water and cautiously acidify it with concentrated hydrochloric acid. Thoroughly triturate the residue, which frequently is an oil, with 5 mL of 5% sodium carbonate solution. Finally, collect the solid by suction filtration and recrystallize it from aqueous alcohol, or methanol, or ethanol, or acetone/petroleum either. (Ref. 1, pp. 246, 247; Ref. 2, p. 249; Ref. 3, p. 262; Ref. 4, pp. 156, 157.)

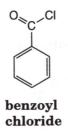

**benzoyl
chloride**

Note

1. Tertiary alcohols are esterified poorly under these conditions.

30.2 Hydrogen 3-Nitrophthalates of Alcohols

Derivative for primary and secondary alcohols (Note 1).

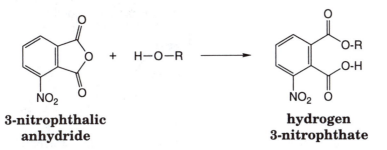

**3-nitrophthalic
anhydride**   **hydrogen
3-nitrophthate**

Procedure

Heat for 2 hours on the steam bath a mixture of 0.3 mL of the alcohol, 0.3 gram of 3-nitrophthalic anhydride, and 0.5 mL of pyridine. Pour the mixture onto ice, acidify with concentrated hydrochloric acid, and isolate the solid ester either by filtration or by extraction with benzene or chloroform. Recover the acid phthalate by extraction into dilute sodium hydroxide followed by acidification of the extract and suction filtration. Recrystallize from water, alcohol/water, or toluene. (Ref. 1, p. 248, Ref. 4, p. 158.)

Note

1. Tertiary alcohols undergo elimination under these conditions.

30.3 Phenyl- and α-Naphthylurethans

Derivatives for primary and secondary alcohols (Note 1) and phenols.

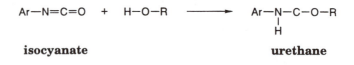

isocyanate   urethane

Procedure

Add a solution of 0.5 gram of the alcohol (Note 2) in 5 mL of ligroin (b.p., 80–100°C) to a solution of 0.5 gram of phenylisocyanate or α-naphthylisocyanate (Note 3) in 10 mL of the same solvent. Heat the mixture for 1–3 hours on the steam bath, filter it while hot (Note 2), and allow the filtrate to cool. Collect the product by suction filtration, and recrystallize it from ligroin or carbon tetrachloride (Ref. 1, p. 246; Ref. 2, p. 252; Ref. 3, pp. 264, 683; Ref. 4, p. 355.)

Notes

1. The urethans of tertiary alcohols are difficult to prepare by this method.
2. The alcohol must be dry; water leads to the formation of the very insoluble diphenylurea or di-α-naphthylurea. The hot filtration will remove most of this if it is formed, but some will crystallize with the product.
3. If a phenol is used, the α-naphthylurethan is the preferred derivative and the reaction should be catalyzed by the addition of a few drops of pyridine.

30.4 Methone Derivatives of Aldehydes

Derivative for low-molecular-weight aldehydes.

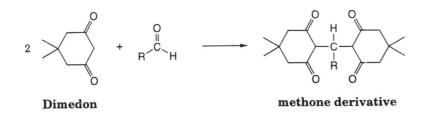

Dimedon   **methone derivative**

Procedure

Dissolve 300 mg of Dimedon (2.1 mmole) in 4 mL of 50% aqueous ethanol. Add to this 1 mmole of the aldehyde and boil the mixture for about 30 seconds. Allow it to cool and stand for crystallization for at least 4 hours. Collect the product by suction filtration and recrystallize from a minimum amount of methanol and water. (Ref. 1, p. 254; Ref. 2, p. 262; Ref. 3, p. 332; Ref. 4, p. 180.)

30.5 2,4-Dinitrophenylhydrazones

Derivative for aldehydes and ketones.

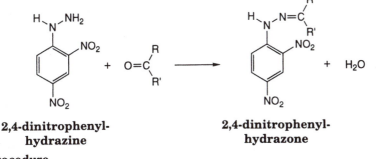

2,4-dinitrophenyl-   **2,4-dinitrophenyl-**
hydrazine   **hydrazone**

Procedure

Preparation of a solution of 2,4-dinitrophenylhydrazine in 30% perchloric acid. Dissolve 1.2 grams (6 mmole) of 2,4-dinitro-phenylhydrazine in a mixture of 16 mL of 60% perchloric acid plus 34 mL of water at room temperature.

Preparation of the derivative. Take 4 mL of the perchloric acid solution of 2,4-dinitrophenylhydrazine and dilute it with 8 mL of water; stir well. Quickly add to this well-stirred mixture 0.5 mmole of the carbonyl compound as a 10-20% solution in ethanol. Collect the resulting 2,4-dinitrophenylhydrazone by suction filtration, and recrystallize it frommethanol, ethyl acetate, dioxane, dioxane/water, or ethanol/water. (Ref. 1, p. 253; Ref. 2, p. 260; Ref. 3, p. 344; Ref. 4, p. 179.)

30.6 Semicarbazones

Derivative for aldehydes and ketones.

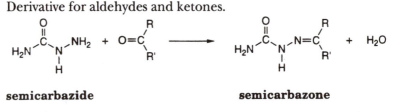

semicarbazide   **semicarbazone**

Procedure

Preparation of an alcoholic solution of semicarbazide acetate. Grind 1 gram of semicarbazide hydrochloride with 1 gram of anhydrous sodium acetate in a mortar. Transfer the mixture to a flask, boil it with 10 mL of absolute ethanol, and filter the suspension while hot.

Preparation of the derivative. Add to the freshly prepared solution of semicarbazide acetate about 0.2 gram of the carbonyl compound, and heat the mixture under reflux on the steam bath for 30–60 minutes. Dilute the hot solution with water to incipient cloudiness, and allow it to cool slowly to room temperature. Collect the semicarbazone by suction filtration and recrystallize it from aqueous alcohol or alcohol. (Ref. 1, p. 253; Ref. 2, pp. 262, 320; Ref. 3 p. 344, Ref. 4, p. 179.)

30.7 Oximes

Derivative for higher-molecular-weight aldehydes and ketones.

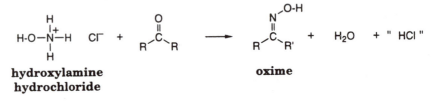

hydroxylamine   **oxime**
hydrochloride

Procedure

Heat for 2 hours under reflux on the steam bath a mixture of 0.5 gram of the carbonyl compound, 0.5 gram of hydroxylamine hydrochloride, 3 mL of pyridine, and 3 mL of absolute ethanol. Remove the solvent by evaporation (by heating the mixture while drawing a current of air over it or with a rotary evaporator; Section 37), and recrystallize the residue from methanol or methanol/water. (Ref. 1, p. 289; Ref. 2, p. 319; Ref. 3, p. 721; Ref. 4, p. 181.)

30.8 Carboxylic Acids by Hydrolysis of Primary Amides and Nitriles

Derivative for primary amides and nitriles.

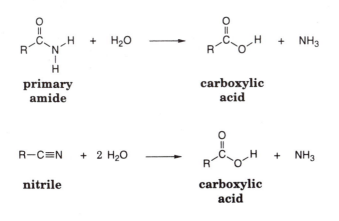

primary amide → **carboxylic acid**

nitrile → **carboxylic acid**

Basic hydrolysis

Procedure A (*easily hydrolyzed amides or nitriles*)

Boil under reflux 10 mmole of the substance with 0.8 gram (20 mmoles) of sodium hydroxide in 3 mL of water until ammonia evolution can be detected no longer (4-10 hours; Notes 1 and 2).

Procedure B (*more difficulty hydrolyzed amides or nitriles*)

Boil under reflux 10 mmole of the substance with 1.2 grams (20 mmoles) of potassium hydroxide in 4 mL of mono-, di-, or triethyleneglycol until ammonia evolution has ceased (about 5 hours; Note 1). At this time, dilute the mixture with 10 mL of water.

Workup for both procedures A and B

Acidify the aqueous solution with 20% sufuric acid. Collect the precipitated acid by suction filtration, wash it with water, and recrystallize it from water or aqueous methanol.

If the acid is a liquid or is relatively soluble in water, it should be converted to a phenacyl ester by the procedure of Section 30.19.

Acidic hydrolysis: for nitriles that resist basic hydrolysis

Procedure

Heat in an oil bath at 160°C for 30 minutes, with stirring and using a reflux condenser, a mixture of 5 mL of 75% sulfuric acid, 200 mg of sodium chloride, and 10 mmole of the nitrile. Raise the temperature of the bath to 190°C and continue to heat and stir for another 30 minutes. Cool the mixture to room temperature and pour it onto 20 grams of ice. Collect the precipitate by suction filtration and dissolve it in 5 mL of 10% sodium hydroxide solution. Any insoluble amide should be removed by suction filtration and, if it is a major product, recrystallized from water or aqueous methanol. Acidify the filtrate to obtain the acid. Collect it by suction filtration and recrystallize it from water, toluene, acetone/water, or alcohol/water. (Ref. 1, p. 291; Ref. 2, p. 321; Ref. 3, pp. 404, 410, 798, 805; Ref. 4, p. 282.)

Notes

1. Ammonia can be detected by the method described in Section 30.1.
2. If the amide or nitrile solidifies in the condenser, 0.5 mL of ethanol can be added to help to redissolve it; the alcohol should be removed by distillation at the end of the heating period.

30.9 9-Acylamidoxanthenes from Amides

Derivative for unsubstituted amides.

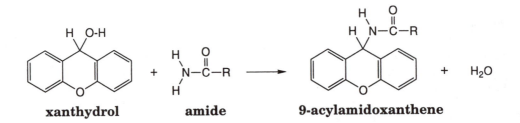

xanthydrol   amide   9-acylamidoxanthene

Procedure

Dissolve 2 mmoles (400 mg) of xanthydrol in 5 mL of glacial acetic acid (Note 1). Add to this 1-1.5 mmole of the unsubstituted amide (Note 2), heat the mixture in a beaker of water at 85°C for 20–30 minutes, and then allow it to cool for crystallization. Collect the product by suction filtration and recrystallize it from 2:1 dioxane: water or 2:1 ethanol: water. (Ref. 1, p. 256; Ref. 2, p. 269; Ref. 3, p. 405; Ref. 4, p. 284.)

Notes

1. If the solution is not clear, allow it to stand for a few minutes and decant or filter the clear liquid from the insoluble material.

2. If the amide is not soluble in glacial acetic acid, it can be added to the reaction mixture as a solution in 2 mL of ethanol; if this is done, add 1 mL of water to the mixture after heating and before cooling.

30.10 Hydrolysis of *N*-Substituted Amides

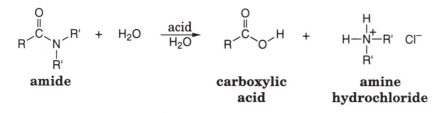

| amide | | carboxylic acid | amine hydrochloride |

Procedure A

Boil under reflux for 2 hours a mixture of 0.5-1 gram of the amide and 10–20 mL of concentrated hydrochloric acid. Cool the mixture and collect any solid by suction filtration. The solid may be the acid corresponding to the amide, the hydrochloride salt of the amine corresponding to the amide, or unchanged amide. If the solid is soluble in water, it is probably the salt of the amine. If it is insoluble in water but soluble in 5% aqueous sodium bicarbonate, it is probably the acid. The melting point as well as solubility properties will indicate unchanged starting material.

The amine can be recovered by making the solution alkaline (use care in neutralizing a strongly acidic solution) and then either filtering with suction or extracting with either, depending on the nature of the amine. It can then be purified by recrystallization or converted to a derivative by procedures suitable for primary and secondary amines.

The acid can be isolated from the solution by reacidification and extraction (if it were not soluble, it would have separated from the solution originally) and converted to a derivative.

Procedure B (*for amides that resist Procedure A*) (Note 1)

Boil under reflux for 30 minutes to 1 hour a mixture of 1 gram of the amide in 10 mL of 70% sulfuric acid (Note 2). Cautiously pour 10 mL of water down the condenser and allow the solution to cool. Remove any solid by suction filtration and recover the amine as in Procedure A. (Ref. 1, p. 255; Ref. 2, p. 267, Ref. 3, p. 801.)

Notes

1. Benzanilide and related compounds, for example.
2. Prepared by pouring 8 mL of concentrated sulfuric acid onto 6 grams of ice.

30.11 Substituted Acetamides from Amines

Derivative for water-insoluble primary and secondary amines.

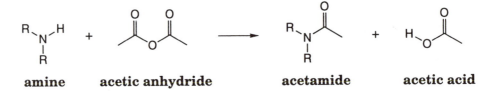

amine   acetic anhydride   acetamide   acetic acid

Procedure

Dissolve 0.5 gram of the amine in 25 mL of 5% hydrochloric acid. Add 5% sodium hydroxide solution in small portions until the mixture becomes cloudy (because of precipitation of the amine). Redissolve the precipitate by adding a little more 5% hydrochloric acid.

Add about 10 grams of ice and then 5 mL of acetic anhydride. Next, while stirring well and all in one portion add a previously prepared solution of 5 grams of sodium acetate trihydrate in 5 mL of water. Cool the mixture in an ice bath and allow it to stand (possibly overnight) for crystallization. (Compare the preparation of acetanilide, Section E42).

Collect the product by suction filtration and wash it well with water. Recrystallize from ethanol or ethanol/water or, if the crude product has been thoroughly dried, from toluene/cyclohexane. (Ref. 1, p. 259; Ref. 3, pp. 576, 652; Ref. 4, p. 232.)

30.12 Substituted Benzamides from Amines

Derivative for primary and secondary amines.

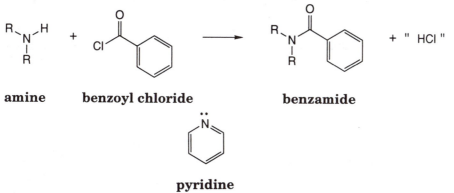

amine   **benzoyl chloride**   **benzamide**   + " HCl "

pyridine

Procedure

Add dropwise 0.5 mL of benzoyl chloride to a solution of 0.5 gram of the amine in 5 mL of dry pyridine and 10 mL of dry toluene. Heat the resulting mixture in a beaker of hot water at 60–70°C for 30 minutes. Pour the mixture into 100 mL of water, separate the toluene layer, extract the aqueous layer with one 10-mL portion of toluene, and combine this with the toluene layer. Wash the combined toluene layers with water followed by 5% sodium carbonate solution, dry over anhydrous magnesium sulfate, and, after removing the magnesium sulfate by filtration, concentrate to a volume of 3 or 4 mL. Stir about 20 mL of hexane into the concentrated toluene solution to precipitate the derivative, collect the product by suction filtration, and wash it with hexane. Recrystallize the benzamide from cyclohexane/hexane or cyclohexane/ethyl acetate. (Ref. 1, p. 260; Ref. 2, p. 272; Ref. 3, p. 652; Ref. 4, p. 233.)

30.13 *p*-Toluenesulfonamides from Amines

Derivative for primary and secondary amines (see Note 1).

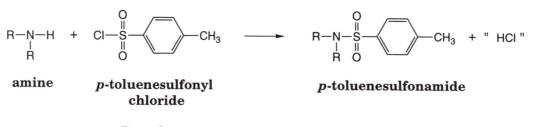

amine   ***p*-toluenesulfonyl**   ***p*-toluenesulfonamide**
                   chloride

Procedure

Boil for 30 minutes under reflux a mixture of 0.5 g of the amine, 1–1.5 g (5–8 mmoles) of *p*-toluenesulfonyl chloride (Note 2), and 3 mL of pyridine. Pour the reaction mixture into 5 mL of cold water and stir until the product crystallizes. Collect the sulfonamide by suction filtration, wash it well with water, and recrystallize it from alcohol or aqueous alcohol. (Ref. 1, p. 261; Ref. 2, p. 274; Ref. 3, p. 653; Ref. 4, p. 233.)

Notes

1. Compare with Hinsberg's test, Section 29.2.
2. An equivalent amount of benzenesulfonyl chloride can be used; the *p*-toluenesulfonamides are said to be more satisfactory derivatives.

30.14 Phenylthioureas and α-Naphthylthioureas

Derivatives for primary and secondary amines.

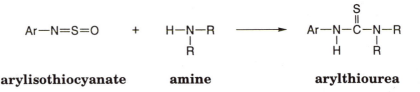

arylisothiocyanate   **amine**   **arylthiourea**

Procedure

Dissolve 0.2 gram of the amine in 5 mL of ethanol (Note 1) and add to this a solution of 0.2 g (0.18 mL; 1.5 mmole) of phenylisothiocyanate or 0.28 g (1.5 mmole) of α-naphthylisothiocyanate in 5 mL of ethanol. If no reaction takes place at room temperature, heat the mixture for 1 or 2 minutes. If no crystals separate upon cooling and scratching (as with aromatic amines), reheat the mixture for another 10 minutes and then cool again (Note 2). Collect the product by suction filtration and recrystallize it from ethanol. (Ref. 1, p. 261; Ref. 2, p. 275; Ref. 3, p. 422; Ref. 4, p. 234.)

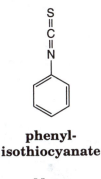

**phenyl-
isothiocyanate**

Notes

1. These reagents are not reactive toward water (compare the analogous isocyanates, Section 30.3), and so an aqueous solution of the amine can be used if it is difficult to obtain the pure amine.

2. If no product can be obtained with an aromatic amine, the reaction can be tried without a solvent with heating for 10 minutes. The product can be isolated at the end of the reaction by the addition of 50% aqueous ethanol.

30.15 Picrates

Derivative for tertiary amines, aromatic ethers, and aromatic hydrocarbons (Note 1).

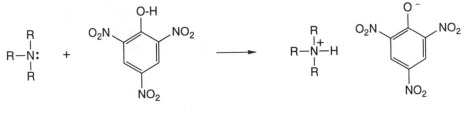

amine   **picric acid**   **picrate salt**

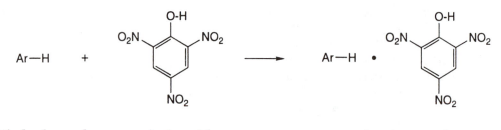

aromatic hydrocarbon   **picric acid**   **picrate complex**

Procedure A (*for tertiary amines; also for aromatic ethers and hydrocarbons that form relatively stable picrates*)

Dissolve 2 mmoles of the substance in 5 mL of 95% ethanol, and add to this solution a solution of 2.2 mmoles of picric acid (500 mg) in 10 mL of 95% ethanol. Heat the mixture to boiling on the steam bath, and allow it to cool slowly for crystallization. Collect the product by suction filtration and recrystallize it from ethanol (Note 2).

Procedure B (*for substances that form relatively unstable picrates*)

Dissolve 5 mmoles of the substance in about 5 mL of boiling chloroform and add a solution of 5 mmoles (1.15 grams) of picric acid in 3 mL of boiling chloroform. Swirl the resulting mixture and allow it to cool for crystallization. Collect the product by suction filtration and recrystallize it from a minimum amount of chloroform (Notes 2 and 3). For amine picrates see Ref. 1, p. 263; Ref. 2, p. 277; Ref. 3, p. 422; Ref. 4, p. 236. For ether and hydrocarbon picrates, see Ref. 1, p. 277; Ref. 2, pp. 298, 315; Ref. 3, pp. 518, 672; Ref. 4, p. 309.)

Notes

1. Certain other aromatic compounds such as primary and secondary amines and halides will form picrates.
2. The picrates of some substances cannot be recrystallized because they dissociate to too large a degree in solution.
3. The melting point should be determined immediately because some picrates decompose.

30.16 Quaternary Ammonium Salts: Methiodides and *p*-Toluenesulfonates

Derivatives for tertiary amines.

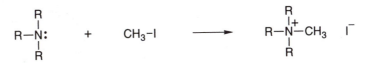

tertiary amine   **methyl iodide**   **quaternary methiodide**

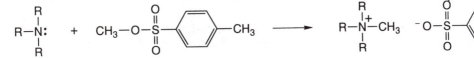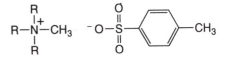

tertiary amine   **methyl *p*-toluenesulfonate**   **quaternary *p*-toluenesulfonate**

Procedure

Dissolve 0.5 gram of the tertiary amine in twice its volume of nitromethane, acetonitrile, or alcohol (Note 1); do the same for 1 gram of the quaternizing agents (methyl iodide or methyl *p*-toluenesulfonate). Combine the solutions, allow the mixture to stand at room temperature for 1 hour, and then heat it for 30 minutes on the steam bath. If the quaternary salt crystallizes out, it should be collected by suction filtration; if not, the solvent should be removed under vacuum (Section 37). The crude product or residue should be recrystallized from ethyl acetate/ethanol. (Ref. 1, p. 262; Ref. 2, p. 277; Ref. 3, p. 660; Ref. 4, p. 235.)

Note

1. The solvents named are listed in decreasing order of suitability for the reaction.

30.17 Carboxylic Acid Amides

A procedure for deriving amides from carboxylic acids. Lower-molecular-weight amides, which are relatively soluble in water, are difficult to isolate by this procedure.

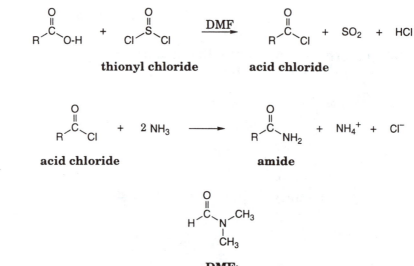

thionyl chloride　　　　　　**acid chloride**

acid chloride　　　　　　　　**amide**

DMF;
dimethylformamide

Procedure

Boil under reflux on the steam bath, using a calcium chloride drying tube in the condenser, a mixture of 1 gram of the carboxylic acid, 5 mL of thionyl chloride, and 1 drop of dimethylformamide for 15-30 minutes. Pour the mixture *cautiously* into 15 mL of ice-cold concentrated ammonium hydroxide. Collect the crude amide by suction filtration and recrystallize it from water or aqueous alcohol. (Ref. 1, p. 235; Ref. 3, p. 361; Ref. 4, p. 271.)

30.18 Anilides, *p*-Toluidides, and *p*-Bromoanilides of Carboxylic Acids

Derivatives for carboxylic acids, acid halides, and anhydrides.

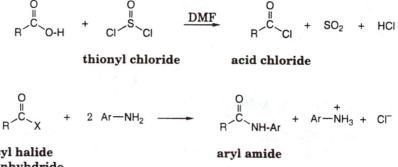

thionyl chloride　　　　　　**acid chloride**

acyl halide　　　　　　　　　**aryl amide**
or anhyhdride

aniline

Procedure

Preparation of the acid chloride. Heat for 30 minutes 1 mmole of the acid with 0.8 mL (1.3 g; 11 mmole) of thionyl chloride, and 1 drop of dimethylformamide on the steam bath under reflux, using a calcium chloride drying tube on the condenser.

Treatment of the acid halide or anhydride with the aromatic amine. Add to the acid chloride just prepared, or to 1 mmole of the acid halide or anhydride, dissolved in dry toluene if it is a solid, a solution of 5 mmoles of the aromatic amine (aniline, *p*-toluidine, or *p*-bromoaniline) dissolved in 20 mL of toluene. Heat the mixture under reflux on the steam bath for about 15 minutes. Cool the mixture to room temperature, transfer it to a separatory funnel, and extract it with 2 mL of water, 5 mL of 5% hydrochloric acid, 5 mL of 5% sodium hydroxide, and 2 mL of water. Evaporate the toluene solution to dryness, and recrystallize the residual amide from aqueous alcohol, methanol, or ethanol. (Ref. 1, p. 236; Ref. 2, p. 238; Ref. 3, p. 361; Ref. 4, p. 272.)

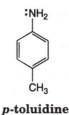

p-toluidine

30.19 Phenacyl and Substituted Phenacyl Esters of Carboxylic Acids

Derivative for carboxylic acids.

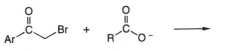
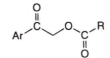

phenacyl bromide   **phenacyl ester**

phenacyl bromide

Procedure A (*for pure acids*)

Dissolve 100 mg (1 mmole) triethylamine in 2 mL of dry acetone, and neutralize the solution by the addition of the carboxylic acid. To this solution, add a solution of 0.5 mmole of the phenacyl bromide (Note 1) (phenacyl bromide, *p*-chlorophenacyl bromide, *p*-bromophenacyl bromide, or *p*-phenylphenacyl bromide) in 3 mL of dry acetone. Allow the mixture to stand for 3 hours at room temperature (a precipitate of triethyl ammonium bromide will form after a short time); then dilute it with 10 mL of water and collect the precipitate by suction filtration. After thoroughly washing the crude phenacyl ester with 5% sodium bicarbonate solution and finally with water, recrystallize it from aqueous alcohol.

Procedure B (*for salts or aqueous solutions of the acid*)

Dissolve 150 mg of the salt in 2 mL of water and add 2 drops of 5% hydrochloric acid; *or* add 5% hydrochloric acid to 2 mL of a basic aqueous or aqueous alcoholic solution of the acid containing about 100 mg of the acid until it is neutral to litmus, and then add 2 more drops of 5% hydrochloric acid. To this slightly acidic solution of the sodium salt of the acid (Note 2), add a solution of 200 mg of the phenacyl bromide (Notes 1 and 3) in ethanol and heat the mixture under reflux—1 hour for a monocarboxylic acid; 2 hours for a dicarboxylic acid; 3 hours for a tricarboxylic acid. Occasionally, a solid separates from solution during the reflux period; it should be brought into solution by the addition of a few milliliters of ethanol. At the end of the heating period, allow the mixture to cool to room temperature and collect the crude phenacyl ester by suction filtration. It should be washed and recrystallized as in Procedure A. (Ref. 1, p. 235; Ref. 2, p. 243; Ref. 3, p. 362; Ref. 4, p. 270.)

Notes

1. The phenacyl halides are lachrymatory and tend to irritate the skin. They should be treated with respect.

2. The reaction mixture should not be alkaline; alkali causes the hydrolysis of the phenacyl bromide to the phenacyl alcohol.

3. *p*-Bromophenacyl bromide is converted to the relatively insoluble *p*-bromophenacyl chloride (m.p., 117°C) by chloride ion. If the mixture contains large amounts of chloride ion, another phenacyl halide should be used.

30.20 *p*-Nitrobenzyl Esters of Carboxylic Acids

Derivative for carboxylic acids.

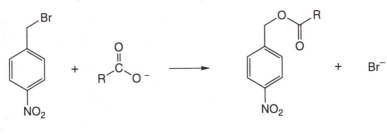

p-nitrobenzyl bromide   **p-nitrobenzyl ester**

Procedure

The procedure for the preparation of phenacyl ester (Section 30.19) may be used with the following exceptions:

- *Procedure A:* *p*-Nitrobenzyl bromide is used in place of the phenacyl bromide; ethanol is used as solvent in place of acetone.

- *Procedure B:* *p*-Nitrobenzyl bromide is used in place of the phenacyl bromide.

- *Procedure A and B:* If *p*-nitrobenzyl chloride is substituted for *p*-nitrobenzyl bromide, 10 mg of sodium iodide should be added.

The benzyl halides are lachrymatory and tend to irritate the skin; they should be treated with respect. (Ref. 1, p. 235; Ref. 2, p. 244; Ref. 3, p. 362; Ref. 4, p. 270.)

30.21 *N*-Benzylamides from Esters

Derivative for esters.

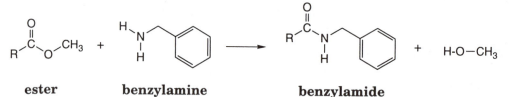

| ester | benzylamine | benzylamide |

Procedure

Heat under reflux for about 1 hour a mixture of 0.5 g of the methyl or ethyl ester (Note 1), 1.5 mL of benzylamine, and 50 mg of ammonium chloride. Cool the mixture and stir it with a little water and finally a little dilute hydrochloric acid (Note 2). Collect the solid by suction filtration and recrystallize it from alcohol/water or acetone/water. (Ref. 1, p. 272; Ref. 2, p. 290; Ref. 3, p. 394; Ref. 4, p. 300.)

Notes

1. Higher esters must first be transesterified to the methyl ester by heating 0.6-1 gram under reflux for 30 minutes in 10 mL of anhydrous methanol in which 0.1 g of sodium metal or sodium methoxide has been dissolved. The residue obtained by evaporation of the excess methanol can then be used in the procedure described.

2. Avoid an excess of hydrochloric acid, as it may dissolve the product as well.

30.22 3,5-Dinitrobenzoates from Esters

Derivative for the alcohol fragment of an ester (Note 1).

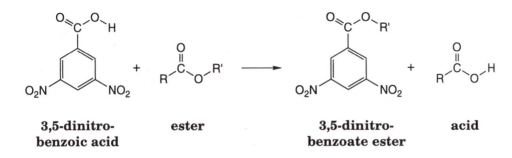

3,5-dinitro-   ester   3,5-dinitro-   acid
benzoic acid   benzoate ester

Procedure

Mix 0.5 gram of the ester with 0.5 gram of powdered 3,5-dinitrobenzoic acid, add a small drop of concentrated sulfuric acid, and heat the mixture. Heat under reflux if the ester boils below 150°C, and at 150°C in an oil bath if it boils above 150°C; heat until the 3,5-dinitrobenzoic acid dissolves and then for an additional 30 minutes. Cool the mixture, dissolve it in 30 mL of ether, and extract it twice with 5% sodium bicarbonate solution to remove 3,5-dinitrobenzoic acid and sulfuric acid (*caution:* foaming from CO_2 evolution). Finally, after washing the ether extract with water, evaporate it to dryness and recrystallize the residue by dissolving it in a minimum of hot ethanol, adding water to incipient cloudiness, and allowing the solution to cool. It may be necessary to induce crystallization by scratching. (Ref. 1, p. 273; Ref. 2, p. 292; Ref. 3, p. 393; Ref. 4, p. 300.)

Note

1. This procedure is not satisfactory for esters of alcohols that are unstable to concentrate sulfuric acid, such as tertiary alcohols or certain unsaturated alcohols. Higher-molecular-weight esters react slowly or not at all.

30.23 Hydrolysis of Esters

$$\underset{\text{ester}}{R-\overset{O}{\overset{||}{C}}-O-R'} \;+\; H-O^- \longrightarrow R-\overset{O}{\overset{||}{C}}-O^- \;+\; H-O-R$$

Procedure

Boil under reflux a mixture of 1 gram of the ester and 10 mL of 1 *M* sodium hydroxide until a clear solution has been obtained (Note 1).

Some of the solution can be used to identify the acid corresponding to the ester by means of Procedure B in Section 30.19 (Note 2). The remainder of the solution can be saturated with potassium carbonate to salt out the alcohol, which can then be extracted and identified by one of the procedures described in Sections 30.1-30.3. (Ref. 2, p. 287; Ref. 3, p. 391; Ref. 4, p. 293.)

Notes

1. If the alcohol that corresponds to the ester is not soluble in water, it will be present as an oil after hydrolysis.
2. If the acid is relatively insoluble in water, it can be isolated by acidification of the hydrolysate, followed by suction filtration.

30.24 Bromination of Aromatic Ethers

Derivative for aromatic ethers.

$$Ar—H \quad + \quad Br—Br \quad \longrightarrow \quad Ar—Br \quad + \quad H—Br$$

Procedure

Add 1 mmole of the aromatic ether to the calculated amount of a 1% (by volume) solution of bromine in acetic acid (Note 1) that has been cooled in an ice bath; continue cooling during the addition. After a few more minutes, remove the mixture from the ice bath and allow it to stand at room temperature for 15 minutes. Isolate the product by diluting the mixture with water and collecting the precipitate by suction filtration. Remove traces of bromine from the product by washing it with dilute sodium bisulfite solution and then water. it can be recrystallized from alcohol or alcohol/water. (Ref. 1, p. 276; Ref. 2, p. 297; Ref. 4, p. 308.)

Note

1. A solution of 1 mL of bromine in 100 mL of acetic acid will contain 1 millimole of bromine per 5.2 mL of solution. Depending on whether a mono-, di-, or tribromo derivative is expected, a 10-20% excess over 1, 2, or 3 millimoles of bromine should be used.

30.25 *S*-Alkylthiuronium Picrates

Derivative for primary and secondary alkyl chlorides, bromides, and iodides.

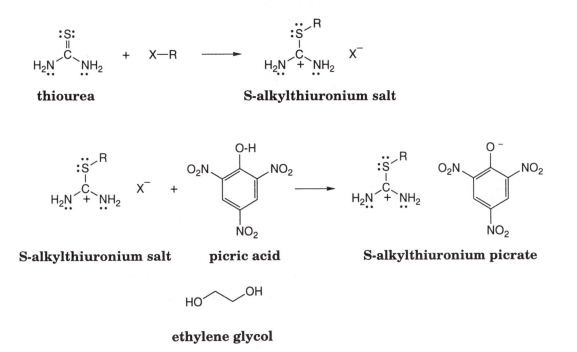

thiourea   S-alkylthiuronium salt

S-alkylthiuronium salt   picric acid   S-alkylthiuronium picrate

ethylene glycol

Procedure (Ref. 4)

Place in a 10-mL standard-taper Erlenmeyer flask 5 ML of ethylene glycol, 300 mg of thiourea, 6 drops of the alkyl halide, and two boiling stones. Fit the flask with a reflux condenser and clamp the flask in position in an oil bath that has been preheated to 117°C. After the flask has been held at this temperature for 30 minutes (Note 1), add down the condenser 1 mL of a saturated solution of picric acid in ethanol (Note 2). Keep the flask in the oil bath for another 15 minutes; then remove it, cool it to room temperature, and add 5 mL of water. Finally, cool the mixture in an ice bath for 15 minutes, with occasional swirling. Collect the *S*-alkylthiuronium picrate by suction filtration. The derivative can be recrystallized from methanol. [Ref. 1, p. 280; Ref. 2, p. 302; Ref. 3, p. 291; Ref. 4, p. 217. See also H. M. Crosby and J. B. Entrikin, *J. Chem. Educ.* **41**, 360 (1964).]

Notes

1. If decomposition occurs (odor of mercaptan; red color), repeat the experiment with the oil bath held at 65°C.
2. The solubility of picric acid is reported to be 1 gram per 12 mL of ethanol.

30.26 *o*-Aroylbenzoic Acids from Aromatic Hydrocarbons

Derivative for aromatic hydrocarbons (Note 1).

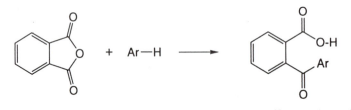

phthalic anhydride   ***o*-aroylbenzoic acid**

Procedure

Add 2.5 grams of powdered anhydrous aluminum chloride to a mixture of 0.5 gram of the hydrocarbon and 0.6 gram of phthalic anhydride in 2 or 3 mL of dry dichloromethane while the mixture is cooled in an ice bath. After the initial reaction has subsided, either allow the mixture to stand at room temperature or heat it under reflux until the evolution of hydrogen chloride has occurred (about half an hour); if the mixture has been heated, allow it to cool. Then add 5 grams of ice and 5 mL of concentrated hydrochloric acid. When the addition product has been hydrolyzed, collect the product by suction filtration and wash it well with water (Note 2). The aroylbenzoic acid should be purified by dissolving it by heating in 5 mL of concentrated sodium carbonate solution, boiling the solution with activated carbon for 5 minutes, filtering, and acidifying to a pH of 3 by the addition of 20% hydrochloric acid. The precipitated acid should then be collected by suction filtration, washed with water, and recrystallized from aqueous alcohol or (after drying) from toluene/petroleum ether. (Ref. 1, p. 285; Ref. 2, p. 519; Ref. 4, p. 235.)

Notes

1. Certain aryl halides can be characterized in this manner also.

2. If the product does not crystallize immediately, it should be allowed to stand overnight.

30.27 Aromatic Acids by Oxidation by Permanganate

Derivative for aromatic hydrocarbons and aryl bromides and chlorides.

$$Ar-R \xrightarrow{\text{KMnO}_4} Ar-\overset{\overset{\displaystyle O}{\|}}{C}-O\text{-}H$$

Procedure

Add 1 gram of the hydrocarbon to a solution of 3 grams of potassium permanganate and 1 gram of sodium carbonate in 75 mL of water, and heat the mixture under reflux until the permanganate color has disappeared (in 15 minutes to 4 hours; Note 1). Cool the solution to room temperature and acidify it cautiously by the addition of 50% sulfuric acid (Note 2). Remove the manganese dioxide by the addition of a concentrated solution of sodium bisulfite (while stirring well and possibly heating on the steam bath) and, after cooling the mixture thoroughly in an ice bath, collect the acid by suction filtration. It can be recrystallized from water. (Ref. 1, p. 285; Ref. 2, pp. 315, 326; Ref. 3, p. 520; Ref. 4, p. 253.)

Notes

1. Potassium permanganate can be detected by dipping a stirring rod into the mixture and touching the rod to a piece of filter paper; a pink color in the ring around the dark spot of managanese dioxide indicates the presence of permanganate.

2. 50% sulfuric acid can be prepared by cautiously pouring 5 mL concentrated sulfuric acid over 10 grams ice.

30.28 Anilides, *p*-Toluidides, and α-Naphthalides from Alkyl Halides

Derivatives for alkyl halides and aryl bromides and iodides.

$$R-X \ + \ Mg \xrightarrow{\text{dry ether}} R\text{-}Mg\text{-}X$$

Grignard reagent

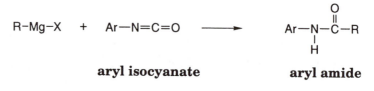

aryl isocyanate   **aryl amide**

Procedure

Preparation of the Grignard reagent. Add a solution of 1 mL of the halide in 5 mL of absolute ether to 0.4 gram of magnesium turnings and a crystal of iodine in a dry flask.

Preparation of the derivative. To the freshly prepared Grignard reagent, add a solution of 0.5 mL of the isocyante (phenyl-, *p*-tolyl-, or α-naphthylisocyanate) in 10 mL of absolute ether. After swirling the mixture and allowing it to stand for 10 minutes, add 20 grams of crushed ice and 1 mL of concentrated hydrochloric acid. After hydrolysis of the solid, separate the ether layer; then, after drying over anhydrous magnesium sulfate, evaporate the ether to give the derivative as a solid residue. The residual amide can be recrystallized from methanol, ethanol, or aqueous alcohol. (Ref. 1, p. 279; Ref. 3, p. 290; Ref. 4, p. 214.)

30.29 2,4,7-Trinitrofluorenone Adducts of Aromatic Hydrocarbons

Derivative for aromatic hydrocarbons.

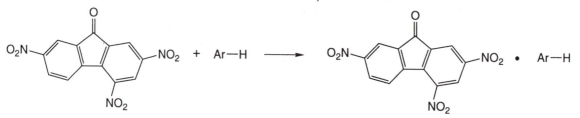

2,4,7-trinitrofluorenone   **2,4,7-trinitrofluorenone adduct**

Procedure

Dissolve 100 mg of 2,4,7-trinitrofluorenone (0.3 mmole) in a mixture of 10 mL of absolute methanol or ethanol and 2 mL of benzene. To this hot solution, add a solution of an equivalent amount (0.3

mmole) of the aromatic hydrocarbon in the minimum amount of 2:
1 alcohol: benzene. Heat the mixture for 30 seconds and then allow
it to cool for crystallization. Collect the product by suction filtration.
It can be recrystallized from absolute ethanol. (Ref. 2, p. 315.)

30.30 Bromination of Phenols

Derivative for phenols.

$$Ar-O-H \quad + \quad \xrightarrow{\text{bromine}} \quad \text{brominated phenol} \quad + \quad HBr$$

Procedure

Prepared a solution of 7.5 grams of potassium bromide and 5 grams
(1.6 mL; 31 mmoles) of bromine in 50 mL of water. Add this solution, dropwise, to a well-stirred solution of 0.5 gram of the phenol
dissolved in water, dioxane, ethanol, or acetone until a weak yellow
color persists. Precipitate the brominated phenol by stirring in 25
mL of cold water and collect the crude product by suction filtration.
Wash the product with dilute sodium bisulfite solution to remove
free bromine, and then recrystallize the solid from ethanol or aqueous alcohol. (Ref. 1, p. 298; Ref. 4, p. 354.)

30.31 Aryloxyacetic Acids from Phenols

Derivative for phenols.

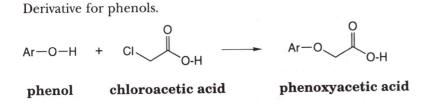

phenol   **chloroacetic acid**   **phenoxyacetic acid**

Procedure

Add 1.25 grams of monochloroacetic acid to a mixture of 1 gram of
the phenol and 4 mL of 10 M sodium hydroxide solution. Add 1 or 2
mL of water as necessary to give a homogeneous solution. Heat the
mixture for 1 hour on the steam bath, cool it to room temperature,
dilute it with 10 or 15 mL of water, and acidify it to a pH of 3 with

dilute hydrochloric acid. Extract the product with 50 mL of ether, wash the extract with 10 mL of water, and then extract the product from the ethereal solution by means of 25 mL of 5% aqueous sodium carbonate solution. Precipitate the product by acidifying the sodium carbonate solution with dilute hydrochloric acid (CO_2 evolution) and collect it by suction filtration. It can then be recrystallized from water. (Ref. 1, p. 298; Ref. 2, p. 331; Ref. 3, p. 682; Ref. 4, p. 354.)

Disposal

Sink: aqueous acids; aqueous bases; acetic acid; acetone; ethanol

Non-halogenated liquid organic waste: all non-aqueous solutions that contain no chlorine, bromine, or iodine.

Halogenated liquid organic waste: all non-aqueous solutions that do contain either chlorine, bromine, or iodine.

Non-halogenated solid organic waste: all solids that contain no chlorine, bromine, or iodine.

Halogenated soid organic waste: all solids that do contain either chlorine, bromine, or iodine.

References

1. R. L. Shriner, R. C. Fuson, and D. Y. Curtin, *The Systematic Identification of Organic Compounds*, 5th edition, Wiley, New York, 1965.

2. N. D. Cheronis and J. B. Entrikin, *Identification of Organic Compounds*, Interscience, New York, 1963.

3. A. I. Vogel, *A Textbook of Practical Organic Chemistry*, 3rd edition, Wiley, New York, 1957.

4. R. L. Shriner, R. C. Fuson, D. Y. Curtin, and T. C. Morrill, *The Systematic Identification of Organic Compounds*, 6th edition, Wiley, New York, 1980. This revision of the book cited in Reference 1 places a greater emphasis on physical methods of analysis. Its Tables of Derivatives are reproduced from Reference 1.

5. A. I. Vogel, *Vogel's Textbook of Practical Organic Chemistry*, 4th edition, revised by Furniss et al., Longman, London and New York, 1978. This revision of *Vogel* is less comprehensive than its predecessor, cited in Reference 3.

6. Z. Rappoport, *Handbook of Tables for Organic Compound Identification*, 3rd edition, The Chemical Rubber Co., Cleveland, Ohio, 1967.

Apparatus and Techniques For Chemical Reactions

A chemical reaction, in which a substance of one structure is converted to a substance of a different structure, typically includes

- combining the materials that are to react, usually with a solvent or a catalyst

- maintaining the mixture at a specified temperature for a certain length of time, with stirring if necessary, and then

- isolating the desired product by the separation procedures described in Sections 8 through 16.

The next sections describe some of the ways by which these various phases of the operation can be carried out in the laboratory. Industrial-scale and micro-scale reactions usually require different procedures.

31.　Assembling the Apparatus

The apparatus in which a reaction is to be carried out can be as simple as a single test tube or flask. More often, however, it is made up of several glass components, such as a boiling flask and condenser, plus sometimes a drying tube to exclude moisture, a separatory funnel by which a liquid reagent or solution can be added, a trap to absorb harmful gases that may be evolved, or any of a number of other parts.

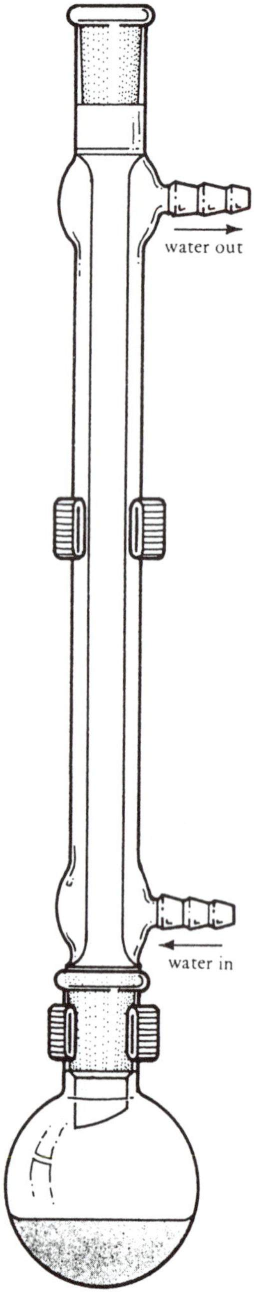

Semi-micro-scale reactions

Figures 31-1 and 31-2 illustrate two of the assemblies that are most often used for carrying out reactions on a semi-micro or larger scale. The simple arrangement, shown in Figure 31-1, is used when a mixture is to be boiled for a while. The condenser, arranged to return the condensed vapors to the reaction flask, ensures that the material in the flask is not lost while the mixture is boiled.

The more elaborate apparatus shown in Figure 31-2 is used when you want to add a liquid reagent or a solution to a boiling reaction mixture. The Claisen adapter, in effect, converts a one-necked flask to a two-necked flask.

There is no doubt that the nicest and most convenient components from which the apparatus can be assembled are those that can be connected by ground-glass joints. The joints should normally be put together without a lubricant. Exceptional situations would be when concentrated solutions of strong bases are used, or when the apparatus must hold a good vacuum, as during a distillation under reduced pressure.

As indicated in Figures 31-1 and 31-2, the weight of the apparatus should be supported by clamps so that the tendency to twist is minimized.

Micro-scale reactions

Micro-scale reactions can be done in micro-scale versions of the kind of glassware shown in Figures 31-1 and 31-2. On the other hand, many micro-scale reactions can be run in a test tube, an inexpensive alternative to component glassware. Because only a small amount of material is used in a micro-scale reaction, the walls of the test tube will often suffice as a condenser, especially when the heating period is short.

If a longer heating period is called for, or if the solvent is especially volatile, a second test tube can be suspended inside the first to act as a condenser, as indicated in Figure 31-3. The second, inner, test tube can be filled with cold water or ice, as required. When the ice melts and the water gets warm, the water can be removed by means of a Pasteur pipette and replaced by more cold water or ice.

Figure 31-1. Apparatus for carrying out a reaction under reflux.

water out

water in

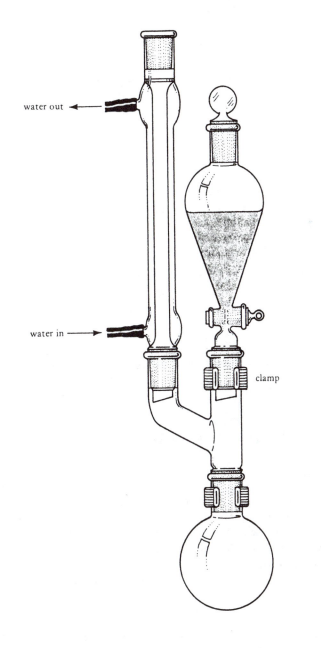

water out

water in

clamp

Figure 31-2. Apparatus that allows heating under reflux and addition of a liquid reagent or solution.

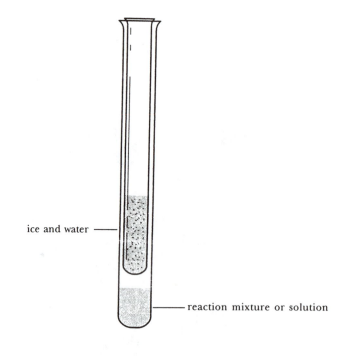

ice and water ——

—— reaction mixture or solution

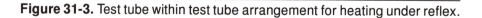

Figure 31-3. Test tube within test tube arrangement for heating under reflex.

32. Temperature Control

Control the rate

A reaction should be carried out at a definite and constant temperature for two reasons. The first reason is that the rate of a reaction increases sharply with increasing temperature, and a reaction will proceed at a convenient or desirable rate only within a relatively narrow temperature range.

The rate increases with temperature according to the relationship

$$\text{Rate} \propto e^{-\Delta G/RT}$$

where ΔG is the free energy of activation of the reaction, R is the gas constant, and T is the absolute temperature.

The exponential dependence on temperature means that a small change in temperature will give a large change in rate. For example, if the free energy of activation is such that the reaction will be half done in 1 minute at 25°C ($T = 298$ K), running the reaction at 35°C ($T = 308$ K) will decrease the half time for the reaction to about 20 seconds, and at 50°C ($T = 323$ K), the half time will be only 5 seconds. In general, an increase in reaction temperature of 10 degrees will make the reaction go between 2 and 3 times faster.

Minimize side reactions

The second reason is that the rates of competing reactions, or *side reactions*, increase faster with an increase in temperature than the rate of the desired reaction. This happens because the free energies of activation of the side reactions are greater than the free energy of reaction of the desired reaction. Thus, the fraction of side products will be greater at higher temperatures. The choice of the optimum temperature at which to carry out a reaction, therefore, always represents a compromise between a desire to carry out the reaction as fast as possible and a desire to minimize side reactions. The exponential dependence of rate on temperature makes it necessary to control the temperature over a narrow range.

Heat at the boiling point of the solvent

One of the easiest ways to maintain a reaction mixture at an approximately constant temperature is to heat the mixture until it boils, using a condenser to condense the vapor, and to return the condensate to the flask (to heat under reflux). As long as the composition of the mixture does not change, the boiling point will remain constant. Because the composition must change as the reaction progresses, an excess of one of the reactants can be used as a solvent so that the boiling point will change less during the course of the reaction. Sometimes the solvent is chosen so that the mixture will boil at the desired temperature. This is the usual procedure if the reactants are solids.

Heat under reflux

The reaction mixture can be boiled by heating the flask with a flame, an oil, steam, or sand bath, or a heating mantle. These methods of heating are described in Section 33.1.

Control of highly exothermic reactions

A highly exothermic reaction presents an additional problem because, as the rate of a reaction increases, the rate of heat evolution will increase. If the rate of heat evolution is greater than the rate at which heat can be removed by condensation of the vapors of the boiling mixture and all other cooling methods, the reaction will proceed at an ever- increasing rate as the temperature of the mixture rises until it boils over or blows out the condenser.

By rate of addition of a reactant. One way to control a highly exothermic reaction is to add one of the reactants to the mixture at such a rate that the reaction can be maintained at the desired temperature (the boiling point or below) by the heat evolved in the reaction. The heat of the reaction may be the only source of heat, or, if the reaction is only moderately exothermic or if the temperature is relatively high, an external source of heat can be used as well. More often it is necessary to remove heat by external cooling. External cooling will always be necessary if the temperature is to be kept below the boiling point of the mixture. When the mixture is not to be allowed to boil, a cooling bath must be used, and the rate of addition of the reactant will be determined by how efficiently heat can be removed from the mixture. Different types of cooling baths are described in Section 33.2.

> **Warning:** *The danger of this method of controlling* **exothermic reactions** *is that the reactant may unintentionally be added faster than it reacts. It is especially easy to add too much reactant at the beginning, when the temperature may be too low for the reaction to proceed at a reasonable rate. If too much reactant accumulates in the mixture, the reaction may accelerate out of control when it finally starts. The only way to be sure the reaction is under way and under control is to determine that the temperature rises when the reactant is added and starts to fall when the addition is stopped or slowed down.*

By micro-scale reactions. The second method of controlling an exothermic reaction is to run the reaction in small batches, or on a small scale, in a large flask so that a large surface area is available for efficient cooling. When the temperature of the reaction starts to climb, the flask can be plunged into a cold bath and swirled for a short time; remove it when the temperature rise has been checked. The use of a large amount of solvent and a still larger flask, so that the mixture can be cooled efficiently in a cold bath, can serve to make the reaction more easily controlled, both by decreasing the rate of bimolecular reactions by dilution, and by increasing the total heat capacity of the mixture. This second method requires a little more judgment and skill on the part of the experimenter, and it is dangerous if large amounts are used.

By trying new reactions on a small scale. From this discussion of exothermic reactions, it should be obvious that new or unfamiliar reactions should be carried out on a small scale until their exothermicity has been determined. Another, less obvious consequence is that, as a reaction is scaled up, more attention must be paid to the problem of heat transfer. If the volume of the reaction, and thus of the apparatus, is increased by a factor of 8, the surface area of the apparatus (through which the heat must flow) is increased by a factor of only $8^{2/3} = 4$. Whereas a cooling bath might not have been needed for the small-scale reaction, it might be required for the large-scale reaction.

33. Methods of Heating and Cooling

The typical chemical reaction is not carried out at room temperature. The next two sections describe methods by which reaction mixtures can be heated or cooled.

33.1 Heating

Heating with gas: the bunsen burner

A traditional source of heat in the organic laboratory is a Bunsen burner flame. The height of the flame and the distance of the flask from the flame control the rate of heating (see Figure 33.1-1).
The advantages of using a burner as a source of heat are that it can be set up very quickly and easily and that the mixture can be brought up to temperature very rapidly. However, a burner has three important disadvantages. The first is that it must not be left unattended. It might go out, creating a gas leak that could lead to an explosion, or the height of the flame might change, providing either too little heat to maintain boiling or so much heat that the mixture boils over and catches fire. The second disadvantage is that if the flask should crack or break, or if the reaction mixture should boil over, the burner could also set the mixture on fire. The third disadvantage is that a burner flame is an exceedingly uneven source of heat. The glass at the bottom of the flask will be very much hotter than the rest of the flask, and undesired reactions may occur. If there is a solid in the reaction mixture, boiling may be uneven with lots of bumping. However, if relatively small amounts are involved, if the heating period is short, and if the mixture will be watched, heating with a flame will often be quite satisfactory.

Heating with electricity

The heating mantle. Three methods of electrical heating are often used. In the first, an electrically heated jacket, or *heating mantle*, is placed around the lower half of the flask; sometimes a second part is used around the upper half also. Figure 33.1-2 illustrates the use of a

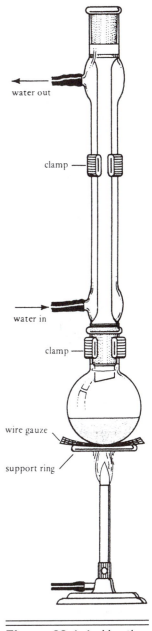

Figure 33.1-1. Heating with a burner flame. Rate of heating is determined by the size of the flame and the height of the flask above the flame. The wire gauze moderates and spreads the heat of the flame.

heating mantle. The rate of heating is controlled by the voltage supplied to the mantle from the variable voltage transformer. A similar device, the Thermowell heater, does not require a variable voltage transformer. The rate of heating of the Thermowell device is set by a proportional controller, which determines the fraction of the time that current is allowed to flow through the heater.

The advantages of this method are that the apparatus can more safely be left unattended, it is a very constant and easily controlled source of heat, and it is not nearly as likely to result in ignition of flammable materials as is a flame.

There are several disadvantages, however. One is that the glass of the flask is likely to be at a considerably higher temperature than the contents of the flask. As in heating with a flame, this can cause local superheating and possible decomposition, as well as uneven boiling. The temperature of the mantle must be determined by a thermocouple, which is somewhat inconvenient. Thus, most chemists use a heating mantle without bothering to determine its temperature. Also, the level of the liquid in the flask should not fall below the top of the mantle. If it does, the glass between the top of the liquid and the top of the mantle will become much hotter than the rest of the flask. Material that splashes onto this part of the flask will be superheated. For this reason, heating with a mantle is not entirely satisfactory for a distillation in which the pot will end up nearly empty. Another disadvantage is that a heating mantle takes quite a long time to warm up and reach thermal equilibrium, perhaps an hour or more. Also, if you wish to turn off the heat in a hurry, it is not enough to turn off the electricity, because the mantle is hotter than the flask and has a high heat capacity. The mantle itself must be removed. For this reason, you should never set up an apparatus using a heating mantle with the mantle resting on the desk top. An additional disadvantage is that the mantle must approximately fit the flask, and a different one may be required for each size of flask. You can buy at least a dozen burners for the price of a mantle and transformer. However, for a large-scale reaction, a heating mantle is more satisfactory than any other apparatus. A heating mantle will also be most useful when you will have to run the same reaction a number of times, and a relatively high temperature or long reaction time is required.

The immersion heater and oil bath. A second method of electrical heating uses an *immersion heater,* supplied by a variable autotransformer, in a bath of a liquid heat-transfer medium, or *"oil bath."* The immersion heater can be of the commercially available Calrod type, or can be easily made from a power resistor, a piece of lamp cord, and a plug (see Figure 33.1-3). The bath liquid can be a mixture of hydrocarbons (mineral oil), a silicone oil, or a polyethylene glycol fraction that freezes just below room temperature (Carbowax 600). The latter is most useful because it is inexpensive, can be used at

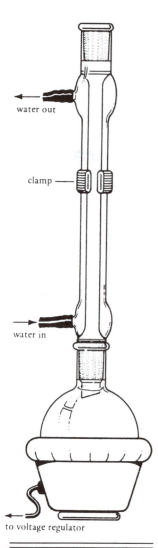

water out

clamp

water in

to voltage regulator

Figure 33.1-2. Heating with a heating mantle. The mantle should be supported on an iron ring or other removable support.

temperatures up to about 175 to 180°C without appreciable decomposition, and has the great advantage of being completely soluble in water, which makes cleaning up much easier. The disadvantage of polyethylene glycol is that it is somewhat hygroscopic and, if exposed for a long time to high humidity, it can boil below the desired temperature; the water will boil out with continued heating.

The advantages of an electrically heated oil bath are that the temperature of the bath (and thus of the contents of the flask) can be determined by a thermometer in the bath; no superheating or local decomposition is possible; and, when a power resistor is used as the heating element, the heat capacity of the heater is so small that the rate of heating can be changed almost instantly by adjusting the voltage from the variable transformer. The electrically heated oil bath has the advantage over the heating mantle that it can be rapidly brought up to temperature, the rate of heating is more quickly and easily adjusted, the contents of the flask are more visible, and a magnetic stirrer can easily be used (Section 34) with it.

The disadvantages of the oil bath are that it is a nuisance to store, it starts to smoke at elevated temperatures (about 150°C for mineral oil, acrid smell; about 180°C for polyethylene glycol, burnt sugar smell), and it presents the danger of spilling hot oil if the container is tipped or broken during use. For small flasks (up to about 500 mL), at temperatures below 150 to180°C, and for short reaction times, an electrically heated oil bath is most useful. For the constant and exact heat control required for a distillation, the electrically heated oil bath is unsurpassed. Of course, the immersion heater, like the heating mantle, requires a variable autotransformer.

The hot plate. A third method of electrical heating makes use of an *electric hot plate.* The hot plate tends to be rather inefficient because of poor contact between the top of the hot plate and the bottom of the flask, and it takes a while to warm up. Because the surface of the hot plate will be much hotter than the contents of the flask, local superheating is possible, just as in heating with a flame.

The hot plate with oil bath. An oil bath heated on a hot plate avoids the problem of superheating and generally provides a much more even and constant source of heat for the flask. The temperature of the bath can be determined with a thermometer.

The stirring hot plate. Occasionally a mixture needs to be both stirred and heated for a long time. In these instances, a *stirring hot plate,* described in Section 34, can be most suitable.

Heating with steam: the steam bath

A steam bath, illustrated in Figure 33.1-4, is often used to heat a solution that boils below about 90°C, or to heat a mixture to approxi-

Advantages of an oil bath

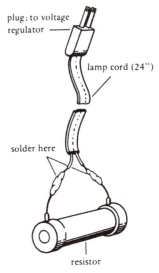

plug; to voltage regulator

lamp cord (24")

solder here

resistor

Figure 33.1-3. An inexpensive immersion heater made from a 125-ohm 5-watt resistor, lamp cord, and plug.

mately 100°C. Most laboratories are supplied with steam from a central boiler; when this is the case, use of a steam bath eliminates the hazards of a flame. The steam bath heats the contents of a flask rapidly, but it has the disadvantage of being good for only one temperature, approximately 100°C.

Heating with hot water

A useful device for heating at temperatures below 100°C is a hot water bath, which can be simply a beaker or steam bath full of hot

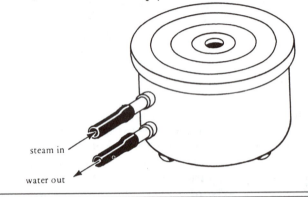

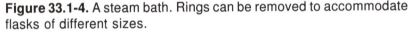

Figure 33.1-4. A steam bath. Rings can be removed to accommodate flasks of different sizes.

water. Sometimes hot water from the faucet will be hot enough; sometimes it may be necessary to heat and reheat the water in the beaker or steam bath with a flame. A steam bath or hot water bath should be used with flammable substances whenever possible.

Heating with hot sand: the sand bath

A method for heating that is quite convenient on the small scale is a dish of hot sand, or *sand bath*. The sand can be heated in a dish on a hot plate, or sand can be heated in a heating mantle. When the sand is hot, the flask to be heated is pushed down into the sand. The sand at the bottom will be hotter than the sand at the top, and the rate of heating and the temperature can be adjusted by raising or lowering the flask in the sand bath. Unlike a steam bath or a water bath, a sand bath can provide temperatures that range from near room temperature at the surface up to 200°C and above well down into the sand.

A sand bath warms up much more slowly than a steam bath; you should turn on the sand bath before you need it.

33.2 Cooling

An ice bath

Depending on the temperature of the flask to be cooled, a cooling bath may be simply a beaker of water or a specially prepared cold bath. For cooling, exposure to the air is very inefficient compared with immersion in a cold liquid, because a liquid has a much higher heat capacity. For cooling below room temperature and down to about 0°C, a mixture of ice and water can be used (Figure 33.2-1). It is

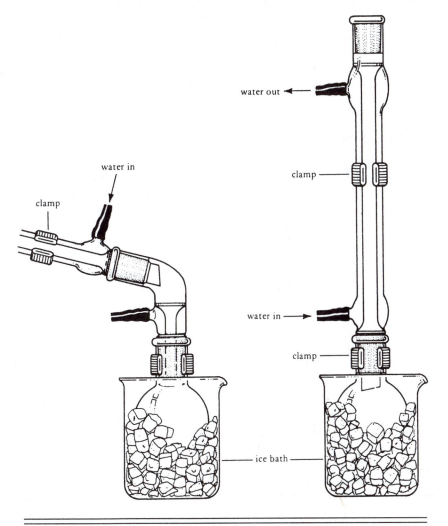

Figure 33.2-1. Use of an ice bath.

much more satisfactory for the ice to be in small pieces than as ice cubes. For cooling to a few degrees below zero, or for more efficient cooling to 0°C, a freezing mixture can be used. This mixture is made by thoroughly mixing 1 part sodium chloride and 3 parts of snow or finely chopped ice (by weight). Ice/salt baths, however, are fairly short-lived and inefficient.

A dry ice/acetone bath

For efficient cooling for extended periods, a Dry Ice/acetone bath can be used. This is prepared by adding lumps of Dry Ice (do not handle with bare hands) to acetone until the desired temperature is reached. The temperature can be maintained at any given level by adding Dry Ice as needed. Do not add the Dry Ice too rapidly, or the carbon dioxide that is evolved will cause the bath to overflow. The ultimate temperature that can be reached with such a bath is about −78°C; use an alcohol or toluene thermometer, as mercury freezes at −40°C.

A crystallizing dish is the most conveniently shaped container for hot or cold baths, because it has a flat bottom and straight sides, is wider than it is tall, and comes in several sizes. Don't forget that it is glass and can be broken. The life of a low-temperature bath can be greatly extended by using a Dewar flask (like a Thermos bottle) as a container, or by making the bath in a beaker that is nested inside a large beaker with a folded towel in between for insulation.

Efficient cooling   For most efficient cooling, the reaction mixture or solution should be in a large flask, to provide a large area of cold surface, and the mixture should be swirled or stirred continuously.

34.   Stirring

"Swirling"

A reaction mixture usually must be stirred. Stirring serves to mix two phases, to mix a reagent as it is gradually added, or to promote efficient heat transfer to a cooling bath or from a heating bath, thus providing for a constant temperature throughout the mixture. If the reaction time is short, swirling the flask by hand may be an acceptable method of mixing. If the mixture is homogeneous and the reaction is neither very endothermic nor very exothermic, so that the transfer of heat does not have to be very efficient, occasional swirling in the heating or cooling bath may be adequate. If constant mixing or stirring is required for more than a short time, however, a stirring mechanism is almost a necessity. The boiling action of a vigorously boiling mixture can often provide adequate mixing, but boiling will be smoother and superheating can be minimized if the mixture is stirred.

Mechanical stirring

Two stirring mechanisms are commonly used. The first makes use of a paddle or wire agitator, which extends on a shaft into the mixture from above the reaction vessel. The shaft is usually a glass sleeve with a ground joint that fits into the center ground glass–jointed neck of the flask. The vigor of the stirring is determined by the speed of the motor, which is governed by a rheostat. Figure 34-1 illustrates a *mechanical stirrer.*

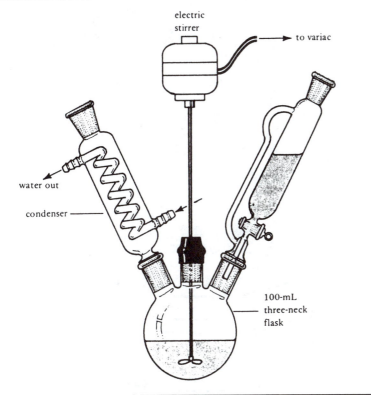

electric stirrer

to variac

water out

condenser

100-mL three-neck flask

Figure 34-1. Three-necked boiling flask set up with condenser, mechanical stirrer, and addition funnel.

Magnetic stirring

The second method makes use of a glass- or Teflon-covered bar magnet, the *stirring bar,* which is placed in the reaction flask. The flask is positioned over a motor, which turns a magnet fastened to its shaft, a *magnetic stirrer. A*s the motor-driven magnet turns, the magnet inside the flask turns and stirs the mixture. Simultaneous heating and stirring are often carried out by use of a combination magnetic stirrer–hot plate. If the flask is heated by the hot plate via an oil bath, the magnetic stirrer can be used to stir both the bath and the mixture. Figure 34-2 shows another arrangement in which a magnetic stirrer is used to stir both an electrically heated oil bath and the contents of a reaction flask. The magnet in the bath should be larger than the one in the flask.

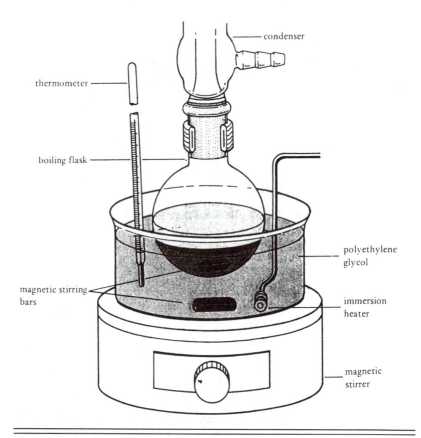

Figure 34-2. Single-necked boiling flask set up with condenser, magnetic stirrer, and oil bath.

Direct mechanical stirring has an advantage over magnetic stirring in that mechanical stirring can provide the greater torque necessary to stir suspended solids or viscous mixtures or very large volumes. The mechanical stirrer can also be used when heating with a flame or with a heating mantle, since the stirrer enters from the top of the flask. On the other hand, a magnetic stirrer is much easier to set up, and is generally much more convenient if the mixture needs only to be stirred and not heated or cooled at the same time. It is most useful when used in conjunction with an electrically heated oil bath, as the stirring motor can stir both the bath and the reaction mixture. This combination is especially suited for heating and stirring during a distillation, particularly a vacuum distillation. Use of a small magnetic stirrer in the pot will essentially eliminate superheating and bumping.

Magnetic stirring is somewhat more convenient when the apparatus is not at atmospheric pressure, or when an inert atmosphere is being used, as magnetic stirring does not require a flask opening. Both types of stirring can be used when a heating or cooling bath is being employed, but with the magnetic stirrer, the flask must be positioned near the bottom of the bath, and the bath container must be made of nonferrous material.

Small air or water turbine-driven magnetic stirring motors are also available. These are useful for stirring "under water" in a constant-temperature bath.

Oil bath and magnetic stirrer

35. Addition of Reagents

Many chemical reactions are carried out by combining the reagents (and solvent and catalyst, if necessary) all at once and keeping the mixture for a length of time at a certain temperature. Sometimes, however, it is necessary to add one or more reagents gradually over a period of time. This would be done, for example, if the reaction were strongly exothermic and the rate could be controlled easily only by adding one reagent a little at a time. Or it would be done if one reagent underwent an undesirable bimolecular reaction with itself; if this reagent were added slowly to an excess of the other materials to keep its concentration low, the rate of its bimolecular reaction with itself would be decreased relative to its bimolecular reaction with another substance. A gaseous reagent must usually be added slowly because of limited solubility; after the solution is saturated, it can be added no faster than it can be consumed.

Addition of solids

If a solid must be added gradually, it is usually added as a solution. If you must add a solid, portions can often be dropped right into the flask or down the condenser; sometimes it will be necessary to poke a length of glass rod down the condenser to dislodge material that has stuck to the walls.

Addition of liquids and solutions

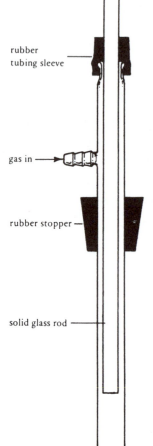

Liquids or solutions can be added by means of a separatory funnel, as shown in Figure 31-2. If it is necessary to add a liquid to a closed system, as when an inert atmosphere is being used, a pressure-equalizing addition funnel is required, as illustrated in Figure 34-1. The rate of addition is controlled by means of the stopcock. When the stopcock is adjusted for a very low rate of flow, it often becomes clogged and you must open it a bit and then reset it for the low rate of addition.

Addition of gases

A gas can be added to a reaction mixture through a tube that dips below the surface of the liquid. If the gas is only slightly soluble, it must be passed in only as quickly as it is consumed, in order not to waste it. The solubility of a gas decreases with increasing temperature, and stirring promotes more rapid solution. The gas should not be so soluble or reactive that the solution is sucked up the tube, and it should not react to produce a solid that will clog the addition tube.

If a gas is very soluble or very reactive, the addition tube must be open-ended and of wide bore so that the solution cannot be drawn up into it very far before the level falls below the end of the tube. Alternatively, the gas can be led just to the surface of the mixture in the vortex formed by the stirrer.

If the product of the reaction with the gas is a solid that tends to clog the tube, as in the treatment of a Grignard reagent with carbon dioxide, a solid rod can be incorporated into the wide-bore addition tube, as shown in Figure 35-1, so that the solid can be dislodged from time to time by pushing the rod down the addition tube.

Measuring gases. If the weight of the gas added is large enough, compared with the weight of the cylinder, the amount added can be determined by the weight loss of the cylinder. A lecture bottle is a good source of gas, from this point of view. Sometimes it may be possible to pass the gas in at a steady rate. If so, the use of a flowmeter

rubber tubing sleeve

gas in →

rubber stopper —

solid glass rod —

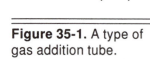

Figure 35-1. A type of gas addition tube.

in the line will allow the amount of gas added to be determined from the length of time of addition. If the gas is not too low-boiling, it can be passed into a tube cooled in a Dry Ice/acetone bath and measured by volume when the density of the liquid at the bath temperature can be estimated. When the desired volume has been condensed, the gas can be passed into the reaction mixture as it vaporizes. Experiment E102 describes this procedure for a chlorination reaction. Of course, if an excess of an inexpensive gas is not objectionable, the amount added need be estimated only very roughly.

36. Control of Evolved Gases

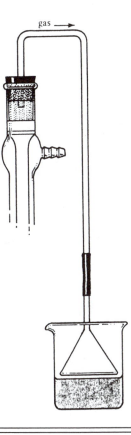

Use the hood

Reactions in which a poisonous or irritating gas is formed are best carried out in a ventilated enclosure, or hood.

Absorb the gas

If hood space is limited, it is possible, when the gas is readily soluble in water and not too poisonous, to set up or modify the apparatus in such a manner that *the gas can be absorbed,* thus preventing it from escaping into the atmosphere. One way to do this is to lead the gas from the apparatus to a beaker of water or an appropriate aqueous solution, where the gas will dissolve rather than escape into the room, as illustrated in Figure 36-1. This method is satisfactory if the rate of gas evolution is not too great and if small amounts are involved. The capacity and efficiency of this system can be increased by using a dilute sodium hydroxide solution (for acidic gases such as HCl or HBr) or a dilute solution of sulfuric acid (for basic gases such as ammonia) in the beaker.

Use the aspirator

A "T" tube fit to the top of the condenser and connected to the vacuum produced by the water aspirator provides another way to dispose of a water-soluble gas. This arrangement is illustrated in Figure 36-2. Although easy to set up, this method uses a lot of water.

    Of course, gases that are not soluble in water, such as carbon monoxide, cannot be controlled by either of these methods.

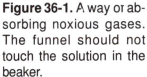

Figure 36-1. A way of absorbing noxious gases. The funnel should not touch the solution in the beaker.

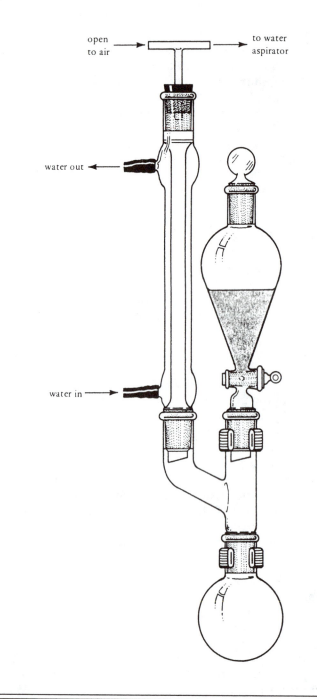

Figure 36-2. Another way to control water-soluble noxious gases. A vacuum adapter can be used in place of the cork and "T" tube.

37. Concentration; Evaporation

To leave a nonvolatile solid or liquid residue, it is often necessary to either concentrate a solution or completely remove the solvent. This is most often done to recover material after an extraction, to evaporate chromatographic fractions, or to concentrate the filtrate after recrystallization.

Removal of solvent by distillation

Possibly the most often used method of concentration or evaporation is to remove the solvent by *distillation*. If the source of heat is a steam or oil bath, this method is entirely satisfactory for either partial or complete removal of the solvent.

Complete removal of solvent by distillation requires two steps. In the first step, the solution is boiled at a bath temperature of 25 to 50°C above the boiling point of the solvent until distillation has slowed almost to a complete stop. In the second step, the residue is heated under vacuum, usually at the same temperature at which the bulk of the solvent was removed, until the remainder of the solvent is gone, as illustrated in Figure 37-1. Evaporation can be judged to be complete when the residue shows no appreciable loss in weight between successive periods of heating under vacuum. If only a small amount of a volatile solvent is present, it can be removed by distillation, using an aspirator to remove the vapors, as illustrated in Figure 37-2.

Beware of frying your product. Although you can concentrate a solution by distillation using a flame or heating mantle as a source of heat, you should not attempt to completely evaporate a solution by heating by either of these methods. If you attempt to remove the last traces of solvent by heating with a flame, you run the risk of decomposing the residue when it becomes viscous, or when a solid separates after the solution becomes more concentrated.

If a large volume of solvent is to be evaporated to give a small residue, it may be appropriate to fit a small flask with a separatory funnel so that solution can be added as the distillation progresses. In this way, the small residue will end up in a small flask, and mechanical losses can be minimized. Or, the bulk of the solvent can be removed using a large flask and the remainder of the solution can be transferred, with rinsing, to a small flask for completion of the evaporation.

Removal of solvent by evaporation in the hood

If large amounts of a volatile solvent must be evaporated, it may be convenient to place the solution in a large shallow dish (evaporating dish) and keep the dish in the hood with the draft on until evaporation is complete. This method is slower, but it requires no attention.

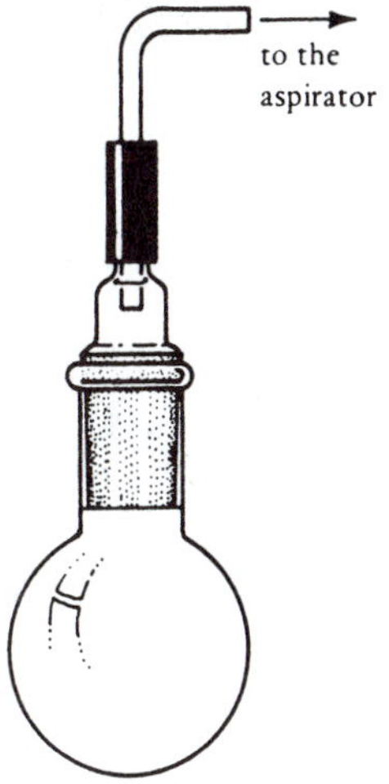

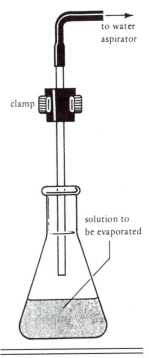

Figure 37-1. Removal of a solvent under vacuum. Heat the flask in a beaker of hot water or on the steam bath.

Figure 37-2. A way to rapidly concentrate a solution. You can heat the flask on the steam bath.

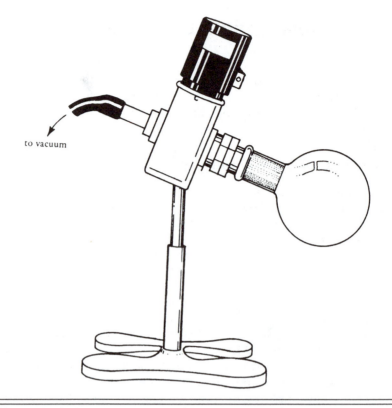

Figure 37-3. A rotary evaporator. You can heat the flask with warm water or with steam.

Removal of solvent with the "rotary evaporator"

It is possible to evaporate solutions very rapidly by using a rotary evaporator, a device that can apply an aspirator vacuum to a flask while rotating it in a heating bath (see Figure 37-3). Rotation of the flask minimizes superheating and bumping by keeping the contents well mixed and provides a large surface area for evaporation by constantly rewetting the walls of the flask.

38.   Use of an Inert Atmosphere

Keep out water

Certain reactions, for instance the Grignard reaction, are better carried out in the absence of oxygen or water vapor because these substances either take part in or catalyze undesired side reactions. If the acceptable levels of oxygen and water vapor are not too low, simply flushing out the air of the reaction flask with dry nitrogen and maintaining a slight positive pressure of nitrogen will be adequate. Figure 38-1 illustrates the essential features of an arrangement to do this: the tank of nitrogen connected to the apparatus through a wash bottle

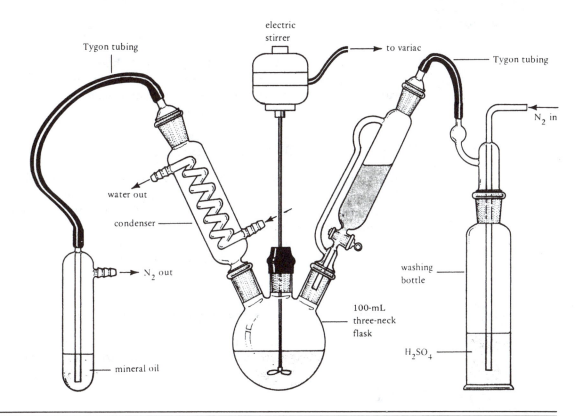

Figure 38-1. An apparatus for carrying out a reaction in an atmosphere of nitrogen. If magnetic stirring is used rather than the mechanical stirring shown, a three-necked flask is not needed. Instead, a single-necked flask with a Claisen adapter can be used. The condenser will go in one neck of the adapter, and the pressure-equalizing addition funnel in the other.

for removal of traces of water, and the bubbler in the exit line to maintain the slight positive pressure and to indicate the flow rate. The apparatus is flushed initially at a moderate flow rate and is then maintained at a slight positive pressure throughout the reaction at a minimum flow rate.

If the acceptable levels of water or oxygen are rather low, usually because their action is catalytic, much more elaborate procedures must be used; these might include working on a vacuum line, recrystallization and redistillation of the reactants and solvents under nitrogen, and heating and pumping out of the reaction vessels.

39. Working Up the Reaction; Isolation of the Product

After the reaction has taken place, the product must be separated from the reaction mixture by one or more of the separation procedures described in Sections 8 through 16. Although each reaction mixture presents a unique problem in separation, most reactions can be "worked up", which means the product can be isolated according to one of a few general schemes.

Isolation of a solid by suction filtration

Insoluble solid product

If the product is a solid, it may crystallize or separate spontaneously from the reaction mixture. When this happens, the solid need only be collected by suction filtration and washed with the appropriate solvents. Depending on the degree of purity required and on the nature of any impurities, the solid can be further purified by recrystallization, sublimation, or some other method, perhaps column chromatography.

Isolation of a liquid by fractional distillation

If the product is a liquid, it can occasionally be isolated simply by fractional distillation of the entire reaction mixture. The initial product fraction can then be further purified, if necessary, by redistillation.

The standard "aqueous workup"

The most common workup will call for the addition of water, or water and ice, to the reaction mixture, or vice versa.

Aqueous workup and isolation of a solid. If the addition of water produces a water-insoluble solid, the solid can then be collected by suction filtration, washed with appropriate solvents, dried, and recrystallized. As an alternative, the solid can be extracted with ether or dichloromethane, and the solid then recovered from the organic solution in the same way that a liquid would be recovered, as explained next.

Neutral liquid or solid product: see flow diagram, Figure 39-1

Aqueous workup and isolation of a neutral liquid or solid. If the product is a neutral substance, an organic solvent such as ether or dichloromethane can be added to dissolve it. The resulting ethereal or dichloromethane solution is then separated from the aqueous phase, using the separatory funnel (Section 14.3).

The solution of the neutral substance in the organic solvent is next washed with portions of either aqueous base (such as dilute sodium hydroxide or dilute sodium bicarbonate) or aqueous acid (such as dilute hydrochloric acid or dilute sulfuric acid). If the organic solution is washed with both acid and base, the washings must be done in succession. Usually, the order of washing with both acid and base is not important. Washing with base will remove acidic impurities, and washing with acid will eliminate basic contaminants, as described in Section 14.2. A final wash with water will then remove traces of the previous aqueous solution.

The organic solution of the neutral substance is then dried over a suitable drying agent (Section 16.2), filtered by gravity to remove the hydrated drying agent (Section 8.1), and finally evaporated to remove the solvent and to leave the neutral organic substance as a residue.

If the residue is a solid, it can be recrystallized. If the residue is a liquid, it can be fractionally distilled. These operations are summarized in the flow diagram in Figure 39-1.

See Figure 39-1

Aqueous workup and isolation of an acidic liquid or solid. If the product is a water-insoluble acid, a typical aqueous workup will involve extraction from the organic solvent into a basic aqueous solution such as dilute sodium hydroxide. After the organic phase has been removed by using the separatory funnel (Section 14.3), the basic aqueous solution of the salt of the acid is acidified, and the free acid is then extracted into a fresh portion of the organic solvent. After the organic solution of the free acid has been separated from the aqueous layer, the organic solution is washed with a small portion of water (to remove traces of the previous aqueous solution) and dried over a suitable drying agent (Section 14.2). The hydrated drying agent is then removed by gravity filtration, and the organic solution is evaporated to remove the solvent if the product is a solid, or distilled if the product is a liquid. This procedure is summarized in the flow diagram of Figure 39-2.

Acidic liquid or solid product: see flow diagram, Figure 39-2

If the product to be purified is a water-insoluble base, an aqueous workup may involve extraction of the base from the organic solvent into aqueous acid, basification of the aqueous solution, and reextraction by a fresh portion of the organic solvent. After the organic solution is dried, the product can be isolated either by evaporation or fractional distillation. These operations are summarized in the flow diagram of Figure 39-3.

Basic liquid or solid product: See flow diagram, Figure 39-3

Questions

1. In the preparation of methyl benzoate by the first procedure in Section E36, the ethereal solution of the crude product is washed first with water and then with aqueous sodium carbonate solution. What is removed, and why, in each extraction?

2. In preparing *N,N*-diethyl-*m*-toluamide ("Off") from *m*-toluic acid and diethylamine, described in Section E43, the toluene solution of the crude product is extracted first with dilute sodium hydroxide solution and then with dilute hydrochloric acid solution. Explain how these extractions remove unreacted starting materials.

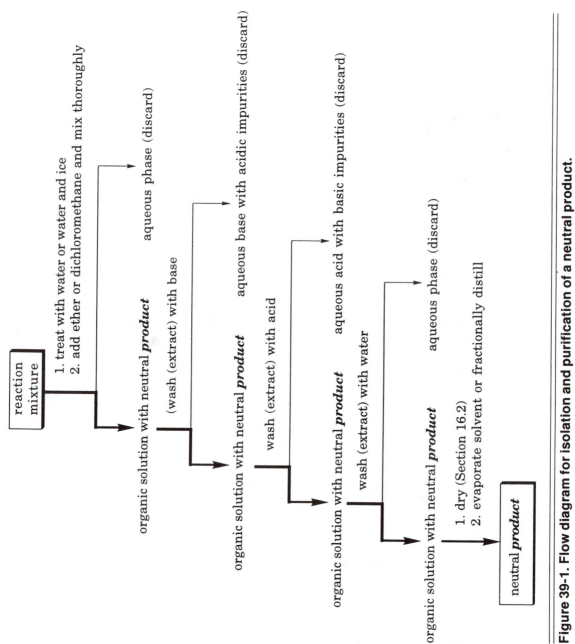

Figure 39-1. Flow diagram for isolation and purification of a neutral product.

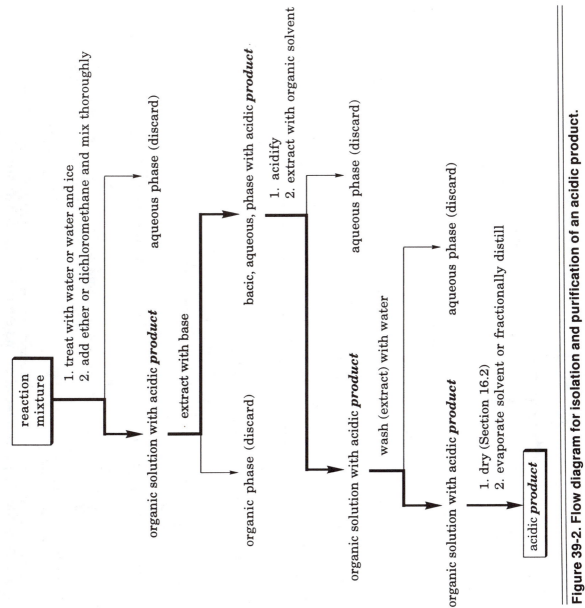

Figure 39-2. Flow diagram for isolation and purification of an acidic product.

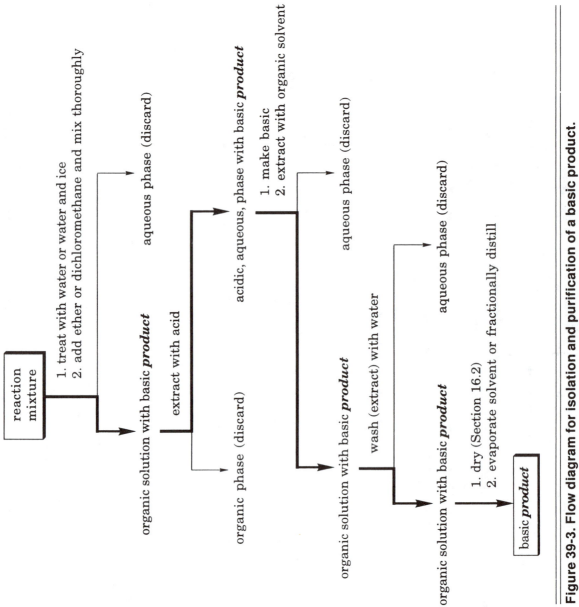

Figure 39-3. Flow diagram for isolation and purification of a basic product.

PART **2**

Experiments

➤ **Isolations and Purifications**

➤ **Transformations**

➤ **Synthetic Sequences: Synthesis Experiments that Use a Sequence of Reactions**

Part 1 of this book presents techniques for the three types of experiments that make up Part 2: isolations and purifications, one-step transformations, and synthetic sequences.

Most of the *isolations and purifications* are straightforward applications of the techniques described in Sections 8 through 16, in which the substance to be isolated is already present as a component of a mixture. The isolation of ibuprofen from ibuprofen tablets is an example.

Several simple experiments that illustrate a single technique are presented in Part 1 as exercises in the sections on recrystallization, distillation, and extraction. These experiments include:

- Recrystallization of *endo*-norbornene-2,3-dicarboxylic acid from water: Exercise 1, Section 9, on page 61. Both semi-micro-scale and micro-scale versions are presented.

- Recrystallization of acetanilide, *m*-nitroaniline, or *p*-nitroanaline: Exercises 2, 3, and 4 , Section 9, on page 62.

- Distillation of methanol: Exercise 1, Section 10, on page 79.

- Distillation of a mixture of methanol and water: Exercise 2, Section 10, on page 80.

- Separation by extraction of a mixture of an acid, a base, and a neutral compound: Exercise 3, Section 14, on page 103.

- Separation by extraction of a mixture of aspirin, phenacetin, and caffeine: Exercise 4, Section 14, on page 106.

- Paper chromatography of spinach leaves, carrots, food coloring, or ink: Exercises 1, 2, 3, and 4, Section 15.6, on pages 121 and 122.

Most of the *transformations* are a one-step synthesis of a compound, followed by its isolation and purification.

The *synthetic sequences* are preparations that must be carried out by a series of reactions. Usually the crude product of the preceding experiment can be used without purification.

All of the experiments have been designed for use in a three-hour laboratory period. Some can be done easily in less time, but a few will require more than three hours for many workers. The estimates of time required, which are given at the end of each procedure, are of the time needed by students who are taking their first course in organic chemistry. The estimates do not include the time needed for any optional crystallizations.

Isolations and Purifications

The separation procedures described in this chapter include the isolations of cholesterol from gallstones (Experiment 1), lactose from powdered milk (Experiment 2), acetylsalicylic acid from aspirin tablets (Experiment 3), ibuprofen from ibuprofen tablets (Experiment 4), caffeine from tea, cola, or NoDoz (Experiment 5), piperine from black pepper (Experiment 6), trimyristin from nutmeg (Experiment 7), clove oil from cloves (Experiment 8), eugenol from clove oil (Experiment 9), and R-(+)-limonene from grapefruit or orange peel (Experiment 10).

The two other experiments in this group involve the isolation of enantiomeric molecules. In the first of these (Experiment 11), the R and S isomers of carvone are isolated, one from oil of caraway and one from oil of spearmint, and their odors, which are different, can be compared. In the second (Experiment 12), the R and S forms of α-phenylethylamine are separated from the racemic mixture by a chiral agent, R,R-(+)-tartaric acid.

E1. Isolation of Cholesterol from Gallstones

In this experiment, cholesterol is isolated from human gallstones by a simple crystallization process. On the average, the cholesterol content of human gallstones is about 75%. The infrared spectrum of cholesterol is shown in Figure E1-1.

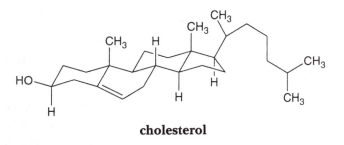

cholesterol

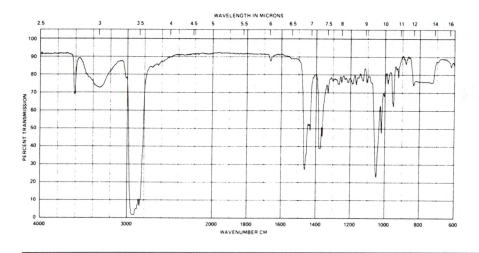

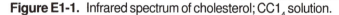

Figure E1-1. Infrared spectrum of cholesterol; CCl_4 solution.

Scale: 2 grams of gallstones.

Procedure (Reference 1)

Place a 2-gram sample of crushed gallstones in a 50-mL Erlenmeyer flask. To this, add 15 mL of 2-butanone (Note 1), and heat the mixture on the steam bath with occasional swirling for about 5 minutes. During this time, most of the solid in the flash should dissolve.

Gravity filtration, page 44

While the mixture is still hot, remove the brown residue of bile pigments by gravity filtration, collecting the filtrate in a 50-mL Erlenmeyer flask.

Decolorizing the solution, page 50

Dilute the filtrate with 10 mL of methanol, add some decolorizing carbon, and warm the mixture on the steam bath for 1–2 minutes.

While the mixture is still hot, remove the carbon by gravity filtration and collect the filtrate in a 50-mL Erlenmeyer flask.

Reheat the filtrate on the steam bath, add 1 mL of hot water, and swirl to dissolve any precipitated solid. (If it does not all dissolve, add a little methanol and heat while swirling until it does.) Allow the solution to come to room temperature, during which time cholesterol will separate as colorless plates.

Suction filtration, page 45

Collect the crystals of cholesterol by suction filtration, and wash them with a few milliliters of cold methanol. Leave the vacuum on for a few minutes to suck the crystals as dry as possible.

Cholesterol can be recrystallized from methanol, using 35 mL per gram. Pure cholesterol melts at 149°C.

Cholesteryl benzoate, the ester of cholesterol and benzoic acid, is converted to a "liquid crystal" when heated to a temperature between 147°C and 180°C. Section E.32 tells how to prepare cholesteryl benzoate and how to observe the liquid crystal phenomenon.

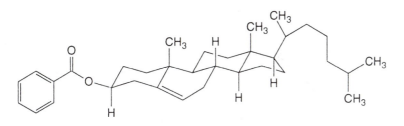

cholesteryl benzoate

Note

1. The 10 mL of dioxane called for in previous editions has been replaced by 15 mL of 2-butanone, so as to avoid exposure to dioxane. Other alternatives are 15 mL of isopropyl alcohol or 10 mL of diethyl ether.

Time: less than 3 hours.

Disposal

> **Solid waste:** filter paper
> **Nonhalogenated liquid organic waste:** methanol; 2-butanone
> **Solid organic waste:** gallstones; cholesterol

Reference

1. L.F. Fieser, *Organic Experiments,* Heath, Boston, 1964, p. 70.

E2. Isolation of Lactose from Powdered Milk

In this experiment, lactose is recovered from the whey that remains after precipitation and removal of the milk proteins. Lactose is a disaccharide that, upon hydrolysis, yields a molecule of glucose and a molecule of galactose. Dry milk is approximately one-half lactose.

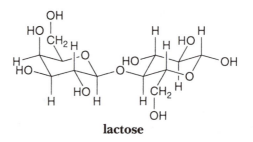

lactose

Scale: 25 grams of nonfat dry milk powder

Procedure

Place 25 grams of nonfat dry milk powder in a 250-mL beaker (Note 1). Add 75 mL of warm water and stir to mix. Adjust the temperature of the mixture to between 40 and 50°C by heating or cooling. Add about 10 mL of 10% acetic acid solution (Note 2) and stir the mixture to coagulate the casein. Precipitation can be judged to be complete when the liquid changes from milky to clear.

Gravity filtration, page 44

Remove the precipitated casein by filtering the mixture by gravity through cheesecloth (Note 3). Collect the filtrate in a 250-mL beaker. Add about 2 grams of calcium carbonate powder to the filtrate, stir it well, and boil the suspension for about 10 minutes (Note 4). Add to the hot mixture as much decolorizing carbon as would cover a nickel, stir the mixture thoroughly, and filter it by suction through a layer of wet filter aid on a Büchner funnel (Note 5).

Transfer the filtrate to a 250-mL beaker and concentrate it to a volume of about 30 mL by boiling over a low flame with a wire gauze between the flame and the beaker (Note 6). When the volume has been reduced to 30 mL, turn off the burner and add 125 mL of 95% ethanol and about the same amount of decolorizing carbon as used before. Stir the mixture well and filter it through a layer of wet filter aid on a Büchner funnel (Note 5).

Allow the clear filtrate to stand for crystallization for at least 24 hours in a stoppered Erlenmeyer flask.

Suction filtration, page 45

Collect the crystals of lactose by suction filtration. They may be washed with a small amount of 95% ethanol. Yield: between 2.5 and 4.5 grams.

Notes

1. The powdered milk is most easily measured by volume using a beaker; 25 grams is about 100 mL.

2. Ten percent acetic acid is prepared by diluting 10 mL of glacial acetic acid to 100 mL. More than 10 mL of 10% acetic acid seems to be required if the dry milk is not fresh.

3. Use two pieces of cheesecloth about 12 inches square. Lay them over a conical funnel large enough to contain all the curds. First, decant as much as possible of the liquid from the coagulated casein into the funnel and then carefully transfer the wet protein mass onto the cheesecloth. After most of the liquid has drained through, a bit more can be recovered by wrapping the cheesecloth around the curds and squeezing.

4. The mixture will foam occasionally.

5. Add about 5 grams of filter aid (we have used Celite) to about 25 mL of water in a small beaker. Wet the filter paper on the Büchner funnel with water, fit the funnel to the suction flask, and apply a gentle suction. Swirl the mixture of filter aid and water to suspend the solid, and pour the suspension all at once into the Büchner funnel. When the water has been drawn through the funnel, a damp pad of filter aid should be in place on top of the filter paper. Remove the funnel momentarily and pour the water out of the filter flask. Be careful not to disturb the layer of filter aid when pouring the mixture to be filtered into the funnel.

6. Some bumping and foaming occur.

Time: 3 hours plus $\frac{1}{2}$ hour to isolate the product after crystallization is complete.

Disposal

Sink: acetic acid; ethanol

Solid waste: powdered milk; cheesecloth; filter paper; casein; calcium carbonate

Solid organic waste: lactose

Reference

1. J. Cason and H. Rapoport, *Laboratory Text in Organic Chemistry*, Prentice-Hall, New York, 1950, p. 113.

E3. Isolation of Acetylsalicylic Acid from Aspirin Tablets

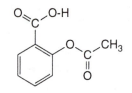

acetylsalicylic acid; aspirin

Aspirin tablets consist of acetylsalicylic acid plus a binder that helps keep the tablet from breaking up. In this experiment, the acetylsalicylic acid is recovered by a crystallization process. A procedure for the preparation of acetylsalicylic acid from salicylic acid is described in Section E.72. The infrared and NMR spectra of acetylsalicylic acid are shown in Figures E3-1 and E3-2.

Traditional Scale Experiment

Scale: ten 5-grain aspirin tablets (3.25 grams of acetylsalicylic acid)

Procedure

Place ten 5-grain aspirin tablets (3.25 grams acetylsalicylic acid) in a 50-mL Erlenmeyer flask. Add to the flask 10 mL of 95% ethanol and heat the mixture to boiling on the steam bath. Swirl the hot mixture

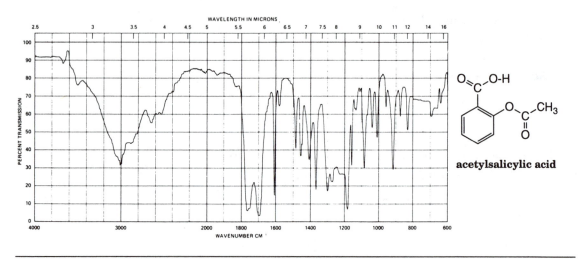

Figure E3-1. Infrared spectrum of acetylsalicylic acid; CHCl$_3$ solution.

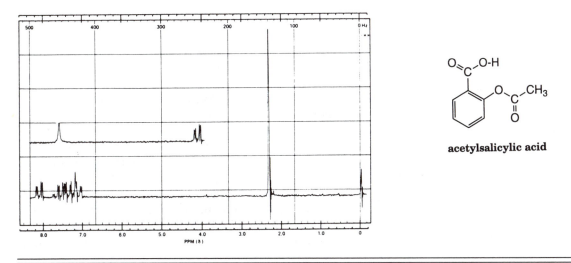

Figure E3-2. NMR spectrum of acetylsalicylic acid; CDCl$_3$ solution.

until the tablets disintegrate. The acetylsalicylic acid will dissolve, but a white residue, the binder, will remain.

Filter the hot solution by gravity (Note 1) to remove the insoluble material, collecting the filtrate in a 125-mL Erlenmeyer flask. Add to the filtrate 50 mL of cool water and swirl the mixture. Acetylsalicylic acid will begin to crystallize almost immediately.

Gravity filtration, page 44

Allow crystallization to take place for about 10 minutes at room temperature and then for about 5 minutes in an ice bath.

Collect the acetylsalicylic acid by suction filtration (Note 2) and wash the product with a little cold water. Recovery: about 2.4 grams (75%).

Suction filtration, page 45

The melting point of acetylsalicylic acid is reported to be 135-136°C.

Notes

1. For a filter, use a stemless conical glass funnel fit with a fluted filter paper. Collect the filtrate in a second 125-mL Erlenmeyer flask.
2. Use a porcelain Hirsch or Büchner funnel with a flat circle of filter paper.

Time: less than 2 hours.

Micro-scale Experiment

Scale: one 5-grain aspirin tablet (325 millligrams of acetylsalicylic acid)

Procedure

Place one 5-grain aspirin tablet (325 milligrams acetylsalicylic acid) in a small test tube. Add to the test tube 1.0 mL of 95% ethanol and heat the mixture to boiling on the steam bath. Poke the tablet with a stirring rod until the tablet disintegrates. The acetylsalicylic acid will dissolve, but a white residue, the binder, will remain.

Micro-scale gravity
filtration, page 45

Filter the hot solution by gravity (Note 1) to remove the insoluble material, collecting the filtrate in a second small test tube. Add to the filtrate 5.0 mL of cool water and stir to mix. Acetylsalicylic acid will begin to crystallize almost immediately.

Allow crystallization to take place for about 10 minutes at room temperature and then for about 5 minutes in an ice bath.

Collect the acetylsalicylic acid by suction filtration (Note 2) and wash the product with a little cold water. Recovery: about 240 milligrams (75%).

The melting point of acetylsalicylic acid is reported to be 135-136°C.

Notes

1. For a filter, use a Pasteur pipette fit with a bit of cotton, as described in Section 8.2. Transfer the solution to the filter with a second Pasteur pipette and collect the filtrate in a second small test tube.
2. Use a porcelain Hirsch or Büchner funnel with a flat circle of filter paper.

Time: about 2 hours.

Disposal

Sink: ethanol
Solid waste: filter paper
Solid organic waste: aspirin tablets; acetylsalicylic acid

E4. Isolation of Ibuprofen from Ibuprofen Tablets

Although it is not a new compound, ibuprofen (Advil, Motrin, and at least 57 other brand names) has recently become a popular analgesic, right up there with acetylsalicylic acid (Aspirin, Ecotrin, Emprin, and about 40 other brand names) and acetaminophen (Tylenol, and again at least 57 other brand names).

In this experiment, you will recover the active ingredient from ibuprofen tablets.

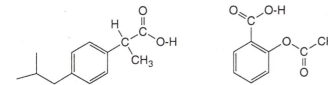

ibuprofen   **acetylsalicylic acid**

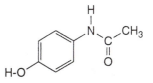

acetaminophen

Scale: 5 ibuprofen tablets (1.0 gram of ibuprofen).

Procedure (Reference 1)

Peel the coating from 5 coated ibuprofen tablets (Notes 1 and 2) by picking at the coating with a metal spatula or your fingernails. Place the peeled tablets in a 100-mL boiling flask, and then add 30 mL of ordinary (ethyl) ether to the flask. Fit the flask with a condenser, and then gently heat the flask on the steambath for about 10 minutes or until the peeled tablets have disintegrated and the ibuprofen appears to have dissolved.

Remove the undissolved binder by filtering the ethereal ibuprofen solution by gravity into a clean 50-mL Erlenmeyer flask.   Gravity filtration, page 44

Fit the 50-mL Erlenmeyer flask with a distillation adapter, thermometer, and condenser. Add a boiling stone to the flask and then, by heating the flask on the steam bath, strip off all the ether (Notes 3 and 4). As an alternative to distillation, you can use a rotary evaporator to remove the solvent.   Evaporation, page 317

When all of the ether has been removed, allow the residue of ibuprofen to cool and to crystallize (Note 5). Yield: 0.80 grams (80%) of crude product. Melting point: 70–75°C.   Suction filtration, page 45

To recrystallize the crude product, add about 4 mL of hexane to the flask, and then cautiously heat the flask on the steam bath until the ibuprofen has dissolved. Transfer the hot solution, using a Pasteur pipette, to a 10-mL Erlenmeyer flask. Use another 1 mL of hexane to rinse the 50-mL flask, and add the rinsings to the ibuprofen solution in the smaller flask.

Add a tiny boiling stone to the smaller flask and gently heat the smaller flask on the steam bath until the solution boils, and continue heating until the volume of the solution has been reduced to 3 to 4 mL. Remove the flask from the heat, cover the top with a piece of aluminum foil, and set the flask aside to cool for crystallization (Note 6). Collect the crystals of ibuprofen by suction filtration. Yield: 0.42 grams (53% recovery) from the crude product; 42% overall from ibuprofen tablets.

Notes

1. Each tablet contains 200 mg (0.001 mole) of ibuprofen.
2. If there is no coating, or if the coating does not peel off easily, simply crush each tablet by squeezing it between a hard object and the desk top. The fractured tablets will disintegrate as the mixture is boiled.
3. Collect the ether and put it in the container reserved for waste ether.
4. When the residue stops bubbling, you can assume that the ether is gone.
5. It may be necessary to scratch the residue to induce crystallization.
6. Ibuprofen can also be extracted using hexane. Add 5 crushed uncoated tablets to 20 mL of hexane. Heat under reflux on the steam bath for 10 minutes. Filter to remove the binder. Strip off the hexane to reduce the volume to about 4 mL, then allow the solution to stand overnight. Yield: 0.45 grams (45%).

Time: less than 3 hours.

Disposal

    Solid waste: filter paper; coating from tablets
    Nonhalogenated liquid organic waste: ether; hexane
    Solid organic waste: ibuprofen tablets; ibuprofen

Reference

1. Brandi Langsdorf, Cornell College 1993, private communication.

Questions

1. Which is the better solvent for ibuprofen, ether or hexane? How do you know?
2. Why bother fitting the flask with a condenser while you are heating to dissolve the ibuprofen?

E5. Isolation of Caffeine from Tea and NoDōz

In these experiments, caffeine is isolated from tea or No Dōz tablets by extraction into water followed by extraction into chloroform, evaporation, and recrystallization from ethanol. It may also be purified by sublimation (Section 13).

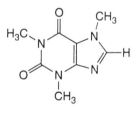

caffeine

Caffeine is described as a cardiac, respiratory, and psychic stimulant, and as a diuretic. It is reported to have a melting range of 235–236°C. The infrared spectrum is shown in Figure E5-1 and the NMR spectrum in Figure E5-2.

Caffeine from Tea

Procedure (Reference 1)

Place 125 mL of water in a 500-mL Erlenmeyer flask. Add 12.5 grams of tea and 12.5 grams of powdered calcium carbonate (Note 1).

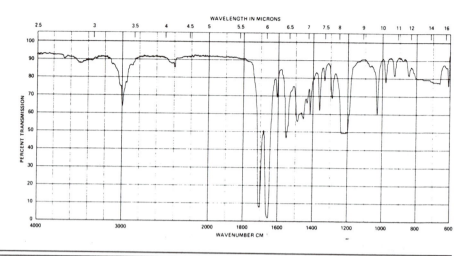

Figure E5-1. Infrared spectrum of caffeine; $CHCl_3$ solution.

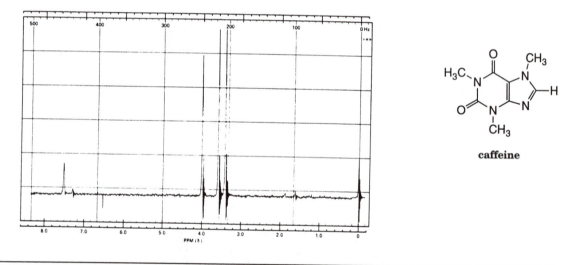

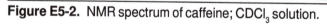

Figure E5-2. NMR spectrum of caffeine; $CDCl_3$ solution.

Boil the contents of the flask with constant stirring for about 20 minutes.

Meanwhile, filter the hot mixture with suction, and press out the liquid from the tea leaves with a large cork. Transfer the filtrate to a 250-mL Erlenmeyer flask, and add 100 mL of dichloromethane. Swirl or gently stir the two layers together for about 10 minutes (Note 2) and then separate them, using a separatory funnel.

Distill the chloroform layer until only about 10 mL remain, transfer the remaining solution to a tared (weighed) 25-mL Erlenmeyer flask (filtering if necessary), and evaporate to dryness on the steam bath. Recrystallize the solid residue of caffeine from 95% ethanol, using 5 mL per gram (Note 3).

Caffeine from NoDōz

Procedure

Place 100 mL of water in a 250-mL Erlenmeyer flask. Add two NoDōz tablets that have been crushed and weighed. Heat the suspension to boiling to ensure solution of the caffeine in the NoDōz tablets; the binder will remain in suspension.

Cool the mixture, add 100 mL of dichloromethane, and gently swirl or stir the two layers together for about 10 minutes (Note 2) and then separate them, using a separatory funnel.

Distill the dichloromethane layer until only about 10 mL remain, transfer the remaining solution to a tared (weighed) 25-mL Erlenmeyer flask (filtering if necessary), and evaporate to dryness on the steam bath. Recrystallize the solid residue of caffeine from 95% ethanol, using 5 mL per gram (Note 3).

Notes

1. R. H. Mitchell, W. A. Scott, and P. R. West suggest using 12.5 grams of tea bags and 9 grams of sodium carbonate [*J. Chem. Educ.* **51,** 69 (1974)]. This variation allows you to omit the suction filtration.

2. More vigorous extraction procedures often result in very troublesome emulsions. Gentle stirring of the mixture with a magnetic stirrer is quite satisfactory. The NoDōz extraction is less likely to form an emulsion.

3. One recrystallization from ethanol is sufficient and yields caffeine in the form of small needles. A reasonable recovery of recrystallized caffeine is about 25–50% of the total present, depending upon the thoroughness of the extraction. The greenish color in the product of the extraction of tea can be completely removed by careful washing of the crystals with ethanol.

Time: about 3 hours.

Disposal

> **Solid waste:** tea leaves; calcium carbonate; filter paper
> **Nonhalogenated liquid organic waste:** ethanol
> **Halogenated liquid organic waste:** dichloromethane
> **Solid organic waste:** NoDōz tablets; caffeine

Reference

1. G. K. Helmkamp and H. W. Johnson, Jr., *Selected Experiments in Organic Chemistry,* 2nd edition, W. H. Freeman, San Francisco, 1968, p. 157.

E6. Isolation of Piperine from Black Pepper

Piperine, a very weakly basic substance, can be isolated from a variety of peppers by extraction with alcohol. Although the aroma of pepper cannot be due to the presence of piperine (it is not volatile), the taste of pepper may be due in part to this substance because piperine wet with alcohol produces an immediate, sharp sensation when it is placed on the tongue. Piperine, along with a small amount of its Z,E isomer, accounts for about 10% of the weight of black pepper.

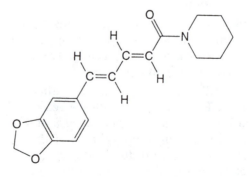

piperine

Scale: 15 grams of ground black pepper.

Procedure

Place 15 grams of ground black pepper and 1 gram of powdered calcium carbonate (Note 1) in a 250-mL boiling flask, add 100 mL of isopropyl alcohol, and, after fitting the flask with a reflux condenser, boil the mixture for about 1 hour on the steam bath.

Heating under reflux, page 303

At the end of the heating period, filter the mixture by gravity into a 125-mL Erlenmeyer flask, clean the 250-mL boiling flask, and return the filtrate to the boiling flask. Fit the boiling flask with a distillation adapter and condenser (see Figure 10.2-1), and boil off all but about 10 mL of the isopropyl alcohol (Note 2). Transfer the residual solution from the boiling flask to a 25-mL Erlenmeyer flask, and set the flask aside to cool for crystallization of piperine (Notes 3 and 4). Collect the product by suction filtration, using small portions of methyl alcohol to rinse the flask and wash the product. Yield: about 0.5 gram.

Gravity filtration, page 44

Suction filtration, page 45

Notes

1. The addition of calcium carbonate should prevent the extraction of acidic components of pepper.

2. Isopropyl alcohol boils at 80°C, only a little below the maximum temperature attainable on the steam bath. To make the distillation proceed quickly, clamp the boiling flask so that it is well down in the rings of the steam bath, and drape a towel over the flask and the steam bath to make a tent that will hold steam around the top of the flask.

3. Crystallization occurs slowly, and the flask must be allowed to stand for at least 24 hours.

4. Alternatively, add 25 mL of water to the isopropyl alcohol solution of piperine, allow the mixture to stand for at least 24 hours so that precipitation will be complete, collect the solid by suction filtration, and recrystallize it from either isopropyl alcohol or acetone.

Time: About 3 hours of working time; crystallization takes at least 24 hours.

Disposal

Solid waste: pepper; filter paper; calcium carbonate
Nonhalogenated liquid organic waste: isopropyl alcohol
Solid organic waste: piperine

Question

1. Piperine is one member of a set of stereoisomers. What is the total number of isomers in this set, and what are the relationships among these stereoisomers?

E7. Isolation of Trimyristin from Nutmeg

The seeds of plants often contain substantial amounts of *triglycerides,* the tri-ester between *glycerol* and three long-chain monocarboxylic acid molecules, often called *fatty acids.*

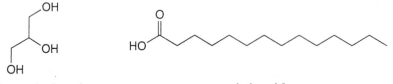

glycerol; glycerine   **myristic acid**

The triglyceride content of nutmegs, the seeds of the fruit of the nutmeg tree (*Myristica fragrans*), native to the Spice Islands, consists almost entirely of the glyceryl tri-ester of the 14-carbon, straight-chain, saturated monocarboxylic acid *myristic acid,* and is called *trimyristin:*

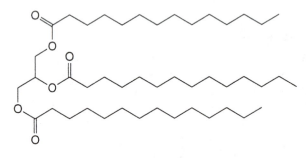

trimyristin; a triglyceride

In this experiment, you will use ether to extract the trimyristin from ground nutmeg seeds and then recrystallize the crude trimyristin from acetone.

Scale: 5 grams of ground nutmeg.

Procedure

Place 5.0 grams of ground nutmeg in a ⚕ 100-mL boiling flask, add 50 mL of ether (Note 1), fit the flask with a condenser, and heat the mixture on the steam bath to a *gentle* boil under reflux for about 15 minutes.

    Allow the mixture to cool a bit and then filter it by gravity through a fluted filter paper into a second ⚕ 100-mL boiling flask. Use another 10 mL of ether to wash solids that remain in the first boiling flask and solids that remain on the filter; let these washings run through the filter into the second ⚕ 100-mL boiling flask that now contains the original 50 mL of ether.

    Fit the second ⚕ 100-mL boiling flask with a distillation adapter and a condenser, add a couple of boiling stones, and then remove by distillation as much of the ether as you can by heating the flask on the steam bath. (Collect the ether and put it in the container reserved for waste ether.) As an alternative to distillation, you can use a rotary evaporator to remove the solvent.

    Cool the residue in the boiling flask to make the oil crystallize (Note 2).

    Now add 25 mL of acetone to the residue in the flask, and heat the mixture to dissolve the solid. When the solid is all dissolved, transfer the hot solution to a 50-mL Erlenmeyer flask, and then let the solution cool for crystallization at room temperature for 10 minutes. Complete the crystallization by placing the flask in an ice bath for another 10 minutes.

    Collect the colorless crystals of trimyristin by suction filtration. Trimyristin can be recrystallized from 95% ethanol using 10 mL per gram. The melting point of trimyristin is reported to be 56–57°C.

Heating under reflux, page 303

Gravity filtration, page 44

Technique of distillation, page 75

Suction filtration, page 45

Notes

1. Fifty mL of dichloromethane can be used instead of the ether.
2. If necessary, crystallization can be induced by rubbing the oil with a stirring rod and cooling the flask in an ice bath.

Time: less than 3 hours.

Disposal

    Sink: ethanol
    Solid waste: nutmeg; filter paper

Nonhalogenated liquid organic waste: ether; acetone

Solid organic waste: trimyristin

Questions

1. Why is it important to fit the flask with a condenser during the extraction?

2. Why is crystallization allowed to take place for 10 minutes at room temperature before the flask is placed in an ice bath?

3. Trimyristin is more soluble in which solvent: ether or acetone? How did you decide?

E8. Isolation of Clove Oil from Cloves

In this experiment, oil of cloves is isolated from cloves by steam distillation followed by extraction of the distillate. The extraction solution can be further treated to yield oil of cloves, or it can be used for the isolation of eugenol, as described in the procedure of Section E9.

Scale: 75 grams of cloves.

Procedure (Reference 1)

Steam distillation, page 87

Place 75 grams of cloves (Note 1) and 250 mL of water in a 500-mL boiling flask. Clamp the flask so that it can be heated with a burner flame, and then fit it with a Claisen adapter. Fit the center neck of the Claisen adapter with a separatory funnel and the side neck with a distillation adapter carrying a thermometer and a condenser set for downward distillation (Figure E8-1). Distill the mixture rapidly by heating the flask with a burner until no more oil can be observed to form in the condensate (Note 2).

Technique of extraction, page 97

Extract the oil of cloves from the distillate with 50 mL of petroleum ether, batchwise if necessary, as described in Section 14.3.

Technique of distillation, page 75

The oil of cloves can be isolated by drying the extract over anhydrous magnesium sulfate (Section 16.2) and removing the petroleum ether by distillation on the steam bath (Section 37).

Notes

1. Whole cloves are much more satisfactory than ground cloves.

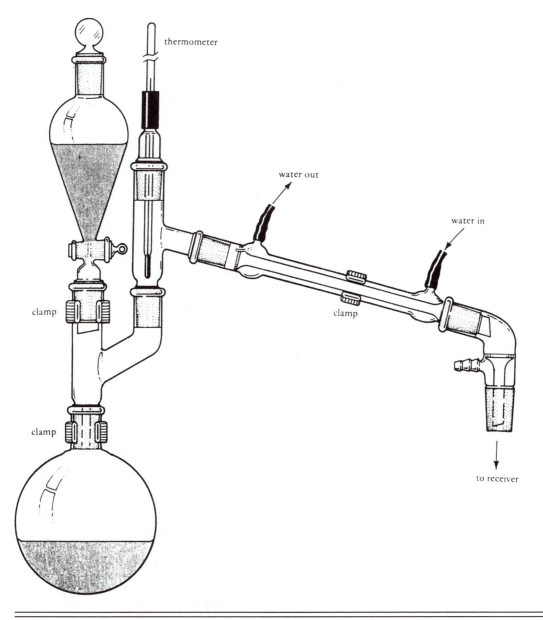

Figure E8-1. Apparatus for the steam distillation of cloves.

2. Steam distillation for 1 hour will remove most of the oil of cloves.

Time: about 3 hours, depending upon how rapidly the steam distillation is done.

Disposal

    Sink: water

    Solid waste: cloves

    Nonhalogenated liquid organic waste: oil of cloves; petroleum ether

    Solid inorganic waste: magnesium sulfate

E9. Isolation of Eugenol from Clove Oil

In this experiment, eugenol is isolated from oil of cloves by extraction with base. Eugenol has the characteristic odor of cloves. Its infrared spectrum is shown in Figure E9-1.

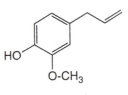

eugenol

Scale: 5 mL of oil of cloves.

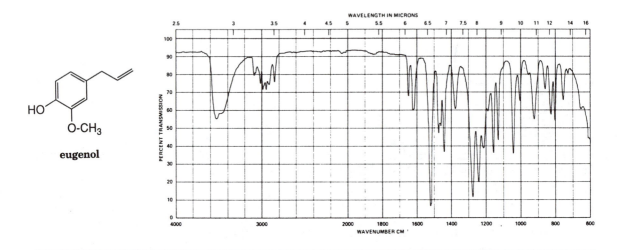

Figure E9-1. Infrared spectrum of eugenol; thin film.

Procedure (Reference 1)

Dissolve 5 mL of oil of cloves in 50 mL of dichloromethane and extract the solution with three 50-mL portions of 5% aqueous potassium hydroxide solution. Combine the basic extracts and acidify them to litmus with 5% aqueous hydrochloric acid; about 100 mL will be needed.

Extract the free eugenol from the aqueous mixture with 50 mL of dichloromethane, dry the organic extract over anhydrous magnesium sulfate, and remove the carbon tetrachloride to give a residue of eugenol by distillation using a steam bath.

Extraction, page 92

Time: about 2 hours.

Disposal

>**Sink:** potassium hydroxide solution; aqueous hydrochloric acid
>**Nonhalogenated liquid organic waste:** oil of cloves; eugenol
>**Halogenated liquid organic waste:** dichloromethane
>**Solid inorganic waste:** magnesium sulfate

Reference

1. Information Division, Unilever Ltd., Unilever House, Blackfriars, London.

E10. Isolation of (R)-(+)-Limonene from Grapefruit or Orange Peel

The major constituent of the steam-volatile oil of grapefruit or orange peel is (R)-(+)-limonene.

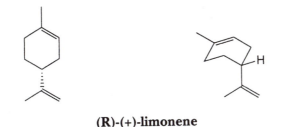

(R)-(+)-limonene

Limonene can be isolated as a material of about 97% purity by a simple steam distillation of citrus fruit peels. It is responsible for the characteristic smell of citrus peel. The infrared spectrum of a commercial sample of (R)-(+)-limonene is shown in Figure E10-1.

Scale: peel from one grapefruit or two oranges.

Procedure

Steam distillation, page 87

Obtain the peel from a grapefruit or from two oranges. Cut the peel into pieces about 5–10 mm square, and put the peel into a 500-mL boiling flask along with about 250 mL of water. Clamp the flask so that it can be heated with a burner flame, and then fit it with a distillation adapter carrying a condenser set for downward distillation (Figure E10-2). Distill the mixture rapidly by heating the flask with a burner until you have collected about 50 mL of distillate.

Technique of extraction, page 97
Drying of solutions, page 132
Technique of distillation, page 75

Extract the limonene from the distillate with 20 mL of either pentane or dichloromethane, separate the extract by means of a separatory funnel, and dry the extract over anhydrous magnesium sulfate for a few minutes (Note 1). Isolate the limonene by filtering the extract by gravity from the magnesium sulfate into a tared (previously weighed) flask and removing the solvent by distillation on the steam bath. Yield: between 1 and 1.5 mL.

Physical properties reported for purified (R)-(+)-limonene include b.p. = 175.5–176°C, d_4^{21} = 0.8403, n_4^{21} = 1.4743, and $[\alpha]_D^{19.5}$ = +124.2°.

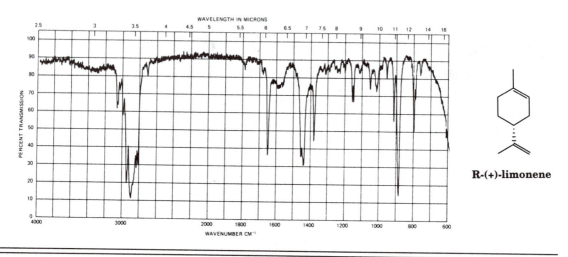

R-(+)-limonene

Figure E10-1. Infrared spectrum of R-(+)-limonene; thin film.

Note

1. As an alternative, you can collect the distillate in two 20 × 150 mm test tubes. The limonene will separate to form a clear layer above the water, and most of the limonene can be removed easily using a Pasteur pipette.

Time: less than 3 hours.

Disposal

   Sink: water

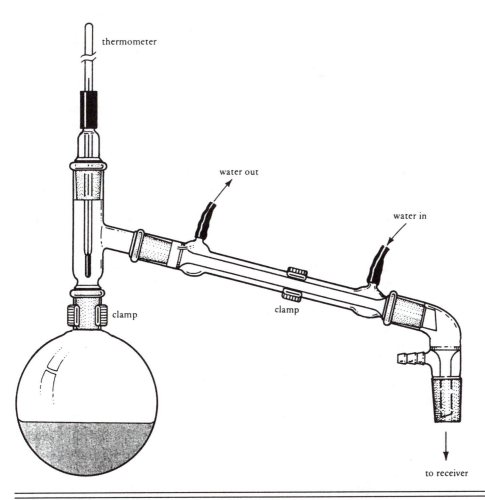

Figure E10-2. Apparatus for the steam distillation of grapefruit or orange peel.

Solid waste: fruit peel

Nonhalogenated liquid organic waste: pentane; limonene

Halogenated liquid organic waste: dichloromethane

Solid inorganic waste: magnesium sulfate

Reference

1.F. H. Greenberg, *J. Chem. Educ.* **45,** 537 (1968).

E11. Isolation of (R)-(–)- or (S)-(+)-Carvone from Oil of Spearmint or Oil of Caraway

Enantiomeric molecules have identical properties, with two exceptions. Enantiomers rotate the plane of polarization of plane-polarized light in opposite directions, and they interact differently with chiral molecules. The latter difference is like the difference in the ways that enantiomeric members of a pair of gloves will interact with a right hand. Enzymes, which are chiral molecules present in every living cell, generally react exclusively with one member of a pair of enantiomers. It is therefore not surprising that enantiomers have often been reported to differ in taste or smell.

One report, in which the evidence for a difference in smell is especially convincing (Reference 1), describes the enantiomers of carvone. (R)-(–)-Carvone, isolable from oil of spearmint; smells like spearmint, while (S)-(+)-carvone, isolable from oil of caraway, smells like caraway. About nine persons out of 10 can distinguish these enantiomers by smell (Reference 2).

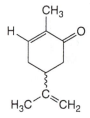

carvone

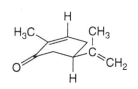

(R)-(–)-carvone (spearmint)
$[\alpha]_D^{20} = -62.5°$

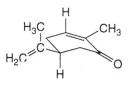

(S)-(+)-carvone (caraway)
$[\alpha]_D^{20} = +61.2°$

The experiment in this section involves the isolation of the enantiomeric forms of carvone, one from each oil. The isolation can be effected by distillation at atmospheric pressure or by distillation under reduced pressure. As you can see from Figures E11-1 and E11-2, the infrared spectra of the enantiomeric forms are identical.

The lower-boiling components of the oils are mostly isomers of molecular formula $C_{10}H_{16}$.

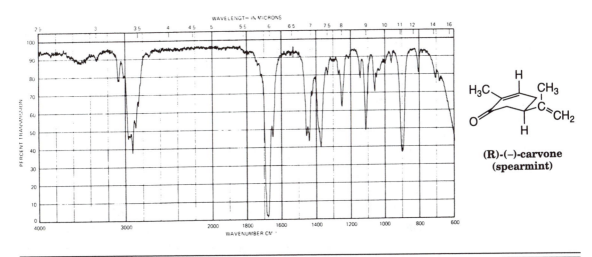

Figure E11-1. Infrared spectrum of (R)-(–)-carvone; thin film.

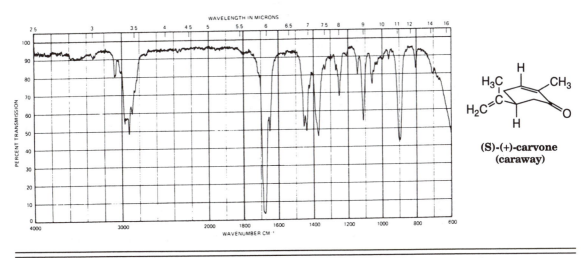

Figure E11-2. Infrared spectrum of (S)-(+)-carvone; thin film.

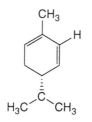

(S)-(+)-α-phellandrene

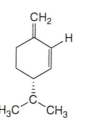

(S)-(+)-β-phellandrene

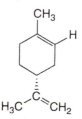

(R)-(+)-limonene

Distillation at Atmospheric Pressure

Scale: 10 mL of oil.

Procedure

Fractional distillation, page 70

Put 10 mL of one of the oils and one or two boiling stones in a 25-mL round-bottom flask; clamp the flask into position in a heating mantle. Place a Claisen adapter in the neck of the flask (Note 1), and put a stopper in the center neck of the Claisen adapter. If you want, you may poke a small amount of stainless steel sponge into the curved part of the Claisen adapter to increase the efficiency of the separation. Then place a distillation adapter in the side neck of the Claisen adapter. Fit a thermometer in the top neck of the distilling adapter and connect the side neck directly to a vacuum adapter. The distillate will be collected as it drips from the center tube of the vacuum adapter. This apparatus is illustrated in Figure E11-3.

Heat the flask with the heating mantle until the oil boils, and continue to heat the flask strongly until material begins to distill. Then adjust the heat so that the distillate accumulates at a rate of 1 drop every 3–5 seconds (Note 2). Collect the distillate in at least 3 fractions: $C_{10}H_{16}$ isomers, intermediate fraction, and carvone fraction.

Atmospheric-pressure boiling points for carvone and the isomers of molecular formula $C_{10}H_{16}$ have been reported as follows. Carvone: various values between 224 and 231°C; limonene: 176°C; α-phellandrene: 176°C; β-phellandrene: 172°C.

Notes

1. By omitting the Claisen adapter and using a 10-mL flask, you can distill as little as 5 mL of oil. If the distillation is done carefully, the separation is quite good.

2. To drive the carvone over, you will probably have to insulate the apparatus by wrapping the adapters with a single layer of aluminum foil.

Time: less than 3 hours.

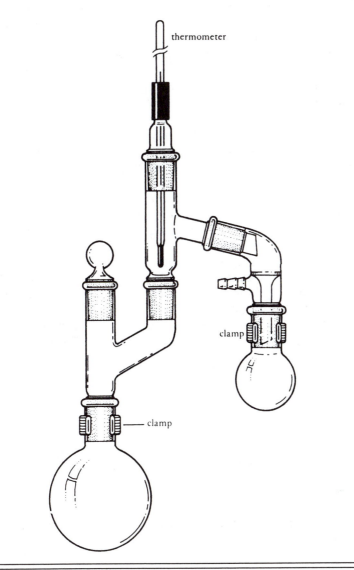

Figure E11-3. Apparatus for the fractional distillation of oil of spearmint or oil of caraway.

Distillation at Reduced Pressure

Scale: 10 mL of oil.

Procedure (Reference 3)

Put 10 mL of one of the oils and one or two boiling stones in a 25-mL round-bottom flask; clamp the flask into position in a heating mantle. If possible, you should use an apparatus like that shown in Figure 11.2-2. Connect the side arm of the adapter to a source of vacuum (Section 11.3) and a manometer (Section 11.4). It is best to use a magnetically stirred oil bath for heating, as shown in Figure 34-2. If magnetic stirring is available, place a very small magnetic stirring bar in the flask instead of boiling stones. If the flask must be heated with a flame, add a few boiling stones and clamp the flask in position on a wire gauze supported by an iron ring. Reduce the pressure of the system by turning on the water or oil pump, and allow the pressure to stabilize at 15–30 Torr. It is vital to the success of the distillation that the pressure of the system does not fluctuate. Heat the flask until the oil starts to boil (Note 1); continue heating until distillation begins. At this time, adjust the rate of heating so that the distillate accumulates at the rate of 1 drop every 3–5 seconds (Section 10.6). Collect the distillate in at least three fractions: $C_{10}H_{16}$ isomers, an intermediate fraction, and the carvone fraction.

The expected boiling points will have to be estimated from the reported reduced pressure boiling points (Section 11.1: Carvone: 115°C (20 Torr), 100°C (10 Torr), 91°C (6 Torr); limonene: 64°C (15 Torr), α-phellandrene: 61°C (11 Torr); β-phellandrene: 57°C (11 Torr).

Note

1. The air-saturated oil will degas under reduced pressure long before it starts to boil.

Time: less than 3 hours.

Disposal

Nonhalogenated liquid organic waste: oils of spearmint and caraway; carvone

Questions

1. The dextrorotatory, (+),-isomers of limonene and the two phellan-

drenes are illustrated. However, although the configurations at the chiral centers are similar, the phellandrene isomers are assigned the S configuration, while the limonene isomer is assigned the R configuration. Explain.

2. What advantage is there to distilling slowly?

References

1. G. F. Russell and J. I. Hills, *Science* **172**, 1043 (1971).
2. L. Friedman and J. G. Miller, *Science* **172**, 1044 (1971).
3. S. L. Murov and M. Pickering, *J. Chem. Educ.* **50**, 74 (1973).

E12. Resolution of α-Phenylethylamine by (R),(R)-(+)-Tartaric Acid

In this experiment, the enantiomeric forms of α-phenylethylamine react with the chiral reagent (R),(R)-(+)-tartaric acid to give diastereomeric salts that differ greatly in their solubility in methanol.

α-phenylethylamine   tartaric acid

The *enantiomeric reactants* have the same structure but different configurations, being *mirror images* of one another; they have identical physical and chemical properties, except for the direction of rotation of plane-polarized light and the interaction with chiral molecules. The *diastereomeric products* have the same structure but different configurations; they are *not mirror images* of one another. The (R)-(+)-amine (R),(R)-(+)-tartrate is quite soluble in methanol, whereas its diastereomeric companion (S)-(-)-amine (R),(R)-(+)-tartrate is relatively insoluble in methanol. This great difference in solubility of diastereomeric salts is very unusual.

The infrared spectrum of the racemic amine is shown in Figure E12-1.

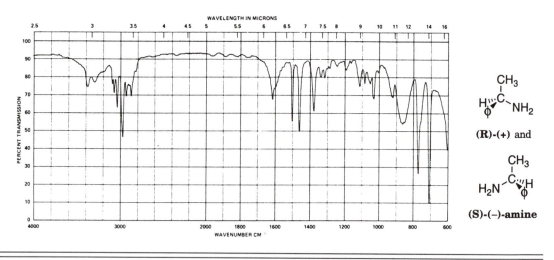

Figure E12-1. Infrared spectrum of (R),(S)-α-phenylethylamine; thin film.

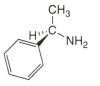 + + 2 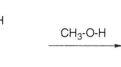 $\xrightarrow{CH_3\text{-}O\text{-}H}$

(R)-(+)amine   (S)-(–)-amine   (R),(R)-(+)-tartaric acid

enantiomers

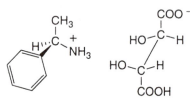 +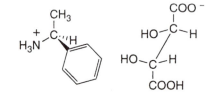

(R)-(+)-amine (R),(R)-(+)-tartrate   (S)-(–)-amine (R),(R)-(+)-tartrate
(more soluble; remains in solution)   (less soluble; crystallizes from solution)

diasteromers

As the equation indicates, when a hot methanolic solution of the racemic amine and an equimolar portion of (R),(R)-(+)-tartaric acid is allowed to cool, the less soluble of the diastereomeric salts crystallizes from solution.

Isolation of this salt followed by treatment with excess sodium hydroxide converts the salt to the free (S)-(-)-amine and disodium tartrate.

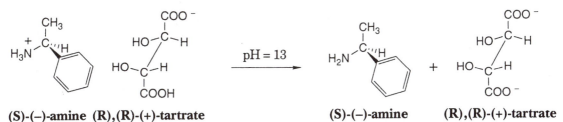

(S)-(–)-amine (R),(R)-(+)-tartrate   **(S)-(–)-amine**   **(R),(R)-(+)-tartrate**

When the methanolic solution that remains after crystallization of the (S)-(–)-amine (R),(R)-(+)-tartrate is evaporated, the residue consists of the (R)-(+)-amine (R),(R)-(+)-tartrate, which did not crystallize, and that part of the (S)-(–)-amine (R),(R)-(+)-tartrate that remained in solution after crystallization. The amine recovered from this residue by treatment with sodium hydroxide will therefore contain more (R)-(+)-amine than (S)-(–)-amine.

Part of the excess (R)-(+)-amine can be isolated by treating a hot ethanolic solution of this amine mixture with an amount of sulfuric acid in methanol just great enough to convert the excess R-(+)-amine to its neutral sulfate salt:

(R)-(+)-amine + (S)-(–)-amine + H$_2$SO$_4$ $\xrightarrow{\text{ethanol}}$

   x + y mole   x mole   1\2 y mole

[(R)-(+)-amine+]$_2$SO$_4$$^=$ + (R)-(+)-amine + (S)-(–)-amine

   1/2 y mole   x mole   x mole
   crystallizes out

The (R)-(+)-amine is then recovered from its neutral sulfate salt by treatment with excess sodium hydroxide.

(S)-(–)-α-Phenylethylamine

Scale: 12.0 grams (0.100 mole) of α-phenylethylamine.

Procedure

Add 15.0 grams of (R),(R)-(+)-tartaric acid (0.100 mole) to 220 mL of methanol in a 500-mL Erlenmeyer flask, and heat the mixture almost to boiling. To the hot solution, *cautiously* add 12.0 grams of racemic α-phenylethylamine (12.8 mL, 0.100 mole); too rapid an addition will cause the mixture to boil over. Stopper the flask and set the solution aside for crystallization. Because crystallization occurs slowly, the solution must be allowed to stand at room temperature for about 24 hours. During this time, the (–)-amine-(+)-hydrogen tartrate separates as prismatic crystals (Note 1).

Isolate the product by carefully decanting the liquid from the crystals, and then rinse the crystals by adding 20 mL or so of cold methanol to the crystals in the original flask, swirling methanol in the flask, and then decanting the additional methanol (Note 2). Yield: 8.7 grams (65%; Note 2).

Technique of extraction, page 97

Place the product (8.7 grams, or 10.5 grams if a second crop has been obtained) in a 100-mL Erlenmeyer flask and add about 45 mL of water; the amine tartrate salt will partially dissolve. Add to this mixture 6.0 mL of 50% sodium hydroxide solution (0.115 mole NaOH) to convert the amine salt to the free base. Then add about 25 mL of ethyl ether and swirl the mixture to extract the free amine into the ether. Separate the ethereal extract using a separatory funnel, and then dry the extract over anhydrous magnesium sulfate for about 10 minutes.

Drying of solutions, page 132

Remove the drying agent from the ethereal extract by gravity filtration, collecting the filtrate in a 100-mL Ŧ round-bottom flask. Add a boiling stone to the flask, fit the flask with a distilling adapter with thermometer and condenser, and heat the flask with the steam bath, to remove most of the ether by distillation.

Technique of distillation, Page 97

Transfer the residue to a 25-mL Ŧ round-bottom flask, using a little ether for rinsing. Add a boiling stone to the flask and set up an apparatus for distillation like that illustrated in Figure 10.2-1. Remove the last of the ether by heating the flask strongly with the steam bath. Then, using a heating mantle as the source of heat, distill the residue to obtain (S)-(–)-α-phenylethylamine (Note 3). Yield: 3.3 grams (55%); b.p., 184–186°C; d_4^{25} 0.953; $[\alpha]_D^{27}$, –38.2° (lit.: $[\alpha]_D^{22}$ –40.3°)

Optical activity, Chapter 21, page 163

Notes

1. Sometimes a substance separates in the form of very fine needles. In this event, warm the mixture to dissolve the crystals, and allow it to cool again. The needlelike crystals dissolve much more rapidly than the prismatic crystals. If possible, seed the solution with prismatic crystals. Workup of the needlelike crystals gives α-phenylethylamine with an $[\alpha]_D^{25}$ of –19 to –21°.

2. A second crop (1.8 grams) can be obtained by concentrating the combined mother liquor and washings to 110 mL and allowing crystallization to proceed at room temperature for another 24 hours.

3. Years ago, there were occasional reports of a problem with foaming. The conditions that cause foaming are not known.

Time: 3 hours plus $\frac{1}{2}$ hour of a previous laboratory period to prepare the initial solution.

Disposal

Nonhalogenated liquid organic waste: methanol; α-phenylethylamine

Solid organic waste: tartaric acid

Solid inorganic waste: sodium hydroxide; magnesium sulfate

Questions

1. Why do you suppose the original reaction mixture will boil over if the amine is added too rapidly?

2. What would be the result of the experiment if (S),(S)-(−)-tartaric acid were used in place of (R),(R)-(+)-tartaric acid?

3. What might be the molecular composition of the needle-like crystals?

4. What experiments might be done in order to determine the conditions for foaming (see Note 4)?

5. How would you expect the infrared spectrum of either separate enantiomer of α-phenylethylamine to compare with that of the racemic mixture?

(R)-(+)-α-Phenylethylamine

Procedure

Allow the methanolic solution remaining from the isolation of the (−)-amine-(+)-hydrogen tartrate to evaporate to dryness (this is most easily done by leaving the solution in an evaporating dish in the hood overnight), and recover the remainder of the amine by treating the residual salt with sodium hydroxide, extracting with ether, and distilling as previously described. Determine the specific rotation of the recovered amine.

Isolate the (+)-amine by treating a hot ethanolic solution of the recovered amine with an amount of sulfuric acid in ethanol slightly greater than that necessary to convert the excess (+)-amine to the neutral sulfate salt. The resulting crystals of (+)-α-phenylethylamine sulfate yield the (+)-amine (Note 1).

$$\textbf{(R)-(+)-amine} + \textbf{(S)-(-)-amine} + \textbf{H}_2\textbf{SO}_4 \xrightarrow{\text{ethanol}}$$

x + y mole   x mole   1½ y mole

$$\textbf{[(R)-(+)-amine+]}_2\textbf{SO}_4^= + \textbf{(R)-(+)-amine} + \textbf{(S)-(-)-amine}$$

½ y mole   x mole   x mole
crystallizes out

Dissolve 12.5 grams (0.103 mole) recovered amine ($[\alpha]_D^{27}$; +24.5°) in 88 mL of 95% ethanol. (This solution contains 0.062 mole excess (+)-amine, based on a $[\alpha]_D^{22}$ of −40.7° for the pure (+)-amine and 0.041 mole racemic amine.) Heat the solution to boiling and add to it a solution of concentrated sulfuric acid (3.2 grams of 98% H_2SO_4; 0.032 mole H_2SO_4) in 180 mL of 95% ethanol. After allowing the mixture to cool slowly to room temperature, collect the crystalline (+)-amine sulfate by suction filtration and wash it thoroughly with ethanol. Yield: 7.8 grams (74%).

Isolate the free (+)-amine as described above (treatment with sodium hydroxide, extraction, and distillation), using 40 mL of water and 5 mL of 50% sodium hydroxide for each 10 grams of the sulfate salt. Yield: 4.4 grams (59%); b.p., 184–186°C; $[\alpha]_D^{27}$, +38.3° (lit.: $[\alpha]_D^{22}$ +40.67°)

Note

1. You will have to calculate the amount of concentrated sulfuric acid you will need in your experiment, as the amount required depends both on the yield of the recovered amine and its optical purity. The directions given illustrate one possibility.

Time: 3 hours to recover the amine enriched in the (R)-(+)-isomer and to prepare the solution. Two hours to isolate the (+)-isomer.

Disposal

Sink: diluted sulfuric acid; ethanol

Nonhalogenated liquid organic waste: methanol; α-phenyl-ethylamine

Solid inorganic waste: sodium hydroxide; magnesium sulfate

Questions

1. If 1.00 mole "recovered amine" had $[\alpha]_D^{25} = +20.35°$:

   a. How many moles of (+)-amine and how many moles of (–)-amine would there be in the sample?

   b. How many moles of excess (+)-amine and how many moles of racemic amine would there be in the sample? (The excess of one enantiomer beyond the racemate expressed as a percent of the total is often called the enantiomeric excess, or ee.)

   c. How much sulfuric acid (both moles and grams) would be needed to convert the excess (+)-amine to its neutral sulfate salt?

2. a. Are the (+)-amine sulfate and (–)-amine sulfate salts of α-phenylethylamine enantiomeric or diastereomeric?

   b. Are their solubilities different in 95% ethanol? Explain.

   c. Why is it possible to isolate the (+)-amine in this experiment?

   d. What might happen if more sulfuric acid is used than the amount necessary to convert the excess (+)-amine to its neutral sulfate salt?

References

1. A. Ault, *J. Chem. Educ.* **42**, 269 (1965).
2. A. Ault, *Organic Syntheses*, Vol. 49, Wiley, New York, 1969, p. 93.

Transformations

In most of these experiments, a substance is prepared by a single reaction. The experiments are grouped by type of reaction, and, in general, reactions typical of aliphatic compounds are presented before the reactions of aromatic compounds.

Often, several examples of each kind of reaction are set forth. By comparing the procedures, you can see which features are essential to that type of reaction and which apply to the particular compounds involved. Amounts of materials are given in moles as well as in the units normally used to measure them so that comparisons between procedures can be made more easily.

Occasionally, more than one procedure is given for the preparation of a particular compound. Sometimes your choice of procedure will be determined by the starting materials, time, or equipment available. At other times, however, experience and understanding will be used to decide which procedure, or combination of procedures, will provide the desired yield or purity with the least effort.

ISOMERIZATIONS

These first two experiments are isomerizations, processes in which one substance is converted to a second substance that has the same molecular formula. In the first of these, one constitutional isomer of molecular formula $C_{10}H_{16}$ is converted to another, more stable, constitutional isomer of the same molecular formula (*constitutional isomerism*). In the second experiment, a more subtle change takes place: the starting materials and products differ only in molecular configuration (*configurational isomerism*).

$$\xrightarrow{\text{AlCl}_3}$$

endo-tetrahydrodicyclopentadiene
$C_{10}H_{16}$

adamantane
$C_{10}H_{16}$

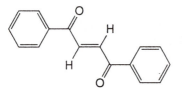

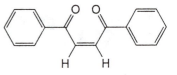

trans-**1,2-dibenzoylethylene**
m.p., 113–114°C
yellow

cis-**1,2-dibenzoylethylene**
m.p., 137–138°C
colorless

E13. Adamantane from *endo*-Tetra-hydrodicyclopentadiene via the Thiourea Clathrate

In this experiment, adamantane is formed by the aluminum chloride-catalyzed rearrangement of *endo*-tetrahydrodicyclopentadiene. Because of its symmetry and freedom from angle strain, adamantane is the constitutional isomer of molecular formula $C_{10}H_{16}$ that would be expected to predominate under equilibrating conditions.

AlCl₃

endo-**tetrahydrodicyclopentadiene**
$C_{10}H_{16}$

adamantane
$C_{10}H_{16}$

The adamantane is recovered from solution in hexane by treatment with a methanolic solution of thiourea to form a crystalline inclusion complex in which the molecules of adamantane occupy channels in the crystal lattice of thiourea. When the inclusion complex is shaken with a mixture of water and ether, it breaks down. The thiourea dissolves in the water and the adamantane in the ether, from which it is recovered by evaporation.

The infrared spectrum of adamantane is shown in Figure E13-1.

Scale: 10 grams (0.073 moles) of *endo*-tetrahydrodicyclopentadiene.

Procedure (Reference 1)

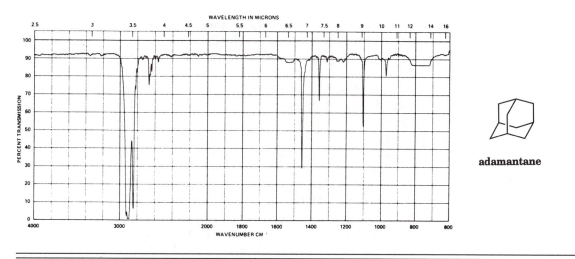

Figure E13-1. Infrared spectrum of adamantane; $CHCl_3$ solution.

Weigh out 4 grams of aluminum chloride and store it until ready for use in a stoppered sample vial.

Place 10.0 grams of *endo*-tetrahydrodicyclopentadiene in a 50-mL ⊈ 19/22 Erlenmeyer flask, and fit the flask with a well-greased inner joint that will serve as an air condenser. Clamp the flask so that it can be heated by a small Bunsen burner flame. Heat the flask gently to melt the tetrahydrodicyclopentadiene. When it has melted, add about one-fourth of the aluminum chloride down the air condenser, and suspend a thermometer in the mixture so that the bulb dips into the liquid as far as possible but does not touch the bottom of the flask. Adjust the flame of the burner so that the mixture is heated just barely to boiling (160–185°C) and is maintained in that condition. Add the remaining aluminum chloride down the condenser in 3 or 4 portions at about 5-minute intervals, and continue the heating for a total of 1 hour. During this time, aluminum chloride will sublime into the condenser tube. It should be pushed down into the reaction mixture from time to time with a spatula.

At the end of the heating period, remove the thermometer and allow the black mixture to cool for an hour. During this time, prepare a solution of 10 grams of thiourea in 150 mL of methanol (Note 1).

At the end of the cooling period, extract the product from the flask by adding hexane, swirling the mixture, and decanting, being careful not to pour out any of the black, tarry material; use a total of 70–80 mL of hexane in 4 or 5 portions. Add 1 gram of chromatography-grade alumina to the combined hexane extracts, and swirl occasionally for about 1 minute. Filter the extract by gravity into the methanolic solution of thiourea, using 10–20 mL of hexane to rinse the flask and filter paper. Stir the two layers together for 2 or 3 minutes to permit the beautifully crystalline inclusion complex to complete its formation (Note 2). Collect the complex by suction filtration and wash it with about 20 mL of hexane. Yield: (after drying to constant weight) 6–7 grams.

Transfer the crystals (which need not be dry) to a 125-mL separatory funnel, and add 80 mL of water and 40 mL of ether. Shake the funnel vigorously for about 5 minutes, releasing the pressure occasionally, and then allow the layers to separate. If any solid remains, draw off as much of the lower layer as is convenient, add 50 mL more water, and shake vigorously until all the solid has disappeared (Note 3). Dry the ethereal extract over anhydrous magnesium sulfate and evaporate it to dryness in a tared (weighed) 50-mL Erlenmeyer flask. The last of the ether can be removed by keeping the flask on its side for about 20 minutes. The yield is 1.45–1.6 grams (14.5–16%) of material with a melting range of 258–265°C (sealed capillary).

The product can be recrystallized from isopropyl alcohol, using 13 mL per gram, with a recovery of about 60%, to give an overall yield of 0.93 gram (9.3%) of adamantane with a melting point of 268–270°C (sealed capillary).

Notes

1. The process of dissolution can be greatly hastened by stirring with a magnetic stirrer or by heating on the steam bath.
2. Saturation of a similar solution with hexane does not result in crystal formation.
3. If a larger separatory funnel is available, all the water can be used at once.

Time: 4 hours.

Disposal

> **Solid waste:** filter paper
>
> **Nonhalogenated liquid organic waste:** ether; methanol; hexane; isopropyl alcohol
>
> **Solid inorganic waste:** thiourea; magnesium sulfate
>
> **Solid organic waste:** tarry residue; tetrahydrocyclopentadiene; adamantane
>
> **Thiourea solution:** allow to evaporate; discard residue with solid organic waste.

Aluminum chloride: add cautiously to water; discard residue after evaporation with solid inorganic waste.

Reference

1. A. Ault and R. Kopet, *J. Chem. Educ.* **46,** 612 (1969).

E14. *cis*-Dibenzoylethyelene from *trans*-1,2-Dibenzoylethylene

In this experiment, the colorless *cis* isomer can be prepared from the yellow *trans* isomer if the hot solution of the *trans* isomer is allowed to stand in the sun during recrystallization. If the solution is not placed in the sun, the *trans* isomer recrystallizes without change.

sunlight

trans-1,2-dibenzoylethylene   *cis*-1,2-dibenzoylethylene
m.p., 113–114°C   m.p., 137–138°C
yellow   colorless

A procedure for the preparation of *trans*-1,2-dibenzoylethylene is given in Reference 1.

Recrystallization of *trans*-1,2-Dibenzoylethylene

Scale: 0.3 grams (0.0013 moles) of *trans*-1,2-dibenzoylethylene.

Procedure (Reference 2)

Place 0.3 grams of *trans*-1,2-dibenzoylethylene in a 25-mL Erlenmeyer flask and add 10 mL of 95% ethanol. Heat the mixture almost to boiling on the steam bath and swirl the flask to dissolve the solid. Remove the flask from the heat and add as much decolorizing carbon as would cover a nickel. Filter the hot solution by gravity and collect the filtrate in another 25-mL Erlenmeyer flask.

Reheat the filtrate if necessary to dissolve any crystals that may have formed during the filtration. Set the flask aside to cool slowly. After several hours, the long, canary-yellow needles of the *trans* isomer can be collected by suction filtration, using a small amount of ethanol for rinsing and washing. Recovery: 0.22 grams (73%). Melting point: 113–115°C (Reference 2).

trans isomer: yellow needles;
m.p. = 113 - 115 °C

Time: 1 hour plus $\frac{1}{2}$ hour to collect the product, not including the time required for the material to stand for crystallization.

Disposal

> **Solid waste:** filter paper
> **Nonhalogenated liquid organic waste:** ethanol
> **Solid organic waste:** dibenzoylethylene

The infrared and proton NMR spectra of *trans*-1,2-dibenzoylethylene are shown in Figures E14-1 and E14-2. In the NMR spectrum, the singlet due to the resonance of the olefinic protons appears at $\delta = 8.0$.

Formation of *cis*-1,2-Dibenzoylethylene

Scale: 0.3 grams (0.0013 moles) of *trans*-1,2-dibenzoylethylene.

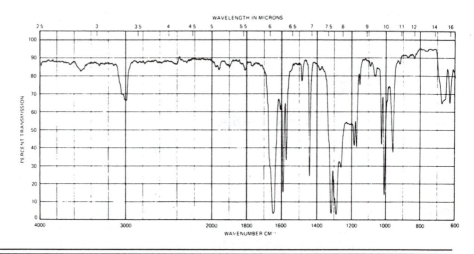

Figure E14-1. Infrared spectrum of *trans*-1,2-dibenzoylethylene; CHCl$_3$ solution.

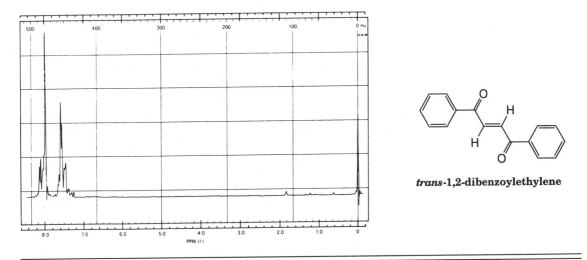

trans-1,2-dibenzoylethylene

Figure E14-2. NMR spectrum of *trans*-1,2-dibenzoylethylene; CDCl$_3$ solution.

Procedure (Reference 2)

Prepare a solution of 0.3 grams of *trans*-1,2-dibenzoylethylene in 10 mL of 95% ethanol as described in the previous procedure. After removing the decolorizing carbon by gravity filtration and reheating to dissolve any crystals that may have formed during the filtration, place the loosely stoppered flask in bright sunlight (Note 1). After several hours, very fine, colorless needles start to form (Note 2). When the flask appears to be about two-thirds full of crystals, they can be collected by suction filtration and washed with a little ethanol. Yield: 0.22 grams (73%). Melting point: 137–138°C (Reference 2).

cis isomer: colorless needles m.p. = 137 - 138 °C

Notes

1. A 275-watt sun lamp can be used instead of the sun (References 1 and 2).
2. If yellow needles of the *trans* isomer form, they will redissolve and give way to the colorless needles of the *cis* isomer on further irradiation.

Time: 1 hour plus $\frac{1}{2}$ hour to collect the product, not including the time required for irradiation and crystallization.

Disposal

Solid waste: filter paper
Nonhalogenated liquid organic waste: ethanol
Solid organic waste: dibenzoylethylene

The infrared and proton NMR spectra of *cis*-1,2-dibenzoylethylene are shown in Figures E14-3 and E14-4. In the NMR spectrum, the singlet due to the resonance of the olefinic protons appears at $\delta = 7.1$.

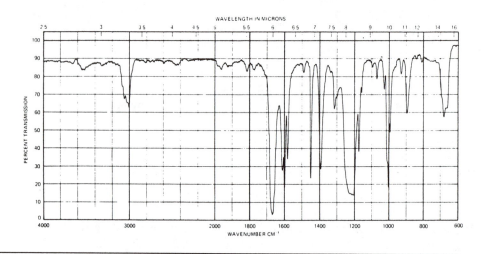

Figure E14-3. Infrared spectrum of *cis*-1,2-dibenzoylethylene; $CHCl_3$ solution.

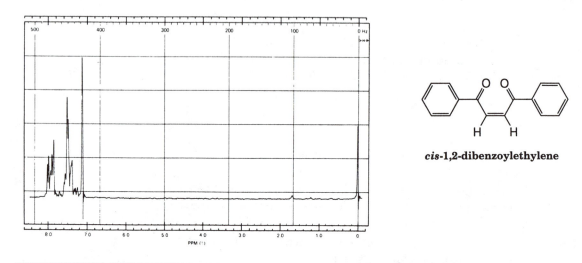

cis-1,2-dibenzoylethylene

Figure E14-4. NMR spectrum of *cis*-1,2-dibenzoylethylene; $CDCl_3$ solution.

References

1. D.J. Pasto, J.A. Duncan, and E. F. Silversmith, *J. Chem. Educ.* **51**, 227 (1974).
2. E. F. Silversmith and F. C. Dunson, *J. Chem. Educ.* **50**, 568 (1973).

PREPARATION OF CYCLOHEXANOL

Two common methods for the preparation of an alcohol are addition of water to an alkene, and reduction of the corresponding aldehyde or ketone. In the first of the next two experiments, cyclohexanol is formed by the acid-catalyzed addition of water to cyclohexene.

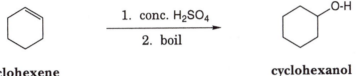

cyclohexene   **cyclohexanol**

In the second, cyclohexanol is formed by the reduction of cyclohexanone by sodium borohydride.

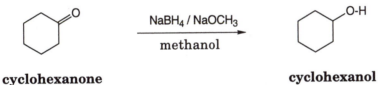

cyclohexanone   **cyclohexanol**

E15. Cyclohexanol from Cyclohexene

In this procedure, cyclohexene is converted first to cyclohexyl hydrogen sulfate, the product of the addition of sulfuric acid to the carbon–carbon double bond of cyclohexene. The cyclohexyl hydrogen sulfate is then hydrolyzed to the alcohol, cyclohexanol, by heating with water.

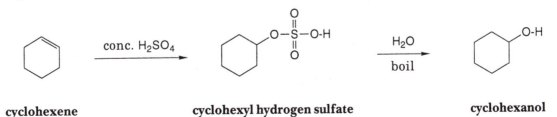

cyclohexene   **cyclohexyl hydrogen sulfate**   **cyclohexanol**

Scale: 0.10 mole (10.1 mL; 8.2 grams) of cyclohexene.

Procedure (Reference 1)

Cautiously add 7.0 mL (0.126 mole) of concentrated sulfuric acid to 3.4 mL (3.4 g; 0.19 mole) of water in a 50-mL ground-glass stoppered Erlenmeyer flask. Cool the solution to room temperature. Add 10.1 mL (8.2 g; 0.10 mole) of cyclohexene. Stopper the flask and shake to mix; the mixture should be shaken or stirred until a clear homogeneous solution is formed (Note 1).

This is a steam distillation; page 87

At this point, pour the mixture into a 250-mL boiling flask and rinse the Erlenmeyer flask with a total of about 120 mL of water, adding the rinsings to the boiling flask. Fit the flask with a distillation adapter and a condenser set for distillation. Heat the mixture to hydrolyze the intermediate and to distill the product. Continue the distillation until 50 or 60 mL of distillate has been collected.

Saturate the distillate with sodium chloride and separate the cyclohexanol by extracting it with ether. Dry the ethereal extract over anhydrous potassium carbonate, filter, and remove the ether by distillation on the steam bath. Distill the residue and collect the fraction boiling between 155 and 162°C as cyclohexanol.

Note

1. Some time can be saved by allowing the mixture to stand between laboratory periods.

Time: 3 hours.

The infrared spectrum of cyclohexanol is shown in Figure E15-1.

Disposal

Sink: sulfuric acid (diluted)

Solid waste: filter paper

Nonhalogenated liquid organic waste: cyclohexene; ether; cyclohexanol

Solid inorganic waste: sodium chloride; potassium carbonate

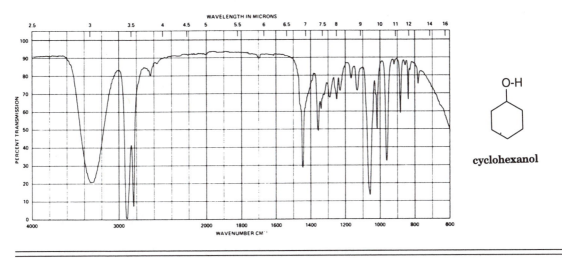

Figure E15-1. Infrared spectrum of cyclohexanol; thin film.

Reference

1. G. H. Coleman, S. Wawzonek, and R. E. Buckles, *Laboratory Manual of Organic Chemistry,* 2nd edition, Prentice-Hall, Englewood Cliffs, N.J., 1962, p. 52.

Questions

1. Write balanced equations for the reactions that take place in this experiment.
2. At which point in the experiment does each reaction take place?
3. Compare this procedure with that for the preparation of cyclohexene from cyclohexanol (Experiment E17). Explain why the reactions go in opposite directions under the different conditions.

E16. Cyclohexanol from Cyclohexanone

Sodium borohydride is an efficient and specific reducing agent for aldehydes, ketones, and acid chlorides, converting these substances to the corresponding alcohols. Other reducible groups such as carbon–carbon double or triple bonds, nitro or cyano groups, and other carbonyl-containing functional groups such as esters or lactones and amides are not reduced under conditions whereby aldehydes and ketones can be reduced. Sodium borohydride is fairly soluble in water, methanol, and ethanol. In these solvents, in the presence of base, it hydrolyzes to form hydrogen very slowly; acidification, however, leads to rapid evolution of hydrogen. The first step of the reaction is hydride transfer from the borohydride ion to the carbonyl group to give an intermediate borate ester.

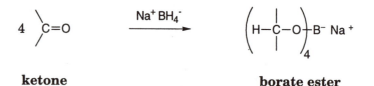

ketone   **borate ester**

The borate ester can then be hydrolyzed by reaction with water at room temperature.

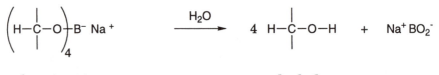

borate ester   **alcohol**

Scale: 0.10 mole (10.3 mL; 9.2 grams) of cyclohexanone.

Procedure

Dissolve 9.8 g (10.2 mL; 0.10 mole) cyclohexanone in 25 mL of methanol. To this solution, add a solution of about 0.5 g sodium methoxide and 1.0 g (0.026 mole) sodium borohydride in 25 mL of methanol. After about 5 minutes, pour the mixture into 100 mL of ice water and add 10 mL of dilute hydrochloric acid. Extract the aqueous mixture with 50 mL of ether, and wash the ethereal extract with three 15-mL portions of water. After drying the extract over anhydrous magnesium sulfate, remove the drying agent by gravity filtration and isolate the product by distillation. Collect as cyclohexanol the portion that boils between 155 and 165°C.

The infrared spectrum of cyclohexanol is shown in Figure E15-1, and the infrared spectrum of cyclohexanone is shown in Figure E16-1.

Time: less than 3 hours.

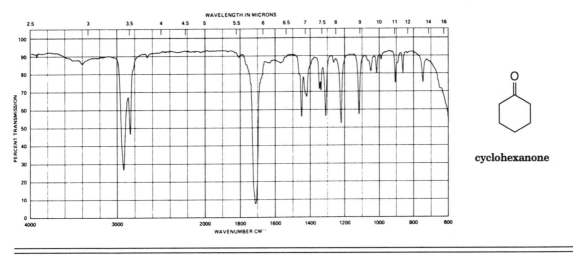

WAVELENGTH IN MICRONS

cyclohexanone

Figure E16-1. Infrared spectrum of cyclohexanone; thin film.

Disposal

Sink: aqueous washes; dilute HCl; sodium methoxide (after addition to water)

Solid waste: filter paper

Nonhalogenated liquid organic waste: cyclohexanone; cyclohexanol; methanol; ether

Solid inorganic waste: magnesium sulfate

Sodium borohydride: add to water; discard residue after evaporation with solid inorganic waste.

Questions

1. Write balanced equations for the reactions that take place in this experiment.
2. At which point in the experiment does each reaction take place?

REACTIONS OF CYCLOHEXANOL

Cyclohexanol, an alcohol, undergoes many of the reactions typical of alcohols, and the next experiments illustrate three of these reactions. These reactions are the acid-catalyzed elimination of water to form an alkene, the acid-catalyzed conversion to the corresponding alkyl bromide, and oxidation to cyclohexanone, the corresponding ketone.

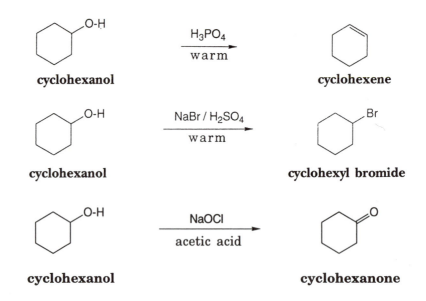

E17. Cyclohexene from Cyclohexanol

In this procedure, cyclohexanol is converted to cyclohexene by heating the alcohol with phosphoric acid. Because the product, cyclohexene, is more volatile than the alcohol, cyclohexanol, the product can be selectively removed by slowly distilling the reaction mixture.

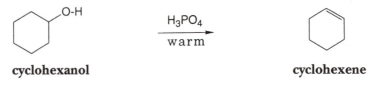

cyclohexanol   cyclohexene

The reaction probably takes place via the conjugate acid of cyclohexanol, which then loses a molecule of water in the rate-limiting step to form the secondary carbocation, which then transfers a proton to the solvent to give the alkene.

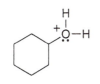

conjugate acid
of cyclohexanol

cyclohexyl
carbocation

Scale: 5.0 grams (0.050 mole) of cyclohexanol.

Procedure (Reference 1)

Introduce 5.0 g (5.2 mL; 0.050 mole) of cyclohexanol, 1 mL of 85% phosphoric acid (Note 1), and a boiling stone into a $50-mL boiling flask. Swirl the flask to mix the layers. Attach to the flask a fractionating column (Note 2) fitted with a distilling adapter, thermometer, and condenser as shown in Figure E17-1.

Heat the flask to bring the contents to a gentle boil. After gently boiling the mixture for 5 minutes or so, slowly (Note 3) distill the mixture until the volume of the residue has been reduced to 1 to 1.5 mL. To reduce loss by evaporation of the product, cool the receiver in an ice bath.

Pour the total distillate into a small separatory funnel, add about 5 mL of water, put the stopper in the top of the separatory funnel, and shake the funnel to mix the contents. After allowing the contents of the separatory funnel to separate into two layers, draw off the lower, aqueous layer and then decant the upper, organic layer into a small, dry Erlenmeyer flask.

Dry the organic material by adding a small amount of anhydrous calcium chloride to the Erlenmeyer flask with the organic material. Allow the mixture to stand with occasional swirling for about 10 minutes.

Carefully decant the organic liquid from the drying agent into a small, clean, dry boiling flask.

Fit the flask containing the dried organic material with a distillation adapter and condenser, and carefully distill the organic material. Collect as cyclohexene the material that boils near 83°C. Cool the collecting vessel in ice to avoid loss by evaporation. The yield of distilled product should be about 3 grams.

The infrared spectrum of cyclohexene is shown in Figure E17-2.

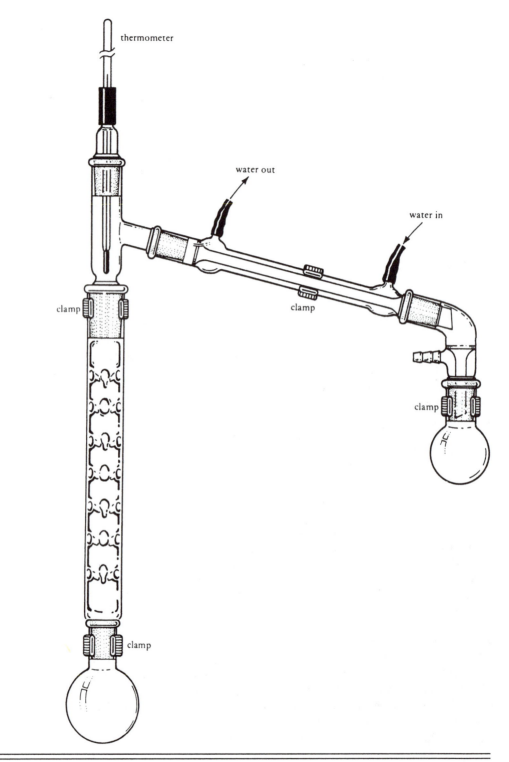

Figure E17-1. Apparatus for the dehydration of cyclohexanol.

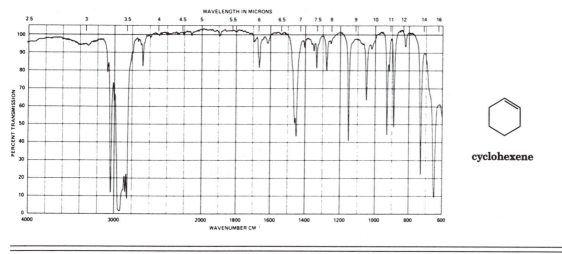

WAVELENGTH IN MICRONS

cyclohexene

Figure E17-2. Infrared spectrum of cyclohexene; thin film.

Notes

1. Sulfuric acid can be used in place of phosphoric acid, but it causes some oxidation of the organic materials.
2. A Vigreux column or a tube packed loosely with stainless steel sponge is suitable.
3. Try to adjust the rate of heating so that the temperature of the distilling vapors does not rise above about 95 to 105°C.

Time: 3 hours.

Disposal

> **Sink:** phosphoric acid (after dilution); residue from boiling flask (after dilution); aqueous washings
> **Solid waste:** filter paper
> **Nonhalogenated liquid organic waste:** cyclohexanol; cyclohexene
> **Solid inorganic waste:** calcium chloride

Questions

1. Why do you suppose that the dehydration of cyclohexanol to

cyclohexene has been the traditional experiment to illustrate the acid-catalyzed dehydration of an alcohol to an alkene?

2. Suppose 2-methylcyclohexanol were subjected to the conditions of this experiment instead of cyclohexanol.

 a. What alkenes would you expect to be formed?

 b. In what relative amounts would you expect them to be formed?

3. When 3,3-dimethyl-2-butanol is boiled with 85% H_3PO_4 according to the procedure of this section, an elimination reaction takes place. If a carbocation mechanism is involved and the initial carbocation undergoes elimination before rearrangement, the product should contain only 3,3-dimethyl-1-butene.

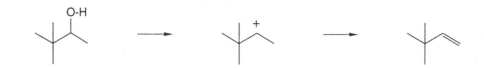

3,3-dimethyl-2-butanol   **3,3-dimethyl-1-butene**

If the initial carbocation should rearrange by methyl migration, two other alkenes could be formed.

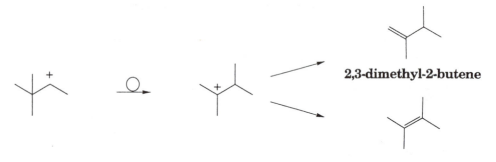

2,3-dimethyl-2-butene

2,3-dimethyl-2-butene

Figure E17-3 shows the proton NMR spectrum of the product mixture obtained in this elimination reaction. The integral of the multiplet at d = 4.6 is 4.2 units, and the integral of the remainder of the spectrum is 124 units. From these data, estimate the relative amounts of the three products formed, and interpret, the results in terms of the relative stability of the products and the mechanism of the reaction.

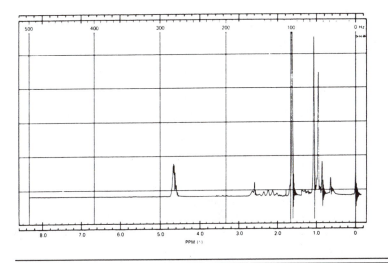

Figure E17-3. NMR spectrum of a mixture of alkenes produced by boiling 3,3-dimethyl-2-butanol with 85% H_3PO_4. The peak at d = 1.6 was recorded at one-tenth the amplitude used to record the rest of the spectrum.

Reference

1. L. F. Fieser, *Experiments in Organic Chemistry*, 3rd edition, revised, Heath, Boston, 1955, p. 62.

E18. Dehydration of 2-Methylcyclo-hexanol: A Variation

When 2-methylcyclohexanol is heated with 85% phosphoric acid, a mixture of alkenes is produced.

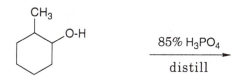

In contrast to the dehydration of cyclohexanol, which can give only a single alkene upon dehydration, you could expect to get a mixture of as many as four alkenes from the dehydration of 2-methylcyclohexanol.

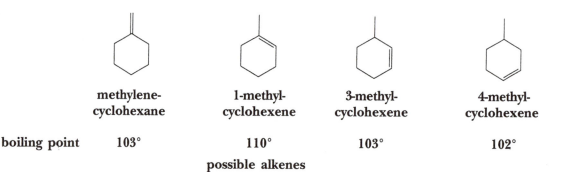

methylene-cyclohexane	1-methyl-cyclohexene	3-methyl-cyclohexene	4-methyl-cyclohexene
boiling point 103°	110°	103°	102°

possible alkenes

In this experiment, you will heat 2-methylcyclohexanol with 85% phosphoric acid to produce a mixture of alkenes that will be subjected to gas chromatographic analysis. After you have determined the relative amounts of the four alkenes present in your product mixture, you should propose a mechanism of reaction that will account for the observed results.

Scale: 5.7 grams (0.050 mole) of 2-methylcyclohexanol.

Procedure

Introduce 5.7 g (6.1 mL; 0.050 mole) of 2-methylcyclohexanol, 1 mL of 85% phosphoric acid, and a boiling stone into a 100-mL boiling flask. Swirl the flask to mix the layers. Attach to the flask a fractionating column (Note 1) fitted with a distilling adapter, thermometer, and condenser (Figure E17-1).

Heat the flask to bring the contents to a gentle boil. After gently boiling the mixture for 5 minutes or so, slowly (Note 2) distill the mixture until the volume of the residue has been reduced to 1 to 1.5 mL. To reduce loss by evaporation of the product, cool the receiver in an ice bath.

Pour the total distillate into a small separatory funnel, add about 5 mL of water, put the stopper in the top of the separatory funnel, and shake the funnel to mix the contents. After allowing the contents of the separatory funnel to separate into two layers, draw off the lower, aqueous layer and then decant the upper, organic layer into a small, dry Erlenmeyer flask.

Dry the organic material by adding a small amount of anhydrous calcium chloride to the Erlenmeyer flask with the organic material. Allow the mixture to stand with occasional swirling for about 10 minutes.

Carefully decant the organic liquid from the drying agent into a small product vial. A typical yield is about 3 grams. The composition of the product mixture will be determined by gas chromatography.

After determining the composition of the mixture of alkenes formed, propose mechanisms to account for the composition of the mixture.

Notes

1. A Vigreux column or a tube packed loosely with stainless steel sponge is suitable.
2. Try to adjust the rate of heating so that the temperature of the distilling vapors does not rise above about 95 to 105°C.

Time: less than 3 hours.

Disposal

> **Sink:** phosphoric acid (after dilution); residue from boiling flask (after dilution); aqueous washings
>
> **Solid waste:** filter paper
>
> **Nonhalogenated liquid organic waste:** 2-methylcyclohexanol; methylcyclohexenes
>
> **Solid inorganic waste:** calcium chloride

Reference

1. Tabor and Champion, *J. Chem. Educ.* **44,** 620 (1967)

Questions

1. Why should you "slowly" distill the reaction mixture?
2. Why should you stop distilling the mixture when the volume has been reduced to about 1 mL?
3. In the gas chromatographic analysis of the products of this reaction,
   a. How does one determine which peak corresponds to which substance?
   b. Does it matter whether you assume that the ratio of peak areas corresponds to the weight ratio or the mole ratio of the alkenes? Explain.
   c. What experiments could you do to show that the ratio of peak areas is a valid measure of product composition?

4. What experiments could you do to tell whether the product composition is due to relative rates of formation or to relative stabilities?

5. a. How many stereoisomeric forms can there be of 2-methylcyclohexanol?

   b. To what extent would you expect that these stereoisomers could be separated by fractional distillation or by gas chromatography? Explain.

   c. Which isomers would you expect to give *exactly the same* results in this experiment if each isomer could be tested separately? Explain.

   d. Which isomers might give *different* results in this experiment if each isomer could be tested separately? Explain.

6. Why do you suppose the dehydration of cyclohexanol is the "traditional" practical example of the acid-catalyzed dehydration of an alkene?

E19. Cyclohexyl Bromide from Cyclohexanol

If an alcohol is heated with a strong acid in the presence of a good nucleophile such as chloride ion or bromide ion, the alcohol can be converted to the corresponding alkyl halide. In this experiment, cyclohexyl bromide is formed by treating an aqueous mixture of cyclohexanol and sodium bromide with concentrated sulfuric acid.

 $\xrightarrow[\text{warm}]{\text{NaBr} / \text{H}_2\text{SO}_4}$

cyclohexanol   **cyclohexyl bromide**

The first step of the reaction is formation of the conjugate acid of the alcohol.

conjugate acid of cyclohexanol

The second step is then formation of cyclohexyl bromide either by an S_N2 process or an S_N1 process.

In an S_N2 process, the conjugate acid of the alcohol reacts directly with bromide ion to give cyclohexyl bromide.

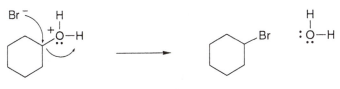

S_N2 process

In an S_N1 process, the conjugate acid of the alcohol undergoes a unimolecular dissociation to give the cyclohexyl carbocation, which then combines with bromide ion to give the product.

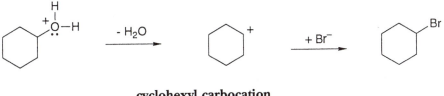

cyclohexyl carbocation
S_N1 process

The infrared spectrum of cyclohexyl bromide is shown in Figure E19-1.

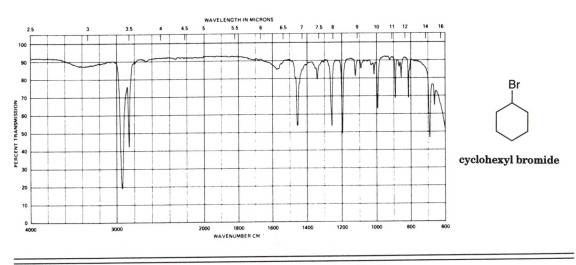

Figure E19-1. Infrared spectrum of cyclohexyl bromide; thin film.

Scale: 5.0 grams (0.050 mole) of cyclohexanol.

Procedure

Apparatus:
Figure 31-2, page 301

Be careful!

Place a magnetic stirring bar in a 100-mL boiling flask, and add to the flask 10 mL of water, 10 g (0.1 mole) of sodium bromide, and 5.0 g (5.2 mL; 0.05 mole) of cyclohexanol. Fit the flask with a Claisen adapter and put the bottom joint of the separatory funnel in the center joint of the adapter. Add 10 mL (18 g; 0.18 mole) of concentrated sulfuric acid to the separatory funnel.

Support the flask over the magnetic stirring motor, and adjust the speed of the motor so that the contents of the flask are stirred quite vigorously. While the contents of the flask are being stirred, add the sulfuric acid drop by drop but otherwise as rapidly as possible; the addition should take about 2 minutes.

Caution!

After completing the addition of the acid, let the mixture stir for 5 more minutes. At this time, pour the hot (Note 1) mixture into the separatory funnel and, after allowing sufficient time for separation of the layers, draw off the aqueous phase. To the cyclohexyl bromide, which remains in the separatory funnel, add a solution of about 2.5 grams potassium carbonate in about 25 mL of water. Mix the layers thoroughly, and, after separating the layers, dry the crude cyclohexyl bromide over a mixture of anhydrous potassium carbonate and anhydrous magnesium sulfate for about 10 minutes.

Filter the crude product from the drying agent into a 25-mL boiling flask, add a pinch of anhydrous potassium carbonate, and distill (it is not necessary to use a condenser; Figure 10.7-1). Collect as product the material that boils between 155° and 165°C. Yield: about 4.5 grams.

Note

1. Be very careful when working with the hot reaction mixture, as it contains sulfuric and hydrobromic acids: it will cause painful burns if it is spilled on your hands, and it will burn holes in your clothes.

Time: 3 hours.

Disposal

Sink: sulfuric acid (after dilution); reaction residue (after dilution)

Solid waste: filter paper

Nonhalogenated liquid organic waste: cyclohexanol

Halogenated liquid organic waste: cyclohexyl bromide

Solid inorganic waste: sodium bromide; potassium carbonate; magnesium sulfate

Questions

1. We assumed in the discussion of the mechanism of this reaction that cyclohexyl bromide is not formed via elimination to give cyclohexene followed by addition of hydrogen bromide to the cyclohexene.

   a. Is there any evidence to support this assumption?

   b. What kinds of experiments might be done to test this assumption?

E20. Cyclohexanone from Cyclohexanol

One of the more common methods for the preparation of a ketone is oxidation of the corresponding secondary alcohol.

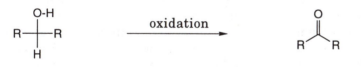

secondary alcohol   **ketone**

A traditional oxidizing agent for this reaction was a sulfuric acid solution of a chromate or a dichromate salt. Because chromate salts are relatively expensive and somewhat hazardous, alternative oxidizing agents have been tried. One of the more popular of these agents is an aqueous solution of sodium hypochlorite, also known as "bleach," which is inexpensive and readily available at the grocery store. In this experiment, we will prepare cyclohexanone by the sodium hypochlorite oxidation of cyclohexanol.

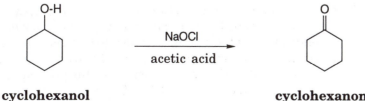

cyclohexanol   **cyclohexanone**

The infrared spectra of cyclohexanol and cyclohexanone are presented in Figures E15-1 and E16-1.

Scale: 5.0 grams (0.050 mole) of cyclohexanol.

Procedure (Reference 1)

Place 5.0 grams (5.2 mL; 0.050 mole) of cyclohexanol, 12 mL of glacial acetic acid, and a magnetic stirring bar in a 125-mL Erlenmeyer flask. Clamp the flask into position on a magnetic stirring motor. Place 80 mL of a 5.25% solution of sodium hypochlorite (0.055 mole NaOCl; Note 1) in an addition funnel (Note 2), and clamp the funnel into position over the neck of the 125-mL Erlenmeyer flask so that the bleach solution can be gradually added while the mixture in the flask is stirred by the magnetic stirrer. Position a thermometer in the Erlenmeyer flask so that the bulb touches the bottom of the flask but does not interfere with the magnetic stirring bar. While the solution is magnetically stirred, add the bleach at such a rate that the temperature of the reaction mixture does not rise much above 35°C.

Bleach should be in excess

After the addition is complete, test the reaction mixture with potassium iodide–starch test paper (Note 3). If the test is *not* positive (*not* blue), add an additional 2 mL of bleach and stir the reaction mixture at room temperature for another 15 minutes; an excess of bleach solution should then be present, as indicated by a greenish-yellow color, and the oxidation of all the alcohol is thereby assured. Reduce the excess bleach by adding concentrated sodium bisulfite until the greenish-yellow color is gone and the solution no longer gives a positive potassium iodide–starch test (Note 3; between 1 and 5 mL will be needed).

This is a steam distillation; page 87

Transfer the contents of the Erlenmeyer flask to a 250-mL boiling flask, and distill the solution until about 25 mL of distillate have been collected in a 125-mL Erlenmeyer flask (Note 4). Add about 3.5 grams of anhydrous sodium carbonate to the distillate to neutralize the acetic acid that has distilled along with the cyclohexanone and water; swirl or stir the mixture until all of the sodium carbonate has dissolved.

Transfer this solution from the Erlenmeyer flask to a separatory funnel, rinse the flask with 15 mL of ether, and add the ether to the separatory funnel. Extract the cyclohexanone into the ether and separate the layers. Dry the ethereal extract over anhydrous magnesium sulfate, filter, and distill. Collect as cyclohexanone the portion of the distillate that boils between 150° and 155°C. Yield: about 4 grams (80%).

Notes

1. A commercial bleach, available at the supermarket, can be used.
2. You should use your separatory funnel as an addition funnel.
3. When sodium hypochlorite is present, it will oxidize iodide to iodine. In the presence of starch, iodine gives an intense blue color. Perform the test by transferring a drop of the reaction mixture to the test paper with a stirring rod.
4. This is a steam distillation.

Time: about 3 hours.

Disposal

Sink: aqueous sodium hypochlorite (bleach); acetic acid

Solid waste: potassium iodide–starch test paper; filter paper

Nonhalogenated liquid organic waste: cyclohexanol; cyclohexanone; ether

Solid inorganic waste: sodium bisulfite; sodium carbonate

Questions

1. What is reduced in this reaction? What is the product of reduction? (You must know this to write a balanced equation for this reaction.)
2. The reaction mixture must be slightly acidic for the oxidation to take place at a reasonable rate. Suggest a mechanism for the oxidation that is consistent with this dependence of the rate on acidity.

Reference

1. N. M. Zuczek and P. S. Furth, *J. Chem. Educ.* **58**, 824 (1981).

ADDITION OF DICHLOROCARBENE TO ALKENES BY PHASE TRANSFER CATALYSIS

One method for the formation of a cyclopropane ring is the addition of a carbene to a carbon–carbon double bond. If the carbene bears substituents, such as a pair of chlorine atoms, a substituted cyclopropane ring can be formed, and the addition of dichloro-carbene to an alkene can be represented in general as

Although carbenes are uncharged, the valence shell of the car-bon atom is electron deficient because it contains only six electrons. For this reason, carbenes act as powerful electrophiles, but the pres-ence of an unshared pair of electrons also gives carbenes nucleo-philic properties.

$$R-\overset{..}{C}-R$$

**a carbene; trivalent; electron-deficient
both electrophilic and nucleophilic**

To avoid reaction with water, a nucleophile, carbene reactions are usually run under anhydrous conditions. In the procedure for the next experiment, however, dichlorocarbene is generated in a water-free environment in an unusual way. The reaction mixture consists of two phases: an aqueous phase containing sodium hydroxide and a catalytic amount of tetra-*n*-butylammonium bromide, and an organic phase containing chloroform and cyclohexene.

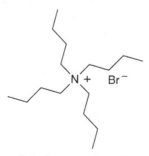

tetra-*n*-butylammonium bromide

Apparently, tetra-*n*-butylammonium ions and hydroxide ions mi-grate into the organic phase as ion pairs, where the hydroxide ion removes a proton from a chloroform molecule to give the trichloro-methide anion.

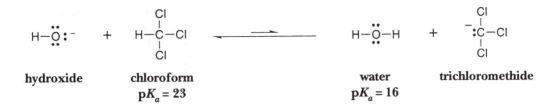

hydroxide   **chloroform**   **water**   **trichloromethide**
$pK_a = 23$ $pK_a = 16$

The trichloromethide anion then undergoes an alpha elimination, in which a chloride ion is lost, to form dichlorocarbene:

trichloromethide   **dichlorocarbene**

The dichlorocarbene, formed in the organic layer where the water concentration is very low and the cyclohexene concentration very high, will usually react to give dichloronorcarane. The chloride ion formed and the tetra-*n*-butylammonium ion then migrate as an ion pair into the water layer, where the ammonium cation can again pick up a hydroxide ion to take into the organic layer. The catalytic role of the tetra-*n*-butylammonium cation can be summarized in this way:

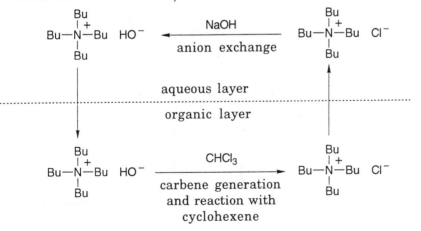

The catalytic function of the tetra-*n*-butylammonium salt in this reaction has been called *phase transfer catalysis*. Other quaternary ammonium salts have the same effect, and no reaction takes place in the absence of the salt.

The next experiment gives a procedure for the addition of dichlorocarbene to cyclohexene, and the two experiments that follow are procedures for the addition of dichlorocarbene to styrene and to 1,5-cyclooctadiene.

E21. Addition of Dichlorocarbene to Cyclohexene

The addition of dichlorocarbene to cyclohexene gives 2,2-dichlorobicyclo[4.1.0]heptane, or dichloronorcarane. The reaction can be represented in this way:

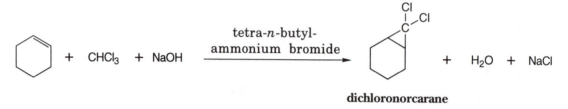

dichloronorcarane

The infrared spectrum of the product, dichloronorcarane, is shown in Figure E21-1.

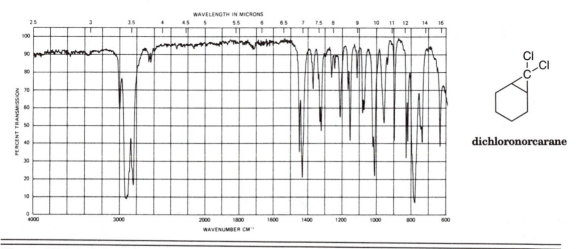

dichloronorcarane

Figure E21-1. Infrared spectrum of dichloronorcarane; thin film.

Scale: 4.1 grams (5.1 mL; 0.050 mole) of cyclohexene.

Procedure

Put a magnetic stirring bar in a 100-mL round-bottom boiling flask. Place a steam bath on a magnetic stirrer, and clamp the flask in position in the rings of the steam bath so that the contents of the flask can be heated by steam and magnetically stirred at the same time. Add to the flask 12 mL of chloroform (17.9 grams; 0.150 mole), 5.1 mL of cyclohexene (4.1 grams; 0.050 mole), and 250 mg (0.00078 mole) of tetra-*n*-butylammonium bromide. Fit the flask with a reflux condenser and then pour, in one portion, down the condenser into the flask, a mixture of 20 mL of 50% sodium hydroxide (0.382 mole) and 15 mL of water. Gently heat and vigorously stir the resulting mixture for 30 minutes (Note 1).

At the end of the heating period, add ice to cool the mixture to room temperature. Then add to the flask 30 mL of ether, and vigorously stir the resulting mixture for three minutes (Note 2). Then transfer the mixture to a separatory funnel, allow the layers to separate, and draw off the lower, aqueous, layer, returning it to the original 100-mL boiling flask. Save the ether extract in a 125-mL Erlenmeyer flask.

Add 15 mL of ether to the aqueous layer in the boiling flask and again vigorously stir the mixture for 3 minutes. Again, using the separatory funnel, separate the layers and then dry the combined ether extracts over anhydrous magnesium sulfate. Filter the dried ethereal solution by gravity into a round-bottom boiling flask, and remove the ether by distillation on the steam bath. Distill the residue at atmospheric pressure using a flame. Collect the fraction that boils between 192 and 197°C as dichloronorcarane. Yield: about 4.7 grams (about 55%).

Notes

1. Heat the mixture just to boiling so that an occasional drop falls from the reflux condenser.
2. The yield is critically dependent on the thoroughness of the extraction.

Time: 3 hours.

Disposal

Sink: spent aqueous phase

Solid waste: filter paper

Nonhalogenated liquid organic waste: cyclohexene; ether

Halogenated liquid organic waste: chloroform; tetra-*n*-butylammonium bromide; dichloronorcarane

Solid inorganic waste: magnesium sulfate

Question

1. What product(s) would you expect to get by a similar treatment of
   a. styrene?
   b. 1,5-cyclooctadiene?

References

1. A. Ault and B. Wright, *J. Chem. Educ.* **53,** 486 (1976).
2. M. Makosza and M. Wawrzyniewicz, *Tetrahedron Lett.* 4659 (1969).
3. G. C. Joshi, N. Singh, and L. M. Pande, *Tetrahedron Lett.* 1461 (1972).
4. K. Isagawa, Y. Kimura, and S. Kwon, *J. Org. Chem,* **39,** 3171 (1974).

E22. Addition of Dichlorocarbene to Styrene: A Variation

The addition of dichlorocarbene to styrene should give 1,1-dichloro-2-phenylcyclopropane:

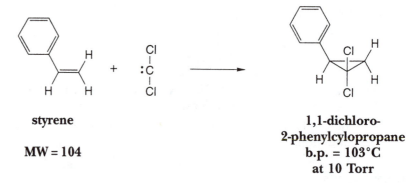

styrene

MW = 104

**1,1-dichloro-2-phenylcylopropane
b.p. = 103°C
at 10 Torr**

An obvious way to attempt to prepare 1,1-dichloro-2-phenyl-cyclopropane is to use the procedure of Section E21, substituting an equimolar amount of styrene for the cyclohexene.

As the equation implies, the "double bonds" of the aromatic ring of styrene will be inert to dichlorocarbene.

Scale: 5.2 grams (5.7 mL; 0.050 mole) of styrene.

Procedure

Use the procedure of Section E21, substituting 5.7 mL (0.050 mole) of styrene for 5.1 mL (0.050 mole) of cyclohexene.

Time: 3 hours.

Disposal

> **Sink:** spent aqueous phase
> **Solid waste:** filter paper
>
> **Nonhalogenated liquid organic waste:** styrene; ether
>
> **Halogenated liquid organic waste:** chloroform; tetra-*n*-butylammonium bromide; 1,1-dichloro-2-phenylcyclopropane
>
> **Solid inorganic waste:** magnesium sulfate

E23. Addition of Dichlorocarbene to 1,5-Cyclooctadiene: Another Variation

When dichlorocarbene is generated in the presence of 1,5-cyclo-octadiene, two equivalents of the carbene should be consumed per mole of the diene, with formation of a *bis*-adduct:

1,5-cyclooctadiene   *bis*-**adduct**

Thus, in adapting the procedure of Section E21 to the use of 1,5-cyclooctadiene, the cyclohexene should be replaced by only half as many moles of the diene.

The reaction of two equivalents of dichlorocarbene with 1,5-cyclooctadiene can, in principle, take place in two ways. Both reactions can take place at, say, the top face of the ring, to form a *cis-bis*-adduct, or one reaction can take place at the top face and the other at the bottom face, to form a *trans-bis*-adduct.

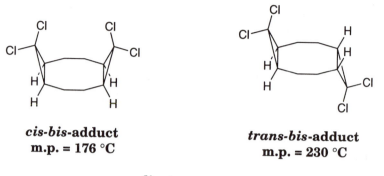

cis-bis-adduct
m.p. = 176 °C

trans-bis-adduct
m.p. = 230 °C

diastereomers

Another way to view these possibilities is to consider the addition of the second equivalent of dichlorocarbene to the product of the addition of the first. The second equivalent can attack either the same face as that attacked by the first, or the face opposite that attacked by the first. Since the diastereomeric products have quite different melting points, it should be easy to determine which isomer is formed in the reaction.

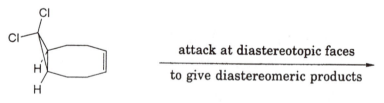

attack at diastereotopic faces
—————————————————————
to give diastereomeric products

Scale: 2.7 grams (3.1 mL; 0.025 mole) of 1,5-cyclooctadiene.

Procedure

Use the procedure of Section E21, substituting 3.1 mL (0.025 mole) of 1,5-cyclooctadiene for 5.1 mL (0.050 mole) of cyclohexene up to the point where the ether used for extraction of the product has been removed to leave the product as a residue.

After the ether that was used for the extraction has been removed by heating on the steam bath, continue to heat the residue, using the steam bath and wrapping a towel around the flask so that it forms a kind of tent to contain the steam, until sufficient volatile material has been driven off that the residue starts to crystallize. Then cool the flask in cold water to complete the crystallization of the residue.

Wash the crude product by adding 5 mL of methanol, crushing the lumps, and decanting the liquid. Repeat the washing with a second 5-mL portion of methanol. Recrystallize the crude product from ethyl acetate and determine its melting point.

Time: 3 hours.

Disposal

>**Sink:** spent aqueous phase
>
>**Solid waste:** filter paper
>
>**Nonhalogenated liquid organic waste:** 1,5-cyclooctadiene; ether; methanol; ethyl acetate
>
>**Halogenated liquid organic waste:** chloroform; tetra-*n*-butylammonium bromide
>
>**Solid inorganic waste:** magnesium sulfate
>
>**Halogenated solid organic waste:** *bis*-adduct

Questions

1. Which of the two possible *bis* adducts was actually formed in this reaction?
2. What is the maximum amount of the other *bis* adduct that could have been formed in the reaction?
3. Account for the formation of the observed product; what influenced the second molecule of dichlorocarbene to attack one of the two diastereotopic faces and not the other?

ALKYL HALIDES FROM ALCOHOLS

In Section E19 we explained how treatment of an alcohol with a strong acid in the presence of chloride or bromide ion could serve to convert the alcohol to the corresponding alkyl halide.

With primary alcohols, the reaction involves establishment of an equilibrium between the alcohol and its conjugate acid,

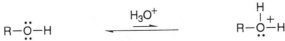

alcohol   conjugate acid of alcohol

followed by a rate-limiting transfer of the alkyl group of the alcohol to the nucleophilic halide ion.

halide ion   **alkyl halide**

With tertiary alcohols, the reaction involves establishment of an analogous equilibrium between the alcohol and its conjugate acid,

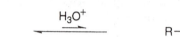

alcohol   **conjugate acid of alcohol**

followed by a rate-limiting loss of water from the conjugate acid of the alcohol to give a carbocation,

conjugate acid of alcohol   **carbocation**

and then combination of the carbocation with halide ion to form the alkyl halide.

alkyl halide

The next four experiments illustrate these reactions.

E24. Isoamyl Bromide from Isoamyl Alcohol

Isoamyl bromide can be prepared by heating isoamyl alcohol with a mixture of sodium bromide and concentrated sulfuric acid. The reaction is probably a nucleophilic displacement of water (the leaving group) by bromide ion (the nucleophile) from the conjugate acid of the alcohol by an S_N2 mechanism.

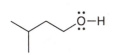

isoamyl alcohol

conc. H_2SO_4

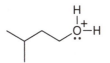

conjugate acid of isoamyl alcohol

proton transfer; a rapidly established equilibrium

conjugate acid of isoamyl alcohol

Br^-

isoamyl bromide

alkyl transfer

rate of reverse reaction is very low under the conditions of the experiment

Isoamyl bromide can be converted to isoamyl acetate, one of the components of the alarm pheromone of the honey bee, by the procedure of Section E30.

Scale: 6.6 grams (8.2 mL; 0.075 moles) of 3-methyl-1-butanol (isoamyl alcohol).

Procedure

Add to a 100-mL ᛋ boiling flask 9 grams (0.087 moles) of sodium bromide, 10 mL of water, and 8.2 mL (6.6 grams; 0.075 moles) of isoamyl alcohol (3-methyl-1-butanol). Cool this mixture in an ice bath and then add, while swirling and cooling, 8.0 mL (14.7 grams; 0.150 mole) of concentrated sulfuric acid.

Fit the flask with a condenser and heat the mixture (Note 1) under reflux for one-half hour (Notes 2 and 3).

Apparatus:
Figure 31-1, page 300

At the end of the heating period, remove the condenser, fit the flask with a distillation adapter, and reconnect the condenser to the distillation adapter for downward distillation. Heat the flask and distill the mixture until no more water-insoluble material comes over (Note 4).

This is a steam distillation; page 87

Pour the distillate into a separatory funnel and add another 25 mL of water to the mixture in the separatory funnel. If necessary, add a pinch of sodium bisulfite to the contents of the funnel to reduce any free bromine (indicated by yellowish color). Now stopper, invert, and shake the funnel gently to mix the contents.

Support the separatory funnel by a clamp or a ring and, after the layers have separated, draw off the lower, organic, layer (Note 5).

Remove (and save, just in case) the upper, aqueous layer, and return the lower, organic, layer to the separatory funnel.

Now add to the organic layer 7 mL of *cold concentrated sulfuric acid. Cautiously* swirl and mix the contents of the separatory funnel and, after allowing the layers to separate, draw off (and save, just in case) the lower, sulfuric acid, layer.

Now add 10 mL of *water* to the organic layer that remains in the separatory funnel. After swirling and mixing, allow the layers to separate and then draw off the lower, organic layer (Note 6).

Remove (and save, just in case) the upper, aqueous layer, and return the lower, organic, layer to the separatory funnel.

Be careful!   Now add 8 mL of a *10% aqueous sodium carbonate* solution to the organic layer that has been returned to the separatory funnel. After swirling and mixing, allow the layers to separate and then draw off the lower, organic layer into a 25-mL Erlenmeyer flask (Note 6).

Dry the organic layer, which is the crude isoamyl bromide, by adding a few pellets of anhydrous calcium chloride. Pour the dried organic layer into a small boiling flask and distill the liquid. Collect as isoamyl bromide the material that boils from about 118°C to about 121°C.

Notes

1. Use a heating mantle to heat the flask.
2. A longer heating time increases the yield only a little.
3. Some hydrogen bromide is evolved. Usually the amount is not objectionable, but it can be trapped by absorption in water.
4. If you look closely at the distillate as it flows down the condenser, you can see droplets of water-insoluble material flowing along in the water. When you can no longer see these droplets, the water-insoluble organic material, which is what you want, is no longer coming over.

5. Although from the difference in density we expect the desired organic layer to be the lower layer at this time, there is no guarantee that this will be so. Dr. Orgo says, *"Always save both layers."*

6. Except for the wash with concentrated sulfuric acid, the desired, organic, layer will be the bottom layer. Thus, the bottom layer will have to be removed, the separatory funnel emptied and rinsed with water, and the organic layer returned to the separatory funnel. Losing material in these transfers is easy.

Time: 3 hours.

Disposal

> **Sink:** sulfuric acid (diluted with water); aqueous layers
> **Nonhalogenated liquid organic waste:** isoamyl alcohol
> **Halogenated liquid organic waste:** isoamyl bromide
> **Solid inorganic waste:** sodium bromide; calcium chloride

Questions

1. Why is concentrated sulfuric acid required for this reaction? Why not just use sodium bromide and water?

2. Why is the crude product washed with concentrated sulfuric acid? (What is removed by this wash?)

3. Why is the crude product next washed with water? (What is removed by this wash?)

4. Why is the crude product finally washed with 10% aqueous sodium bicarbonate solution? (What is removed by this wash?)

5. At the start of the experiment, there is only one liquid layer. During the course of the experiment, a second liquid layer appears. Why does this happen? (What is the composition of the second liquid layer?)

6. In this experiment, an alcohol has been converted to an alkyl halide. It is also possible, however, to convert an alkyl halide to an alcohol. What determines the direction of this reaction?

E25. *n*-Butyl Bromide from *n*-Butyl Alcohol: A Variation

n-Butyl bromide can be prepared by boiling *n*-butyl alcohol with a mixture of sodium bromide, water, and concentrated sulfuric acid. The reaction most likely proceeds by way of formation of the conjugate acid of the alcohol,

alcohol   **conjugate acid of alcohol**

followed by a rate-limiting transfer of the alkyl group of the alcohol to the nucleophilic halide ion.

halide ion   **alkyl halide**

The infrared spectrum of *n*-butyl bromide is shown in Figure E25-1.

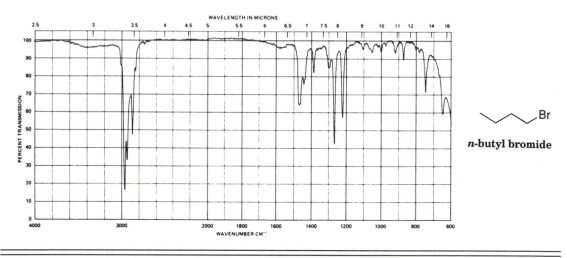

n-butyl bromide

Figure E25-1. Infrared spectrum of *n*-butyl bromide; thin film.

Scale: 5.6 grams (6.9 mL; 0.075 moles) of 1-butanol (*n*-butyl alcohol).

Procedure

Use the procedure of Section E24, substituting 6.9 mL of 1-butanol for the 8.2 mL of 3-methyl-1-butanol, and collecting as product the fraction that boils between 99° and 103°C.

Time: 3 hours.

Disposal

> **Sink:** sulfuric acid (diluted with water); aqueous layers
> **Nonhalogenated liquid organic waste:** *n*-butyl alcohol
> **Halogenated liquid organic waste:** *n*-butyl bromide
> **Solid inorganic waste:** sodium bromide; calcium chloride

E26. *tert*-Butyl Chloride from *tert*-Butyl Alcohol

tert-Butyl chloride is prepared by treating *tert*-butyl alcohol with concentrated hydrochloric acid.

tert-butyl alcohol
conc. HCl
room temperature

tert-butyl chloride

The first step of the reaction is establishment of an equilibrium concentration of the conjugate acid of the alcohol.

tert-butyl alcohol
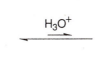
H_3O^+

conjugate acid of tert-butyl alcohol

The next step is a rate-limiting loss of a water molecule to give the *tert*-butyl carbocation,

tert-butyl carbocation

followed by combination of the carbocation with a chloride ion to form *tert*-butyl chloride.

tert-butyl chloride

Apparently, the *tert*-butyl carbocation is formed at a convenient rate near room temperature and without additional Lewis acid catalysis.

The infrared and proton NMR spectra of *tert*-butyl chloride are shown in Figures E26-1 and E26-2.

Scale: 7.4 grams (9.3 mL; 0.0100 mole) of 2-methyl-2-propanol (*tert*-butyl alcohol).

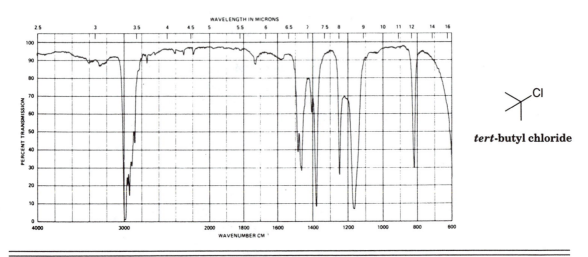

Figure E26-1. Infrared spectrum of *tert*-butyl chloride; thin film.

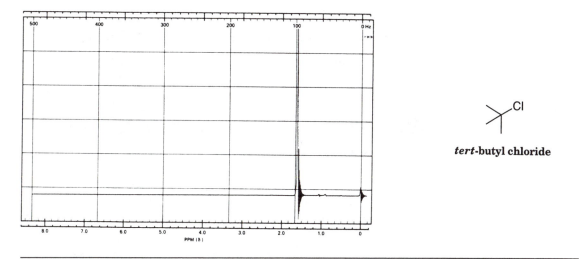

Figure E26-2. NMR spectrum of *tert*-butyl chloride; neat.

Procedure

Place 30 mL (35.4 g; 0.36 mole) concentrated hydrochloric acid, which has been cooled in an ice bath, in a 125-mL separatory funnel. To this, add 9.3 mL (7.4 g; 0.100 mole) *tert*-butyl alcohol.

Shake the mixture occasionally over 20 minutes, relieving the internal pressure by inverting the funnel and cautiously opening the stopcock.   First one layer; then two

Allow the mixture to stand until the layers have separated cleanly, and then remove and discard the lower, hydrochloric acid layer.

Wash the product with 10 mL of water and then with 10 mL of aqueous sodium bicarbonate solution.

Dry the product over anhydrous calcium chloride and distill, collecting the fraction that boils between 49° and 52°C as *tert*-butyl chloride. Cool the distillation receiver in an ice bath so as to minimize loss of product by evaporation.

Time: less than 3 hours.

Disposal

> **Sink:** concentrated HCl (diluted with water); aqueous wash; aqueous sodium bicarbonate
> **Solid waste:** filter paper

Nonhalogenated liquid organic waste: *tert*-butyl alcohol

Halogenated liquid organic waste: *tert*-butyl chloride

Solid inorganic waste: calcium chloride

E27. *tert*-Amyl Chloride from *tert*-Amyl Alcohol: A Variation

Treatment of *tert*-amyl alcohol by the same procedure used for the conversion of *tert*-butyl alcohol to *tert*-butyl chloride should serve to convert *tert*-amyl alcohol to *tert*-amyl chloride.

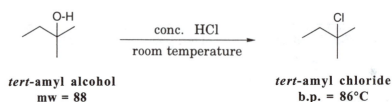

tert-amyl alcohol
mw = 88

tert-amyl chloride
b.p. = 86°C

Scale: 8.8 grams (10.9 mL; 0.100 moles) of 2-methyl-2-butanol (*tert*-amyl alcohol).

Procedure

Use the procedure of Experiment E26, substituting 10.9 mL of 2-methyl-2-butanol for the 9.3 mL of 2-methyl-2-propanol, and collecting as product the fraction that boils at about 86°C.

Time: less than 3 hours.

Disposal

Sink: concentrated HCl (diluted with water); aqueous washes

Solid waste: filter paper

Nonhalogenated liquid organic waste: *tert*-amyl alcohol

Halogenated liquid organic waste: *tert*-amyl chloride

Solid inorganic waste: calcium chloride

E28. Competitive Nucleophilic Substitution of Butyl Alcohols by Bromide and Chloride Ion

The relative reactivities of two species can sometimes be determined by allowing them to compete for a limited amount of a second reagent. In this experiment, bromide ion and chloride ion compete to form the corresponding alkyl halides from one of the isomeric butyl alcohols. The relative nucleophilicity of the two ions is inferred from the relative amounts of the two alkyl halides formed.

Procedure (Reference 1)

Cautiously pour 25 mL of concentrated sulfuric acid over 30 grams of ice. Pour the resulting solution into a 100-mL boiling flask. To this, add 6.69 g (0.125 mole) ammonium chloride and 12.24 g (0.125 mole) ammonium bromide. Fit the flask with a reflux condenser, and bring the solids into solution by swirling and warming the flask on the steam bath. Now add 0.10 mole of one of the isomeric butyl alcohols (Note 1) and boil the mixture very gently for 15–30 minutes, depending on your estimate of the reactivity of the alcohol.

Cool the reaction mixture in an ice bath, and transfer it to a separatory funnel. Save the organic layer and wash it twice with 10-mL portions of cold concentrated sulfuric acid, once with 25 mL of water, and once with aqueous sodium bicarbonate solution. Dry the organic phase over anhydrous calcium chloride.

Determine the ratio of alkyl bromide to alkyl chloride by one of the following methods:

1. Determination of the index of refraction (Section 20.1), assuming a linear relation between the refractive index and molar composition of the sample (Note 2).

2. Determination of the density (Section 19.1), assuming a linear relation between the density and molar composition of the sample (Note 2).

3. Vapor-phase chromatography on a Carbowax column (Sections 15.7 and 15.8), assuming that the ratio of the peak areas equals the molar ratio.

Notes

1. Normal, secondary, and tertiary butyl alcohol appear to work satisfactorily in this experiment. When isobutyl alcohol was used, the index of refraction of the product was greater than that of isobutyl bromide.

2. The following values have been reported for the density and the index of refraction of isomeric butyl chlorides and bromides. (The data are from A. I. Vogel, *Practical Organic Chemistry*, 3rd edition, Wiley, New York, 1957, pp. 293 and 294. The values in parentheses were estimated from the other values.)

		n-Butyl	i-Butyl	s-Butyl	t-Butyl
d_4^{20}	Cl	0.886	0.881	0.874	0.846
d_4^{20}	Br	1.274	1.253	1.256	(1.226)
n_D^{20}	Cl	1.402	1.398	1.397	1.386
n_D^{20}	Br	1.440	1.435	1.437	(1.424)

Time: 3 hours.

Disposal

> **Sink:** concentrated sulfuric acid (diluted with water); aqueous washes
>
> **Nonhalogenated liquid organic waste:** alcohols
>
> **Halogenated liquid organic waste:** alkyl halides
>
> **Solid inorganic waste:** ammonium chloride; ammonium bromide; calcium chloride

Questions

1. What effect would the following procedural variations have on the numerical value of the observed ratio, compared with what the value would be if the variation had not occurred?

   a. The mixture was boiled so vigorously that some of the product escaped from the condenser.

   b. Some of the product evaporated during work-up.

c. The product contained substances other than the two alkyl halides. (Consider each method of analysis.)

d. The alcohol was added before the ammonium salts had completely dissolved.

2. In interpreting the results, it will be assumed that the observed ratio of R-Br to R-Cl is the result of the relative rates of formation of the two compounds and not of their relative stability. How could you show experimentally whether or not this is a valid assumption?

3. One possible result is that the ratios would be the same for each alcohol. How would this result be interpreted?

4. Another result might be that the ratio of R-Br to R-Cl would be different for each alcohol. What would be a reasonable variation in this ratio with the structure of the alcohol?

5. Assuming that class results are available from the reaction of normal, secondary, and tertiary butyl alcohols, interpret them by the presently accepted mechanisms for this type of reaction.

6. What difference would it make if the initial concentrations of ammonium chloride and ammonium bromide were not equal? How would you take this into account?

7. What effect would the following factors have on the numerical value of the observed product ratio, compared with what the value would be otherwise? Assume an observed ratio of 2.0.

   a. 0.10 mole of each ammonium salt was used.

   b. 0.05 mole of alcohol was used.

   c. The reaction was not complete (some alcohol remained unreacted).

   d. Some of the alcohol was converted to the alkene.

8. Answer Question 7 assuming an observed ratio of 1.0.

9. Answer Question 7 assuming an observed ratio of 0.5.

10. What would be the result in this experiment if 0.25 mole of alcohol were used?

11. It should now be apparent that taking the observed ratio of R-Br to R-Cl as a measure of the relative nucleophilicity of bromide and chloride ions in this reaction is an approximation. In fact, in an experiment using 0.125 mole each of bromide and chloride and 0.10 mole of alcohol, a relative nucleophilicity of 2.80 is required to account for an observed ratio of 2.0. If only 0.05 mole of alcohol had been used in an experiment that was otherwise the same, a relative nucleophilicity of 2.23 would account for an observed ratio of 2.0. Derive an exact expression, verify the examples given, and show under what conditions the exact expression reduces to the approximation.

Reference

1. G. K. Helmkamp and H. W. Johnson, *Selected Experiments in Organic Chemistry,* 2nd edition, Freeman, San Francisco, 1968, p. 59.

E29. Kinetics of the Hydrolysis of *tert-*Butyl Chloride

When *tert*-butyl chloride is dissolved in aqueous acetone, it reacts to give *tert*-butyl alcohol and hydrogen chloride.

tert-butyl chloride   *tert*-butyl alcohol

The mechanism of this reaction is slow loss of chloride ion from the halide, to give the *tert*-butyl carbocation, followed by a more rapid reaction of the carbocation with water to form the conjugate acid of the alcohol.

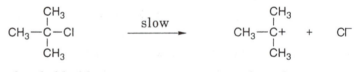

tert-butyl chloride   carbocation

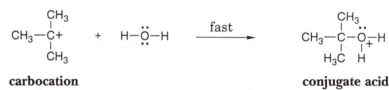

carbocation   **conjugate acid
of the alcohol**

Although water and *tert*-butyl alcohol are similar in basic strength, the fact that there is a very large excess of water present means that most of the molecules of the conjugate acid of *tert*-butyl alcohol will transfer a proton to water.

conjugate acid   **water**   ***tert*-butyl**   **hydronium**
of the alcohol   **alcohol**   **ion**

Because, according to this mechanism, the rate-limiting step of the reaction is the ionization of *tert*-butyl chloride to form the *tert*-butyl carbocation, the rate of the reaction should be proportional to the concentration of *tert*-butyl chloride,

$$\text{rate} = \frac{d[\text{R-Cl}]}{dt} \propto [\text{R-Cl}] \qquad \text{(E29.1)}$$

or

$$\text{rate} = \frac{d[\text{R-Cl}]}{dt} = -k[\text{R-Cl}] \qquad \text{(E29.2)}$$

where k is the proportionality constant, or rate constant, and the minus sign indicates that the concentration of *tert*-butyl chloride decreases with time. Separation of the variables

$$\frac{d[\text{R-Cl}]}{[\text{R-Cl}]} = -k\, dt \qquad \text{(E29.3)}$$

followed by integration gives

$$\log[\text{R-Cl}] = \frac{-kt}{2.303} + C \qquad \text{(E29.4)}$$

where C is the constant of integration of the indefinite integral.

Because, at the start of the experiment, when $t = 0$, the concentration of *tert*-butyl chloride will have its initial value, $[\text{R-Cl}]_0$, the value of the constant of integration can be determined to be $[\text{R-Cl}]_0$,

$$\log[\text{R-Cl}]_0 = -k(0) + C = C$$

Thus, Equation E29.4 can be written as

$$\log[\text{R-Cl}] = \frac{-kt}{2.303} + \log[\text{R-Cl}]_0 \qquad \text{(E29.5)}$$

or as

$$\log\frac{[R-Cl]}{[R-Cl]_0} = \frac{-kt}{2.303} \tag{E29.6}$$

Because $[R-Cl]/[R-Cl]_0$ is the fraction of *tert*-butyl chloride remaining at any time t, Equation E29.6 states that a graph of the log of the fraction of *tert*-butyl chloride remaining at any time t versus t will give a straight line whose slope equals $-k/2.303$. The rate constant, k, which is a measure of the intrinsic rate of a particular reaction under specified conditions of temperature and solvent composition, can be determined from the slope of this line.

In this experiment, the fraction of *tert*-butyl chloride remaining at any particular time will be determined by following the production of acid during the course of the reaction. Measured amounts of 0.010 M sodium hydroxide solution will be added to the mixture to make it basic. Because hydrogen chloride is produced during the course of the reaction, the base will be consumed; when the base is gone, the solution will become acidic. The change in color of the indicator, bromthymol blue, from blue (basic) to yellow (acidic) will signal the time at which the base has been just used up. The original concentration of *tert*-butyl chloride will be proportional to the total volume of base consumed over the entire course of the reaction, V_∞:

$$\log[R-Cl]_0 = (\text{constant})V_\infty \tag{E29.7}$$

and the concentration of *tert*-butyl chloride remaining at the intermediate time t will be proportional to the *difference* between this total volume of base and the volume of base that has been consumed up to the intermediate time:

$$[R-Cl] = (\text{constant})(V_\infty - V_t) \tag{E29.8}$$

Thus, the fraction of *tert*-butyl chloride remaining at any time will equal the following ratio:

$$\frac{[R-Cl]}{[R-Cl]_0} = \frac{(\text{constant})(V_\infty - V_t)}{(\text{constant})\ V_\infty} = \frac{V_\infty - V_t}{V_\infty} \tag{E29.9}$$

Substituting in Equation E29-6 then gives

$$\log\frac{V_\infty - V_t}{V_\infty} = \frac{-kt}{2.303} \tag{E29.10}$$

which states that a graph of the log of the volume of base remaining to be consumed divided by the total that is ultimately consumed versus the time elapsed to that point will be a straight line with a slope equal to $-k/2.303$.

The details of the ways by which the rate constant k can be calculated will be presented after the procedure.

Materials

- Approximately 0.095 *M* *tert*-butyl chloride in acetone, prepared by adding 0.52 mL of *tert*-butyl chloride to 50 mL of reagent grade acetone (1 mL per run)
- Approximately 0.010 *M* sodium hydroxide solution (10 mL per run)*
- Water:acetone, 9:1, by volume (50 mL per run)*

Bromthymol blue indicator solution, 0.04%

125-mL Erlenmeyer flask

Shallow glass or plastic pan

Three 5-mL plastic syringes, with needle (21 gauge)

Thermometer

Magnetic stirring bar

Magnetic stirring motor

Timer

Procedure

Add 50 mL of the 9:1 water:acetone solution to the 125-mL Erlenmeyer flask; add a magnetic stirring bar. Adjust the temperature of the solvent to the desired value (Note 1). Clamp the flask in position over a magnetic stirring motor, and start the stirrer. Fill each of two 5-mL plastic syringes with exactly 5.0 mL of 0.010 *M* sodium hydroxide solution (Note 2). Add 3 drops of the bromthymol blue indicator solution to the flask, and then introduce as accurately as possible 2.0 mL of 0.010 *M* sodium hydroxide from the first syringe; the solution should now have a pale blue color. Next, using a third plastic sy-

*The 0.010 *M* sodium hydroxide solution and the 9:1 water:acetone solution must be prepared ahead of time from boiled, distilled water; they must be carbonate free.

ringe, add exactly 1.0 mL of the solution of *tert*-butyl chloride in acetone (Note 2). Express the *tert*-butyl chloride solution from the syringe in less than 1 second, if possible, and simultaneously start the timer.

Record the time (without stopping the timer) at which the solution becomes acidic, as indicated by a change in color of the indicator from blue to yellow, and then quickly add another 2.0-mL portion of base from the first syringe; the solution should then become light blue.

Again, record the time at which the color changes from blue to yellow, and then add the final 1.0 mL of sodium hydroxide solution from the first syringe; again the color of the solution should revert to blue. After recording the time of the third change in color from blue to yellow, add portions of base from the second syringe until a total of 9.0 mL of base has been added.

After each addition, the solution should be blue, and in each case the time at which the color change from blue to yellow occurs should be recorded. Three 1-mL portions followed by two 0.5-mL portions would be appropriate. Three or 4 minutes after the last change in color from blue to yellow, add base drop by drop from the syringe until the solution just turns blue; if it fades to yellow, add another drop of base. When the blue color from the addition of 1 drop of base persists for 30 seconds (Note 3), the reaction can be assumed to be essentially complete. If the solutions have exactly the concentrations specified, the addition of a total of 9.5 mL of base will be required (Note 4).

Notes

1. It is easiest to run the reaction at room temperature. If the solvent has been prepared at least a day ahead of time, it will be at room temperature and it can be used at the temperature at which it comes from the container. Determine this temperature with your thermometer.

   If you want to run the reaction at a temperature that is only a few degrees above or below that of the lab (at 25°C, for example), you can adjust the temperature of the solvent in the 125-mL flask by swirling the flask in a pan of water (warm or cool, depending on whether you want to raise or lower the temperature) until your thermometer indicates that the solvent has attained the desired temperature.

If you want to run the reaction at a temperature that differs more than about 5° from that of the room, you will have to clamp the flask in place within a shallow glass or plastic dish, which is then supported by the magnetic stirrer. The dish should contain water of the desired temperature. The water in the dish serves as a thermal buffer between the Erlenmeyer flask and the air of the room.

2. Fill the syringe by drawing in somewhat more than the desired volume. While holding the syringe with the needle pointing up, withdraw the plunger slightly to remove the liquid from the needle, tap the barrel to cause all the air bubbles to rise to the top, and then slowly push the plunger in until liquid just starts to come out the tip of the needle. Adjust the volume of the solution in the syringe to that desired by slowly sliding the plunger in, expressing the excess solution into a spare beaker.

3. A longer waiting period will be required at lower temperatures, because the reaction will proceed more slowly.

    The reaction will be approximately 99.9% complete when the total elapsed time is 10 times that required to reach the 5-mL point. For example, if the color change from blue to yellow that corresponds to the final addition from the first syringe takes place after an elapsed time of 50 seconds, the reaction should be about 99.9% complete after a total elapsed time of 500 seconds, or about 8.5 minutes.

4. A volume of 1.0 mL of 0.095 M tert-butyl chloride contains 0.095 mmole of *tert*-butyl chloride, and the hydrogen chloride produced by its hydrolysis will consume the base contained in 9.5 mL of 0.010 M sodium hydroxide solution. However, the concentrations of the solutions probably will not be exactly equal to those specified, and therefore the volume of base required probably will not be exactly 9.5 mL. Because the results of the experiment are interpreted in terms of the fractional extent of reaction, the actual concentrations of *tert*-butyl chloride and of base need not be accurately known.

Disposal

Sink: water:acetone solution; sodium hydroxide solution

Halogenated liquid organic waste: *tert*-butyl chloride solution

Analysis of results

One way to approach the analysis of the results is to prepare a table for each run in which each row corresponds to a time, and each column to a value measured at that time or calculated for that time. The headings for such a table could be these:

Time	Volume of Base Added	Volume of Base Remaining	Fraction of Base Remaining	Log Fraction of Base Remaining
t	V_t	$V_\infty - V_t$	$(V_\infty - V_t)/V_\infty$	$\log\left[(V_\infty - V_t)/V_\infty\right]$

Then a graph of the values in the last column (log of the fraction of base remaining) versus the value of the first column (time) should give a straight line of a slope equal to $-k/2.303$. The rate constant k can then be calculated by multiplying the slope (rise over run) by -2.303. The units of the rate constant are time^{-1}, and the magnitude of the number indicates that fraction of the amount present that will undergo reaction during the next unit of time, neglecting any decrease in rate due to consumption of starting material.

Alternatively, the time at which only $\frac{1}{2}$ of the initial amount is left can be estimated from the line [the time corresponding to log $(\frac{1}{2})$, or -0.301]. The rate constant can be calculated from this time, the half-life or half-time $t_{1/2}$, by the following relationships:

$$k = \frac{\ln 2}{t_{1/2}} = \frac{0.693}{t_{1/2}} \qquad\qquad (E29.11)$$

For example, if the half-time is 50 seconds, the first-order rate constant is 0.0139 sec^{-1}. This relationship between half-time and rate constant is valid only for first-order reactions.

Questions

1. Given the following data, calculate the half-time and the first-order rate constant for the hydrolysis of *tert*-butyl chloride in 90% water:acetone at 25°C.

Time[a]	Volume of Base Added[b]	Volume of Base Remaining	Fraction of Base Remaining	Log Fraction of Base Remaining
t	V_t	$V_\infty - V_t$	$(V_\infty - V_t)/V_\infty$	$\log[(V_\infty - V_t)/V_\infty]$
0	0			
14	2			
33	4			
46	5			
62	6			
82	7			
112	8			
137	8.5			
177	9.0			
∞	9.6			

[a] in seconds
[b] in mL

2. Derive Equation E29.11 from Equation E29.10.

3. Show that if a first-order reaction runs for 10 half times, only 0.00098 of the original amount of starting material will remain.

4. Sometimes first-order processes are described by a characteristic time, or time constant, t. This number is the reciprocal of the first-order rate constant:

$$\tau = \frac{1}{k} \tag{E29.12}$$

We have seen that a half-time of 50 seconds corresponds to a first-order rate constant of 0.0139 sec^{-1}. According to Equation E29.12, this rate constant corresponds to a characteristic time, or time constant, t, of 72 seconds. What fraction remains after the elapse of the characteristic time?

ISOAMYL ACETATE: A COMPONENT OF THE ALARM PHEROMONE OF THE HONEY BEE

Pheromones

Many organisms communicate with members of their own species and with other species by means of chemical signals (References 1 through 4, Section E31). For example, we all understand the message conveyed by the smell of the skunk, and bees are known to make use of quite an elaborate system of chemical signals. Substances that are used to convey messages between members of a particular species are often called *pheromones,* and several classes of pheromones have been recognized. These include sex attractants, alarm pheromones, trail-marking pheromones, and aggregation pheromones. Some pheromones are quite ordinary compounds. For example, *n*-**undecanal** is a sex attractant for the female greater wax moth, **valeric acid** attracts the male sugar beet wireworm, and **isoamyl acetate** is one component of the alarm pheromone of the honeybee.

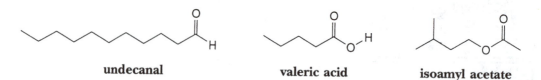

undecanal   **valeric acid**   **isoamyl acetate**

Other pheromones are more complex, and the molecules may include one or more rings and several sites of unsaturation. Examples include **(R)-(+)-limonene**, a substance that triggers a frenzy of biting and snapping among Australian harvester termites, and **phenylacetic acid**, one component of the stink of the stink pot turtle. Long-chain unsaturated, aliphatic alcohols, aldehydes, ketones, and esters seem to be the most common structures of most insect pheromones.

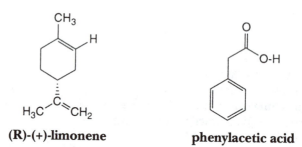

(R)-(+)-limonene   **phenylacetic acid**

One of the most interesting aspects of pheromone communication is the structural and stereochemical specificity of the messenger molecules. For example, the sex attractant of the silkworm moth is **10E,12Z-hexadecadien-1-ol.** Of the four possible diastereomeric forms for this structure, the 10E, 12Z isomer is at least 10^9 times more active than any of the other three isomers.

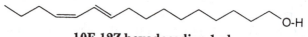

10E,12Z-hexadecadien-1-ol

Another example of specificity is that the New York version of the European corn borer is attracted to E-11-tetradecenyl acetate, while the Iowa version is attracted to the Z isomer.

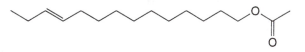

(E)-11-tetradecenyl acetate:
turns on New York corn borers

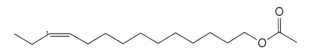

(Z)-11-tetradecenyl acetate:
turns on Iowa corn borers

Specificity also extends to enantiomeric forms of pheromones. Only the (S)-(+) isomer of 4-methyl-3-heptanone functions as an alarm pheromone for Texas leaf-cutting ants, and it appears that the 7R,8S-(+) isomer of *cis*-7,8-epoxy-2-methyloctadecane (disparlure) is far more active as a sex attractant to the female gypsy moth than the enantiomeric 7S,8R-(−) isomer.

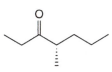

(S)-(+)-4-methyl-3-heptanone

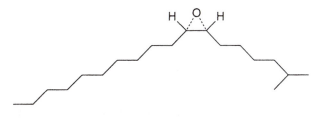

7R,8S-(+)-disparlure

Enantiomeric specificity such as this should not be surprising, however, as many similar examples of this kind of biological specificity are known. Often, we see that only one stereoisomeric form of a drug is physiologically active, or that only one enantiomer of a substance, such as penicillamine, is toxic. Although S penicillamine is toxic, the R isomer is used as a chelating agent to accelerate the elimination from the body of normal heavy metals present in too high, nonphysiological concentrations (Wilson's disease) and of nonphysiological heavy metals (heavy metal poisoning). The R and S forms of amino acids generally have different tastes: only the S form of monosodium glutamate (MSG) has a meaty taste, the enantiomer of glucose is bitter, only the R enantiomer of luciferin reacts enzymatically with the production of light, S-ibuprofen acts more rapidly than R-ibuprofen, and, of course, the R and S isomers of carvone smell like spearmint and caraway, respectively.

Preparation of Esters

Esters of carboxylic acids can be prepared either by forming the alkyl-to-oxygen bond (*alkyl* transfer) or by forming the acyl-to-oxygen bond (*acyl* transfer).

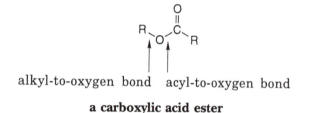

alkyl-to-oxygen bond   acyl-to-oxygen bond

a carboxylic acid ester

Although formation of an ester by alkyl transfer is less common than formation of an ester by acyl transfer, the first synthesis (Section E30) of isoamyl acetate, in which potassium acetate is alkylated by isoamyl bromide, is an example of this approach.

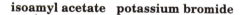

isoamyl bromide potassium acetate   **isoamyl acetate   potassium bromide**

Acylation of an alcohol by an acyl halide, an anhydride, or the carboxylic acid itself is the approach most often followed for the preparation of an ester. The second synthesis (Section E31) of isoamyl acetate from acetic acid and isoamyl alcohol illustrates ester formation by acyl transfer.

isoamyl alcohol   **acetic acid**   **isoamyl acetate**   **water**

Isoamyl acetate has been shown to be one of the active compounds in the alarm pheromone of the honey bee (Reference 5, Section E31). Cotton balls containing freshly excised honey bee stings and placed near the hive entrance were seen to be more frequently stung than control balls. Gas chromatographic and infrared analysis of 700 stings indicated that approximately one microgram of isoamyl acetate can be isolated from each sting. Other substances in addition to isoamyl acetate must be present in the stings, however. Although cotton balls treated with isoamyl acetate do alert and agitate the guard bees, other balls containing an equivalent number of stings incite the bees to sting as well.

E30. Isoamyl Acetate from Isoamyl Bromide and Potassium Acetate

Isoamyl acetate can be prepared by heating a mixture of potassium acetate and isoamyl bromide in dimethylformamide.

isoamyl bromide   potassium acetate   **isoamyl acetate   potassium bromide**

The mechanism of the reaction is an S_N2 displacement of bromide by acetate. The inability of the polar aprotic solvent, dimethylformamide (DMF), to form hydrogen bonds with the oxygen atoms of acetate ion allows the acetate ion to manifest its nucleophilicity.

N,N-dimethylformamide; DMF
polar, aprotic; no hydrogen bonding to the acetate anion

The infrared and proton NMR spectra of isoamyl acetate are shown in Figures E30-1 and E30-2.

Scale: 5.6 mL (7.6 grams; 0.050 mole) of isoamyl bromide.

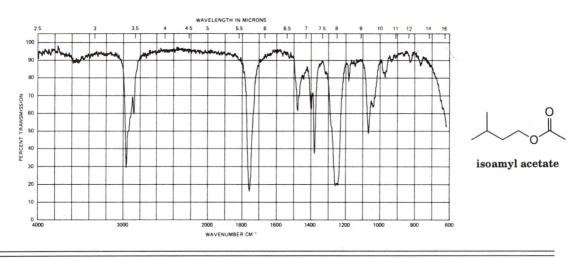

isoamyl acetate

Figure E30-1. Infrared spectrum of isoamyl acetate; thin film.

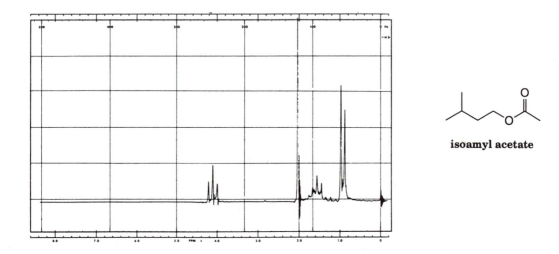

isoamyl acetate

Figure E30-2. NMR spectrum of isoamyl acetate; CCl₄ solution.

Procedure

Place in a 100-mL ᵴ boiling flask 5.6 mL (7.6 grams; 0.050 mole) of isoamyl bromide, 6 grams (0.06 mole) of potassium acetate, 10 mL of dimethylformamide, and a magnetic stirring bar. Fit the flask with a condenser, and support the flask in a steam bath so that the flask can be heated by steam and magnetically stirred at the same time. Heat and stir the mixture for 2 hours (Note 1).

At the end of the heating period, cool the contents of the flask by swirling the flask in cold water, and then add 15 mL of cold water down the condenser. Swirl the mixture to dissolve all of the solid material.

Transfer the mixture to a separatory funnel, and add 15 mL of ethyl ether, using some of the ether to rinse the boiling flask. Thoroughly mix the contents of the separatory funnel, and, after allowing the layers to separate, draw off the lower, aqueous, layer. Wash the remaining ether layer with one 5-mL portion of water.

Transfer the ethereal solution to a 25-mL Erlenmeyer flask, add anhydrous magnesium sulfate to dry the solution, and then filter the dried solution into a 25- or 50-mL boiling flask. Fit the flask with a distillation adapter and condenser, and strip off the ether, using the steam bath as a source of heat and collecting the distillate.

Distill the residual crude isoamyl acetate using a heating mantle as the source of heat.

Collect the fraction that boils between 137° and 145°C as isoamyl acetate. Yield: about 5 grams (about 75%).

Notes

1. If the mixture is stirred well, there will be a time during the reaction when no solid is present. However, solid is always present both at the start and the end of the experiment.
2. You should *not* use a rotary evaporator to strip off the ether; see Question 4.

Time: Less than 4 hours.

Disposal

Sink: aqueous washes

Solid waste: filter paper

Nonhalogenated liquid organic waste: dimethylformamide; ether; isoamyl acetate

Halogenated liquid organic waste: isoamyl bromide

Solid inorganic waste: potassium acetate; magnesium sulfate

Questions

1. What is the solid present at the start of the reaction?
2. What do you suppose the solid is that is present at the end of the reaction?
3. Why is the ethereal extract of the reaction mixture washed with water? (What is removed by this treatment?)
4. Why do you *not* want to use a rotary evaporator to strip off the ether in this experiment?
5. Sometimes a student allows water to get into the reaction mixture. When this happens, the material that is isolated at the end of the experiment is isoamyl bromide and not isoamyl acetate. How might you account for this?

E31. Isoamyl Acetate from Isoamyl Alcohol and Acetic Acid; The Fischer Esterification

Isoamyl acetate, sometimes called pear oil or banana oil, can be prepared by heating together acetic acid and isoamyl alcohol in the presence of concentrated sulfuric acid.

isoamyl alcohol   acetic acid   isoamyl acetate   water

An excess of acetic acid is used to shift the equilibrium toward product formation, because the acid is a little less expensive and is easier to remove from the product.

Scale: 5.4 mL (4.4 grams; 0.050 mole) of isoamyl alcohol.

Procedure

Place in a 100-mL ᵴ round-bottom boiling flask 5.4 mL of isoamyl alcohol (3-methyl-1-butanol; 4.4 grams; 0.050 mole), 11.4 mL of glacial acetic acid (12 grams; 0.20 mole), and 1 mL of concentrated sulfuric acid (1.8 grams; 0.018 mole). Add a few boiling chips, fit the flask with a condenser, and boil the mixture gently, using a heating mantle, for 1 to 2 hours (Note 1).

Setup:
Figure 33.1-2, page 306

At the end of the heating period, cool the contents of the flask by swirling the flask in a beaker of cold water for 2 or 3 minutes and then add 25 grams of ice to the boiling flask (Notes 2 and 3).

Transfer the reaction mixture to a separatory funnel and add to the separatory funnel 20 mL of diethyl ether, using some of the ether to rinse the boiling flask. Thoroughly mix the contents of the separatory funnel, allow the layers to separate, and draw off the lower, aqueous, layer.

Wash the ether layer with one 25-mL portion of cold water, and then extract the ether layer with a solution of 1.5 grams of sodium carbonate dissolved in 25 mL of water, using this solution in two 12.5-mL portions (Note 4).

Dry the ether extract over anhydrous magnesium sulfate, remove the drying agent by gravity filtration, and strip off the ether by distillation on the steam bath (Note 5).

Distill the residue, using a heating mantle, to isolate the isoamyl acetate. The boiling point of isoamyl acetate is reported to be 142°C at 756 Torr. Yield (Note 1): 4 to 4.5 grams (60–70%).

Notes

1. A slightly higher yield is obtained with the longer reaction time.
2. Twenty-five grams of ice is approximately the amount of crushed ice that can be contained in a 50-mL beaker.
3. The point is to cool the mixture well below the boiling point of the solvent that will be used for extraction. Because the solvent will be diethyl ether, b.p. 35°C, the mixture should be cooled to below 25°C.
4. Carbon dioxide will be produced. Take care not to build up excessive pressure in the separatory funnel.
5. You should *not* use a rotary evaporator to strip off the ether; see Question 4.

Time: About 3 hours.

Disposal

> **Sink:** sulfuric acid (after dilution); acetic acid; aqueous washes
>
> **Solid waste:** filter paper
>
> **Nonhalogenated liquid organic waste:** isoamyl alcohol; ether; isoamyl acetate
>
> **Solid inorganic waste:** magnesium sulfate

References

1. J. H. Law and F. E. Regnier, **Pheromones**, *Annu. Rev. Biochem.*, **40,** 533 (1971).
2. D. A. Evans and C. L. Green, **Insect attractants of natural origin**, *Chem. Soc. Rev.* **2,** 75 (1973).
3. William C. Agosta, **Using chemicals to communicate**, *J. Chem. Educ.* **71,** 242 (1994)
4. William C. Agosta, *Chemical Communication: The Language of Pheromones,* Scientific American Library, New York, N.Y., 1992
5. R. Boch, D. A. Shearer, and B. C. Stone, *Nature* **195,** 1018 (1962).

Questions

1. a. List the components of the reaction mixture *before* the solution is boiled for 2 hours.

   b. List the components of the reaction mixture *after* the solution has been boiled for 2 hours.

   c. Explain how the prescribed work-up serves to isolate the desired product from the other components of the reaction mixture. Be specific: tell what is removed, and why, by each action.

2. What is the role of the sulfuric acid? What would be different if you forgot to add the sulfuric acid?

3. We discovered that the yield in this experiment is higher when you cool the reaction mixture somewhat before adding the ice, rather than just adding the ice to the hot reaction mixture. Account for this difference.

4. Why should you *not* use a rotary evaporator to strip off the ether?

LIQUID CRYSTALS

In 1888, an Austrian botanist prepared cholesteryl benzoate and noticed that at about 150°C it "melted" to give a cloudy liquid that became clear only on further heating to above 180°C. Since this first observation of the phenomenon, which has come to be called the formation of a "liquid crystal," many other compounds have been found that show similar behavior.

The liquid crystal phase differs from the ordinary liquid phase as follows. Although the liquid crystal phase takes the shape of its container, its molecules are still ordered to a certain extent, and so the liquid still exhibits some of the properties of the crystalline state. In cholesteryl benzoate, the molecules are associated in layers, with their long axes parallel to one another, and the layers are stacked in a spiral arrangement. This is the so-called *twisted nematic* form, which was earlier called the cholesteric form. All substances that give the twisted nematic form of liquid crystal are made up of chiral molecules; the chirality of the molecules is what leads to the large-scale spiral organization. This spiral organization makes it possible for certain cholesterol derivatives, though colorless in themselves, to selectively scatter light into different colors. As cholesteryl benzoate enters and leaves the liquid crystal phase, reddish-purple colors can be seen if the sample is strongly illuminated from the side. An easy way to see this is described below. Reference 1 (Section E32) discusses the phenomenon of liquid crystals in more detail.

It is often desirable to obtain only one member of a pair of enantiomers. One way to do this is by separating a mixture of enantiomers, as in the resolution of α-phenylethylamine (Section E12). Another is to start the synthesis with only one member of a pair of enantiomers rather than starting with the racemic mixture. However it is done, something chiral must be involved. A particularly attractive possibility is the use of a chiral catalyst, and much research is being done to find ways to form or consume exclusively only one member of a pair of enantiomers in a process that makes use of a chiral catalyst. One recent experiment of this general nature was to heat phenyl ethyl malonic acid in the liquid crystal phase of cholesteryl benzoate to form 2-phenylbutyric acid by thermal decarboxylation (Reference 2, Section E32).

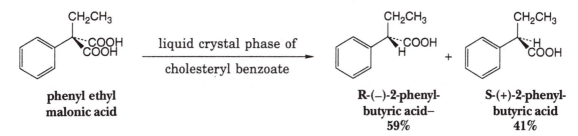

phenyl ethyl malonic acid → liquid crystal phase of cholesteryl benzoate → R-(−)-2-phenyl-butyric acid– 59% + S-(+)-2-phenyl-butyric acid 41%

Apparently, the two carboxyl groups, which reside in enantio-meric environments, react at slightly different rates in the chiral reaction medium, although there is disagreement about the relative importance in this experiment of the chirality of the individual molecules and of the large-scale spiral structure (Reference 3, Section 32).

Although this approach has been applied to synthetic organic chemistry only recently, enzymes have been doing similar things for millions of years. For example, the alcohol dehydrogenases react exclusively with either the R or the S proton in the alpha-methylene group of ethanol.

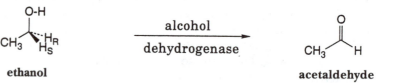

One class of alcohol dehydrogenases removes exclusively H_R, and the other removes exclusively H_S (Reference 4, Section 32).

E32. Cholesteryl Benzoate from Cholesterol

In this experiment, the ester cholesteryl benzoate is prepared by treating the alcohol with an acid chloride in pyridine solution. Upon melting, this material behaves as a liquid crystal.

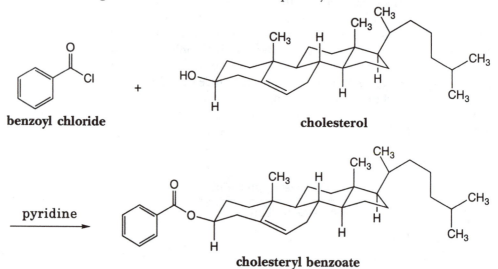

The infrared spectrum of cholesteryl benzoate is shown in Figure E32-1.

Scale: 2.0 grams (0.0026 moles) of cholesterol.

Procedure

Place 1.0 gram (0.0026 mole) of cholesterol in a 50-mL Erlenmeyer flask. To this, add 3 mL of pyridine and swirl the mixture to dissolve the cholesterol. Then add 0.4 mL (0.48 gram; 0.0035 mole) of benzoyl chloride (Note 1), and heat the resulting mixture on the steam bath for about 10 minutes.

At the end of the heating period, cool the mixture somewhat by swirling the flask in a beaker of cold water. Then dilute the mixture with 15 mL of methanol, and collect the solid cholesteryl benzoate by suction filtration, using a little methanol to rinse the flask and to wash the crystals.

Recrystallize the cholesteryl benzoate by heating it in an Erlenmeyer flask with 20 mL of ethyl acetate until it has dissolved, filtering the hot solution by gravity, allowing the filtrate to cool to room temperature, and collecting the crystals by suction filtration (Note 2). Yield: from 0.6 to 0.8 grams (45 to 65%).

Notes

1. Benzoyl chloride is a lachrymator. It should be stored in the hood, and you should take your flask to the hood and

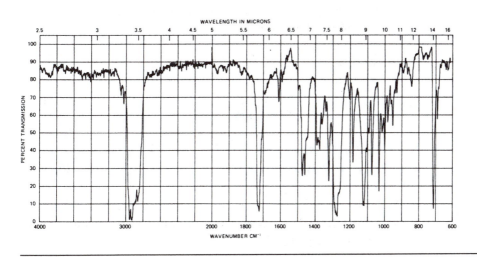

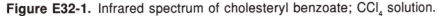

Figure E32-1. Infrared spectrum of cholesteryl benzoate; CCl_4 solution.

work there while making the addition. A graduated pipette should be supplied for use in transferring the benzoyl chloride.

2. The recovery can be improved somewhat by cooling the mixture in an ice bath before collecting the crystals.

Disposal

Solid waste: filter paper

Nonhalogenated liquid organic waste: pyridine; methanol; ethyl acetate

Halogenated liquid organic waste: benzoyl chloride

Solid organic waste: cholesterol; cholesteryl benzoate

The liquid crystal phenomenon

The phenomenon of the formation of the liquid crystal phase of cholesteryl benzoate can be easily seen by placing 100–200 milligrams of the compound on the end of a microscope slide and heating the sample by holding the slide with a pair of tongs above a small burner flame. The solid will turn first to a cloudy liquid and then with further heating to a clear melt. On cooling, the cloudy liquid will first appear, and then it will change to a hard, crystalline solid. With strong lighting from the side, purple and red colors will be seen as the sample changes phases on both heating and cooling. The more cautious the heating, the better you can see the changes. You can repeat the heating and cooling many times with the same sample.

Time: 3 hours.

References

1. L. Verbit, *J. Chem. Educ.* **49,** 37 (1972)

2. L. Verbit, T. R. Halbert, and R. B. Patterson, *J. Org. Chem.* **40,** 1649 (1975).

3. W. H. Pirkle and P. L. Rinaldi, *J. Am. Chem. Soc.* **99,** 3510 (1977).

4. W. L. Alworth, *Stereochemistry and Its Application in Biochemistry,* Wiley-Interscience, 1972, p. 77.

ACETYLATION OF GLUCOSE

Kinetic or Thermodynamic Control

When D-glucose is heated with acetic anhydride and zinc chloride, α-D-glucose pentaacetate is formed.

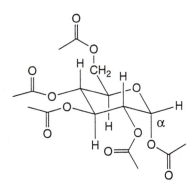

α-D-glucose pentaacetate
m.p. = 112–113°C

If sodium acetate is used in place of zinc chloride, the product is β-D-glucose pentaacetate.

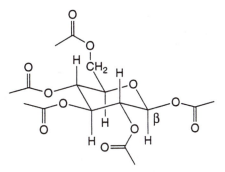

β-D-glucose pentaacetate
m.p. = 132–134°C

The two products differ only in the configuration at C-1, the anomeric carbon.* The following scheme is proposed to account for the different results of these two procedures.

*Because these diastereoisomers differ in configuration at only a single chiral center, they can be called epimers. Epimers of this type, created by formation of a new chiral center upon cyclization of an aldose, can also be called anomers.

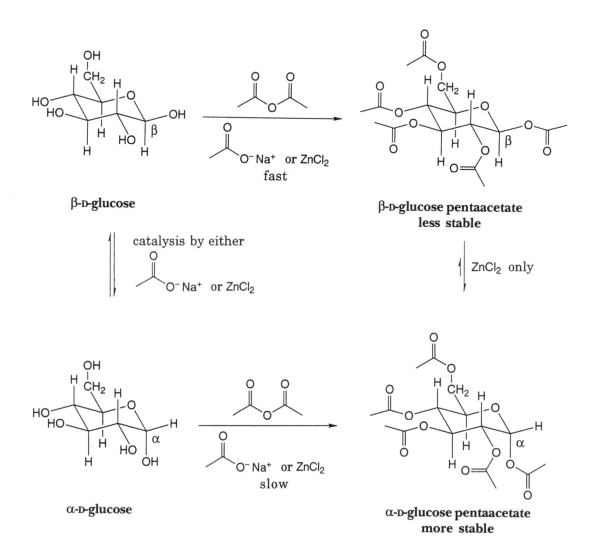

β-D-glucose

β-D-glucose pentaacetate
less stable

catalysis by either

O
‖
CH₃—C—O⁻ Na⁺ or ZnCl₂

ZnCl₂ only

α-D-glucose

α-D-glucose pentaacetate
more stable

In both experiments, β-D-glucose pentaacetate is the initial product, as it is formed more rapidly than α-D-glucose pentaacetate (kinetic control). When zinc chloride is the catalyst, however, the initially formed β-D-glucose pentaacetate is rapidly isomerized to the more stable α isomer (thermodynamic control). Apparently, sodium acetate is not an effective catalyst for this isomerization.

It may be surprising that the α anomer is the more stable because in this isomer the anomeric acetoxyl group has the axial configuration. As required by this scheme, however, when β-D-glucose pentaacetate is treated with acetic anhydride in the presence of zinc chloride, the α isomer can be isolated from the reaction mixture in good yield.

E33. α-D-Glucose Pentaacetate from Glucose

Scale: 2.0 grams (0.011 moles) D-glucose.

Procedure

Place 0.5 g anhydrous zinc chloride (0.0037 mole), 12 mL acetic anhydride (13.0 g; 0.127 mole), and 2.0 g anhydrous D-glucose (0.011 mole) in a 50-mL round-bottom boiling flask. Add a boiling stone, fit the flask with a condenser, and heat the flask cautiously with a burner flame until the contents start to boil. Remove the flask from the heat until the brief exothermic reaction is over, and then, by further heating with the flame, gently boil the mixture for about 2 more minutes.

While it is still hot, pour the solution in a thin stream with good stirring into about 200 mL of a mixture of water and ice. Stir the resulting suspension until the oil has solidified. Break up any lumps and collect the solid by suction filtration.

Recrystallize the crude α-D-glucose pentaacetate from 10 mL of methanol. The melting point of purified α-D-glucose pentaacetate is reported to be 112–113°C. Yield: about 1.9 grams (about 50%).

Time: 2 hours.

Disposal

> **Sink:** glucose; aqueous filtrate
>
> **Solid waste:** filter paper
>
> **Nonhalogenated liquid organic waste:** acetic anhydride; methanol
>
> **Solid inorganic waste:** zinc chloride
>
> **Solid organic waste:** glucose pentaacetate

E34. β-D-Glucose Pentaacetate from Glucose

Procedure

The procedure is the same as that just given for the preparation of α-D-glucose pentaacetate except that 1.2 g anhydrous sodium acetate (0.015 mole) is used in place of the anhydrous zinc chloride.

The melting point of purified β-D-glucose pentaacetate is reported to be 132–134°C. Yield: about 2.4 grams (about 62%).

Time: 2 hours.

Disposal

> **Sink:** glucose; aqueous filtrate; sodium acetate
> **Solid waste:** filter paper
> **Nonhalogenated liquid organic waste:** acetic anhydride; methanol
> **Solid organic waste:** glucose pentaacetate

Questions

1. The scheme proposed to account for these results specifies explicitly that the equilibrium between the α and the β isomers of D-glucose pentaacetate favors the α isomer. Why (or under what conditions) is the position of the equilibrium between the α and the β forms of D-glucose irrelevant?

2. An alternative explanation for the results of these two reactions is that when zinc chloride is used as catalyst, the α isomer is formed more rapidly, but when sodium acetate is the catalyst the β isomer is formed more rapidly. Explain how this interpretation is or is not consistent with the experimental facts.

3. Why is it reasonable that the -OH of the anomeric carbon atom of β-D-glucose is acetylated more rapidly than that of α-D-glucose?

4. Why is it reasonable that β-D-glucose pentaacetate can be converted to the α isomer by zinc chloride catalysis but not by sodium acetate catalysis?

E35. Acetylation of Glucose in *N*-Methylimidazole: A Variation

Scale: 0.5 grams (0.003 moles) of glucose.

Procedure

Wachowiak and Connors (Reference 1) report that when 0.5 gram of glucose is dissolved by heating in 5 mL of *N*-methylimidazole

(Note 1), 1.5 mL of acetic anhydride is added, and, after 15 minutes, 5 mL of water is added, glucose pentaacetate crystallizes from solution during the next 5 to 10 minutes in a yield of more than 90%. They report that they recrystallized their product from ethanol.

Catalysis by *N*-methylimidazole occurs by acyl transfer to *N*-methylimidazole, to give *N'*-acetyl-*N*-methylimidazole, followed by acyl transfer from *N'*-acetyl-*N*-methylimidazole to the -OH groups of glucose, as indicated the following scheme.

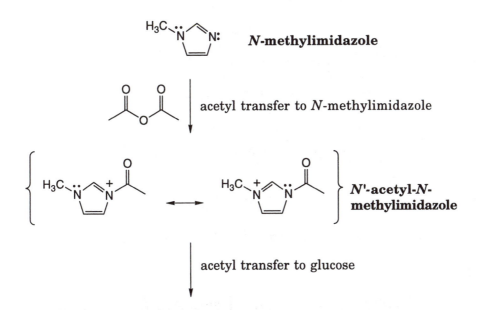

acetylated glucose

The fact that *N*-methylimidazole catalyzes the acetylation of glucose by acetic anhydride implies that (1) *N*-methylimidazole is acetylated by acetic anhydride more rapidly than glucose; (2) the product of acetylation of *N*-methylimidazole, *N'*-acetyl-*N*-methylimidazole, acetylates glucose more rapidly than acetic anhydride; and (3) *N'*-acetyl-*N*-methylimidazole is less stable than the product of the acetylation of glucose. The origins of catalytic effects such as these are not yet understood.

Note

1. *N*-methylimidazole (1-methylimidazole) is available from the Aldrich Chemical Company.

Time: about 2 hours.

Disposal

> **Sink:** glucose; ethanol
>
> **Solid waste:** filter paper
>
> **Nonhalogenated liquid organic waste:** *N*-methylimidazole; acetic anhydride
>
> **Solid organic waste:** glucose pentaacetate

Questions

1. Which isomer, α-D-glucose pentaacetate or β-D-glucose pentaacetate, is formed by this procedure? The melting point of the α isomer is reported to be 112–113°C, while that of the β isomer is reported to be 132–134°C.

2. During the recrystallization of the product from ethanol, the crystals that form upon cooling sometimes disappear if the flask is allowed to stand overnight! What might be the reason for this?

Reference

1. R. Wachowiak and K. A. Connors, *Anal. Chem.* **51,** 27 (1979).

E36. Preparation of Methyl Benzoate, Oil of Niobe

This section presents a standard procedure for the preparation of methyl benzoate from benzoic acid and methanol: treatment of benzoic acid with an excess of methanol in the presence of a strong acid.

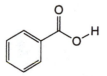 + 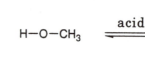 $\rightleftarrows$ 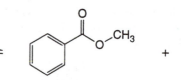 + H—O—H

| benzoic acid | methanol | methyl benzoate | water |
| 0.050 mole | 0.300 mole | | |

The equilibrium is shifted to the right by the six-fold excess of methanol.

The infrared and proton NMR spectra of the product methyl benzoate are shown in Figures E36-1 and E36-2.

Scale: 6.1 grams (0.050 mole) of benzoic acid.

Procedure

Add to a 100-mL boiling flask 6.1 g (0.050 mole) of benzoic acid, 12.1 mL (9.6 g; 0.30 mole) of methanol, and 1 mL concentrated sulfuric acid (Note 1). Fit the flask with a reflux condenser and boil the mixture for about 45 minutes.

Apparatus:
Figure 31-1, page 300

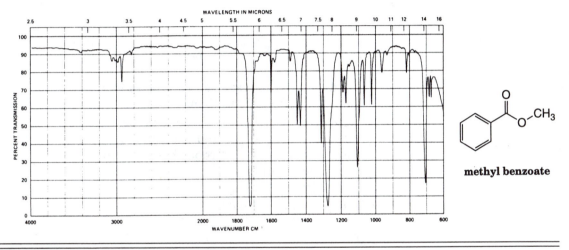

Figure E36-1. Infrared spectrum of methyl benzoate; thin film.

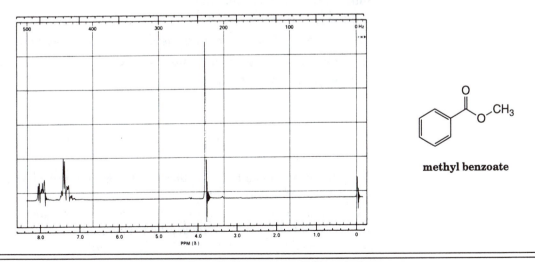

Figure E36-2. NMR spectrum of methyl benzoate; CCl$_4$ solution.

Cool the mixture to room temperature and pour it into a separatory funnel that contains 25 mL of cold water. Rinse the flask with 15 mL of ether, and pour this into the separatory funnel. Mix the contents of the separatory funnel, allow them to settle, and then draw off the aqueous layer. Wash the ethereal layer with 10 mL of water and then 15 mL of 5% aqueous sodium carbonate solution (Note 2).

Dry the ethereal extract over anhydrous magnesium sulfate, filter the solution into a 25-mL boiling flask, and remove the ether by distillation on the steam bath. Use a condenser and collect the ether. Complete the distillation with a flame, collecting as methyl benzoate the material boiling above 190°C. Yield: about 70%.

Notes

1. Add the sulfuric acid cautiously, allowing it to run down the wall of the flask. Swirl to mix.

2. Five percent sodium carbonate solution corresponds to 0.75 grams (0.007 mole) of sodium carbonate in 15 mL of solution. Add the sodium carbonate solution in portions; swirl to mix; watch for foaming.

Time: 3 hours.

Disposal

Sink: sulfuric acid (after dilution); aqueous washes
Solid waste: filter paper
Nonhalogenated liquid organic waste: methyl benzoate; ether
Solid inorganic waste: magnesium sulfate
Solid organic waste: benzoic acid

Questions

1. What is the purpose of washing the ethereal solution with water? (What is removed by washing with water?)

2. What is the purpose of washing the ethereal solution with aqueous sodium carbonate? (What is removed by washing with aqueous sodium carbonate?)

THE GRIGNARD REACTION

One important method for the formation of a carbon–carbon bond is the Grignard reaction. The first step of this three-step reaction is to prepare the Grignard reagent by allowing an ethereal solution of an alkyl halide to react with magnesium.

$$R-X \quad + \quad Mg \quad \xrightarrow{\text{dry ether}} \quad R-Mg-X$$

alkyl halide   **alkylmagnesium halide**
                                                          Grignard reagent

The reaction is strongly inhibited by traces of water, and therefore dry reagents and equipment are essential. Although there is disagreement as to the molecular structure of the active Grignard reagent, it behaves as though it were a source of R: $^-$. Thus, it is a very good nucleophile and a very strong base.

The second step is to treat the ethereal solution of the Grignard reagent with an aldehyde, ketone, or ester. The three equations illustrate the mode of reaction of the reagent with each of these three types of compound.

acetaldehyde
an aldehyde

halomagnesium salt
of a secondary alcohol

acetone
a ketone

halomagnesium salt
of a tertiary alcohol

ethyl acetate
an ester

halomagnesium salt
of a tertiary alcohol

The third step is the hydrolysis of the halomagnesium salt with aqueous acid and isolation of the alcohol produced.

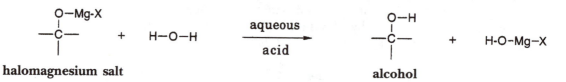

halomagnesium salt   alcohol

The next four sections illustrate some of these possibilities.

E37. Aliphatic Alcohols: Preparation by Grignard Synthesis

In this experiment, you are given a general procedure for the preparation of an aliphatic alcohol by the Grignard synthesis.

For your synthesis, choose a combination of alkyl halide and aldehyde, ketone, or ester from the following lists. Your product should have more than five carbon atoms so that it will not be too soluble in water. Higher-boiling alcohols, especially those that can form a tertiary carbocation, may have to be distilled under reduced pressure to prevent elimination to give an alkene.

- *Alkyl halides:* methyl iodide, ethyl iodide, ethyl bromide, *n*-propyl bromide, *n*-butyl bromide (Note 1)
- *Aldehydes:* acetaldehyde, propionaldehyde, *n*-butyraldehyde, *iso*-butyraldehyde (Note 2)
- *Ketones:* acetone, 2-butanone, 2-pentanone, 3-pentanone, cyclopentanone, cyclohexanone
- *Esters:* ethyl acetate, ethyl propionate

Preparation of the Grignard Reagent

Scale: 0.020 mole of alkyl halide.

Procedure

Place 2.0 g (0.080 mole) of magnesium in a dry 100-mL boiling flask (Note 3). Fit the flask with a Claisen adapter, and fit an addition funnel in the center neck of the adapter and a condenser with a drying tube in the side neck of the adapter, as shown in Figure E37-1. Add to the flask 25 mL of anhydrous ether (Note 1) and 0.020 mole of alkyl halide (Note 1). The reaction should start immediately (Note 4).

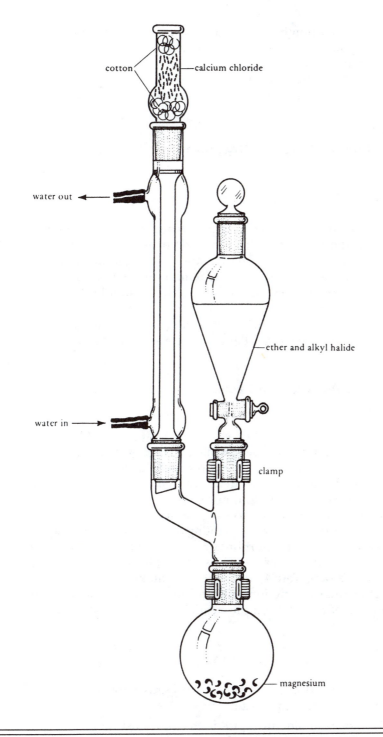

cotton

calcium chloride

water out

ether and alkyl halide

water in

clamp

magnesium

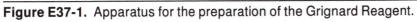

Figure E37-1. Apparatus for the preparation of the Grignard Reagent.

When the reaction has moderated, add via the addition funnel another 0.020-mole portion of the alkyl halide, and repeat with two more 0.020-mole portions.

When the reaction has moderated from the last addition of alkyl halide, gently reflux the mixture on the steam bath for about 15 minutes.

Treatment of the Grignard Reagent with Aldehyde, Ketone, or Ester

Procedure

Cool the flask containing the Grignard reagent with an ice bath. While cooling and swirling or stirring, add drop by drop from the addition funnel a solution of 0.075 mole of the aldehyde or ketone (or 0.038 mole of the ester) in 10 mL of dry ether. The reaction is often quite exothermic, and the addition should therefore be made with caution.

After all the solution has been added, remove the ice bath and allow the mixture to stand at room temperature for about 20 minutes.

Hydrolysis of the Addition Product

Procedure

Pour the reaction mixture slowly, while stirring, onto a mixture of chipped ice and dilute sulfuric acid (prepared by adding 3 mL of concentrated sulfuric acid to about 25 grams of ice and then adding about 25 mL of water). Any addition complex remaining in the reaction flask can be hydrolyzed by pouring the aqueous acid/ether mixture back in and swirling. After separating the ether layer, extract the water layer with two 12-mL portions of ordinary solvent ether.

Combine all the ether solutions, wash with aqueous sodium bisulfite (if necessary; Note 4) and aqueous sodium bicarbonate, and dry over anhydrous magnesium sulfate. Isolate the product alcohol by evaporation of the ether on the steam bath and distillation of the residue.

Notes

1. Ether and alkyl halide can be dried satisfactorily over anhydrous magnesium sulfate.
2. Unless fresh aldehydes are used, they must be purified by distillation.

3. If the humidity is not high, the flask need not be dried by any special method. If desired, the flask can be dried by heating it with a flame, fitting it with a calcium chloride drying tube, and then allowing it to cool to room temperature.

4. If the reaction does not start within a few minutes, the following may be tried:

   a. With a clean, dry stirring rod (not fire polished), crush two or three pieces of the magnesium under the surface of the solution to break the magnesium and expose fresh surfaces. If this is successful, little bubbles will appear where the magnesium has been crushed, and the reaction mixture will become slightly cloudy.

   b. Add a tiny crystal of iodine. If this is done, the ethereal solution of the final reaction product should be treated with a solution of sodium bisulfite to remove the iodine.

   c. Warm the mixture on the steam bath and see if boiling will continue when the steam bath is removed.

   d. Add about 1 mL of someone else's successfully formed Grignard reagent.

   e. Start over, taking more care to see that the apparatus and reagents are dry.

5. Have a beaker or pan of cold water available in case the reaction mixture boils so vigorously that ether starts to escape from the condenser. If this happens, the reaction can be moderated by immersing the flask in the cold water.

Time: 3 hours to the end of the second step; 3 hours to work up the reaction mixture and isolate the product.

Disposal

> **Sink:** sulfuric acid (diluted with water); aqueous washes
>
> **Solid waste:** filter paper
>
> **Nonhalogenated liquid organic waste:** ether; aldehydes and ketones; product alcohols
>
> **Halogenated liquid organic waste:** alkyl halides
>
> **Solid inorganic waste:** magnesium metal; magnesium sulfate

Questions

1. Which combinations of alkyl halide and aldehyde or ketone could be used to prepare 2-butanol by a Grignard synthesis?

2. Which combinations of alkyl halide and aldehyde or ketone could be used to prepare 2-methyl-2-butanol by a Grignard synthesis?

3. a. What would be the final product of the Grignard synthesis between methyl propionate and two equivalents of ethyl magnesium iodide?

   b. What would ethyl propionate give when treated as in 3a?

Reference

A good general reference on the Grignard reagent and its uses is

1. M. S. Kharasch and O. Reinmuth, *Grignard Reactions of Nonmetallic Substances,* Prentice-Hall, Englewood Cliffs, N.J., 1964.

PREPARATION OF TRIPHENYLMETHANOL

The tertiary alcohol triphenylmethanol can be prepared by any one of three different Grignard syntheses.

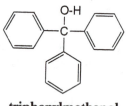

triphenylmethanol

First, triphenylmethanol can be prepared by treating the corresponding ketone, benzophenone, with *one* equivalent of the Grignard reagent derived from bromobenzene.

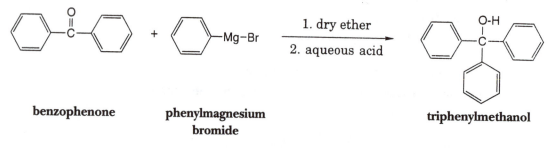

benzophenone   **phenylmagnesium**   **triphenylmethanol**
                          bromide

Second, triphenylmelthanol can be prepared by treating the ester methyl benzoate with *two* equivalents of the Grignard reagent derived from bromobenzene.

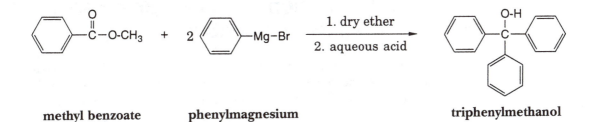

methyl benzoate   **phenylmagnesium**   **triphenylmethanol**
                             bromide

In this second method, attack of the Grignard reagent on the ester produces a tetrahedral intermediate that eliminates methoxide ion to form benzophenone, which then reacts with the second molecule of the Grignard reagent to give the salt of the tertiary alcohol.

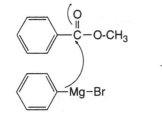
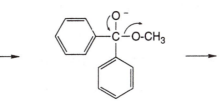
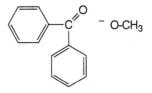

methyl benzoate   **tetrahedral intermediate**   **benzophenone**

   The fact that the ester methyl benzoate can be converted first to benzophenone and then to triphenylmethanol suggests that it might be possible to start with the dimethyl ester of carbonic acid, dimethyl carbonate, and convert this diester to triphenylmethanol by treatment with *three* equivalents of the Grignard reagent derived from bromobenzene.

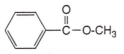

dimethyl carbonate

   The first equivalent of phenylmagnesium bromide would react with the carbonate ester to give methyl benzoate, which would react with a second equivalent of phenylmagnesium bromide to give benzophenone, which would react with the third equivalent of the Grignard reagent to give, after hydrolysis, triphenylmethanol.

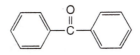

dimethyl carbonate   **methyl benzoate**   **benzophenone**

E38. Preparation of Triphenylmethanol from Benzophenone

This experiment presents the first of the three methods for the preparation of triphenylmethanol. In this experiment, *one* equivalent of the Grignard reagent reacts with one equivalent of the ketone to give, after hydrolysis, the tertiary alcohol.

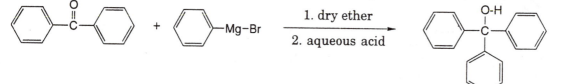

benzophenone   phenylmagnesium   triphenylmethanol
                bromide

Preparation of the Grignard reagent

Scale: preparation of 0.020 mole (5.2 grams) of triphenylmethanol.

Procedure

Place 0.5 g (0.020 mole) of magnesium in a dry 100-mL boiling flask (Note 1). Fit the flask with a Claisen adapter, and fit an addition funnel in the center neck of the adapter and a condenser with a drying tube in the side neck of the adapter, as shown in Figure E37-1. Add to the flask 10 mL of anhydrous ether (Note 2). Add to the addition funnel 2.1 mL (3.1 grams; 0.020 mole) of bromobenzene (Note 2). Now allow about $\frac{1}{4}$ of the bromobenzene to run into the flask with the ether and magnesium metal. The reaction between the magnesium and the bromobenzene should start immediately (Note 3).

When the initial reaction has moderated, add drop by drop from the addition funnel the remaining bromobenzene at such a rate that the vigorous boiling is maintained but not so fast that ether escapes from the top of the condenser (Note 4).

When the reaction has moderated from the last addition of bromobenzene, gently reflux the mixture on the steam bath for about 15 minutes.

Notes

1. If the humidity is not high, the flask need not be dried by any special method. If desired, the flask can be dried by heating it with a flame, fitting it with a calcium chloride

drying tube, and then allowing it to cool to room temperature.

2. Ether and bromobenzene can be dried satisfactorily over anhydrous magnesium sulfate.

3. If the reaction does not start within a few minutes, the following may be tried:

   a. With a clean, dry stirring rod (not fire polished), crush two or three pieces of the magnesium under the surface of the solution to break the magnesium and expose fresh surfaces. If this is successful, little bubbles will appear where the magnesium has been crushed, and the reaction mixture will become slightly cloudy.

   b. Add a tiny crystal of iodine. If this is done, the ethereal solution of the final reaction product should be treated with a solution of sodium bisulfite to remove the iodine.

   c. Warm the mixture on the steam bath and see if boiling will continue when the steam bath is removed.

   d. Add about 1 mL of someone else's successfully formed Grignard reagent.

   e. Start over, taking more care to see that the apparatus and reagents are dry.

4. Have a beaker or pan of cold water available in case the reaction mixture boils so vigorously that ether starts to escape from the condenser. If this happens, the reaction can be moderated by immersing the flask in the cold water.

Reaction with Benzophenone

Procedure

Cool the ethereal solution of phenylmagnesium bromide by swirling or stirring it in a pan of cold water. Then, while continuing to swirl or stir the mixture, add from the addition funnel a solution of 3.64 grams (0.02 mole) of benzophenone in 15 mL of anhydrous ether. Add the solution of benzophenone at such a rate that the reaction mixture boils gently. When the addition is complete, boil the mixture under reflux on the steam bath for about 1–5 minutes.

Hydrolysis of the Addition Product

Procedure

Pour the reaction mixture slowly, while stirring, onto a mixture of chipped ice and dilute sulfuric acid (prepared by adding 3 mL of concentrated sulfuric acid to about 25 grams of ice and then adding about 25 mL of water). Any addition complex remaining in the reaction flask can be hydrolyzed by pouring the aqueous acid/ether mixture back in and swirling. After separating the ether layer, extract the water layer with one 20-mL portion of ordinary solvent ether.

Wash the ethereal solution with aqueous sodium bicarbonate, and dry it over anhydrous magnesium sulfate. Filter the solution into a 50-mL boiling flask, add to the flask 12 mL of hexane, and then concentrate the solution by distilling off the majority of the ether on the steam bath. Condense and collect the ether, and put the recovered ether in the container reserved for it. Continue the distillation until crystals start to appear.

At this time, remove the flask from the distillation apparatus, and set the flask aside to cool to room temperature. Complete the crystallization by cooling the flask in an ice bath and collect the product by suction filtration, using a small amount of cold hexane to rinse the flask and to wash the crystals. The melting point of triphenylmethanol is 160–163°C.

Time: about 3 hours.

Disposal

> **Sink:** aqueous washes
> **Solid waste:** filter paper
> **Nonhalogenated liquid organic waste:** ether; hexane
> **Halogenated liquid organic waste:** bromobenzene
> **Solid inorganic waste:** magnesium metal; magnesium sulfate
> **Solid organic waste:** benzophenone; triphenylmethanol

E39. Preparation of Triphenylmethanol from Methyl Benzoate

This experiment presents the second of the three methods for the preparation of triphenylmethanol. In this experiment, *two* equivalents of the Grignard reagent react with one equivalent of the ester to give, after hydrolysis, the tertiary alcohol.

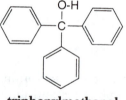

methyl benzoate   phenylmagnesium   triphenylmethanol
                              bromide

Preparation of the Grignard reagent

Scale: preparation of 0.020 mole (5.2 grams) of triphenylmethanol.

Procedure

Place 1.0 g (0.040 mole) of magnesium in a dry 100-mL boiling flask (Note 1). Fit the flask with a Claisen adapter, and fit an addition funnel in the center neck of the adapter and a condenser with a drying tube in the side neck of the adapter, as shown in Figure E37-1. Add to the flask 15 mL of anhydrous ether (Note 2). Add to the addition funnel 4.2 mL (6.3 grams; 0.040 mole) of bromobenzene (Note 2). Now allow about $\frac{1}{4}$ of the bromobenzene to run into the flask with the ether and magnesium metal. The reaction between the magnesium and the bromobenzene should start immediately (Note 3).

When the initial reaction has moderated, add drop by drop from the addition funnel the remaining bromobenzene at such a rate that the vigorous boiling is maintained but not so fast that ether escapes from the top of the condenser (Note 4).

When the reaction has moderated from the last addition of bromobenzene, gently reflux the mixture on the steam bath for about 15 minutes.

Notes

1. If the humidity is not high, the flask need not be dried by any special method. If desired, the flask can be dried by heating it with a flame, fitting it with a calcium chloride drying tube, and then allowing it to cool to room temperature.

2. Ether and bromobenzene can be dried satisfactorily over anhydrous magnesium sulfate.

3. If the reaction does not start within a few minutes, the following may be tried:

    a. With a clean, dry stirring rod (not fire polished), crush two or three pieces of the magnesium under the surface of the solution to break the magnesium and expose fresh surfaces. If this is successful, little bubbles will appear where the magnesium has been crushed, and the reaction mixture will become slightly cloudy.

    b. Add a tiny crystal of iodine. If this is done, the ethereal solution of the final reaction product should be treated with a solution of sodium bisulfite to remove the iodine.

    c. Warm the mixture on the steam bath and see if boiling will continue when the steam bath is removed.

    d. Add about 1 mL of someone else's successfully formed Grignard reagent.

    e. Start over, taking more care to see that the apparatus and reagents are dry.

4. Have a beaker or pan of cold water available in case the reaction mixture boils so vigorously that ether starts to escape from the condenser. If this happens, the reaction can be moderated by immersing the flask in the cold water.

Reaction with Methyl Benzoate

Procedure

Cool the ethereal solution of phenylmagnesium bromide by swirling or stirring it in a pan of cold water. Then, while continuing to swirl or stir the mixture, add from the addition funnel a solution of 2.5 mL (2.7 grams; 0.020 mole) of methyl benzoate in 10 mL of anhydrous ether. Add the solution of methyl benzoate at such a rate that the reaction mixture boils gently. When the addition is complete, boil the mixture under reflux on the steam bath for about 1–5 minutes.

Hydrolysis of the Addition Product

Procedure

Pour the reaction mixture slowly, while stirring, onto a mixture of chipped ice and dilute sulfuric acid (prepared by adding 3 mL of concentrated sulfuric acid to about 25 grams of ice and then adding about 25 mL of water). Any addition complex remaining in the reaction flask can be hydrolyzed by pouring the aqueous acid/ether mixture back in and swirling. After separating the ether layer, extract the water layer with one 20-mL portion of ordinary solvent ether.

    Wash the ethereal solution with aqueous sodium bicarbonate, and dry it over anhydrous magnesium sulfate. Filter the solution into a 50-mL boiling flask, add to the flask 12 mL of hexane, and then concentrate the solution by distilling off the majority of the ether on the steam bath. Condense and collect the ether, and put the recovered ether in the container reserved for it. Continue the distillation until crystals start to appear.

    At this time, remove the flask from the distillation apparatus and set the flask aside to cool to room temperature. Complete the crystallization by cooling the flask in an ice bath and collect the product by suction filtration, using a small amount of cold hexane to rinse the flask and to wash the crystals. The melting point of triphenylmethanol is 160–163°C.

Time: about 3 hours.

Disposal

> **Sink:** aqueous washes
> **Solid waste:** filter paper
> **Nonhalogenated liquid organic waste:** ether; methyl benzoate; hexane
> **Halogenated liquid organic waste:** bromobenzene
> **Solid inorganic waste:** magnesium metal; magnesium sulfate
> **Solid organic waste:** triphenylmethanol

Question

1. Explain how triphenylmethanol could be prepared from bromobenzene and ethyl benzoate. How would the procedure given in this section have to be modified if ethyl benzoate were used?
2. a. What would be the product of reaction of methyl benzoate with two equivalents of ethyl magnesium bromide?
   b. Show how this substance could be prepared by two other Grignard syntheses.

E40. Preparation of Triphenylmethanol from Dimethyl Carbonate

This experiment presents the third of the three methods for the preparation of triphenylmethanol. In this experiment, *three* equivalents of the Grignard reagent react with one equivalent of the diester to give, after hydrolysis, the tertiary alcohol.

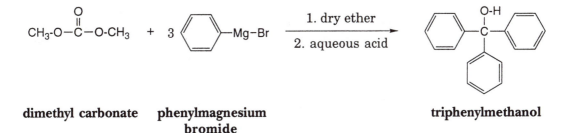

dimethyl carbonate   phenylmagnesium   triphenylmethanol
bromide

Preparation of the Grignard Reagent

Scale: preparation of 0.020 mole (5.2 grams) of triphenylmethanol.

Procedure

Place 1.5 g (0.060 mole) of magnesium in a dry 100-mL boiling flask (Note 1). Fit the flask with a Claisen adapter, and fit an addition funnel in the center neck of the adapter and a condenser with a drying tube in the side neck of the adapter, as shown in Figure E37-1. Add to the flask 25 mL of anhydrous ether (Note 2). Add to the addition funnel 6.3 mL (9.4 grams; 0.060 mole) of bromobenzene (Note 2). Now allow about $\frac{1}{4}$ of the bromobenzene to run into the flask with the ether and magnesium metal. The reaction between the magnesium and the bromobenzene should start immediately (Note 3).

When the initial reaction has moderated, add drop by drop from the addition funnel the remaining bromobenzene at such a rate that the vigorous boiling is maintained but not so fast that ether escapes from the top of the condenser (Note 4).

When the reaction has moderated from the last addition of bromobenzene, gently reflux the mixture on the steam bath for about 15 minutes.

Notes

1. If the humidity is not high, the flask need not be dried by any special method. If desired, the flask can be dried by heating it with a flame, fitting it with a calcium chloride drying tube, and then allowing it to cool to room temperature.

2. Ether and bromobenzene can be dried satisfactorily over anhydrous magnesium sulfate.

3. If the reaction does not start within a few minutes, the following may be tried:

   a. With a clean, dry stirring rod (not fire polished), crush two or three pieces of the magnesium under the surface of the solution to break the magnesium and expose fresh surfaces. If this is successful, little bubbles will appear where the magnesium has been crushed, and the reaction mixture will become slightly cloudy.

   b. Add a tiny crystal of iodine. If this is done, the ethereal solution of the final reaction product should be treated with a solution of sodium bisulfite to remove the iodine.

   c. Warm the mixture on the steam bath and see if boiling will continue when the steam bath is removed.

  d. Add about 1 mL of someone else's successfully formed Grignard reagent.

  e. Start over, taking more care to see that the apparatus and reagents are dry.

4. Have a beaker or pan of cold water available in case the reaction mixture boils so vigorously that ether starts to escape from the condenser. If this happens, the reaction can be moderated by immersing the flask in the cold water.

Reaction with Dimethyl Carbonate

Procedure

Cool the ethereal solution of phenylmagnesium bromide by swirling or stirring it in a pan of cold water. Then, while continuing to swirl or stir the mixture, add from the addition funnel a solution of 1.7 mL (1.8 grams; 0.020 mole) of dimethyl carbonate (Note 1) in 10 mL of anhydrous ether. Add the solution of dimethyl carbonate at such a rate that the reaction mixture boils gently. When the addition is complete, boil the mixture under reflux on the steam bath for about 1–5 minutes.

Note

1. An equivalent amount of diethyl carbonate can be used in place of the dimethyl carbonate.

Hydrolysis of the Addition Product

Procedure

Pour the reaction mixture slowly, while stirring, onto a mixture of chipped ice and dilute sulfuric acid (prepared by adding 3 mL of concentrated sulfuric acid to about 25 grams of ice and then adding about 25 mL of water). Any addition complex remaining in the reaction flask can be hydrolyzed by pouring the aqueous acid/ether mixture back in and swirling. After separating the ether layer, extract the water layer with one 20-mL portion of ordinary solvent ether.

Wash the ethereal solution with aqueous sodium bicarbonate, and dry it over anhydrous magnesium sulfate. Filter the solution into a 50-mL boiling flask, add to the flask 12 mL of hexane and then concentrate the solution by distilling off the majority of the ether on the steam bath. Condense and collect the ether, and put the recovered ether in the container reserved for it. Continue the distillation until crystals start to appear.

At this time, remove the flask from the distillation apparatus and set the flask aside to cool to room temperature. Complete the crystallization by cooling the flask in an ice bath and collect the product by suction filtration, using a small amount of cold hexane to rinse the flask and to wash the crystals. The melting point of triphenylmethanol is 160–163°C.

Time: about 3 hours.

Disposal

Sink: aqueous washes

Solid waste: filter paper

Nonhalogenated liquid organic waste: ether; dimethyl carbonate; hexane

Halogenated liquid organic waste: bromobenzene

Solid inorganic waste: magnesium metal; magnesium sulfate

Solid organic waste: triphenylmethanol

Question

1. Write out reactions for the conversion of dimethyl carbonate to triphenylmethanol.

E41. Preparation of Aniline from Nitrobenzene

Aromatic nitro compounds can be reduced to the corresponding aromatic amine by many reagents. In this experiment, aniline is produced from nitrobenzene by tin and concentrated hydrochloric acid.

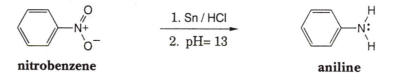

The distillate obtained in the course of the procedure, a mixture of aniline and water, can be used directly for the preparation of acetanilide (Section E42); assume that the mixture contains 0.025 mole of aniline.

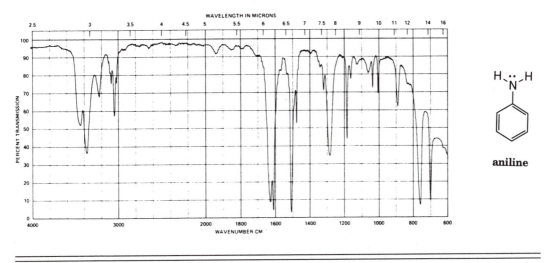

Figure E41-1. Infrared spectrum of aniline; thin film.

The infrared spectrum of aniline is shown in Figure E41-1.

Scale: 3.1 grams (0.025 mole) of nitrobenzene

Procedure

Place 6 g (0.050 mole) of granulated tin and 2.6 mL (3.1 g; 0.025 mole) of nitrobenzene in a 100-mL boiling flask. Make an ice bath ready and add to the tin and nitrobenzene mixture 13 mL (0.16 mole) of concentrated hydrochloric acid. Insert a thermometer and swirl the mixture. Keep the temperature of the reaction between 55 and 60°C by swirling and occasional immersion of the flask in the ice bath.

After 15 minutes, fit the flask with a condenser and heat the reaction mixture on the steam bath with frequent swirling until the condensate in the condenser shows the absence of oily drops of unreacted nitrobenzene (about 15 minutes).

Cool the flask in an ice bath, and while swirling and cooling, slowly add 12 mL 50% aqueous sodium hydroxide (0.23 mole), followed by 20 mL of water.

Fit the flask with a Claisen adapter (Note 1), and fit the center neck with a separatory funnel and the side neck with an adapter and condenser set for downward distillation (Note 2). Steam distill the aniline from the flask by heating it with a flame, adding water by means of the separatory funnel to keep the volume constant. After the distillate begins to come over clear, continue to distill; distill about 10 mL more.

Apparatus:
Figure 12.2-1, page 89

Aniline can be isolated from the distillate by salting it out (using 5 g sodium chloride per 25 mL distillate), extraction with ether or dichloromethane, drying over sodium sulfate, and distilling, collecting the fraction that boils between 180 and 185°C as aniline.

Notes

1. The joints involving the Claisen adapter must be well greased, and the apparatus must be disassembled immediately after the distillation is stopped. Otherwise, the joints will freeze tight (Section 31).
2. The apparatus should look like that shown in Figure 12.2-1.

Time: 3 hours if the aniline is isolated and distilled.

Disposal

Sink: concentrated HCl (after dilution); aqueous washes; sodium chloride

Nonhalogenated liquid organic waste: nitrobenzene; aniline; ether

Halogenated liquid organic waste: dichloromethane

Solid inorganic waste: tin metal

The residue from the steam distillation, which contains tin salts, should be placed in a container reserved for this material.

PREPARATION OF AMIDES

Amines can be converted to the corresponding amides of acetic acid in several ways. One is acetylation by boiling with acetic acid. A second is acetylation by heating with acetic anhydride. A third is acetylation in aqueous solution by treatment with acetic anhydride at a pH of about 5. These three approaches are illustrated by the three procedures presented in Section E42 for the conversion of aniline to acetanilide.

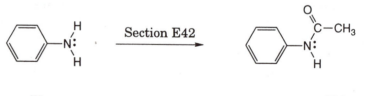

aniline   acetanilide

Then, in Section E43, the conversion of a carboxylic acid to an amide by way of the carboxylic acid chloride is illustrated by the preparation of the mosquito repellent *N,N*-dimethyl-*m*-toluamide.

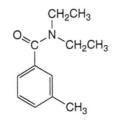

**N,N-dimethyl-*m*-toluamide;
"Off" mosquito repellent**

E42. Acetanilide from Aniline

The acetanilide produced by any of the three procedures that follow should have an infrared spectrum like that shown in Figure E42-1.

Choice of procedures

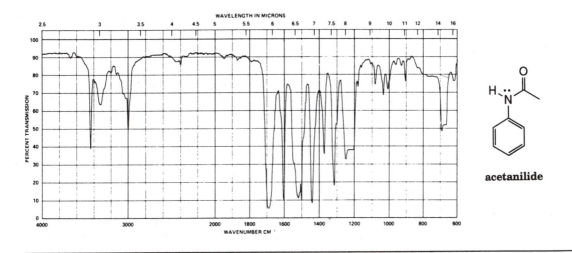

Figure E42-1. Infrared spectrum of acetanilide; $CHCl_3$ solution.

Acetylation with Acetic Acid

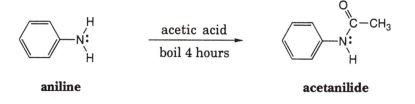

aniline   acetanilide

acetic acid
boil 4 hours

Procedure

Place 2.3 mL (2.3 g; 0.025 mole) aniline, 9.3 mL (9.7 g; 0.160 mole) acetic acid, and a pinch of zinc dust (Note 1) in a 50-mL boiling flask. Fit the flask with an air condenser, and heat the mixture under reflux for about 4 hours.

Pour the hot mixture (Note 2) in a thin stream into 200 mL of cold water. After cooling the mixture in an ice bath for about 10 minutes, collect the product by suction filtration, and wash it with cold water (Note 3). The melting point of acetanilide is reported as 113–115°C.

Notes

1. The zinc reduces the colored impurities in the aniline and helps prevent its oxidation during the reaction.
2. If the mixture is allowed to cool, it will set to a solid cake.
3. Acetanilide can be recrystallized from water.

Acetylation with Acetic Anhydride

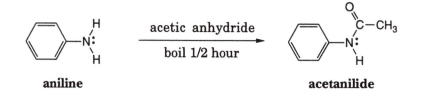

aniline   acetanilide

acetic anhydride
boil 1/2 hour

Procedure

Place 2.3 mL (2.3 g; 0.025 mole) aniline, 2.4 mL (2.6 g; 0.025 mole) acetic anhydride, and a pinch of zinc dust (Note 1) in a 50-mL boiling flask. Fit the flask with a condenser and boil the mixture gently for 30 minutes.

Pour the hot mixture in a thin stream into 100 mL of cold water. After cooling the mixture in an ice bath for about 10 minutes, collect the product by suction filtration and wash it with cold water (Note 3). The melting point of acetanilide is reported as 113–115°C.

Notes

1. The zinc reduces the colored impurities in the aniline and helps prevent its oxidation during the reaction.
2. If the mixture is allowed to cool, it will set to a solid cake.
3. Acetanilide can be recrystallized from water.

Acetylation with Acetic Anhydride and Sodium Acetate in Water

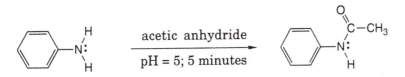

aniline

acetanilide

Procedure

Dissolve 2.3 mL (2.3 g; 0.025 mole) aniline in 65 mL of water and 2.1 mL (0.025 mole) of concentrated hydrochloric acid (Note 1).

Prepare a solution of 2.5 g (0.030 mole) anhydrous sodium acetate (Note 2) in 15 mL of water.

To the solution of aniline hydrochloride, add 2.8 mL (3.1 g; 0.030 mole) acetic anhydride and, as soon as this has been brought into solution by swirling or stirring, add the solution of sodium acetate.

Cool the mixture in an ice bath for about 10 minutes, collect the product by suction filtration, and wash it with cold water. The melting point of acetanilide is reported as 113–115°C.

Notes

1. If the solution is too highly colored, add as much decolorizing carbon as would cover a circle 1 cm in diameter, swirl the mixture to mix it thoroughly, and then remove the carbon by suction filtration. Before pouring your solution into

the Büchner funnel, wet the circle of filter paper, and then apply the vacuum to make sure that the circle of filter paper is pulled down tight over the holes in the funnel.

2. It may seem silly to add *anhydrous* sodium acetate to *water,* but it turns out that anhydrous sodium acetate costs about the same (per *gram*) as the trihydrate, and the anhydrous material dissolves *much faster* than the trihydrate.

Disposal

> **Sink:** acetic acid (after dilution); sodium acetate; aqueous washes
> **Solid waste:** filter paper
> **Nonhalogenated liquid organic waste:** aniline; acetic anhydride
> **Solid inorganic waste:** zinc metal
> **Solid organic waste:** acetanilide

Questions

1. Estimate the pH of the solution at each stage of the reaction:
    a. After the aniline is added to the water in the flask.
    b. After the concentrated hydrochloric acid is added.
    c. After the acetic anhydride is added.
    d. One millisecond after the sodium acetate is added.
    e. One minute after the sodium acetate is added.
2. Indicate the state of the original aniline at each stage of the reaction:
    a. After the aniline is added to the water in the flask.
    b. After the concentrated hydrochloric acid is added.
    c. After the acetic anhydride is added.
    d. One millisecond after the sodium acetate is added.
    e. One minute after the sodium acetate is added.
3. What is the role of the concentrated hydrochloric acid in this reaction?
4. What is the role of the sodium acetate in this reaction?
5. What is the role of the acetic anhydride in this reaction?
6. Do we need to worry about acetate ion reacting with acetic anhydride? Explain.
7. Which is the stronger base, aniline or acetanilide? Explain.

E43. *N,N*-Diethyl-*m*-Toluamide from *m*-Toluic Acid; A Mosquito Repellent: "Off"

One of the more effective insect repellents is the *N,N*-diethylamide of *m*-toluic acid. It can be prepared by converting *m*-toluic acid to the acid chloride with thionyl chloride and then treatment of the product with diethylamine.

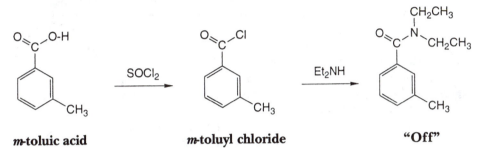

m-toluic acid   **m-toluyl chloride**   "Off"

The amide is an ingredient of a number of commercial insect repellents, but it is probably most familiar as the active ingredient of "Off." Its IR spectrum is shown in Figure E43-1.

Scale: 3.4 grams (0.060 mole) of *m*-toluic acid

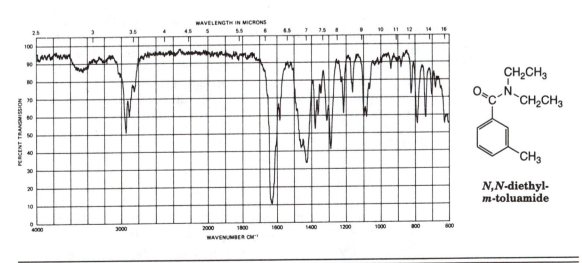

N,N-diethyl-*m*-toluamide

Figure E43-1. Infrared spectrum of *N,N*-diethyl-*m*-toluamide; thin film.

Procedure

Place 3.4 grams (0.060 mole) of *m*-toluic acid in a dry 250-mL boiling flask and add to this 9.0 mL (14.5 grams; 0.122 mole) of thionyl chloride (Note 1). Fit the flask with a reflux condenser, and heat the mixture on the steam bath for about 20 minutes. Because SO_2 and HCl will be evolved during the heating period, you must either work in the hood or make provision for the absorption of these gases (Section 36).

At the end of the heating period, remove the condenser, and add to the flask 40 mL of toluene. Add a magnetic stirring bar to the flask and clamp the flask in position over a magnetic stirring motor. Fit the flask with a Claisen adapter, and place in the center neck a dropping funnel that contains a solution of 20 mL (14.2 grams; 0.196 mole) of diethylamine in 25 mL of toluene. Add the diethylamine solution dropwise while stirring the contents of the flask; the addition should take from 30 to 40 minutes.

When the addition of the diethylamine solution is complete, add to the flask 25 mL of cold water. Continue stirring to thoroughly mix the contents of the flask. Transfer the dark reaction mixture to a separatory funnel and draw off the lower, aqueous layer (Note 2). Wash the toluene solution again with a second 25-mL portion of water. Now wash the toluene solution first with a solution prepared by diluting 5 mL of 50% sodium hydroxide with water to 50 mL (use this solution in two 25-mL portions) and then with a solution prepared by diluting 10 mL of concentrated hydrochloric acid with water to 50 mL (use this solution also in two 25-mL portions). Finally, wash the toluene solution with two more 25-mL portions of water.

Transfer the brown toluene solution of *N,N*-diethyl-*m*-toluamide to a 125-mL Erlenmeyer flask, and dry it for 5 to 10 minutes over anhydrous magnesium sulfate. Filter the dried solution into a 100-mL boiling flask, and remove the toluene by distillation. Because the boiling point of the product is quite high, it should be distilled under reduced pressure (Section 10; Note 3). Although it is colorless when pure, the product is usually obtained as a light brown oil. The boiling point has been reported to be 160°C at 20 Torr and 173°C at 24 Torr. Yield: about 7 grams (60%).

Notes

1. Thionyl chloride is a lachrymator. You should work in the hood while you measure it out and add it to your flask.

2. It may be impossible to distinguish the layers. If so, draw off 25 mL of the liquid.

3. The product can also be purified by chromatography on alumina (B. J.-S. Wang, *J. Chem. Educ.* **51,** 631 (1974).

Time: 4 hours.

Disposal

> **Sink:** aqueous washes
>
> **Solid waste:** filter paper
>
> **Nonhalogenated liquid organic waste:** toluene; diethylamine;
>
> **Solid inorganic waste:** magnesium sulfate
>
> **Solid organic waste:** *m*-toluic acid
>
> **Thionyl chloride:** add cautiously to water; discard the resulting solution in the sink

Reference

1. E. T. McCabe, W. F. Barthels, S. I. Gertler, and S. A. Hall, *J. Org. Chem.* **19,** 493 (1954).

ELECTROPHILIC SUBSTITUTION REACTIONS OF BENZENE DERIVATIVES

Most electrophilic substitution reactions of benzene and of substituted benzenes are thought to take place according to the following general mechanism.

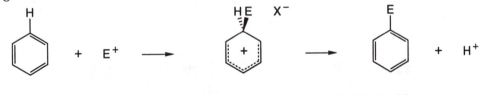

benzene   **sigma complex**   **substituted benzene**

In some cases, the experimental evidence is interpreted best by proposing that E^+, the electrophile, is completely formed as a cation before it reacts with the aromatic compound, as implied by Equation 1. In other cases, the data are best interpreted in terms of a simultaneous attack of the aromatic system on E-X and loss of the leaving group X-:

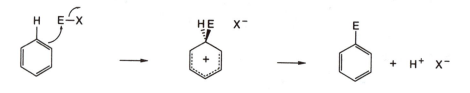

benzene sigma complex substituted benzene

For convenience, however, electrophilic aromatic substitution reactions are often interpreted in terms of attack by previously formed E^+, with the mental reservation that this may not always be the best representation of the reaction.

A complete discussion of the variety of mechanistic possibilities for electrophilic aromatic substitution reactions would have to include the possibility of formation of pi complexes before and after the sigma complex. In some cases, the formation of the first pi complex appears to be the rate-determining step.

The reactivity toward electrophilic substitution of a monosubstituted benzene may be either greater or less than that of benzene itself; the substituent is said to be either activating or deactivating. The new substituent may enter either the *ortho* and *para* positions, or the *meta* position; the original substituent is said to be either *ortho-para* directing, or *meta* directing. Substituents of the general type–A=O have always been found to be deactivating and *meta* directing; examples are the nitro group, $-NO_2$, and the carbonyl group, $-C=O$. Substituents of the general type A: have, with the exception of bromine and chlorine, been found to be activating and *ortho-para* directing. Bromine and chlorine have been found to be *ortho-para* directing but slightly deactivating, presumably because of greater electron withdrawal through an inductive mechanism rather than election release by a resonance mechanism.

E44. Methyl *m*-Nitrobenzoate from Methyl Benzoate

The actual nitrating agent in the nitration of most aromatic compounds is the nitronium ion, NO_2^+. The source of the nitronium ion can be a nitronium salt, such as nitronium fluoroborate, $NO_2^+BF_4^-$, or, as in this experiment, the ion can be derived from nitric acid by the action of a strong acid such as sulfuric acid.

The first step in the conversion of nitric acid to the nitronium ion is proton transfer from the strong acid to the –O–H oxygen of nitric acid.

nitric acid	**conjugate acid of nitric acid**

The second-step is loss of water as the leaving group from the conjugate acid of nitric acid to form the nitronium ion, NO^+_2.

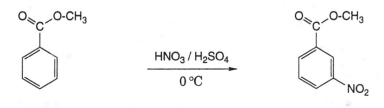

conjugate acid of nitric acid	**nitronium ion**

The conversion of methyl *m*-nitrobenzoate to nitric acid can be summarized in this way.

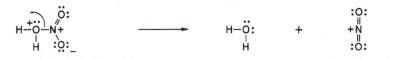

methyl benzoate   **methyl *m*-nitrobenzoate**

Methyl benzoate can be prepared from benzoic acid as indicated in Experiment E36; the infrared spectrum of methyl *m*-nitrobenzoate is shown in Figure E44-1.

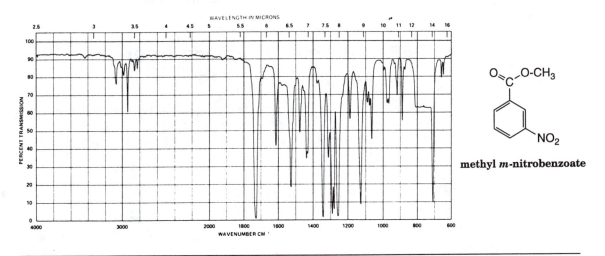

Figure E44-1. Infrared spectrum of methyl *m*-nitrobenzoate; CCl_4 solution.

Scale: 2.7 grams (0.020 mole) of methyl benzoate.

Procedure

Place 5.8 mL (10 g; 0.100 mole) of concentrated sulfuric acid in a 50-mL Erlenmeyer flask, cool it to 0°C in an ice bath, and then add 2.5 mL (2.7 g; 0.020 mole) of methyl benzoate while swirling. Prepare a mixture of 2 mL (3.6 g; 0.036 mole) of concentrated sulfuric acid and 2 mL (2.8 g; 0.031 mole) of concentrated nitric acid, and cool it in an ice bath. Add the cold acid mixture dropwise over a period of about 5 minutes to the methyl benzoate solution, which is constantly swirled in the ice bath.

Allow the resulting mixture to stand at room temperature for an additional 10 minutes with occasional swirling, and then pour it while stirring over about 20 grams of crushed ice.

Collect the resulting solid by suction filtration, and wash it thoroughly with water to remove the acids. Finally, wash the product with two 2-mL portions of ice-cold methanol. The product can be recrystallized from a small amount of methanol.

Time: 2 hours.

Disposal

Sink: sulfuric acid (after dilution with water); nitric acid (after dilution with water)

Solid waste: filter paper

Nonhalogenated liquid organic waste: methyl benzoate; methanol

Solid organic waste: methyl *m*-nitrobenzoate

E45. *p*-Bromoacetanilide from Acetanilide

The electrophilic agent in many electrophilic aromatic bromination reactions is elemental bromine; bromide ion is displaced from the bromine molecule as the electrophilic attack is made upon the aromatic ring.

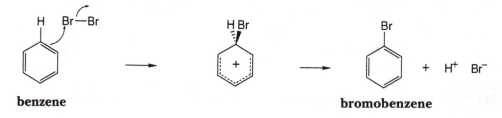

benzene   bromobenzene

The conversion of acetanilide to *p*-bromoacetanilide can be represented in this way.

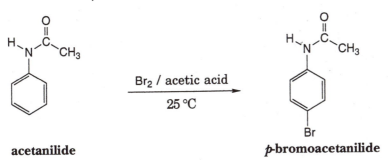

$$\xrightarrow[\text{25 °C}]{\text{Br}_2 \text{ / acetic acid}}$$

acetanilide   ***p*-bromoacetanilide**

Procedure A of this section calls for elemental bromine as one of the reagents. Procedure B provides for the formation of bromine within the reaction mixture. This formation occurs by the reaction of potassium bromate with bromide ion according to the following equation.

$$5 \text{ HBr} \quad + \quad \text{KBrO}_3 \quad \longrightarrow \quad 3 \text{ Br}_2 \quad + \quad \text{KBr} \quad + \quad 3 \text{ H}_2\text{O}$$

Acetanilide can be prepared from aniline by the procedure of Experiment E42. Figure E45-1 shows the infrared spectrum of *p*-bromoacetanilide.

Scale: 2.7 grams (0.020 mole) of acetanilide.

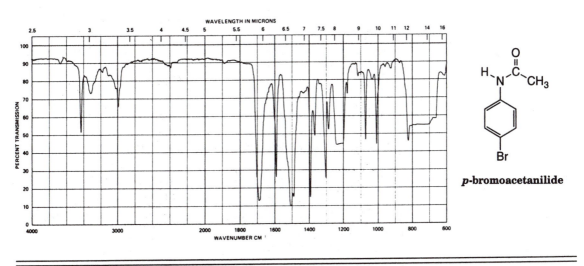

***p*-bromoacetanilide**

Figure E45-1. Infrared spectrum of *p*-bromoacetanilide; CHCl₃ solution.

Procedure A (use of elemental bromine)

Dissolve 2.7 g (0.020 mole) acetanilide in 10 mL of glacial acetic acid in a 125-mL Erlenmeyer flask, and while stirring, slowly add a solution of 1.04 mL (3.24 g; 0.02 mole) of bromine in 2 mL of glacial acetic acid (caution: see Note 1).

Stir the mixture for 2 or 3 minutes more, and while stirring, slowly add 80 mL of water. Add enough concentrated sodium bisulfite solution to discharge the yellow color, and then collect the product by suction filtration, washing it well with water and sucking it as dry as possible. Yield: 4.0 g (96%). The product can be recrystallized from ethanol, using 3.3 mL per gram, with a recovery of 90%; m.p., 168–169°C.

Note

1. See Section 1.7 for precautions to be observed when working with bromine.

Time: less than 2 hours.

Disposal

Sink: acetic acid

Solid waste: filter paper

Nonhalogenated liquid organic waste: ethanol

Halogenated solid organic waste: *p*-bromoacetanilide

Solid inorganic waste: sodium bisulfite

Solid organic waste: acetanilide

Bromine: reduce with sodium bisulfite solution; discard residue after evaporation with solid halogenated inorganic waste

Scale: 1.35 grams (0.010 mole) of acetanilide.

Procedure B (use of potassium bromate, Reference 1)

Dissolve 1.35 grams (0.010 mole) of acetanilide in 6 mL of glacial acetic acid in a 125-mL Erlenmeyer flask. Add to this solution 0.280 grams (0.0017 moles) of potassium bromate. Then add to this well-stirred mixture 1.0 mL (0.009 moles) of 48% hydrobromic acid; a yellow color will appear. Continue to stir the mixture for 30 minutes. At this time, add about 80 mL of water to the reaction flask, and continue to stir for another 15 minutes. At this time, collect the solid that has formed by suction filtration, washing it well with water and then with dilute sodium bisulfite solution, finally sucking the crude product as dry as possible. Yield: 2.0 g (96%). The product can be recrystallized from ethanol, using 3.3 mL per gram, with a recovery of 90%; m.p., 168–169°C.

Time: less than 2 hours.

Disposal

Sink: acetic acid; hydrobromic acid (after dilution with water)

Solid waste: filter paper

Nonhalogenated liquid organic waste: ethanol

Halogenated solid organic waste: *p*-bromoacetanilide

Solid inorganic waste: sodium bisulfite

Solid organic waste: acetanilide

Potassium bromate: oxidizing agent; store separately

Reference

1. Paul Schatz, University of Wisconsin; private communication.

E46. 2,4-Dinitrobromobenzene from Bromobenzene

Nitration of Bromobenzene with a hot mixture of concentrated sulfuric acid and nitric acid gives the product of dinitration.

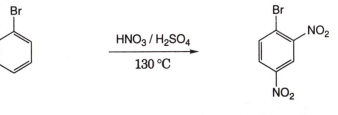

bromobenzene
colorless liquid

2,4-dinitrobromobenzene
pale yellow solid

The 2,4-dinitrobromobenzene produced should have infrared and NMR spectra like those of Figures E46-1 and E46-2.

Scale: 3.1 grams (0.020 mole) of bromobenzene.

Caution: 2,4-dinitrobromobenzene is a mild irritant for some people; See Note 2.

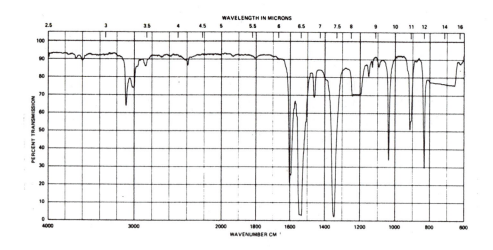

Figure E46-1. Infrared spectrum of 2,4-dinitrobromobenzene; $CHCl_3$ solution.

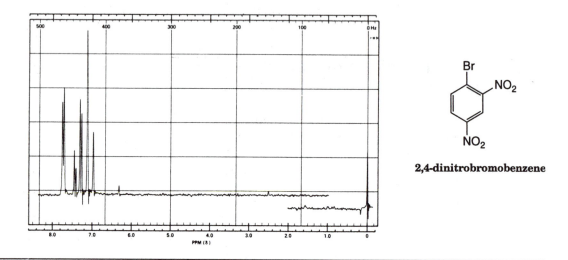

2,4-dinitrobromobenzene

Figure E46-2. NMR spectrum of 2,4-dinitrobromobenzene; $CDCl_3$ solution.

Procedure

Heat a mixture of 15 mL (27 g; 0.27 mole) of concentrated sulfuric acid and 5 mL (7 g; 0.08 mole) of concentrated nitric acid in a 50-mL Erlenmeyer flask to 85–90°C. Add 2.1 mL (3.1 g; 0.020 mole) of bromobenzene in 3 or 4 portions during a minute, and swirl the mixture well after each addition. The temperature will rise to 130–135°C (Note 1).

Allow the mixture to stand for 5 minutes, swirling occasionally, then cool it nearly to room temperature and pour it over about 100 g of ice. Stir the resulting mixture until the product has solidified, crush the lumps, and collect the crude product by suction filtration (Note 2).

Recrystallize the product by dissolving it in 15 mL of hot 95% ethanol and allowing the solution to cool (Note 3). After crystals have started to form, cool the solution in a cold-water bath and then in an ice bath. Crystallization is complete in about 5 minutes. Collect the product by suction filtration (Note 2), and wash the crystals with a small amount of cold ethanol.

Notes

1. The degree to which the temperature will rise depends on the rate of addition and how well the bulb of the thermometer is covered by the reaction mixture. Don't worry if the temperature does not reach 130°C.

2. 2,4-Dinitrobromobenzene is a mild irritant for some people. You should wear gloves whenever you work with 2,4-dinitrobromobenzene. Be sure to wash your hands after you do this experiment. Avoid touching your face or eyes while you are working with 2,4-dinitrobromobenzene. If you should get some 2,4-dinitrobromobenzene on your skin, wash first with a little ethanol and then with a little very dilute aqueous ammonia.

3. The product usually separates as an oil, but vigorous swirling of the mixture when the oil appears will promote crystallization.

Time: 2 hours.

Disposal

Sink: sulfuric acid (after dilution); nitric acid (after dilution); aqueous washings

Solid waste: filter paper

Nonhalogenated liquid organic waste: ethanol

Halogenated liquid organic waste: bromobenzene

Halogenated solid organic waste: 2,4-dinitrobromobenzene

Question

1. Occasionally, a student will use 5 mL of concentrated sulfuric acid and 15 mL of concentrated nitric acid (instead of 15 mL of the former and 5 mL of the latter). When the wrong mixture is used, a colorless solid with a melting point of about 125°C is obtained. What is the colorless solid, and why is it formed? (The story behind this problem is that several students one year obtained a colorless solid of melting point about 125°C. After looking up the melting point of a reasonable alternative product, we determined that the alternative product could be formed by running the experiment with the amounts of the acids reversed.)

NUCLEOPHILIC AROMATIC SUBSTITUTION REACTIONS OF 2,4-DINITROBROMOBENZENE

Certain benzene derivatives undergo nucleophilic substitution reactions. The larger the number of electron-withdrawing substituents on the ring, especially in the positions *ortho* and *para* to the point of substitution, the greater the rate of the reaction. This effect is explained by a two-step addition–elimination mechanism in which the role of the electron-withdrawing substituent is to stabilize the negative charge of the benzene ring in the intermediate.

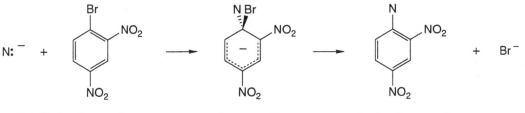

2,4-dinitrobromobenzene   intermediate   substitution product

The same groups, for example the –NO$_2$ group, are activating for nucleophilic aromatic substitution and deactivating for electrophilic aromatic substitution; both effects are explained by the electron-withdrawing ability of the groups.

E47. 2,4-Dinitroaniline

When ammonia is the nucleophile, 2,4-dinitrobromobenzene can be converted to 2,4-dinitroaniline.

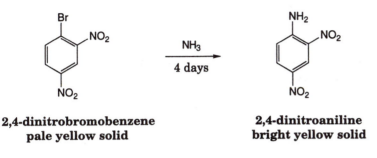

2,4-dinitrobromobenzene
pale yellow solid

2,4-dinitroaniline
bright yellow solid

Scale: 1.84 grams (0.0075 moles) of 2,4-dinitrobromobenzene.

Caution: 2,4-dinitrobromobenzene is a mild irritant for some people; See Note 1.

Procedure

Place 20 mL of ethanol and 1.84 g (0.0075 mole) of 2,4-dinitrobromobenzene (Note 1) in a 125-mL Erlenmeyer flask. Heat the mixture gently to dissolve the solid. Add 15 mL (13.5 g, 0.240 mole) of concentrated ammonium hydroxide. Relieve the resulting cloudiness of the solution by adding another 15 mL of ethanol and warming slightly. Allow the solution to stand for at least 96 hours (Note 2).

After 4 days, collect the yellow solid by suction filtration, and recrystallize it from a mixture of 6 mL of water and 15 mL of ethanol. Yield: about 60%.

Notes

1. 2,4-Dinitrobromobenzene is a mild irritant for some people. You should wear gloves whenever you work with 2,4-dinitrobromobenzene. Be sure to wash your hands after you do this experiment. Avoid touching your face or eyes while you are working with 2,4-dinitrobromobenzene. If you should get some 2,4-dinitrobromobenzene on your skin, wash first with a little ethanol and then with a little very dilute aqueous ammonia.

2. A shorter period of standing will give a lower yield. Heating the mixture seems to speed loss of ammonia more than the desired reaction.

Time: 1 hour plus *4-day reaction time.*

Disposal

> **Sink:** ammonium hydroxide (after dilution)
> **Solid waste:** filter paper
> **Nonhalogenated liquid organic waste:** ethanol
> **Halogenated solid organic waste:** 2,4-dinitrobromobenzene
> **Solid organic waste:** 2,4-dinitroaniline

E48. 2,4-Dinitrophenylhydrazine

When hydrazine is the nucleophile, 2,4-dinitrobromobenzene can be converted to 2,4-dinitrophenylhydrazine. 2,4-Dinitrophenylhydrazine has been widely used to convert liquid aldehydes and ketones to nicely crystalline solids with sharp, characteristic melting points.

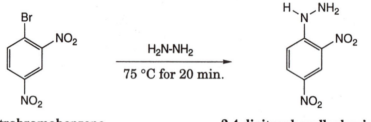

2,4-dinitrobromobenzene   2,4-dinitrophenylhydrazine
pale yellow solid   either orange blades or red-purple prisms

Scale: 0.5 grams (0.002 moles) of 2,4-dinitrobromobenzene.

Caution: 2,4-dinitrobromobenzene is a mild irritant for some people (Note 1).

Procedure

In a 50-mL Erlenmeyer flask, prepare a solution of 0.5 g (0.002 mole) of 2,4-dinitrobromobenzene (Note 1) in 7.5 mL of 95% ethanol. Heat the solution almost to boiling, and add to it a solution of 0.5 mL (0.010 mole) of 64% hydrazine in 2.5 mL of 95% ethanol. Allow the light orange solution, which rapidly turns a deep red-purple, to cool undisturbed for 15–20 minutes.

Collect the resulting crystals by suction filtration and wash them with a little ethanol. The product sometimes crystallizes as red-purple prisms alone, and sometimes as red-purple prisms and orange blades. Recrystallization of either form from boiling ethyl acetate (50 mL per gram) yields a product in the form of orange plates.

Red-purple prisms

Orange plates

Note

1. 2,4-Dinitrobromobenzene is a mild irritant for some people. You should wear gloves whenever you work with 2,4-dinitrobromobenzene. Be sure to wash your hands after you do this experiment. Avoid touching your face or eyes while you are working with 2,4-dinitrobromobenzene. If you should get some 2,4-dinitrobromobenzene on your skin, wash first with a little ethanol and then with a little very dilute aqueous ammonia.

Time: 2 hours.

Disposal

Solid waste: filter paper
Nonhalogenated liquid organic waste: ethanol; ethyl acetate; hydrazine solution
Halogenated solid organic waste: 2,4-dinitrobromobenzene
Solid organic waste: 2,4-dinitrophenylhydrazine

E49. 2,4-Dinitrodiphenylamine

When aniline is the nucleophile, 2,4-dinitrobromobenzene can be converted to 2,4-dinitrodiphenylamine.

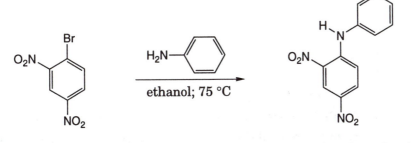

2,4-dinitrobromobenzene
pale yellow solid

2,4-dinitrodiphenylamine
beautiful red needles

Scale: 1.84 grams (0.0075 moles) of 2,4-dinitrobromobenzene.

Caution: 2,4-dinitrobromobenzene is a mild irritant for some people; See Note 1.

Procedure

Place 20 mL of ethanol, 1.84 g (0.0075 moles) of 2,4-dinitro-bromobenzene, (note 1) and 1.4 mL (1.4 g; 0.015 mole) of aniline in a 100-mL boiling flask. Fit the flask with a condenser and, using the steam bath as a source of heat, gently boil the mixture under reflux until all the solid has dissolved.

After allowing the mixture to cool slowly to room temperature (Note 2), collect the red needles by suction filtration.

Long red needles

Recrystallize your product from ethanol. Use a 125-mL ℥ Erlenmeyer flask fit with a condenser. You may need more than 100 mL of ethanol to dissolve all of your product, and it is important to know that 2,4-dinitrodiphenylamine dissolves only slowly in boiling ethanol. Allow the hot solution to cool slowly to room temperature. Your reward for this effort will be a product in the form of beautiful, long red needles.

Notes

1. 2,4-Dinitrobromobenzene is a mild irritant for some people. You should wear gloves whenever you work with 2,4-dinitrobromobenzene. Be sure to wash your hands after you do this experiment. Avoid touching your face or eyes while you are working with 2,4-dinitrobromobenzene. If you should get some 2,4-dinitrobromobenzene on your skin, wash first with a little ethanol and then with a little very dilute aqueous ammonia.

2. The solution should be allowed to stand for at least an hour before filtering.

Time: less than 3 hours.

Disposal

Solid waste: filter paper
Nonhalogenated liquid organic waste: ethanol; aniline
Halogenated solid organic waste: 2,4-dinitrobromobenzene
Solid organic waste: 2,4-dinitrodiphenylamine

E50. 2,4-Dinitrophenylpiperidine

When piperidine is the nucleophile, 2,4-dinitrobromobenzene can be converted to 2,4-dinitrophenylpiperidine.

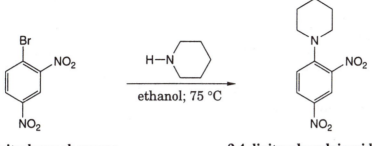

2,4-dinitrobromobenzene
pale yellow solid

2,4-dinitrophenylpiperidine
fine orange needles

Scale: 1.84 grams (0.0075 moles) of 2,4-dinitrobromobenzene.

Caution: 2,4-dinitrobromobenzene is a mild irritant for some people (Note 1).

Procedure (Reference 1)

Place 25 mL of ethanol and 1.84 g (0.0075 moles) of 2,4-dinitrobromobenzene (Note 1) in a 100-mL boiling flask. Fit the flask with a condenser, and heat the mixture gently on the steam bath to dissolve the solid. Then add 2.0 mL (1.72 g; 0.020 mole) of piperidine and boil the mixture under reflux for 10 minutes.

    After allowing the solution to cool slowly to room temperature (Note 2) and finally cooling it in an ice bath, collect the orange needles by suction filtration. The product can be recrystallized from 20 mL of hot ethanol.

Orange needles

Notes

1. 2,4-Dinitrobromobenzene is a mild irritant for some people. You should wear gloves whenever you work with 2,4-dinitrobromobenzene. Be sure to wash your hands after you do this experiment. Avoid touching your face or eyes while you are working with 2,4-dinitrobromobenzene. If you should get some 2,4-dinitrobromobenzene on your skin, wash first with a little ethanol and then with a little very dilute aqueous ammonia.

2. The solution tends to become supersaturated. Crystallization can be induced by adding a seed crystal or by tapping
the flask with a spatula.

Time: 2 hours.

Disposal

Solid waste: filter paper
Nonhalogenated liquid organic waste: ethanol; piperidine
Halogenated solid organic waste: 2,4-dinitrobromobenzene
Solid organic waste: 2,4-dinitrophenylpiperidine

Reference

1. J. W. McFarland, *Organic Laboratory Chemistry*, C. V. Mosby, St. Louis,
1969, p. 209.

E51. 4-Substituted 2,4-Dinitrophenyl-anilines: A Variation

If a *p*-substituted aniline is used instead of aniline in the procedure of
Section E49, the reaction with 2,4-dinitrobromobenzene should give
a 4′-substituted 2,4-dinitrophenylaniline.

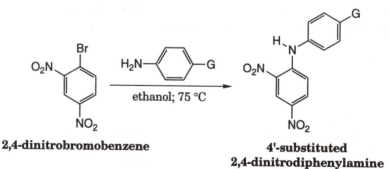

2,4-dinitrobromobenzene   4′-substituted
                                           2,4-dinitrodiphenylamine

Some *p*-substituted anilines that could be used include: *p*-
methylaniline (*p*-toluidine); *p*-bromoaniline; *p*-ethoxyaniline *(p*-phe-
netidine); and *p*-methoxyaniline (*p*-anisidine), which gives a beautifully crystalline deep red product.

Scale: 1.84 grams (0.0075 moles) of 2,4-dinitrobromobenzene.

Caution: 2,4-dinitrobromobenzene is a mild irritant for some people (Note 1).

Procedure

Place 20 mL of ethanol, 1.84 g (0.0075 moles) of 2,4-dinitro-bromobenzene (Note 1), and 0.015 mole of the aromatic amine in a 100-mL boiling flask. Fit the flask with a condenser and, using the steam bath as a source of heat, gently boil the mixture under reflux until all the solid has dissolved.

After allowing the mixture to cool slowly to room temperature (Note 2), collect the product by suction filtration.

Recrystallize your product from ethanol. Use a 125-mL ⚡ Erlenmeyer flask fit with a condenser. You may need as much as 100 mL of ethanol to dissolve all of your product, and it may be true that your product will dissolve only slowly in boiling ethanol. Allow the hot solution to cool slowly to room temperature.

Notes

1. 2,4-Dinitrobromobenzene is a mild irritant for some people. You should wear gloves whenever you work with 2,4-dinitro-bromobenzene. Be sure to wash your hands after you do this experiment. Avoid touching your face or eyes while you are working with 2,4-dinitrobromobenzene. If you should get some 2,4-dinitrobromobenzene on your skin, wash first with a little ethanol and then with a little very dilute aqueous ammonia.

2. The solution should be allowed to stand for at least an hour before filtering.

Time: less than 3 hours.

Disposal

> **Solid waste:** filter paper
> **Nonhalogenated liquid organic waste:** ethanol; aromatic amines
> **Halogenated solid organic waste:** 2,4-dinitrobromobenzene
> **Solid organic waste:** aromatic amines; products

DIAZONIUM SALTS OF AROMATIC AMINES

Diazonium salts are prepared from the corresponding aromatic amine by treatment of an acidic solution of the amine with one equivalent of aqueous sodium nitrite solution.

$$Ar—NH_2 \ + \ HNO_2 \ + \ H\text{-}Cl \ \longrightarrow \ Ar—N_2{}^+ \ Cl^- \ + \ 2 \ H_2O$$

aromatic   nitrous   aryl diazonium

amine   acid   salt

The diazonium ion is best represented as a resonance hybrid of two contributing Lewis structures.

benzenediazonium ion

    The diazotization is carried out as near 0°C as possible to minimize the rate of the reaction of the diazonium salt with water to form the corresponding phenol.

    Any phenol that forms will react with another molecule of the diazonium salt to form a highly colored azo compound (compare Experiments E62 through E65). If the aqueous solution of the diazonium salt is allowed to warm to room temperature, the formation of the phenol can become a significant reaction.

    Because diazotization is an exothermic reaction, you might attempt to minimize the temperature rise and subsequent formation of phenol by very slow addition of the sodium nitrite solution. However, the diazonium salt can react with undiazotized amine to form an azo compound (Experiment E65). Because of this, it is desirable to diazotize as quickly as possible to minimize the time during which the diazonium salt and undiazotized amine are present together. As a

A compromise   compromise, diazotization is usually carried out as fast as possible with good stirring and cooling, but slowly enough that the temperature of the solution does not rise above about 5°C. The lower the temperature is, the slower, also, will be the reaction between the diazonium salt and undiazotized amine.

    An excess of sodium nitrite is to be avoided because the product of a subsequent reaction can react with excess nitrous acid.

E52. Benzenediazonium Chloride from Aniline

In this experiment, aniline is converted to the corresponding diazonium salt, benzenediazonium chloride.

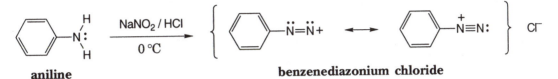

aniline   **benzenediazonium chloride**

Benzenediazonium chloride can then be converted to chlorobenzene by the procedure of Experiment E54.

Scale: 4.7 grams (0.050 mole) of aniline.

Procedure

Dissolve 4.6 mL (4.7 g; 0.050 mole) of aniline in a mixture of 12 mL (0.144 mole) of concentrated hydrochloric acid and 12 mL of water. Cool the solution to 0°C in an ice bath, and, while stirring and cooling (Note 1), add a solution of 3.7 g (0.054 mole) of sodium nitrite in 12 mL of water at such a rate that the temperature of the solution does not exceed 5°C. Add only enough of this solution to give a slight excess of nitrous acid (Note 2). The resulting solution of benzenediazonium chloride must be stored in the ice bath and used fairly quickly.

Notes

1. The efficiency of cooling can be increased by adding pieces of ice to the reaction mixture.

2. Because nitrous acid oxidizes iodide to iodine, the presence of nitrous acid in the mixture can be detected by placing a drop of the solution on a piece of starch potassium iodide test paper. An immediate blue color will be produced if nitrous acid is present. If the quantities of reagents have been carefully measured, one need not start testing for excess nitrous acid until most of the solution of sodium nitrite has been added. The test for nitrous acid should be positive for at least 1–2 minutes after the last addition has been made to

safely assume that slightly more than an equivalent of sodium nitrite has been added.

Time: 1 hour.

Disposal

> **Sink:** hydrochloric acid (after dilution)
> **Nonhalogenated liquid organic waste:** aniline
> **Solid inorganic waste:** sodium nitrite

E53. *p*-Nitrobenzenediazonium Sulfate from *p*-Nitroaniline

p-Nitroaniline can be diazotized by the same procedure that was used
for the diazotization of aniline in Section E52.

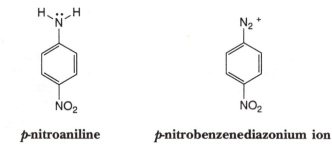

p-nitroaniline   *p*-nitrobenzenediazonium ion

Some of the coupling reactions of the *p*-nitrobenzenediazonium
ion with phenols and aromatic amines to form dyes are presented in
Experiments E62 through E65.

Scale: 1.38 grams (0.010 mole) of *p*-nitroaniline.

Procedure

Make a solution of 2 mL (0.036 mole) of concentrated sulfuric acid
in 10 mL of water. Add to this 1.38 g (0.010 mole) of *p*-nitroaniline,
and heat the mixture gently to bring the amine into solution. Cool
the mixture to about 10°C in an ice bath and while stirring and cooling (Note 1), add to the suspension of *p*-nitroaniline sulfate, a solution of 0.69 g (0.010 mole) of sodium nitrite in 2 mL of water at such
a rate that the temperature does not rise above about 10°C. Store the
resulting solution of *p*-nitrobenzenediazonium sulfate in the ice bath.

Note

1. The efficiency of cooling can be increased by adding pieces of ice to the reaction mixture.

Time: 1 hour.

Disposal

Sink: sulfuric acid (after dilution)
Nonhalogenated liquid organic waste: *p*-nitroaniline
Solid inorganic waste: sodium nitrite

REPLACEMENT REACTIONS OF DIAZONIUM SALTS

The two most useful reactions of aryl diazonium salts are the replacement reactions and the coupling reactions. The procedure of Experiment E54 illustrates one of the replacement reactions, and Experiments E62 through E65 present examples of coupling reactions.

E54. Chlorobenzene from Benzene-diazonium Chloride

When the cuprous salt of benzenediazonium chloride is heated, it decomposes with loss of nitrogen gas to give chlorobenzene.

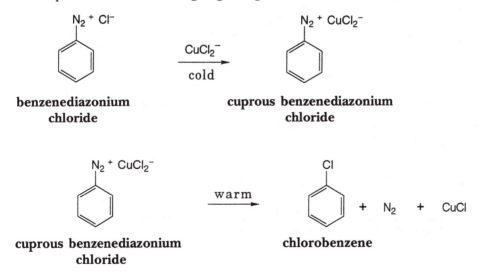

The mechanism of the decomposition of cuprous diazonium chloride is not completely agreed upon, but the reaction does appear to involve free radical intermediates.

Figure E54-1 shows the infrared spectrum of chlorobenzene.

Scale: 4.7 grams (0.050 mole) of aniline.

Procedure

Dissolve 15 g (0.060 mole) of copper sulfate pentahydrate in 50 mL of hot water in a 250-mL boiling flask. Then add and dissolve 5 g (0.087 mole) of sodium chloride. Prepare a solution of sodium sulfite by dissolving 3.5 g (0.034 mole) of sodium bisulfite and 2.9 mL (0.055 mole) of 50% aqueous sodium hydroxide in 50 mL of water.

Reduce the copper by adding the solution of sodium sulfite in several portions over 2–3 minutes and mixing well. Allow the mixture containing the precipitated cuprous chloride to stand and settle in a pan of cold water while you prepare a solution of benzenediazonium chloride as described in Experiment E52.

When the diazotization is complete, decant the supernatant liquid from the precipitated cuprous chloride, and wash the precipitate by adding water, swirling, allowing to settle, and decanting. Then dissolve the cuprous chloride by adding 22 mL (0.26 mole) of concentrated hydrochloric acid and swirling the mixture. Cool this solution in the ice bath and slowly add the solution of benzenediazonium chloride.

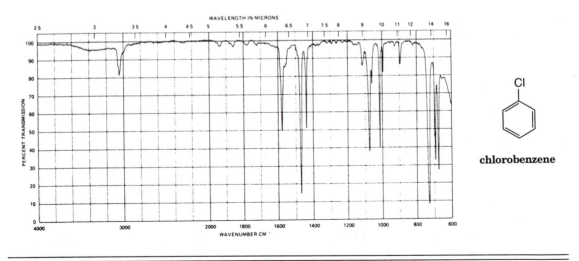

chlorobenzene

Figure E54-1. Infrared spectrum of chlorobenzene; thin film.

Allow the mixture to stand at room temperature, swirling occasionally for about 10 minutes. During this time, the initially formed
cuprous diazonium chloride begins to decompose to give nitrogen
and chlorobenzene. Cautiously heat the mixture to 50°C on the steam
bath to complete the decomposition. Steam distill the mixture to isolate the chlorobenzene (Section 12). Separate the product from the
distillate by extraction with ether, washing with dilute sodium hydroxide solution, drying, and distilling.

The Sandmeyer reaction

Time: 3 hours.

Disposal

> **Sink:** sodium chloride; sodium hydroxide (after dilution)
> **Nonhalogenated liquid organic waste:** ether
> **Halogenated liquid organic waste:** chlorobenzene
> **Solid inorganic waste:** copper sulfate; sodium bisulfite

REACTIONS OF VANILLIN

Vanillin, the key ingredient of the most popular flavor of all time,
contains several functional groups and will undergo a variety of organic reactions.

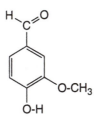

vanillin
an aldehyde; an aromatic ether; a phenol

Gathered here are seven reactions of vanillin that provide examples of the chemistry of vanillin. These include the acetylation of
the phenol hydroxyl group, substitution of the aromatic ring, reduction of the aldehyde function, formation of the oxime and
semicarbazone of the aldehyde function, and the aldol reaction. The
reaction times are usuallly short, and the yields are generally good.

E55. Acetylvanillin from Vanillin

In this procedure, a basic solution of vanillin is treated with the acetylating agent acetic anhydride to form the acetate ester of the phenolic hydroxyl group.

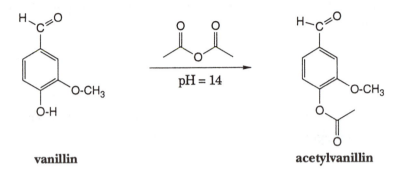

vanillin   **acetylvanillin**

Scale: 3.04 grams (0.020 mole) of vanillin.

Procedure

In a 125 mL-Erlenmeyer flask, dissolve 3.04 grams (0.020 mole) of vanillin in 22 mL of 1 *M* aqueous potassium hydroxide (0.022 mole). Add to this a solution of 2.1 mL (2.27 grams; 0.022 mole) of acetic anhydride in 45 mL of ether.

Stir the resulting mixture vigorously until it loses its yellow color and the lower, aqueous layer appears almost colorless.

At this time, pour the mixture into a separatory funnel and draw off the lower, aqueous layer. Transfer the remaining ethereal solution to a clean Erlenmeyer flask, dry the solution with anhydrous magnesium sulfate, remove the drying agent by gravity filtration into another, ₹, Erlenmeyer flask, and then, using the steam bath, strip off all the ether (Notes 1 and 2), and allow the residue of acetylvanillin to cool and crystallize.

The yield of the crude product is 3.50 grams (90%) with a melting point of 69–75°C.

Recrystallize the crude product from 40 mL of 50% ethanol in water. The yield of recrystallized product is 2.36 grams (67%, 61% overall yield) with a melting point of 76–79°C.

Notes

1. Fit the Erlenmeyer flask with a distillation adapter and condenser to as to condense and recover the ether.
2. When the residue stops bubbling, you can assume that the ether is gone.

Time: less than 2 hours.

Disposal

> **Sink:** potassium hydroxide solution; aqueous washes
> **Solid waste:** filter paper
> **Nonhalogenated liquid organic waste:** acetic anhydride; ether; ethanol
> **Solid inorganic waste:** magnesium sulfate
> **Solid organic waste:** vanillin; acetylvanillin

Questions

1. What is the role of the potassium hydroxide?
2. What is the state of the vanillin after it is added to the aqueous potassium hydroxide solution?
3. Why is the aqueous solution yellow at the start of the reaction?
4. Why does the aqueous solution lose its yellow color during the course of the reaction?
5. Why does the product end up in the ether layer and not in the water layer?
6. In which phase, aqueous or ethereal, does the actual acetyl transfer take place?
7. Why did acetyl transfer take place to the phenolic hydroxyl group and not to the aromatic ring?

E56. 5-Bromovanillin from Vanillin

In this experiment, vanillin is subjected to electrophilic aromatic substitution by bromine. The reaction takes place almost exclusively at the position *ortho* to the phenolic hydroxyl group.

vanillin   5-bromovanillin

Scale: 3.04 grams (0.020 mole) of vanillin.

Procedure

In a 125-mL Erlenmeyer flask, dissolve 3.04 grams (0.020 moles) of vanillin in 11 mL of glacial acetic acid. Now add to the flask 12 mL of a 2 *M* solution of bromine in glacial acetic acid (0.024 moles of bromine; Note 1). After swirling the flask to mix the contents, allow the resulting mixture to stand overnight at room temperature.

Collect the resulting prismatic crystals by suction filtration, and wash them well with cold water. Yield: 3.80 grams (81%). Melting point: 158–164°C. The crude product can be recrystallized from ethanol.

Note

1. Enough bromine solution for eight runs can be prepared by adding 10.2 mL (31.8 grams; 0.20 moles) of bromine to 90 mL of glacial acetic acid.

Time: 1 hour of working time; 24-hour reaction time.

Disposal

Sink: acetic acid (after dilution)

Solid waste: filter paper

Nonhalogenated liquid organic waste: ethanol

Halogenated solid organic waste: 5-bromovanillin

Solid organic waste: vanillin

Bromine: reduce with sodium bisulfite solution; discard residue after evaporation with solid halogenated inorganic waste.

Question

1. Why should bromination take place *ortho* to the phenolic hydroxyl group rather than *para* to the methoxy group?

E57. 5-Nitrovanillin from Vanillin

When vanillin is treated at room temperature with a solution of nitric acid in acetic acid, the product is 5-nitrovanillin.

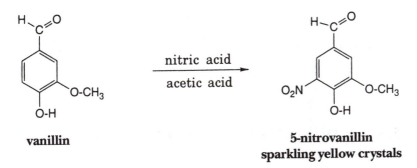

vanillin

5-nitrovanillin
sparkling yellow crystals

Scale: 3.04 grams (0.020 mole) of vanillin.

Procedure

In a 125-mL Erlenmeyer flask in which a small magnetic stirring bar has been placed, dissolve 3.04 grams (0.020 moles) of vanillin in 40 mL of glacial acetic acid. Place the flask in a shallow dish of cold water, and put the pan with the flask on a magnetic stirring motor.

While the solution of vanillin is vigorously stirred, add to it, dropwise, over a period of about 5 minutes, a solution of 1.5 mL of concentrated nitric acid (0.024 moles of HNO_3) in 10 mL of glacial acetic acid. During the addition, the temperature of the solution should remain below 30°C. As the nitric acid is added, the solution will turn yellow and then brown, and a solid will start to separate.

After the addition is complete, stir the mixture for another 5 minutes, and then add 50 mL of cold water. Stir the mixture briefly and then collect the crude 5-nitrovanillin by suction filtration, using water and finally methanol to rinse the flask and to wash the solid.

The yield of crude product, after drying, is about 3.3 grams (0.017 moles; 83%). The crude product can be recrystallized from 80 mL of boiling ethyl acetate to give 2.80 grams of sparkling yellow crystals (71% yield overall; 85% recovery from recrystallization).

Yellow crystals

Time: about 1 hour.

Disposal

>   **Sink:** acetic acid (after dilution); nitric acid (after dilution); aqueous washes

Solid waste: filter paper

Nonhalogenated liquid organic waste: methanol; ethyl acetate

Solid organic waste: vanillin; 5-nitrovanillin

Question

1. Why should nitration take place *ortho* to the phenolic hydroxyl group rather than *para* to the methoxy group?

E58. Vanillin Oxime from Vanillin

One of the standard procedures for the identification of an unknown aldehyde or ketone was conversion to the corresponding oxime, which was usually a nicely crystalline compound whose melting point had been recorded previously. When vanillin is treated with hydroxylamine at a pH of about 5, it is converted to the corresponding oxime.

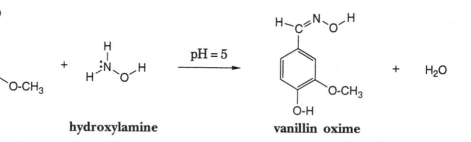

vanillin   hydroxylamine   vanillin oxime

Scale: 3.04 grams (0.020 mole) of vanillin.

Procedure

Dissolve 1.52 grams (0.022 mole) of hydroxylamine hydrochloride and 5.44 grams (0.040 mole) of sodium acetate trihydrate in 20 mL of water while warming on the steam bath.

To this hot solution, add 3.04 grams (0.020 moles) of solid vanillin. Swirl the flask until all of the vanillin has first melted and then completely dissolved, and then continue to heat the solution on the steam bath for another 5 minutes.

Remove the flask from the steam bath, and allow the mixture to stand first at room temperature and then in an ice bath for crystallization.

Collect the product by suction filtration, using a little water for rinsing and washing. Yield: about 3.0 grams (0.18 moles; 90%) The melting point of the oxime of vanillin is reported to be 117°C.

Time: about $1\frac{1}{2}$ hours.

Disposal

> **Sink:** aqueous washes
>
> **Solid waste:** filter paper
>
> **Solid inorganic waste:** hydroxylamine hydrochloride; sodium acetate
>
> **Solid organic waste:** vanillin; vanillin oxime

Questions

1. Why is the reaction run at a pH of about 5?
2. How is a pH of about 5 achieved in this experiment?

E59. Vanillin Semicarbazone from Vanillin

In this experiment, we treat vanillin with semicarbazide at a pH of about 5, thereby converting vanillin to the corresponding semicarbazone, a reaction that illustrates the addition/elimination reactions of aldehydes and ketones with ammonia derivatives.

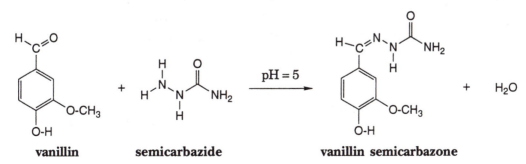

| vanillin | semicarbazide | vanillin semicarbazone |

Scale: 0.76 grams (0.005 mole) of vanillin.

Procedure

Dissolve 0.60 grams (0.0054 mole) of semicarbazide hydrochloride and 1.36 grams (0.010 mole) of sodium acetate trihydrate in a mixture of 16 mL of 95% ethanol and 4 mL of water while warming on the steam bath.

To this hot solution, add a hot solution of 0.76 grams (0.005 moles) of vanillin in 10 mL of 95% ethanol. Swirl the flask to mix the contents, and heat the reaction mixture on the steam bath for about 10 minutes.

After crystals start to form, remove the flask from the steam bath, and allow the mixture to stand at room temperature until crystallization appears to be complete.

Collect the product by suction filtration, using 95% ethanol for rinsing and washing. Yield: about 0.98 grams (94%).

Time: about $1\frac{1}{2}$ hours.

Disposal

> **Sink:** aqueous washes
>
> **Solid waste:** filter paper
>
> **Nonhalogenated liquid organic waste:** ethanol
>
> **Solid inorganic waste:** semicarbazide hydrochloride; sodium acetate
>
> **Solid organic waste:** vanillin; vanillin semicarbazide

Questions

1. Why is the reaction run at a pH of about 5?
2. How is a pH of about 5 achieved in this experiment?
3. Account for the fact that one amino group of semicarbazide is nucleophilic whereas the other is not.

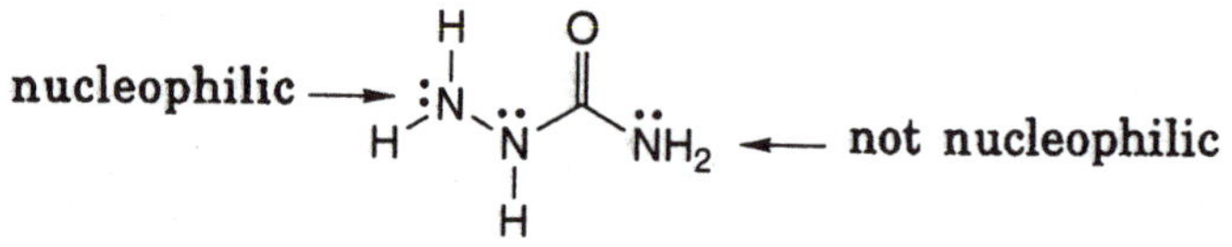

E60. Vanillyl Alcohol from Vanillin

The aldehyde vanillin, the major flavor component of vanilla extract, is easily reduced to the corresponding alcohol by treatment with sodium borohydride in aqueous base.

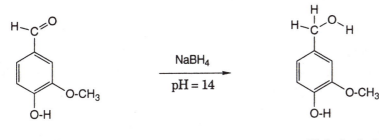

vanillin **vanillyl alcohol**

Scale: 3.04 grams (0.020 mole) of vanillin.

Procedure

Add to a 250-mL Erlenmeyer flask 3.04 grams (0.020 mole) of vanillin and 22 mL of 1 M aqueous NaOH solution (0.022 mole of NaOH). Swirl the flask to dissolve the vanillin, and then cool the contents of the flask to between 10 and 15°C by swirling the flask in an ice bath.

Now, in several portions over a period of 5 minutes and while mixing well, add 0.40 gram (0.010 mole) of sodium borohydride, and then let the flask stand at room temperature for about 30 minutes.

Again cool the contents of the flask by briefly swirling it in the ice bath. Next, add in 0.05-mL portions most of a solution of 2 mL of concentrated HCl (0.024 mole of HCl) in 15–20 mL of water, stopping when the solution in the flask is slightly acidic to litmus (when the pH is on the acidic side of 7).

If at this point, the product has not separated from solution, it can be encouraged to do so by scratching the walls of the flask near the bottom with a stirring rod.

Thoroughly cool the contents of the flask, and then collect the vanillyl alcohol by suction filtration.

After as much water as possible has been sucked from the product, scrape it out of the suction funnel and set it aside to dry. Recrystallize the dried vanillyl alcohol from ethyl acetate, using about 5 mL of ethyl acetate per gram of dried vanillyl alcohol. You may need to filter the hot solution to remove some insoluble material. The melting point of vanillyl alcohol is reported to be 113–115°C.

Time: less than 2 hours.

Disposal

> **Sink:** aqueous sodium hydroxide; hydrochloric acid (after dilution); aqueous washes
> **Solid waste:** filter paper
> **Nonhalogenated liquid organic waste:** ethyl acetate
> **Solid organic waste:** vanillin; vanillyl alcohol
> **Sodium borohydride:** add to water; discard residue after evaporation with solid inorganic waste

Questions

1. What is the role of the sodium hydroxide?
2. What is the role of the sodium borohydride?
3. What is the state of the original vanillin after it has dissolved and before the sodium borohydride has been added?
4. What is the state of the original vanillin after the sodium borohydride has been added and 30 minutes have passed?
5. What is the state of the vanillin after the pH of the solution has been adjusted to 7?
6. The product has two more hydrogen atoms than the starting material. What was the source of each one?

Reference

1. David Todd, *Experimental Organic Chemistry*, Prentice-Hall, Englewood Cliffs, N.J., 1979, p. 137.

E61. Vanillideneacetone from Vanillin

The conjugate base of a ketone can react as a nucleophile with an aldehyde to give the product of addition to the carbon–oxygen bond of the aldehyde. Although this initial product of addition can sometimes be isolated, a longer reaction time allows a subsequent elimination reaction to take place. Thus, when a basic solution of vanillin in acetone is allowed to stand for two days, vanillideneacetone is formed.

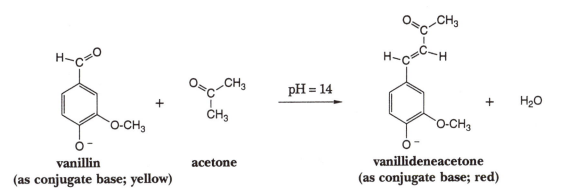

vanillin
(as conjugate base; yellow)

acetone

vanillideneacetone
(as conjugate base; red)

At the pH of the reaction mixture, about 14, a small fraction of the acetone, $pK_a = \sim 20$, is present as the conjugate base of acetone.

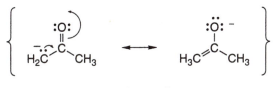

conjugate base of acetone
the enolate anion of acetone

The reaction probably takes place by reaction of the conjugate base of acetone as a nucleophile with the electrophilic carbonyl carbon of vanillin.

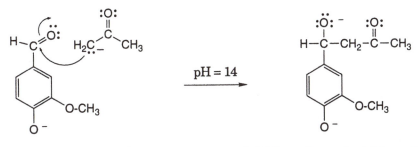

vanillin; as conjugate base

initial product of addition; dianion

Because the pH of the solution is about 14, this intermediate will be present as the monoanion. The solution is basic enough to keep the phenolic proton mainly off but not basic enough to keep more than a small fraction of the alcoholic protons off.

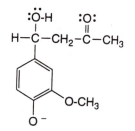

intermediate product; monoanion; keto form

The subsequent elimination reaction takes place via the enol form of the intermediate monoanion.

intermediate product   **vanillideneacetone**
monoanion; enol form   **as conjugate base; red**

Acidification of the reaction mixture after the reaction has become complete converts the ionic conjugate base of vanillideneacetone to the neutral molecule.

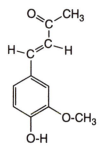

vanillideneacetone; neutral molecule

Scale: 3.04 grams (0.020 mole) of vanillin.

Procedure

Dissolve 3.04 grams (0.020 mole) of vanillin in 12 mL (9.5 grams; 0.164 moles) of acetone in a 125-mL Erlenmeyer flask. To this solution, add 9 mL of 10% aqueous sodium hydroxide (0.025 mole NaOH). Stopper the flask and allow the resulting solution to stand at room temperature for 48 hours (Notes 1 and 2).

At this time, add 50 mL of water and then, while *vigorously stirring* the mixture, acidify it by *slowly adding* 15 mL of 10% aqueous HCl (0.044 moles HCl) (Note 3).

Collect the resulting fine yellow precipitate by suction filtration, using water for rinsing and washing. The yield of dried crude product is 3.14 grams (81%).

The crude product should be recrystallized from 11 mL of 1:1 ethanol:water to give 2.64 grams (68% yield overall) of recrystallized product.

Yellow crystals

Notes

1. At first, the solution will be yellow (the conjugate base of vanillin), but it will gradually darken until it is deep red (the conjugate base of vanillideneacetone).

2. After 24 hours, work-up affords mostly recovered vanillin; after 72 hours the yield of dried crude product is about 3.3 grams.

3. The more slowly the acid is added and the harder the solution is stirred, the more nicely crystalline the product. Dr. Orgo says "Take your time and stir like hell."

Disposal

Sink: sodium hydroxide; hydrochloric acid; aqueous washes
Solid waste: filter paper
Nonhalogenated liquid organic waste: acetone
Solid organic waste: vanillin; vanillideneacetone

Questions

1. Indicate the state of the original vanillin at each of the following stages of the reaction:

a. After the vanillin is dissolved in acetone but before the sodium hydroxide is added.

b. After the vanillin is dissolved in acetone and 1 minute after the sodium hydroxide has been added.

c. After the vanillin is dissolved in acetone and 48 hours after the sodium hydroxide has been added.

d. As in 1c but 1 minute after 50 mL of water has been added.

e. As in 1d but after 15 mL of 10% aqueous HCl has been added.

SYNTHESIS OF DYES

One of the most important reactions of aromatic diazonium salts is the "coupling" reaction with aromatic amines and phenols. The products of these coupling reactions are brightly colored, and many such substances have found extensive use as dyes for cloth, inks, and foods. At this time, nine substances are approved for use as colorants for foods, and of these three are the monoazo dyes shown here.

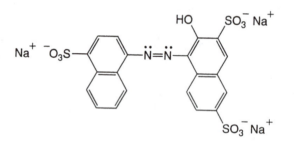

F, D & C Red No. 2
Amaranth

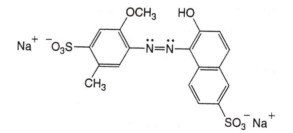

F, D & C Red No. 40
Allura Red

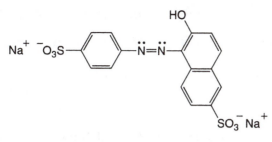

F, D & C Yellow No. 6

The next four experiments illustrate some of the many coupling reactions that benzenediazonium chloride (Experiment E52) and *p*-nitrobenzenediazonium sulfate (Experi-ment E53) will undergo.

E62. Benzenediazonium Chloride and β-Naphthol: 1-Phenylazo-2-Naphthol (Sudan I)

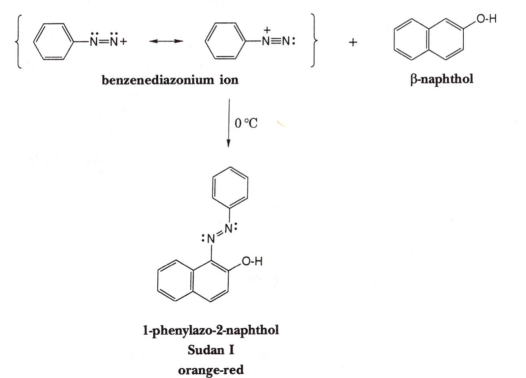

benzenediazonium ion   β-naphthol

0 °C

1-phenylazo-2-naphthol
Sudan I
orange-red

Preparation of benzenediazonium chloride: diazotization

This is the procedure of Section E52, reduced to a 0.010-mole scale.

Scale: 0.93 grams (0.010 mole) of aniline.

Procedure

Dissolve 0.92 mL (0.93 g; 0.010 mole) of aniline in a mixture of 2.4 mL (0.029 mole) of concentrated hydrochloric acid and 12 mL of water. Cool the solution to 0°C in an ice bath, and add, while stirring and cooling (Note 1), a solution of 0.74 g (0.011 mole) of sodium nitrite in 2.4 mL of water at such a rate that the temperature of the solution does not exceed 5°C. Add only enough of this solution to give a slight excess of nitrous acid (Note 2). The resulting solution of benzenediazonium chloride must be stored in the ice bath and used fairly quickly.

Notes

1. The efficiency of cooling can be increased by adding pieces of ice to the reaction mixture.

2. Because nitrous acid oxidizes iodide to iodine, the presence of nitrous acid in the mixture can be detected by placing a drop of the solution on a piece of starch potassium iodide test paper. An immediate blue color will be produced if nitrous acid is present. If the quantities of reagents have been measured carefully, you need not start testing for excess nitrous acid until most of the solution of sodium nitrite has been added. The test for nitrous acid should be positive for at least 1–2 minutes after the last addition has been made to safely assume that slightly more than an equivalent of sodium nitrite has been added.

Preparation of 1-phenylazo-2-naphthol: coupling

Procedure

Prepare a solution of 1.44 g (0.010 mole) of β-naphthol in 5 mL of 3 M sodium hydroxide (0.015 mole). After cooling this solution to 5°C, slowly add while stirring well the solution of benzenediazonium chloride. After allowing the reaction mixture to stand in the ice bath for 15 minutes, collect the product by suction filtration, and wash it thoroughly with cold water. The product can be recrystallized from ethanol or from glacial acetic acid.

Time: 3 hours.

Disposal

> **Sink:** concentrated HCl (after dilution); sodium hydroxide; aqueous washes; acetic acid
>
> **Solid waste:** filter paper
>
> **Nonhalogenated liquid organic waste:** aniline; ethanol
>
> **Solid inorganic waste:** sodium nitrite
>
> **Solid organic waste:** β-naphthol

Questions

1. Why is the coupling reaction done in a weakly basic solution?
2. What must be the two starting materials for the synthesis of F, D & C Red No. 2 by a diazonium coupling reaction?

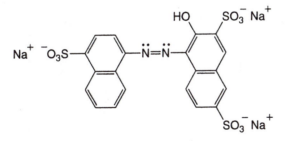

F, D & C Red 3
No. 2
Amaranth

3. What is the function of the sodium sulfonate groups in the F, D & C dyes?

E63. p-Nitrobenzenediazonium Sulfate and Phenol: p-(4-nitrobenzeneazo)-phenol

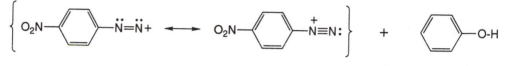

p-nitrobenzenediazonium ion   **phenol**

0 °C

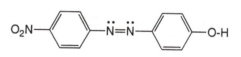

p-(4-nitrobenzeneazo)-phenol
orange-red

Preparation of p-nitrobenzenediazonium chloride: diazotization

This is the procedure of Section E53.

Scale: 1.38 grams (0.010 mole) of p-nitroaniline.

Procedure

Make a solution of 2 mL (0.036 mole) of concentrated sulfuric acid in 10 mL of water. Add to this 1.38 g (0.010 mole) of p-nitroaniline and heat the mixture gently to bring the amine into solution. Cool the mixture to about 10°C in an ice bath and while stirring and cooling (Note 1), add to the suspension of p-nitroaniline sulfate, a solution of 0.69 g (0.010 mole) of sodium nitrite in 2 mL of water at such a rate that the temperature does not rise above about 10°C. Store the resulting solution of p-nitrobenzenediazonium sulfate in the ice bath.

Note

1. The efficiency of cooling can be increased by adding pieces of ice to the reaction mixture.

Preparation of *p*-(4-nitrobenzeneazo)-phenol: coupling

Procedure

Prepare a solution of 0.94 g (0.010 mole) of phenol in 5 mL of 1 *M* sodium hydroxide. Cool this mixture to 5°C and add to it the diazonium salt solution. Stir the mixture for 5 minutes and collect the product by suction filtration. The product can be recrystallized from toluene.

Time: 3 hours.

Disposal

> **Sink:** sulfuric acid (after dilution); sodium hydroxide solution
> **Solid waste:** filter paper
> **Nonhalogenated liquid organic waste:** toluene
> **Solid inorganic waste:** sodium nitrite
> **Solid organic waste:** *p*-nitroaniline; phenol; *p*-(4-nitrobenzeneazo)-phenol

E64. *p*-Nitrobenzenediazonium Sulfate and β-Naphthol: 1-(*p*-nitrophenylazo)-2-naphthol (Para Red; American Flag Red)

p-nitrobenzenediazonium ion   β-naphthol

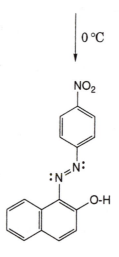

1-(*p*-nitrophenylazo)-2-naphthol
para red; American Flag red
red

Preparation of *p*-nitrobenzenediazonium chloride: diazotization

This is the procedure of Section E53.

Scale: 1.38 grams (0.010 mole) of *p*-nitroaniline.

Procedure

Make a solution of 2 mL (0.036 mole) of concentrated sulfuric acid in 10 mL of water. Add to this 1.38 g (0.010 mole) of *p*-nitroaniline and heat the mixture gently to bring the amine into solution. Cool the mixture to about 10°C in an ice bath and while stirring and cooling (Note 1), add to the suspension of *p*-nitroaniline sulfate, a solution of 0.69 g (0.010 mole) of sodium nitrite in 2 mL of water at such a rate that the temperature does not rise above about 10°C. Store the resulting solution of *p*-nitrobenzenediazonium sulfate in the ice bath.

Note

1. The efficiency of cooling can be increased by adding pieces of ice to the reaction mixture.

Preparation of 1-(*p*-nitrophenylazo)-2-naphthol: coupling

Procedure

Prepare a solution of 1.44 g (0.010 mole) of β-naphthol in 25 mL of 10% sodium hydroxide solution (0.042 mole). When this solution has been cooled to 10°C, pour it into the diazonium salt solution. Acidify the mixture to litmus, and collect the red product by suction filtration. The product can be recrystallized from toluene or acetic acid.

Time: 3 hours.

Disposal

Sink: sulfuric acid (after dilution); sodium hydroxide solution; acetic acid

Solid waste: filter paper

Nonhalogenated liquid organic waste: toluene

Solid inorganic waste: sodium nitrite

Solid organic waste: *p*-nitroaniline; β-naphthol; 1-(*p*-nitrophenylazo)-2-naphthol

E65. *p*-Nitrobenzenediazonium Sulfate and Dimethylaniline: *p*-(4-nitrobenzeneazo)-*N,N*-Dimethylaniline

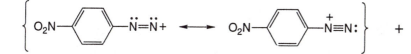

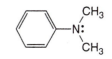

p-nitrobenzenediazonium ion   *N,N*-dimethylaniline

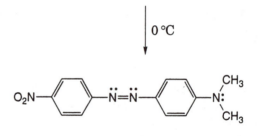

_p_-(4-nitrobenzeneazo)-_N,N_-dimethylaniline
reddish-brown

Preparation of _p_-nitrobenzenediazonium chloride: diazotization

This is the procedure of Section E53.

Scale: 1.38 grams (0.010 mole) of _p_-nitroaniline.

Procedure

Make a solution of 2 mL (0.036 mole) of concentrated sulfuric acid in 10 mL of water. Add to this 1.38 g (0.010 mole) of _p_-nitroaniline and heat the mixture gently to bring the amine into solution. Cool the mixture to about 10°C in an ice bath and while stirring and cooling (Note 1), add to the suspension of _p_-nitroaniline sulfate, a solution of 0.69 g (0.010 mole) of sodium nitrite in 2 mL of water at such a rate that the temperature does not rise above about 10°C. Store the resulting solution of _p_-nitrobenzenediazonium sulfate in the ice bath.

Note

1. The efficiency of cooling can be increased by adding pieces of ice to the reaction mixture.

Preparation of _p_-(4-nitrobenzeneazo)-_N,N_-dimethylaniline: coupling

Procedure

Prepare a solution of 1.21 g (0.010 mole) of _N,N_-dimethylaniline in 1.5 mL of 1 _M_ hydrochloric acid. Cool this solution in the ice bath and add the diazonium salt solution; mix well. Add 10 mL cold 1 *M* sodium hydroxide solution. Collect the product by suction filtration.

Time: 3 hours.

Disposal

> **Sink:** sulfuric acid (after dilution); hydrochloric acid solution; sodium hydroxide solution
>
> **Solid waste:** filter paper
>
> **Solid inorganic waste:** sodium nitrite
>
> **Solid organic waste:** *p*-nitroaniline; *N,N*-dimethylaniline; *p*-(4-nitro-benzeneazo)-*N,N*-dimethylaniline

THE DIELS-ALDER REACTION

The Diels-Alder reaction has been one of the most fascinating reactions in organic chemistry because it can so easily produce a molecule with a complex structure as a single diastereomer.

In its simplest form, the Diels-Alder reaction is a reaction between a conjugated diene and an alkene in which a six-membered ring is formed.

diene dienophile   **Diels-Alder adduct**

The reaction appears to involve only a single step (one transition state).

Many different combinations of diene and dienophile will undergo this reaction, and the following experiments illustrate a few of the great variety of structures that can result.

E66. Butadiene (from 3-Sulfolene) and Maleic Anhydride

A relatively simple Diels-Alder reaction is the combination of the simplest diene of them all, butadiene, with maleic anhydride to give the *meso* isomer of a bicyclic anhydride.

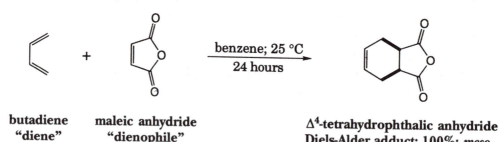

butadiene
"diene"

maleic anhydride
"dienophile"

Δ^4-tetrahydrophthalic anhydride
Diels-Alder adduct; 100%; *meso*

The product can be redrawn in this way to more clearly indicate its configuration:

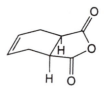

meso Δ^4-tetrahydrophthalic anhydride

Diels-Alder reactions with butadiene must generally be carried out in a pressure vessel, because the temperature at which reaction takes place readily is usually far above the boiling point of butadiene, -3°C. However, Sample and Hatch (Reference 1) described a convenient way to avoid the use of a pressure vessel in the preparation of the Diels-Alder adduct between butadiene and maleic anhydride. In their procedure, butadiene is generated by the thermal decomposition of 3-sulfolene, a sort of reverse Diels-Alder reaction itself.

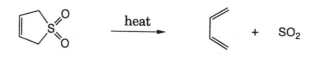

3-sulfolene

butadiene

Apparently, at the temperature of the reaction, butadiene reacts with maleic anhydride more rapidly than it can be lost from the reaction mixture.

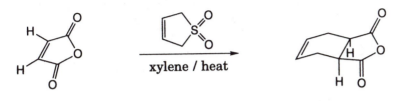

maleic anhydride Δ^4-**tetrahydro-phthalic anhydride**

Scale: 2.45 grams (0.025 mole) of maleic anhydride.

Procedure (Reference 1)

Place 4.25 g, (0.038 mole) of 3-sulfolene, and 2.45 g (0.025 mole) of powdered maleic anhydride in a 125-mL Erlenmeyer flask. Add 2 mL of xylene, fit the flask with a condenser, and stopper the condenser with a plug of absorbent cotton (Note 1). Swirl the flask gently to effect partial solution. Remove the cotton plug, and cautiously heat the mixture to a slow boil (Notes 2 and 3).

After boiling the mixture for about one-half hour, allow it to cool for about 5 minutes (Note 4), and then add 25 mL of toluene (Note 5) and 0.5 g of decolorizing carbon. Heat the suspension on the steam bath and filter it by gravity.

Reheat the filtrate until the product redissolves, and then add, while swirling, 10–12 mL of petroleum ether (Note 5) until a slight cloudiness persists. Rewarm the mixture until it is substantially clear, and finally allow the solution to cool in an ice bath for crystallization.

Notes

1. The mixture will become quite cool. The absorbent cotton will prevent water vapor from entering the flask and causing hydrolysis of the maleic anhydride.

2. Since sulfur dioxide will be evolved, the reaction must be run in the hood, or some provision must be made to absorb the fumes.

3. Because the reaction is exothermic, some care must be taken to avoid overheating at the beginning of the heating period.

4. If the cooling period is too long, the product will separate as a hard cake that will be difficult to redissolve.

5. The solvents should be dry to avoid hydrolyzing the product to the corresponding diacid. Since the diacid is much less soluble, it cannot be removed by crystallization.

Time: 3 hours.

Disposal

> **Solid waste:** filter paper
> **Nonhalogenated liquid organic waste:** xylene; toluene; petroleum ether
> **Solid inorganic waste:** carbon
> **Solid organic waste:** 3-sulfolene; maleic anhydride; product

Reference

1. T. E. Sample, Jr., and L. F. Hatch, *J. Chem. Educ.* **45,** 55 (1968).

E67. Cyclopentadiene and Maleic Anhydride

In all Diels-Alder reactions, the configuration of the product indicates that the reaction takes place with *syn* addition to the double bond of the dienophile. With cyclopentadiene, *syn* addition can take place in two different ways to give products with the same structure but with different configurations, the "*exo*" and the "*endo*" products.

"*exo*" product

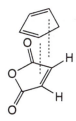 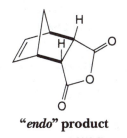

"*endo*" product
m.p. = 165°C

With butadiene (Experiment E66), the two different modes of reaction would give two equivalent (but presumably rapidly interconverting) conformations of the same molecule, and thus there is no way of knowing whether one path was followed in preference to the other. With cyclopentadiene, under the conditions of this experiment, the isomer that corresponds to *endo* addition can be isolated in high yield.

The predominance of the *endo* isomer in Diels-Alder reactions is generally observed, and, because the *exo* isomer is usually the more stable, the predominance of the *endo* isomer must be the result of kinetic control rather than thermodynamic control. The predominance of *endo* addition (Alder Rule) has been interpreted as the result of inductive effects, electrostatic effects of an initial charge transfer between diene and dienophile, "maximum accumulation of unsaturation," and, most recently, orbital symmetry effects.

When the *exo* isomer is more stable, high temperatures and long reaction times can give the *exo* isomer. Sometimes, the *endo* isomer can be converted to the *exo* isomer by heating. Although it would be reasonable to propose that the *exo* isomer is formed via dissociation of the *endo* isomer to diene and dienophile, followed by recombination, the isomerization of the *endo* adduct of cyclopentadiene and maleic anhydride takes place at 190°C in an open flask without observable loss in weight. One would expect free cyclopentadiene to be lost under these conditions, although the success of the procedure of Section E66 indicates that this loss may not necessarily take place.

Cyclopentadiene dimerizes on standing via a Diels-Alder reaction. It must be prepared by thermal decomposition of the dimer.

$$\underset{\substack{\textbf{dicylopentadiene}\\\textbf{b.p., 170°C}}}{}\ \underset{\substack{\text{boil; 170 °C}\\ \text{room temperature}}}{\rightleftharpoons}\ \underset{\substack{\textbf{cyclopentadiene}\\\textbf{b.p., 43°C}}}{}\ +$$

dicylopentadiene
b.p., 170°C

cyclopentadiene
b.p., 43°C

The *endo-cis* diacid can be prepared as described in Exercise 1 at the end of Section 9.

Thermal cracking of dicyclopentadiene to cyclopentadiene

Scale: 15 mL of dicyclopentadiene.

Procedure

Place 15 mL of dicyclopentadiene in a 50-mL boiling flask. Fit the flask with a fractionating column, and connect an adapter with a condenser set for downward distillation to the top of the column. Heat the dicyclopentadiene until it boils, and then continue to heat at such a rate that the cyclopentadiene distills at about its boiling point of 43°C. Cool the receiver in an ice bath. A fair fractionating column and some patience are needed. It can take 30 minutes to collect 6 mL of cyclopentadiene.

Time: about 1 hour.

Scale: 2.45 grams (0.025 mole) of maleic anhydride.

Procedure

Place 8 mL of ethyl acetate in a 50-mL Erlenmeyer flask. Add to the flask 2.45 g (0.025 mole) of maleic anhydride and make the anhydride dissolve by heating the mixture on the steam bath. Add 8 mL of hexane to the solution, and cool the mixture thoroughly in an ice bath (Note 1). Add 2.5 mL (2.0 g; 0.030 mole) of dry cyclopentadiene (Notes 2 and 3), and swirl the mixture in the ice bath until the product separates as a white solid. Then heat the mixture on the steam bath until the solid is all dissolved (Note 4).

Now allow the solution to stand for crystallization; slow crystallization will result in a beautiful display of crystal formation (Note 5).

Notes

1. Some maleic anhydride may crystallize.
2. Cyclopentadiene must be prepared by heating its dimer, dicyclopentadiene, to its boiling point of 170°C. At this temperature, monomeric cyclopentadiene slowly distills from the boiling flask. Upon standing at room temperature, the monomer slowly reverts to dicyclopentadiene, the form in which cyclopentadiene is available for purchase.

Your Laboratory Instructor may have prepared the cyclopentadiene for you.

3. Water in the distilled cyclopentadiene will cause it to be cloudy. The cyclopentadiene can be dried with a little anhydrous calcium chloride.

4. Any diacid formed by water that got into the reaction mixture will remain undissolved at this point; it should be removed by gravity filtration.

5. Once again, Dr. Orgo says that only the least sophisticated chemist would even *think* of placing the flask in an ice bath before crystallization has become complete at room temperature.

Disposal

Solid waste: filter paper

Nonhalogenated liquid organic waste: dicyclopentadiene; cyclopentadiene; hexane

Solid organic waste: maleic anhydride; product

Questions

1. a. How many stereoisomeric forms are possible for the product of this reaction? (This question can be answered in more than one way. State your assumptions and show how your answer follows.)

   b. How many stereoisomers are actually formed in this reaction? What factors limit the number of stereoisomers actually formed?

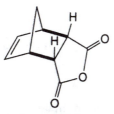

the product of this reaction

2. Why does heating dicyclopentadiene form cyclopentadiene, whereas allowing cyclopentadiene to stand at room temperature leads to dicyclopentadiene? (What must be true of the free energy, enthalpy, and entropy changes in this reaction?) Are these deductions consistent with other things we have said? (Explain.)

3. Propose an alternative procedure for this reaction in which the cyclopentadiene is produced in situ, in analogy with the in situ production of butadiene in Experiment E66.

References

1. W. J. Sheppard, *J. Chem. Educ.* **40,** 40 (1963).
2. L. F. Fieser, *Organic Experiments,* Heath, Boston, 1964, p. 83.

E68. Furan and Maleic Anhydride

In the Diels-Alder reaction between furan and maleic anhydride, the only adduct that has been isolated is one whose melting point has been variously reported as 116–117°C, 122°C, and 125°C with foaming and decomposition. The stereochemistry of this product was shown to correspond to the *exo* mode of addition. It is suspected that this is the thermodynamically more stable isomer.

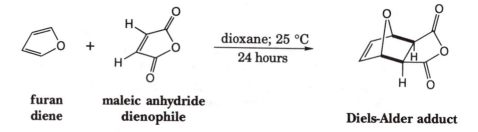

furan   maleic anhydride
diene   dienophile   Diels-Alder adduct

Scale: 2.45 grams (0.025 mole) of maleic anhydride.

Procedure

Place 2.45 grams (0.025 mole) of maleic anhydride in a 50-mL ground-glass-stoppered Erlenmeyer flask. Add 7 mL of dioxane (Note 1), and swirl the mixture until the anhydride has all dissolved. Add to the flask 1.8 mL (1.7 grams; 0.025 mole) of furan, swirl to mix, and allow the solution to stand for at least 24 hours.

Isolate the crystalline adduct by suction filtration, washing it with a little ether. Yield: about 2.5 grams (65%; Note 1).

Note

1. If half as much dioxane is used, the yield will be greater than 80% but the product will separate from solution in hard lumps.

Time: 1 hour plus a reaction time of at least 24 hours.

Disposal

> **Solid waste:** filter paper
> **Nonhalogenated liquid organic waste:** dioxane; furan; ether
> **Solid organic waste:** maleic anhydride; product

E69. α-Phellandrene and Maleic Anhydride

One of the Diels-Alder reactions reported by Diels and Alder was that between the diterpene α-phellandrene and maleic anhydride.

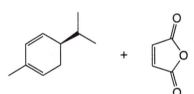 +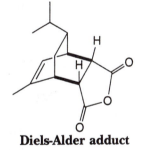

ether ⟶

α-phellandrene   maleic anhydride   Diels-Alder adduct

The reaction takes place as the diene and dienophile are allowed to stand overnight in ether solution to give a product that, in the words of Diels and Alder, ". . . bildet grosse glasglänzende Krystalle von ungewöhnlicher Schönheit. . . ."

α-Phellandrene is obtained by steam distillation of eucalyptus wood and leaves, but the commercial material is only about 60% pure. The contaminant does not interfere with the Diels-Alder reaction, but its presence must be taken into account when measuring the α-phellandrene.

Scale: 0.025 mole of α-phellandrene (5.7 grams of 60% pure commercial material).

Procedure

Place 2.45 grams (0.025 mole) of maleic anhydride in a 50-mL ground-glass-stoppered Erlenmeyer flask. Next, add to the flask 0.025 mole (3.41 grams 100% or 5.7 grams 60%) α-phellandrene. Finally, add 13 mL of anhydrous ether, and swirl the flask to mix the contents. The initial yellow color of the solution, due to a "charge transfer" complex, will fade as the reaction progresses. Fit the flask with a condenser and heat the mixture just to boiling on the steam bath for an hour (Note 1).

Allow the mixture to cool to room temperature, and then finally cool the mixture in an ice bath. Collect the solid product by suction filtration. The yield is about 3.5 grams.

Recrystallize the crude product by dissolving it with heating under reflux in about 40 mL of methanol. Set the hot solution aside to cool slowly for crystallization.

Note

1. Alternatively, let the reaction mixture stand at room temperature for 24 hours and then decant the liquid from the crystals.

Time: about 2 hours working time; 24-hour reaction time.

Disposal

Solid waste: filter paper

Nonhalogenated liquid organic waste: α-phellandrene; ether; methanol

Solid organic waste: maleic anhydride; product

Question

1. a. How many stereoisomeric forms are possible for the product of this reaction? (This question can be answered in more than one way. State your assumptions and show how your answer follows.)

   b. How many stereoisomers are actually formed in this reaction? What factors limit the number of stereoisomers actually formed?

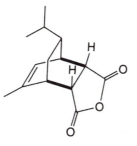

the product of this reaction

References

1. O. Diels and K. Alder, *Justus Liebigs Ann. Chem.* **460,** 116 (1927).
2. David Todd, *Experimental Organic Chemistry,* Prentice-Hall, Englewood Cliffs, New Jersey, 1979, p. 256.

THE WITTIG REACTION

An important method for the synthesis of alkenes is the Wittig reaction. In one version of this reaction, a triphenyl alkyl phosphonium salt is treated with a strong base to form an inner salt, an ylide, which is the Wittig reagent.

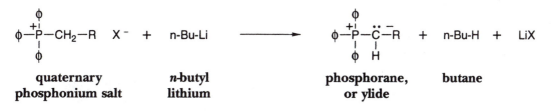

| quaternary | *n*-butyl | phosphorane, | butane |
| phosphonium salt | lithium | or ylide | |

The first step of the Wittig reaction is attack of the nucleophilic carbon of the phosphorane, the carbon connected to phosphorus, on the electrophilic carbon of the aldehyde or ketone.

first intermediate

Just as the electrophilic nature of the carbon of a carbonyl group can be emphasized by writing the aldehyde or ketone as a resonance hybrid,

resonance representation of an aldehyde or ketone

so the nucleophilic nature of the α carbon of the phosphorane is indicated by representing the phosphorane as a resonance hybrid:

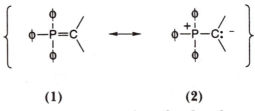

(1)   (2)

resonance representation of a phosphorane

The negative charge on carbon is stabilized both by the positive charge on the adjacent phosphorus atom and by delocalization of the unshared pair of electrons on carbon into empty 3d orbitals on phosphorus, as implied by contributor (1) to the resonance hybrid.

   Thus, the first step of the Wittig reaction could also be represented like this:

**first intermediate
a betaine**

The second and third steps of the Wittig reaction are cyclization of the first intermediate to give a second, cyclic, intermediate,

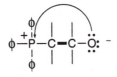

first intermediate   **second intermediate,
                                           an oxaphosphatane**

followed by elimination of triphenylphosphine oxide, to form the alkene and triphenylphosphine

second intermediate	**alkene** **triphenyl-phosphine oxide**

The great stability of the phosphorus–oxygen "double bond" provides the decrease in free energy that drives the reaction.

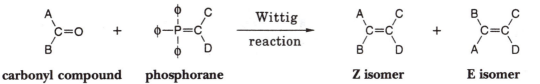

triphenylphosphine oxide

The diastereoselectivity of the Wittig reaction

If both the phosphorane and the carbonyl compound are unsymmetrically substituted, two diastereomeric alkenes can be formed.

carbonyl compound	**phosphorane**	**Z isomer**	**E isomer**

The relative amounts of the Z and E forms of the alkene depend on the relative amounts of the diastereomeric intermediates and their relative rates of elimination of triphenylphosphine oxide.

E70. The Preparation of *trans*-Stilbene

The Phosphonate Variation

A modification of the Wittig reaction has been described by Seus and Wilson (Reference 1) in which a phosphonate-stabilized carbanion acts as the nucleophilic reagent.

For example, as illustrated by the procedure of this experiment, when benzaldehyde is added to a solution of diethyl benzylphosphonate and sodium methoxide in dimethylformamide, *trans*-stilbene is formed in a yield of 85%.

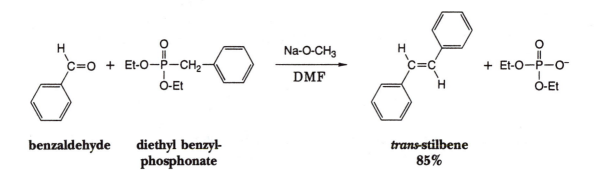

benzaldehyde   diethyl benzyl-
                phosphonate

trans-stilbene
85%

This reaction probably takes place by conversion of the diethyl benzylphosphonate to its resonance-stabilized conjugate base, or carbanion,

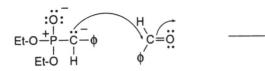

resonance-stabilized carbanion

The carbanion then adds to the carbonyl group of benzaldehyde to give the first intermediate,

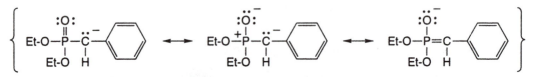

first intermediate

which cyclizes to the second intermediate,

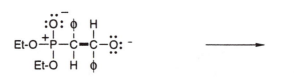

first intermediate

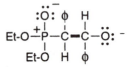

second intermediate

and the second intermediate then eliminates diethyl phosphate to form *trans*-stilbene.

second intermediate   ***trans*-stilbene**

Preparation of diethyl benzylphosphonate from Benzyl Chloride and Triethylphosphate by the Michaelis-Arbuzov Reaction

The most common method for the preparation of an alkyl-phosphonate ester such as diethyl benzylphosphonate is treatment of triethyl phosphite with an alkyl halide, the so-called Michaelis-Arbuzov reaction.

triethyl phosphite   **diethyl alkylphosphonate**

This transformation occurs by alkyl transfer to phosphorus from the alkyl halide by an S_N2 mechanism,

followed by transfer of the ethyl group from oxygen to the halide by another S_N2 process,

diethyl alkylphosphonate

a reaction driven to completion by the loss of the volatile ethyl halide from the reaction mixture.

Scale: 3.2 grams (0.025 mole) of benzyl chloride.

Procedure

Add to a 25-mL round-bottom flask 2.9 mL (3.3 g; 0.025 mole) of benzyl chloride (Note 1) and 4.3 mL (4.15 g; 0.025 mole) of triethyl phosphite. Add a boiling stone to the flask and fit the flask with a reflux condenser. Clamp the flask in position over a wire gauze supported by an iron ring so that the contents can be heated under reflux by a burner flame. Lower a thermometer down the condenser until the bulb rests on the bottom of the boiling flask so that the temperature of the boiling liquid can be determined. By means of a burner flame, heat the mixture to boiling (Note 1) and continue to boil it for 1 hour (Note 2).

At the end of the heating period, remove the burner, the wire gauze, and the iron ring, and allow the flask to cool. After the temperature of the contents of the flask has fallen below 100°C, cooling can be hastened by immersing the flask in a beaker of cold water. The cool diethyl benzylphosphonate is suitable for use in the Wittig reaction. You should use the entire product in the following reaction.

Notes

1. All transfers of reagents and the reaction itself must be carried out in a good hood. Benzyl chloride is a potent lachrymator; it should be dispensed by means of a burette. Triethyl phosphite is not known to be particularly toxic, but it does have an obnoxious smell. Ethyl chloride is produced as a gas during the reaction.
2. A shorter heating period results in a lower yield. The temperature of the boiling liquid is initially about 165°C; at the end of the hour it will be above 220°C.

Time: about 2 hours.

Preparation of *trans*-stilbene from Diethyl Benzylphosphonate by a Wittig Reaction

Scale: 0.025 mole) of diethyl benzylphosphonate.

Procedure

Add to a 125-mL Erlenmeyer flask 1.5 g (0.028 mole) of sodium methoxide (Note 1) and 13 mL of dimethylformamide. Add to this suspension 5.21 mL (5.7 grams; 0.025 mole) of diethyl benzyl-phosphonate (Note 2) and an additional 12 mL of dimethyl-formamide (Note 3). After the temperature of the resulting mixture has been adjusted to 20°C by brief swirling in an ice bath, add 2.5 mL (2.6 g, 0.025 mole) of benzaldehyde (Note 4). Swirl the flask to mix the contents. The temperature of the reaction mixture will start to rise; keep the temperature of the mixture between 30 and 40°C by occasionally swirling the flask in the ice bath.

After about 5 minutes, the temperature will no longer tend to rise. At this point, allow the flask to stand at room temperature for another 5 minutes, and then add 15 mL of water to precipitate the *trans*-stilbene.

Collect the product by suction filtration, using 30 mL of a 1:1 mixture (by volume) of water and methanol for rinsing and washing.

The product should be recrystallized from isopropyl alcohol to give clusters of diamond-shaped prisms (Note 5). The yield, after recrystallization, is about 2.5 grams (about 55%). The melting point of *trans*-stilbene is reported to be 124–126°C, and its infrared spectrum is shown in Figure E70-1.

Notes

1. Sodium methoxide is caustic. Avoid spilling or breathing the finely divided powder. Sodium methoxide from a recently opened bottle should be used.

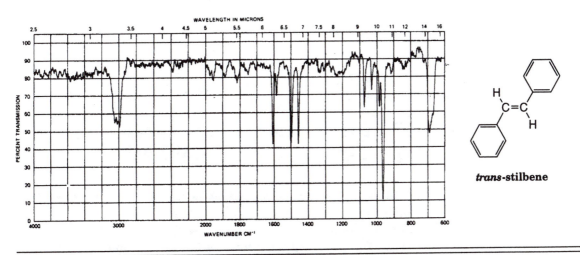

trans-stilbene

Figure E70-1. Infrared spectrum of *trans*-stilbene; $CHCl_3$ solution.

2. Diethyl benzylphosphonate can be either purchased or prepared by the procedure given just above. If you prepare it, use the entire product and assume that it is 0.025 mole.

3. This second portion of dimethylformamide can be used to assist in the transfer of the diethyl benzylphosphonate.

4. Benzaldehyde from a recently opened bottle should be used.

5. The amount of isopropyl alcohol required will be between 30 and 50 mL, depending on how thoroughly the crude product has been sucked dry during suction filtration. Methanol can be used for rinsing and washing when the recrystallized material is collected by suction filtration.

Time: less than 2 hours.

Disposal

Solid waste: filter paper

Nonhalogenated liquid organic waste: triethyl phosphite; dimethylformamide; benzaldehyde; isopropyl alcohol

Halogenated liquid organic waste: benzyl chloride

Solid inorganic waste: sodium methoxide

Solid organic waste: *trans*-stilbene

Reference

1. E. J. Seus and C. V. Wilson, *J. Org. Chem.* **26**, 5243 (1961).

E71. Preparation of *trans,trans*-1,4-Diphenylbutadiene: A Variation

In this variation on the procedure of Section E70, 0.025 mole of cinnamaldehyde is substituted for 0.025 mole of benzaldehyde. Because the aldehyde contains a *trans* double bond and the Wittig reaction produces a *trans* double bond, the 1,4-diphenylbutadiene formed is the *trans,trans* isomer.

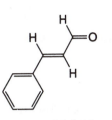 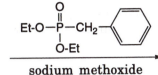 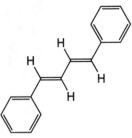

sodium methoxide

trans-cinnamaldehyde

trans,trans-1,4-
diphenylbutadiene

Preparation of *trans,trans*-1,4-diphenylbutadiene from Diethyl Benzylphosphonate by a Wittig Reaction

Scale: 0.025 mole) of diethyl benzylphosphonate.

Procedure

Add to a 125-mL Erlenmeyer flask 1.5 g (0.028 mole) of sodium methoxide (Note 1) and 13 mL of dimethylformamide. Add to this suspension 5.21 mL (5.7 grams; 0.025 mole) of diethyl benzyl-phosphonate (Note 2) and an additional 12 mL of dimethyl-formamide (Note 3). After the temperature of the resulting mixture has been adjusted to 20°C by brief swirling in an ice bath, add 3.15 mL (3.30 g, 0.025 mole) of *trans*-cinnamaldehyde (Note 4). Swirl the flask to mix the contents. The temperature of the reaction mixture will start to rise, the reaction mixture will gradually change in color to deep red, and the product will start to crystallize. Keep the temperature of the mixture between 30 and 40°C by occasionally swirling the flask in the ice bath.

When the temperature of the mixture shows no further tendency to rise, allow the mixture to stand at room temperature for 10 minutes and then add a mixture of 20 mL of water and 10 mL of methanol. Swirl the flask to break up the mass of crystals and mix the contents, and collect the product by suction filtration. Pour the red filtrate back into the reaction flask to help transfer the remainder of the product to the funnel.

Thoroughly wash the crystals with water until the red color is entirely removed, and then wash them with methanol until the washings are almost colorless so as to remove a yellow contaminant. Continue to apply suction to the crude product for about 5 minutes to remove as much methanol as possible. Then remove the product from the funnel, break up the lumps, and spread it out to dry some more.

The dry material can be recrystallized from cyclohexane, using 10 mL per gram, to give light, faintly yellow plates of the purified hydrocarbon. *trans,trans*-1,4-diphenylbutadiene is reported to melt at 151–153°C, and its infrared spectrum is shown in Figure E71-1.

Notes

1. Sodium methoxide is caustic. Avoid spilling or breathing the finely divided powder. Sodium methoxide from a recently opened bottle should be used.

2. Diethyl benzylphosphonate can be either purchased or prepared by the procedure given in Section E70. If you prepare it, use the entire product and assume that it is 0.025 mole.

3. This second portion of dimethylformamide can be used to assist in the transfer of the diethyl benzylphosphonate.

4. Cinnamaldehyde from a recently opened bottle should be used.

Time: less than 2 hours.

Disposal

Solid waste: filter paper

Nonhalogenated liquid organic waste: triethyl phosphite; dimethylformamide; cinnamaldehyde; methanol; cyclohexane

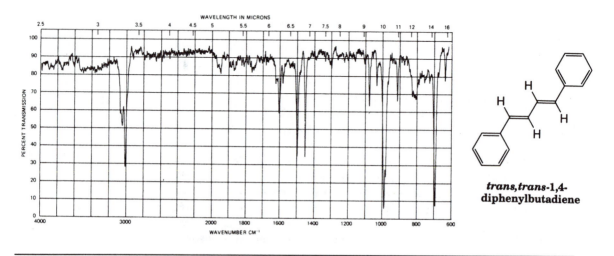

trans,trans-1,4-diphenylbutadiene

Figure E71-1. Infrared spectrum of *trans,trans*-1,4-diphenylbutadiene; CCl_4 solution.

Halogenated liquid organic waste: benzyl chloride
Solid inorganic waste: sodium methoxide
Solid organic waste: *trans,trans*-1,4-diphenylbutadiene

Question

1. In how many stereoisomeric forms could 1,4-diphenylbutadiene exist?

Reference

1. E. J. Seus and C. V. Wilson, *J. Org. Chem.* **26,** 5243 (1961).

ANALGESICS: ASPIRIN, PHENACETIN, AND TYLENOL

Acetylsalicylic acid, *p*-acetamidophenol, and *p*-ethoxyacetanilide are used extensively as antiinflamatories, analgesics (pain relievers), and antipyretics (fever reducers).

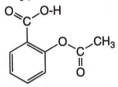 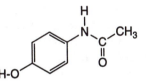 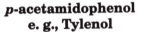 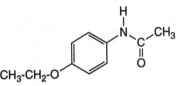

acetylsalicylic acid aspirin   **p-acetamidophenol e. g., Tylenol**   **p-ethoxyacetanilide phenacetin**

The next four sections present procedures by which these compounds can be prepared.

E72. Acetylsalicylic Acid from Salicylic Acid: Preparation of Aspirin

The industrial method of production of acetylsalicylic acid is similar to the method given in this section, which is an acid-catalyzed acetyl transfer from acetic anhydride to the phenolic –O-H group of salicylic acid.

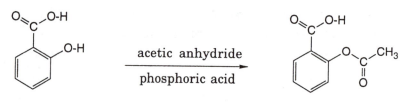

salicylic acid   **acetylsalicylic acid
                                                                Aspirin**

The current industrial production in the United States is about 60 million pounds, or about 300 5-grain tablets for each man, woman, and child in the United States.

The infrared and proton NMR spectra of acetylsalicylic acid are shown in Figures E3-1 and E3-2.

Scale: 1.38 grams (0.010 mole) of salicylic acid.

Procedure

Add to a 50-mL Erlenmeyer flask 1.38 grams (0.010 mole) of salicylic acid and 2.8 mL (3.1 grams; 0.030 mole) of acetic anhydride. To this mixture, add 3 drops of 85% phosphoric acid and swirl the flask to mix the contents. Fit the flask with a reflux condenser (Note 1), and heat the mixture on the steam bath for about 5 minutes.

Without cooling the mixture, add 1 mL (0.056 mole) of water in one portion down the condenser. The excess acetic anhydride will hydrolyze, and the contents of the flask will come to a boil (Note 2).

When the exothermic hydrolysis reaction has ended, add 25 mL of cold water to the flask, cool the mixture to room temperature, stir and rub the mixture with a stirring rod if necessary to induce crystallization, and finally allow the mixture to stand in the ice bath for 10 minutes or so to complete crystallization.

Collect the product by suction filtration and wash it with a little water. The product can be recrystallized from water. Acetylsalicylic acid is reported to melt at 135°C with rapid heating.

Notes

1. You need not fit the condenser with hoses; the idea is to minimize the escape of acetic anhydride vapors.
2. Acetic anhydride has an unpleasant smell, and this operation converts the excess acetic anhydride to acetic acid. You have to pay good attention if you want to see the mixture boil.

Time: About 1 hour.

Disposal

    Sink: phosphoric acid (after dilution); aqueous washes
    Solid waste: filter paper
    Nonhalogenated liquid organic waste: acetic anhydride
    Solid organic waste: salicyclic acid; acetylsalicylic acid

Questions

1. What is the purpose of the phosphoric acid?

2. Why does the acetyl group end up on the phenolic –O-H group and not the –O-H of the carboxyl group? Is the acyl group ever transferred to the "wrong" –O-H group? If so, then what can happen, and why? If not, why not?

3. When 1 mL of water is added down the condenser to hydrolyze the excess acetic anhydride, why do we not worry about hydrolysis of the product? Or should we worry about hydrolysis of the product? What experiment might we do to determine whether the product is hydrolyzed by this operation?

E73. *p*-Acetamidophenol from *p*-Amino-phenol: Preparation of Tylenol

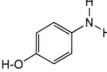

p-aminophenol

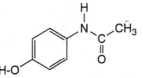

p-acetamidophenol
e.g., Tylenol

pH = about 5

Scale: 1.1 gram (0.010 mole) of *p*-aminophenol.

Procedure

Place 1.1 grams (0.010 mole) of *p*-aminophenol (Note 1) in a 50-mL Erlenmeyer flask. To this flask, add 10 mL of water and 0.9 mL (0.011 mole) of concentrated HCl. Swirl the flask to dissolve the amino phenol (Note 2).

Why does the aminophenol dissolve?

Prepare a solution of 1.0 grams (0.012 mole) of anhydrous sodium acetate (Note 3) in 6 mL of water.

To the solution of *p*-aminophenol in HCl, add 1.1 mL (1.2 grams; 0.012 mole) of acetic anhydride. Swirl the mixture to dissolve almost all of the acetic anhydride, add the solution of sodium acetate, swirl the mixture to mix it thoroughly, and set it aside for crystallization (Note 4).

Collect the product by suction filtration, and wash it with a little cold water. Yield: about 1.1 grams; about 75%.

Notes

1. Commercial *p*-aminophenol is often quite dark. It can be cleaned up somewhat by suspending it in about twice its volume of methanol and then recovering it by suction filtration.

2. If the solution is too highly colored, add as much decolorizing carbon as would cover a circle 1 cm in diameter, swirl the mixture to mix it thoroughly, and then remove the carbon by suction filtration. Before pouring your solution into the Büchner funnel, wet the circle of filter paper, and then apply the vacuum to make sure that the circle of filter paper is pulled down tight over the holes in the funnel.

3. It may seem silly to add *anhydrous* sodium acetate to *water*, but anhydrous sodium acetate costs about the same (per *gram*) as the trihydrate, and the anhydrous material dissolves *much faster* than the trihydrate.

4. The solution can be allowed to stand overnight or longer, or it can be cooled more quickly after crystals have started to form by immersion first in cold water and then in an ice bath.

Time: About 1 hour of working time.

Disposal

Sink: concentrated HCl (after dilution); aqueous washes
Solid waste: filter paper
Nonhalogenated liquid organic waste: acetic anhydride
Solid inorganic waste: sodium acetate
Solid organic waste: *p*-aminophenol; *p*-acetamidophenol

Questions

1. Estimate the pH of the solution at each of the following stages of the reaction:

   a. After the *p*-aminophenol is added to the water in the flask.

   b. After the concentrated hydrochloric acid is added.

   c. After the acetic anhydride is added.

   d. One millisecond after the sodium acetate is added.

   e. One minute after the sodium acetate is added.

2. Indicate the state of the original *p*-aminophenol at each of the following stages of the reaction:

   a. After the *p*-aminophenol is added to the water in the flask.

   b. After the concentrated hydrochloric acid is added.

   c. After the acetic anhydride is added.

   d. One millisecond after the sodium acetate is added.

   e. One minute after the sodium acetate is added.

3. What is the role of the concentrated hydrochloric acid in this reaction?

4. What is the role of the sodium acetate in this reaction?

5. What is the role of the acetic anhydride in this reaction?

6. Do we need to worry about acetate ion reacting with acetic anhydride? Explain.

7. Which is the stronger base, *p*-aminophenol or *p*-acetamidophenol? Explain.

8. Why does the acetylation of *p*-aminophenol with 1 equivalent of acetic anhydride in a solution of pH = 4 give the product of acetyl transfer to nitrogen rather than to oxygen?

9. What product would you expect if *p*-aminophenol were acetylated with 2 equivalents of acetic anhydride at a pH of 11?

10. What product would you expect if *p*-aminophenol were acetylated with 1 equivalent of acetic anhydride at a pH of 11?

E74. *p*-Ethoxyacetanilide from *p*-Ethoxy-aniline: Preparation of Phenacetin

In this preparation of *p*-ethoxyacetanilide (phenacetin), the aromatic amine *p*-ethoxyaniline is converted to the corresponding amide of acetic acid by the action of acetic anhydride by the same procedure presented in Experiment E73 for the conversion of *p*-aminophenol to *p*-acetamidophenol.

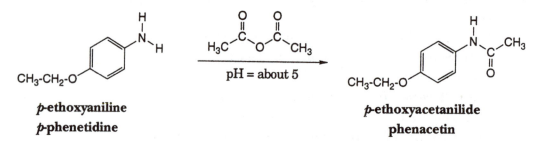

p-ethoxyaniline
p-phenetidine

p-ethoxyacetanilide
phenacetin

In the next section, Experiment E75, phenacetin is prepared from *p*-acetamidophenol (Tylenol) by converting the phenolic –O-H group of *p*-acetamidophenol to the corresponding ethyl ether.

The infrared and proton NMR spectra of *p*-ethoxyacetanilide are shown in Figures E74-1 and E74-2.

p-acetamidophenol
e.g., Tylenol

p-ethoxyacetanilide
phenacetin

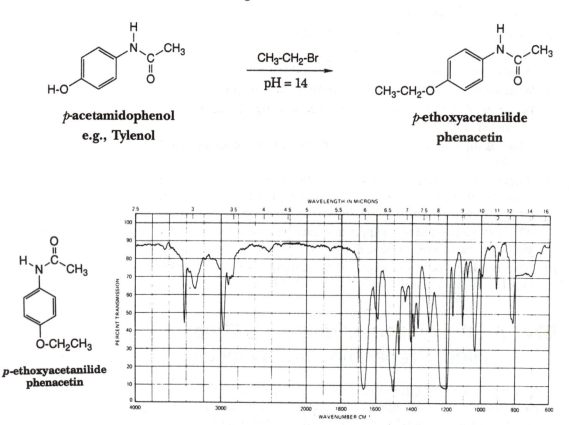

Figure E74-1. Infrared spectrum of *p*-ethoxyacetanilide; $CHCl_3$ solution.

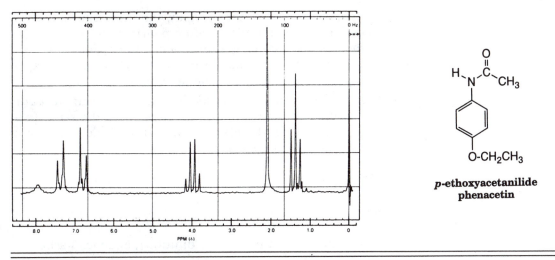

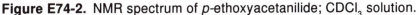

Figure E74-2. NMR spectrum of *p*-ethoxyacetanilide; CDCl$_3$ solution.

The *Merck Index* describes phenacetin as an analgesic and antipyretic.

Scale: 1.4 grams (0.010 mole) of *p*-ethoxyaniline.

Procedure

Add 28 mL of water and 1.3 mL (1.4 grams; 0.010 mole) of *p*-ethoxy-aniline (Note 1) to a 50-mL Erlenmeyer flask. Dissolve the amine by adding 0.9 mL (0.011 mole) of concentrated hydrochloric acid (Note 2).

Why does the amine dissolve?

　　Prepare a solution of 1.0 grams (0.012 mole) of anhydrous sodium acetate (Note 3) in 6 mL of water.

　　To the solution of *p*-ethoxyaniline hydrochloride, add 1.1 mL (1.2 g; 0.012 mole) of acetic anhydride. Swirl the mixture for a few seconds and then add the solution of sodium acetate. *p*-Ethoxy-acetanilide forms and precipitates immediately.

　　Collect the crude product by suction filtration and wash it with cold water.

　　Recrystallize the damp material from about 7 mL of 95% ethanol. On slow cooling, *p*-ethoxyacetanilide will crystallize as long spars. Yield after recrystallization: 1.4 grams (78%); m.p., 137–138°C.

Notes

1. *p*-Ethoxyaniline is also called *p*-phenetidine.
2. If the solution is too highly colored, add as much decoloriz-ing carbon as would cover a circle 1 cm in diameter, swirl

the mixture to mix it thoroughly, and then remove the carbon by suction filtration. Before pouring your solution into the Büchner funnel, wet the circle of filter paper, and then apply the vacuum to make sure that the circle of filter paper is pulled down tight over the holes in the funnel.

3. It may seem silly to add *anhydrous* sodium acetate to *water*, but anhydrous sodium acetate costs about the same (per *gram*) as the trihydrate, and the anhydrous material dissolves *much faster* than the trihydrate.

Time: About 2 hours.

Disposal

Sink: concentrated HCl (after dilution); aqueous washes

Solid waste: filter paper

Nonhalogenated liquid organic waste: *p*-ethoxyaniline; acetic anhydride

Solid inorganic waste: sodium acetate

Solid organic waste: *p*-ethoxyacetanilide

1. Estimate the pH of the solution at each of the following stages of the reaction:
   a. After the *p*-ethoxyaniline is added to the water in the flask.
   b. After the concentrated hydrochloric acid is added.
   c. After the acetic anhydride is added.
   d. One millisecond after the sodium acetate is added.
   e. One minute after the sodium acetate is added.

2. Indicate the state of the original *p*-ethoxyaniline at each of the following stages of the reaction:
   a. After the *p*-ethoxyaniline is added to the water in the flask.
   b. After the concentrated hydrochloric acid is added.
   c. After the acetic anhydride is added.
   d. One millisecond after the sodium acetate is added.
   e. One minute after the sodium acetate is added.

3. What is the role of the concentrated hydrochloric acid in this reaction?

4. What is the role of the sodium acetate in this reaction?

5. What is the role of the acetic anhydride in this reaction?

6. Do we need to worry about acetate ion reacting with acetic anhydride? Explain.

7. Which is the stronger base, *p*-ethoxyaniline or *p*-ethoxyacetanilide? Explain.

E75. *p*-Ethoxyacetanilide from *p*-Acetamidophenol: Another Preparation of Phenacetin

In this second method for the preparation of *p*-ethoxyacetanilide (phenacetin), the product is prepared from *p*-acetamidophenol (e.g., Tylenol) by converting the phenolic –O-H group of *p*-acetamidophenol to the corresponding ethyl ether.

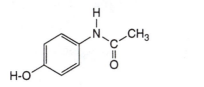

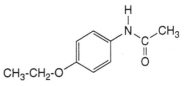

$$CH_3\text{-}CH_2\text{-}Br$$
$$pH = 14$$

p-acetamidophenol

e.g., Tylenol

p-ethoxyacetanilide

phenacetin

The *Merck Index* describes phenacetin as an analgesic and antipyretic agent.

The infrared and proton NMR spectra of *p*-ethoxyacetanilide are shown in Figures E74-1 and E74-2.

Scale: 1.5 grams (0.010 mole) of *p*-acetamidophenol.

Procedure

Place 1.5 grams (0.010 mole) of *p*-acetamidophenol and 10 mL of methanol in a 50-mL boiling flask. To this mixture, add 0.63 mL (0.012 mole) of 50% sodium hydroxide solution; swirl the mixture to dissolve the *p*-acetamidophenol.

Why does the *p*-acetamidophenol dissolve?

Fit the flask with a reflux condenser, and clamp the apparatus in position on a steam bath.

Now add 1.5 mL (2.2 g; 0.020 mole) (Note 1) of ethyl bromide down the condenser, and boil the mixture gently for about 2 hours (Note 2).

At the end of the heating period, add 20 mL of hot water down the condenser; crystals should begin to appear as the last of the water is added.

Remove the condenser, and set the flask aside for crystallization (Note 3).

Collect the glistening, colorless crystals by suction filtration, washing them with portions of cold water. Yield: about 3 grams; about 80%.

Notes

1. Only 1 equivalent of ethyl bromide is required. We use 2 equivalents to decrease the time needed for the reaction

2. A reaction time of 1 hour gives a yield of about 60%.

3. The mixture can be allowed to stand overnight or longer, or crystallization can be hastened by immersing the flask first in cold water and then in an ice bath.

Time: less than 2 hours working time plus 2 hours reaction time.

Disposal

Sink: sodium hydroxide (after dilution); aqueous washes
Solid waste: filter paper
Nonhalogenated liquid organic waste: methanol
Halogenated liquid organic waste: ethyl bromide
Solid organic waste: *p*-acetamidophenol

Questions

1. Why does the *p*-acetamidophenol dissolve in the reaction mixture?
2. Why is the product not soluble in base?

NITRATION OF PHENACETIN

E76. 2-Nitrophenacetin from Phenacetin

In this experiment, phenacetin is converted to the corresponding 2-nitro derivative by treatment with concentrated nitric acid in acetic acid.

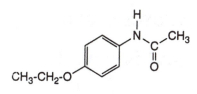

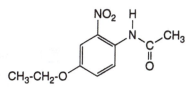

p-ethoxyacetanilide
phenacetin
colorless

2-nitro-4-ethoxyacetanilide
2-nitrophenacetin
golden yellow

Scale: 0.90 grams (0.005 mole) of *p*-ethoxyacetanilide.

Procedure

In a 125-mL Erlenmeyer flask, place 0.90 grams (0.005 mole) of *p*-ethoxyacetanilide (phenacetin). Add to the flask 20 mL of glacial acetic acid, and swirl the mixture to dissolve the solid. Add a magnetic stirring bar to the flask, and clamp the flask into position on a magnetic stirrer.

While stirring the solution, add dropwise 1.0 mL (0.015 mole) of concentrated nitric acid. After adding the nitric acid, warm the solution on the steam bath for about 10 minutes.

Now swirl the flask in cold water to cool the contents to about room temperature, and then add to the flask about 60 mL of cold water. Place the flask in an ice bath until crystallization appears to be complete.

Collect the crude product by suction filtration and wash it with cold water. The crude product can be recrystallized from water.

Golden yellow crystals

Time: less than 2 hours.

Disposal

> **Sink:** acetic acid; nitric acid (after dilution); aqueous washes
> **Solid waste:** filter paper
> **Solid organic waste:** phenacetin; 2-nitrophenacetin

Reference

1. R. A. Russell, R. W. Switzer, R. W. Longmore, B. Hoang Dutton, and L. Harland, *J. Chem. Educ.* **67,** 168 (1990).

ESTERIFICATION OF SALICYCLIC ACID

E77. Preparation of Methyl Salicylate: Oil of Wintergreen

Salicylic acid is both a carboxylic acid and an aromatic alcohol, or phenol. In Experiment E72, we saw that salicylic acid could be converted to an ester through the phenolic OH group by acetylation with acetic anhydride to form acetylsalicylic acid, or aspirin. In this experiment, we see that salicylic acid can also be converted to an ester through the carboxyl group by boiling salicylic acid with an excess of methanol in the presence of concentrated sulfuric acid, the same procedure that served to convert benzoic acid to methyl benzoate in Experiment E36.

salicylic acid   acetylsalicylic acid   methyl salicylate
                              aspirin   oil of wintergreen

Methyl salicylate has the familiar wintergreen smell, and it is the material responsible for both the natural smell of wintergreen (*Gaultheria*), and the wintergreen flavor of candies. Methyl salicylate is also an ingredient of liniments such as Icy Hot and Flex-all, which are used to relieve mild body aches and pains. Methyl salicylate is an active ingredient that is absorbed through the skin and then hydrolyzed to salicylic acid.

The infrared and proton NMR spectra of methyl salicylate are shown in Figures E77-1 and E77-2.

Scale: 1.4 grams (0.010 mole) of salicylic acid.

Procedure

Place 1.4 grams (0.010 mole) of salicylic acid and 4.0 mL (0.100 mole) of methyl alcohol in a 25-mL boiling flask. Swirl the mixture in the flask to dissolve the salicylic acid. Then, while swirling, add dropwise 1.5 mL (0.027 mole) of concentrated sulfuric acid. Fit the flask with

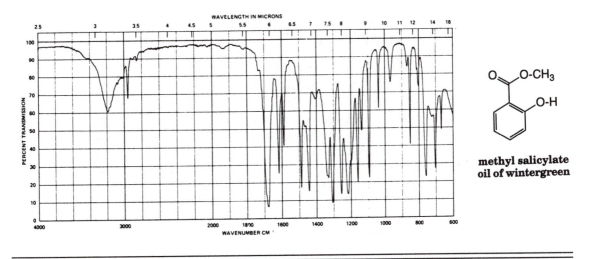

Figure E77-1. Infrared spectrum of methyl salicylate; thin film.

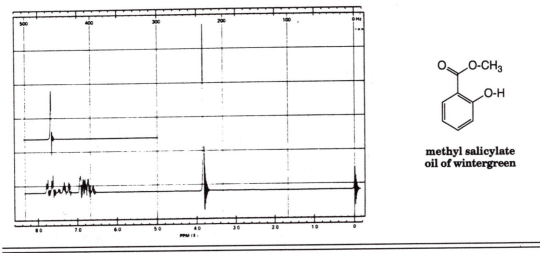

Figure E77-2. NMR spectrum of methyl salicylate; $CDCl_3$ solution.

a condenser, add a boiling stone, and heat the mixture on the steam bath under reflux for about and hour and a half.

At the end of the heating period, swirl the flask in cold water to cool the contents to about room temperature. Then pour the reaction mixture into a small separatory funnel that contains about 50 mL of water. Add 20 mL of dichloromethane to the separatory funnel, and mix the contents of the funnel to extract the methyl salicylate into the dichloromethane layer. Draw off the lower, organic layer, pour out and temporarily reserve the aqueous layer, and return the organic layer to the separatory funnel. Extract the organic layer with a solution of sodium bicarbonate prepared by dissolving about 1 gram (about 0.01 moles) of sodium bicarbonate in 20-mL of water. Draw off the lower, organic layer of dichloromethane plus methyl salicylate, dry it over anhydrous magnesium sulfate, and filter it into a 25-mL boiling flask. Add a boiling stone and drive off the dichloromethane by distillation on the steam bath to give a residue of methyl salicylate.

Smells like wintergreen

Methyl salicylate boils at about 222°C, and it is hard to successfully distill a small amount of material with a boiling point as high as this. You might, however, want to combine the residues from several preparations and carry out a distillation with this larger amount.

Time: less than 3 hours.

Disposal

> **Sink:** sulfuric acid (after dilution); aqueous washes
>
> **Solid waste:** filter paper
>
> **Nonhalogenated liquid organic waste:** methyl salicylate; methanol
>
> **Halogenated liquid organic waste:** dichloromethane
>
> **Solid inorganic waste:** sodium bicarbonate; magnesium sulfate
>
> **Solid organic waste:** salicylic acid

COCONUT ALDEHYDE; γ-NONANOLACTONE

An early and continuing motivation for organic synthesis has been the search for compounds useful as flavors and fragrances. Sometimes the aim is to duplicate at lower cost a substance that occurs in nature, sometimes it is to prepare a similar or related compound with the hope that it will have an interesting property, and sometimes it is to produce a less costly substitute of unrelated structure. One such substance is γ-nonanolactone, whose synthesis is presented in the next two sections.

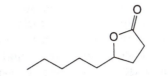

γ-nonanolactone; "coconut aldehyde"

γ-Nonanolactone has been described as having "a strong coconut odor and flavor with a very faint anisaldehydelike bynote" and "an odor indistinguishable from that of coconut." Up to this time, however, it has not been found to occur naturally, though it is often present in suntan lotions.

Physical properties reported for this compound include: boiling point, 137–138°C/17 mm; $d_4^{19.5}$, 0.9672; $n_4^{19.5}$ 1.4462. The infrared spectrum is shown in Figure E79-1.

An interesting book about the sources, properties, and chemistry of flavors and fragrances is P. Z. Bedoukian, *Perfumery and Flavoring Synthetics*, American Elsevier, New York, 1967.

E78. 3-Nonenoic Acid from Heptaldehyde and Malonic Acid

The reaction between an aldehyde and malonic acid to give an unsaturated acid is a variation of the Knovenagle reaction. In this reaction,

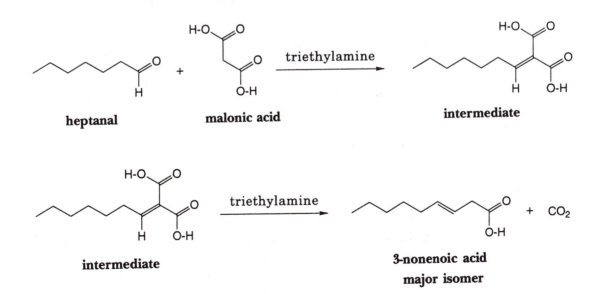

a β-dicarboxylic acid is an intermediate, but it loses one of the two carboxyl groups to give an unsaturated monocarboxylic acid. Although you might expect the conjugated acid to be the major product, it appears that in this case the unconjugated isomer predominates. The actual ratio of isomers may be irrelevant for us because the isomers are likely to be in equilibrium under the conditions of the following reaction.

Scale: 6.7 mL (0.050 mole) of heptanal.

Procedure

Place 5.2 grams (0.050 mole) of malonic acid and 6.7 mL (5.70 grams; 0.050 mole) of heptanal in a 125-mL Erlenmeyer flask. Add to the flask 10 mL (7.2 grams; 0.072 mole) of triethylamine. Heat this mixture on the steam bath, and swirl the flask until the solid has dissolved. Heat the solution on the steam bath for about 1 hour, or until the evolution of carbon dioxide has ceased.

At the end of the heating period, cool the contents of the flask by swirling it in cold water, and then pour the reaction mixture into a 125-mL separatory funnel. Next, rinse the reaction flask with 30 mL of ethyl ether, and pour the rinsings into the separatory funnel. Now pour 8 mL (0.096 mole) of concentrated HCl over about 25 g of ice, stir the mixture until no more ice appears to dissolve, and then pour the cold acid into the separatory funnel (Note 1). Thoroughly mix the contents of the separatory funnel, and then draw off and discard the lower, aqueous layer. Wash the ethereal layer, which has been retained in the separatory funnel, with one 25-mL portion of water.

Transfer the ethereal solution to a 50-mL Erlenmeyer flask, and remove most of the ether by heating the flask on the steam bath (Note 2).

The residue of crude 3-nonenoic acid can be converted to coconut aldehyde by treatment with 85% sulfuric acid as described in the next section.

Notes

1. The reaction of HCl with the material in the separatory funnel is only slightly exothermic.
2. Evaporate the ether in the hood, or fit the flask with a distillation adapter connected to a condenser so that the ether is condensed and ether fumes are not released into the room.

Time: about 3 hours.

Disposal

> **Sink:** concentrated HCl (after dilution); aqueous washes
> **Solid waste:** filter paper
> **Nonhalogenated liquid organic waste:** heptanal; triethyl-amine; ether; 3-nonenoic acid
> **Solid organic waste:** malonic acid

Question

1. Why is the reaction mixture treated with hydrochloric acid?

E79. Coconut Aldehyde from 3-Nonenoic Acid

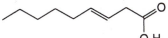

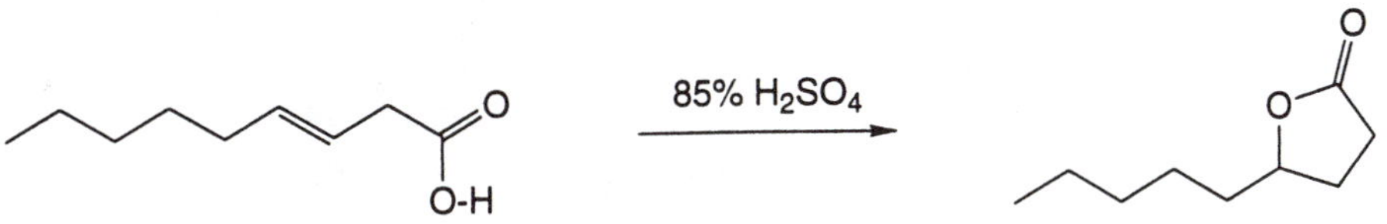

3-nonenoic acid $\xrightarrow{85\% \text{ H}_2\text{SO}_4}$ γ-nonanolactone "coconut aldehyde"

Scale: 6.7 mL (0.050 mole) of heptanal.

Procedure

To the 50-mL flask that contains the crude 3-nonenoic acid from the reaction of 0.050 mole of malonic acid with 0.050 mole of heptanal (Note 1), add 10 mL of 85% sulfuric acid (0.16 mole) (Note 2). Swirl the flask to mix the contents (Note 3), and heat the resulting solution on the steam bath for about 1 hour.

At the end of the heating period, use a medicine dropper to transfer the reaction mixture in small portions to a 400-mL beaker that contains a mixture of 20 grams (0.19 mole) of sodium carbonate and about 90 mL of water (Note 4).

After the transfer is complete, stir the mixture well to ensure that all the acid has had a chance to react with the base. Then pour about half of the oily suspension of coconut aldehyde in water into a 125-mL separatory funnel, and extract it with a 30-mL portion of ether. After the layers have separated, draw off the lower, aqueous phase (Note 5), and add the remainder of the suspension of product and water. After thoroughly mixing the contents of the separatory funnel, again allow the layers to separate, and drain off the lower, aqueous layer (Note 5). Then wash the ethereal layer, which remains in the separatory funnel, with about 25 mL of water.

Next, transfer the ethereal phase to a 50-mL Erlenmeyer flask, dry the solution with anhydrous magnesium sulfate (Note 6), and remove the ether by distillation on the steam bath.

Smells like coconut   The residue of coconut aldehyde can be purified by distillation under reduced pressure (Note 7). Yield: about 3 grams (about 40% overall from malonic acid) of a very pale yellow oil that, at low concentrations, smells like coconut.

Notes

1. The preparation of 3-nonenoic acid is described in Experiment E78.

2. 85% sulfuric acid is prepared by adding 85 grams (46 mL) of concentrated sulfuric acid to 15 grams of ice.

3. The reaction mixture quickly turns dark brown.

4. As the acidic reaction mixture is added to the mixture of sodium carbonate and water, carbon dioxide gas is produced. The gas causes the mixture to foam, and the large beaker serves to contain the foam.

5. If salts start to crystallize from the aqueous phase and interfere with removal of the water layer, add more water to the separatory funnel and swirl the contents gently so as to redissolve the salts.

6. Add anhydrous magnesium sulfate to the ethereal solution, allow the suspension to stand for about 10 minutes, and then remove the magnesium sulfate by gravity filtration. If you plan to purify the product by distillation, filter the solution into the boiling flask that you plan to use for the distillation. If the boiling flask is too small to hold all the filtrate at once, you can add part of it, concentrate it by distillation, and then add more.

7. Coconut aldehyde boils at 151°C under 23 Torr of pressure. The product can be distilled at atmospheric pressure, but it is then darker in color and has an acrid note in its smell.

The infrared spectrum of γ-nonanolactone ("coconut aldehyde") is shown in Figure E79-1.

Time: about 3 hours.

Disposal

> **Sink:** sulfuric acid (after dilution); aqueous washes
>
> **Solid waste:** filter paper
>
> **Nonhalogenated liquid organic waste:** 3-nonenoic acid; coconut aldehyde; ether
>
> **Solid inorganic waste:** sodium carbonate; magnesium sulfate

References

1. *Chemical Abstracts* **56,** 8549 (1962).
2. *Chemical Abstracts* **59,** 11265 (1963)

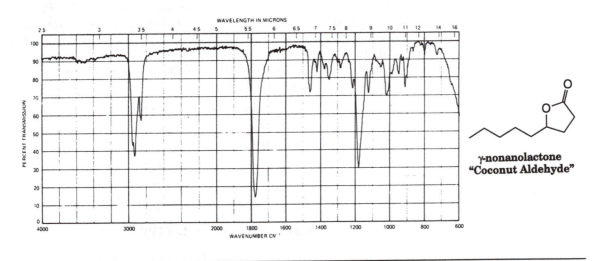

Figure E79-1. Infrared spectrum of γ-nonanolactone; thin film.

CATALYSIS BY THIAMINE

E80. Thiamine-Catalyzed Formation of Benzoin from Benzaldehyde

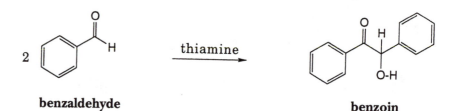

benzaldehyde   benzoin

Thiamine pyrophosphate (TPP) is a necessary coenzyme for a number of enzyme-catalyzed reactions, all of which involve, in effect, the transfer of an acyl group, R-C=O, as a carbanion.

acyl carbanion

Organic chemists try to avoid proposing acyl carbanions as free intermediates because the mechanism of action of thiamine lacks the structural features (electron-accepting groups) that are thought to stabilize such ions. The way in which acyl groups are transferred in biochemical reactions at a pH of 7 and at 37°C, then, remained a mystery until Professor Breslow of Columbia University proposed a role for thiamine (Reference 1).

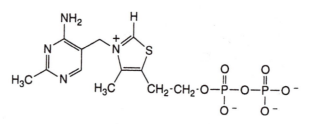

thiamine pyrophosphate (TPP)

Breslow remembered that a famous old reaction of organic chemistry, the cyanide-catalyzed condensation of benzaldehyde to form benzoin (Section E92), appeared to involve the equivalent of an acyl carbanion.

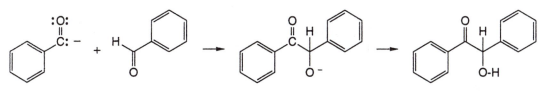

acyl carbanion   **benzaldehyde**   **benzoin**

Cyanide ion had long been known as a specific catalyst for this reaction, and cyanide was thought to act by adding to the carbonyl group of benzaldehyde to form the cyanohydrin,

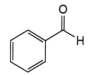 +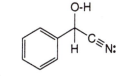

benzaldehyde   **cyanide ion**   **benzaldehyde cyanohydrin**

which could then lose the aldehyde hydrogen to form a resonance-stabilized anion (Reference 2).

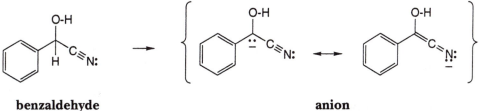

benzaldehyde   **anion**
cyanohydrin   **resonance-stabilized**

This anion could add to the carbonyl group of a second molecule of benzaldehyde to form the cyanohydrin of benzoin, which would then lose HCN to give benzoin and regenerate the catalyst.

anion +   **benzaldehyde**   **benzoin cyanohydrin**   **benzoin**

Breslow also knew that a number of thiazolium salts, including thiamine, had been reported to catalyze the conversion of benzaldehyde to benzoin, and not to give the product originally expected. In addition, he had determined that the C-2 proton of the thiazolium ring of thiamine was rapidly exchanged for deuterium in D_2O at room temperature.

For these and other reasons, he suggested that thiamine could catalyze the benzoin condensation in the following way.

Thiamine loses a proton to the solvent (or to the enzyme).

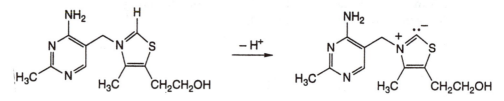

thiamine   **conjugate base of thiamine**

The conjugate base of thiamine adds to benzaldehyde to give thiamine-benzaldehyde.

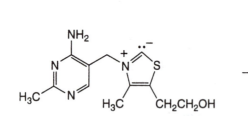

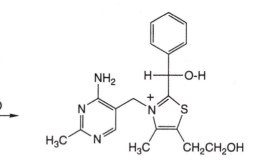

conjugate base of thiamine   **thiamine-benzaldehyde**

Thiamine-benzaldehyde then loses a proton to give a resonance-stabilized equivalent of an acyl carbanion.

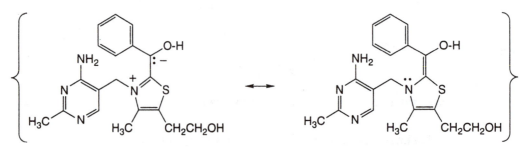

benzaldehyde acyl anion equivalent

The equivalent of the acyl anion of benzaldehyde then adds to a second molecule of benzaldehyde to give a product that eliminates thiamine to give benzoin and to regenerate the catalyst.

benzaldehyde acyl anion equivalent

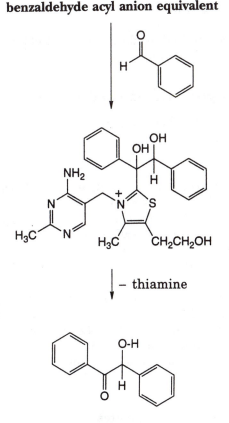

benzoin

Reasoning from this model, Breslow proposed a similar role for thiamine pyrophosphate in the enzyme-catalyzed systems: resonance stabilization of the carbanion equivalent of an aldehyde.

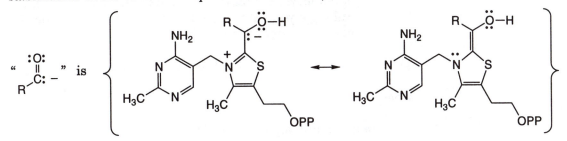

resonance-stabilized acyl anion equivalent

Breslow was very careful not to draw conclusions that were not supported by the evidence. However, his proposal has now come to be accepted as representing the mechanism of thiamine action in enzymatic reactions in which thiamine pyrophosphate participates as a coenzyme. The role of the enzyme itself still remains to be determined.

The following procedure illustrates the use of thiamine as a catalyst in the benzoin condensation.

Thiamine-Catalyzed Benzoin Condensation

Scale: 2.5 mL (0.025 moles) of benzaldehyde.

Procedure

Place 0.84 grams (0.0025 mole) of thiamine hydrochloride in a 25-mL Erlenmeyer flask. Add 2.5 mL of water and swirl the mixture until the thiamine hydrochloride has all dissolved. Then add 6 mL of 95% ethanol, 2.5 mL of 2 M NaOH (0.005 mole), and 2.5 mL (2.6 grams; 0.025 mole) of benzaldehyde (Note 1), swirling the flask during and after each addition to thoroughly mix the contents of the flask.

Allow the resulting mixture to stand at room temperature for at least 6 hours (Note 2).

Collect the resulting crystals of benzoin by suction filtration, washing them with about 6 mL of a cold mixture of 5 mL of ethanol and 1 mL of water. Yield: about 1.8 grams (about 70%).

Notes

1. Always use a fresh bottle of benzaldehyde.
2. The mixture may be allowed to stand for 24 hours or for a week. It is helpful to add a seed crystal. By avoiding supersaturation, you enable crystallization to start sooner, and the resulting crystals are larger and more easily washed on the filter paper after collection by suction filtration.

Time: 2 hours plus time required for reaction to occur.

Disposal

   Sink: ethanol; aqueous NaOH
   Solid waste: filter paper

Nonhalogenated liquid organic waste: benzaldehyde
Solid inorganic waste: thiamine hydrochloride
Solid organic waste: benzoin

Questions

1. Draw the structure of "thiamine hydrochloride" (thiamine chloride hydrochloride; Vitamin B1).

2. The free acyl carbocation is sometimes invoked as an intermediate in the aluminum-chloride-catalyzed acylation of aromatic compounds. Does it have any resonance stabilization? Explain.

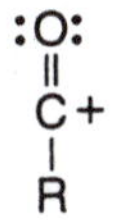

acyl carbocation

3. Thiamine pyrophosphate is a required coenzyme in the enzyme-catalyzed decarboxylation of pyruvic acid to form acetaldehyde. Show in detail how thiamine could be involved in this reaction.

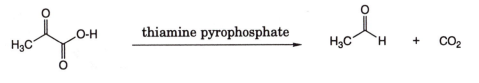

pyruvic acid   **acetaldehyde**

4. Thiamine catalyzes both the self-condensation of acetaldehyde to acetoin and the reaction of pyruvic acid with acetaldehyde to give acetoin. Show in detail how this might occur.

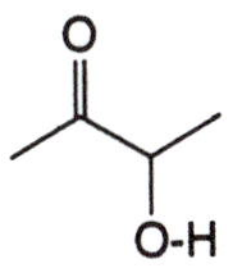

acetoin

References

1. R. Breslow, *J. Am. Chem. Soc.* **80,** 3719 (1958).
2. A. Lapworth, *J. Chem. Soc.* **83,** 995 (1903).
3. T. S. Bruice and S. Benkovic, *Bioorganic Mechanisms,* Volume 2, W. A. Benjamin, Inc., New York, 1966, Chapter 8.

A MODEL FOR THE REDUCING AGENT NADH

Many chemical oxidation-reduction reactions are rather brutal, taking place at high temperatures and at the extremes of pH. In contrast, biochemical redox reactions take place rapidly at about room temperature and at a pH of about 7.

Chemists continue to be interested in trying to understand and reconstruct these gentle biochemical reactions. One approach is to study the reactions of related but less complex compounds, to study model reactions. In this experiment, we will work with a model that has been used to try to understand the mechanism of the biochemical reducing agent NADH (nicotinamide adenine dinucleotide, reduced form). The reducing agent, NADH, can be represented in this way,

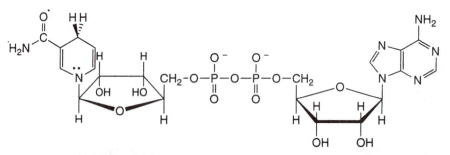

nicotinamide adenine dinucleotide, reduced form; NADH

and its oxidized form, NAD+, by the following structure.

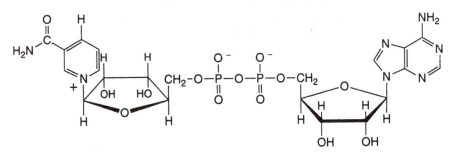

nicotinamide adenine dinucleotide, oxidized form; (NAD⁺)

The model compound to be used in this experiment is one of several that were studied some years ago by Professor Frank Westheimer at Harvard (Reference 1, Section E83).

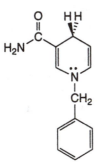

1-benzyldihydronicotinamide

In that research, deuterium-labeling experiments showed that in the reduction of malachite green by 1-benzyldihydronicotinamide, the hydrogen lost by the reducing agent is transferred directly to the hydrogen acceptor.

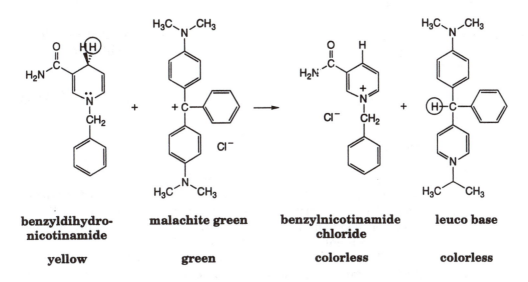

benzyldihydro-nicotinamide	malachite green	benzylnicotinamide chloride	leuco base
yellow	green	colorless	colorless

This series of reactions, suggested by Dr. Bernard Golding (University of Warwick, Coventry, England), includes the two-step preparation of 1-benzyldihydronicotinamide and a demonstration of its ability to reduce malachite green under mild conditions.

The infrared spectrum of 1-benzyldihydronicotinamide is shown in Figure E82-1.

E81. 1-Benzylnicotinamide Chloride from Nicotinamide

In this procedure, nicotinamide is warmed with benzyl chloride in dimethylsulfoxide, thereby transferring the benzyl group from chloride to the ring nitrogen of nicotinamide. The reaction starts when the temperature of the mixture reaches about 135°C, and the crystals of the product separate from the solution as the exothermic reaction occurs even though the temperature of the mixture continues to rise.

Benzyl chloride is a nasty lachrymator, and you should work in the hood while adding the benzyl chloride to your flask.

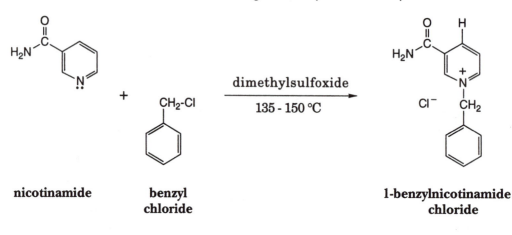

nicotinamide   benzyl chloride   1-benzylnicotinamide chloride

Scale: 1.0 gram (0.008 mole) of nicotinamide and 5.0 mL (0.043 mole) of benzyl chloride.

Procedure

Clamp a 25-mL round-bottom boiling flask into position over a wire gauze on an iron ring so that the flask can be heated with a burner flame. Add to the flask 2.5 mL of dimethylsulfoxide and 1 g (0.008 mole) of nicotinamide. Heat the flask briefly with the burner, and swirl the contents to dissolve the nicotinamide.

Tear gas!
Use the hood.

When the solid has all dissolved, turn off the burner, and add to the flask 5 mL (5.5 g; 0.043 mole) of benzyl chloride (Note 1). Fit the flask with a Claisen adapter and, with a thermometer adapter in the center neck of the Claisen adapter, position a thermometer so that the bulb dips into the solution in the flask. Fit the side arm of the Claisen adapter with a reflux condenser. Using the burner again, heat the flask until crystals begin to separate from the solution (Note 2).

At this time, discontinue heating and allow the exothermic reaction to occur spontaneously (Note 3). After the reaction has taken place and the mixture has cooled to below 60°C (Note 4), remove the condenser, adapters, and thermometer and add 10 mL of isopropyl alcohol. Stopper the flask and mix the contents of the flask by shaking it vigorously. Collect the product by suction filtration, and wash it with a little isopropyl alcohol and then some hexane.

The crude product can be used to prepare 1-benzyl-dihydronicotinamide. It can also be recrystallized by heating it with 5 mL of 95% ethanol, adding just enough water (about 0.5 mL) to cause all the solid to dissolve, and allowing the solution to cool to room temperature.

Notes

1. Because chloride is a lachrymator (it is a tear gas), you should work in the hood while adding it.
2. Crystals of the product start to appear when the temperature reaches about 135°C.
3. The temperature will rise to about 150–155°C before it starts to fall again.
4. When the temperature has fallen somewhat below 100°C, the flask may be placed in a beaker of cold water to cool it faster.

Time: about 1 hour.

Disposal

Solid waste: filter paper

Nonhalogenated liquid organic waste: dimethylsulfoxide; isopropyl alcohol; ethanol

Halogenated liquid organic waste: benzyl chloride (*Caution: lachrymator*)

Solid organic waste: nicotinamide

Question

1. Why does benzyl transfer take place to the ring nitrogen and not the amide nitrogen?
2. Why should the product dissolve better in aqueous methanol than in pure methanol?

E82. 1-Benzyldihydronicotinamide from 1-Benzylnicotinamide Chloride

In this procedure, 1-benzylnicotinamide chloride is reduced to 1-benzyldihydronicotinamide by treatment with sodium dithionite in base.

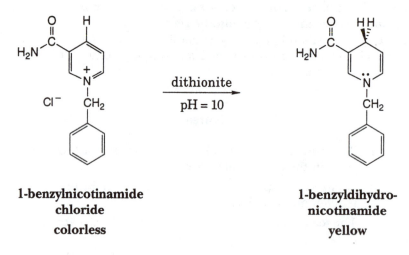

1-benzylnicotinamide
chloride

colorless

1-benzyldihydro-
nicotinamide

yellow

Scale: 2.0 grams (0.008 mole) of 1-benzylnicotinamide chloride.

Procedure

Prepare a solution of 2.76 g of anhydrous sodium carbonate and 5.14 g (about 0.03 mole) of sodium dithionite in 40 mL of water in a 125-mL Erlenmeyer flask. Place a magnetic stirring bar in the flask, and arrange for the solution to be stirred by a magnetic stirrer.

Dissolve 2.0 g (0.008 moles) of 1-benzylnicotinamide chloride in 10 mL of water and add this solution all at once to the well-stirred solution of sodium carbonate and sodium dithionite. The resulting yellow-orange solution will suddenly produce a yellow precipitate in about 60 seconds. Continue to stir the mixture for another 10 minutes, and then collect the yellow solid by suction filtration using several small portions of water for rinsing and washing.

Yellow spars   This product can be used directly for the reduction of malachite green, or it can be recrystallized to give beautiful yellow spars. To recrystallize, dissolve the crude product in 5 mL of hot 95% ethanol, and filter the solution by gravity to remove a small amount of an insoluble impurity. Dilute the filtrate with 5 mL of warm water and, after swirling to mix, allow the solution to stand undisturbed for several hours; it helps to seed the solution. Collect the crystals by suc-

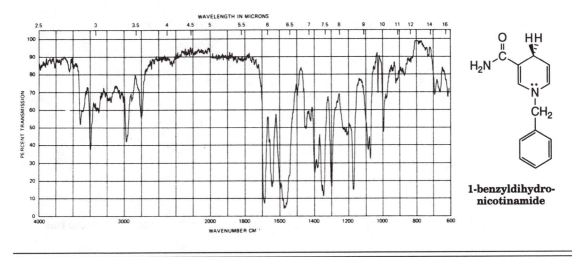

Figure E82-1. Infrared spectrum of 1-benzyldihydronicotinamide; $CHCl_3$ solution.

tion filtration, and wash them with a small amount of ice-cold 50% aqueous ethanol.

The IR spectrum of 1-benzyldihydronicotinamide is shown in Figure E82-1.

Time: less than 1 hour.

Disposal

> **Sink:** aqueous washings
>
> **Solid waste:** filter paper
>
> **Solid inorganic waste:** sodium carbonate; sodium dithionite
>
> **Solid organic waste:** 1-benzylnicotinamide chloride; 1-benzyldihydronicotinamide

E83. Reduction of Malachite Green by 1-Benzyldihydronicotinamide

The progress of this reduction is indicated by the change in color from blue-green (presence of malachite green) to yellow (presence of excess benzyldihydronicotinamide).

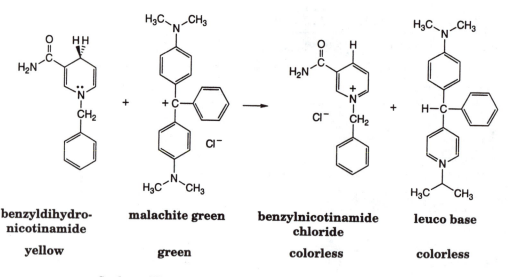

benzyldihydro-   **malachite green**   **benzylnicotinamide**   **leuco base**
nicotinamide   **chloride**

yellow   **green**   **colorless**   **colorless**

Scale: milligram amounts.

Procedure

Dissolve about 10 milligrams (0.03 mmole) of malachite green in 1
mL of 95% ethanol in a small test tube. To this add 20–40 milligrams
of 1-benzyldihydronicotinamide (0.1 to 0.2 mmoles). Swirl the mixture gently to dissolve the crystals. The intense green-blue of the malachite green will give way to the yellow of the excess reducing agent in
2 to 4 minutes.

Time: about 30 minutes.

Disposal

> **Sink:** aqueous reaction mixture
> **Nonhalogenated liquid organic waste:** ethanol
> **Solid organic waste:** malachite green; 1-benzyldihydro-
> nicotinamide

Questions

1. Malachite green is a carbocation salt. What accounts for its great
   stability?

2. Ethyl benzoylformate is not reduced by 1-benzyldihydronicotinamide
   alone at room temperature in the dark in acetonitrile solution. In

the presence of an equimolar amount of magnesium perchlorate, however, racemic ethyl mandelate is formed in quantitative yield (Reference 2). Explain.

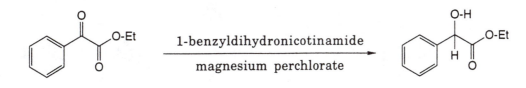

ethyl benzoylformate   **ethyl mandelate**

3. When ethyl benzoylformate is treated with (R)-(−)-*N*-α-methylbenzyl-1-benzyldihydronicotinamide in the presence of an equimolar amount of magnesium perchlorate in acetonitrile solution, the ethyl mandelate produced is 60% R and 40 % S.

   a. Draw formulas indicating the structure and configuration of (R)-(−)-*N*-α-methylbenzyl-1-benzyldihydronicotinamide and of the two enantiomers of ethyl mandelate.

   b. Explain the results of the experiment.

References

1. D. Mauzerall and F. H. Westheimer, *J. Am. Chem. Soc.* **77**, 2261 (1954).

2. Y. Ohnishi, M. Kagami, and A. Ohno, *J. Am. Chem. Soc.* **97**, 4766 (1975).

TWO THERMOCHROMIC COMPOUNDS

Compounds whose color depends on temperature are said to be *thermochromic.* Dixanthylene is a very pale yellow-green solid at room temperature, but it becomes dark blue when melted or heated in solution; at liquid nitrogen temperature it is completely colorless. Dianthraquinone is a bright canary yellow at room temperature, but it becomes a brilliant parrot green when heated in solution. The thermochromism of each compound can be observed very nicely during recrystallization from mesitylene.

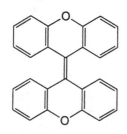

dixanthylene
yellow-green at room temperature
dark blue in hot solution

dianthraquinone
canary yellow at room temperature
parrot green in hot solution

E84. Dixanthylene from Xanthone

Dixanthylene can be prepared by the reductive dimerization of xanthone by zinc and acetic acid.

xanthone   **dixanthylene**

Scale: 1.0 gram (0.005 mole) of xanthone.

Procedure

Place 1.0 g (0.005 mole) of xanthone in a 100-mL boiling flask, add 40 mL of glacial acetic acid, and swirl until most of the solid has dissolved. Add 0.3 g (0.005 mole) of zinc dust and 1 drop of concentrated hydrochloric acid, and boil the mixture under reflux for about 90 minutes.

Every 5 minutes or so, add 1 drop of concentrated hydrochloric acid down the condenser, taking care not to add the acid so often that the solution takes on a red color. Add three more 0.3-g portions of zinc dust, one at a time at intervals of about 20 minutes, adding the first about 20 minutes after the mixture has been brought to a boil.

At the end of the period of boiling under reflux, allow the solution to cool completely to room temperature, collect the solid by suction filtration, and wash it thoroughly with water.

Separate any zinc from the product by boiling it with mesitylene to dissolve the dixanthylene (about 10 mL will be required), filtering the hot, blue solution by gravity, reheating the filtrate if necessary to redissolve any material that may have crystallized, and allowing the filtrate to cool to room temperature. Collect the very pale yellow-green crystals by suction filtration, and wash them with a little toluene. Yield: about 0.36 g (35%).

The thermochromism of dixanthylene can be observed by heating a sample in a melting point capillary over a small burner flame, or, better, by recrystallization from boiling mesitylene, using about 19 mL per gram (recovery: about 90%)

Dark blue in boiling mesitylene

Time: 3 hours.

Disposal

> **Sink:** concentrated HCl (after dilution); acetic acid
> **Solid waste:** filter paper
> **Nonhalogenated liquid organic waste:** mesitylene; toluene
> **Solid inorganic waste:** Zinc dust
> **Solid organic waste:** xanthone; dixanthylene

References

1. A. Schonberg and A. Mustafa, *J. Chem. Soc.* **1944,** 67.
2. G. Gurgenjanz and S. von Kostanecki, *Chem. Ber.* **28,** 2310 (1895).
3. A. Ault, R. Kopet, and A. Serianz, *J. Chem. Educ.* **48,** 410 (1971).

E85. Dianthraquinone from Anthrone via 9-Bromoanthrone

Dianthraquinone can be prepared by bromination of anthrone to give 9-bromoanthrone,

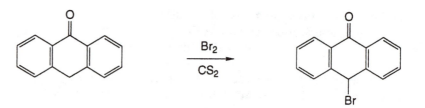

anthrone   9-bromoanthrone

followed by treatment of this product with diethylamine in chloroform.

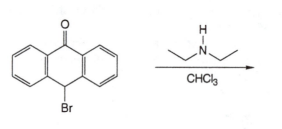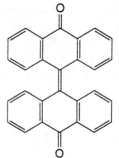

9-Bromoanthrone from Anthrone

Scale: 2.0 grams (0.010 mole) of anthrone.

Procedure (Reference 1)

Suspend 2.0 grams (0.010 mole) of anthrone in 6 mL of carbon disulfide. Add bromine dropwise while stirring over a period of about 15 minutes until the bromine color persists; about 0.5 mL (1.6 grams; 0.010 mole) will be required (Note 1). Collect the tan crystalline product by suction filtration and wash it with a little toluene. Yield: 2.4 grams (85%).

Note

1. Because hydrogen bromide gas is evolved in this reaction, the reaction must be carried out in the hood, or some provision must be made to absorb the gas produced.

Time: 1 hour.

Dianthraquinone from 9-Bromoanthrone

Scale: 2.0 grams (0.007 mole) of 9-bromoanthrone.

Procedure (Reference 2)

Dissolve 2.0 grams (0.0073 mole) of 9-bromoanthrone (Note 1) in 30 mL of chloroform (Note 2). Add 2.0 mL (1.4 g; 0.019 mole) of diethylamine, swirl to mix, and allow the resulting warm solution to stand for 2 hours. During this time, the color of the solution will change from yellow to a dark red-brown.

At this time, add 75 mL of ether slowly while swirling. After 5 minutes, collect the yellow precipitate by suction filtration and wash it thoroughly with ether. Suspend the precipitate in about 10 mL of 95% ethanol, and then collect the canary-yellow dianthraquinone by suction filtration, washing it with a little ethanol. Yield: about 0.8 g (55%).

The thermochromism of dianthraquinone can be observed very nicely upon recrystallization from boiling mesitylene using about 50 mL per gram.

Parrot green in boiling mesitylene

Notes

1. The 9-bromoanthrone must be prepared just before use; otherwise the yield of dianthraquinone is greatly reduced.
2. Any insoluble material should be removed at this point by gravity filtration.

Time: 3 hours.

Disposal

> **Solid waste:** filter paper
>
> **Nonhalogenated liquid organic waste:** carbon disulfide; diethylamine; ether; mesitylene
>
> **Halogenated liquid organic waste:** chloroform
>
> **Halogenated solid organic waste:** 9-bromoanthrone
>
> **Solid organic waste:** anthrone; dianthraquinone
>
> **Bromine:** reduce with aqueous sodium bisulfite; discard residue after evaporation with inorganic solid waste

References

1. L. Barnett, J. W. Cook, and M. A. Matthews, *J. Chem. Soc.* **123,** 1994 (1923); and K. H. Meyer, *Annalen* **379,** 62 (1911).
2. L. Barnett, J. W. Cook, and H. H. Grainger, *J. Chem. Soc.* **121,** 2059 (1922).
3. A. Ault, R. Kopet, and A. Serianz, *J. Chem. Educ.* **48,** 410 (1971).

E86. A Photochromic Compound: 2-(2,4-Dinitrobenzyl)pyridine

Crystals of the compound 2-(2,4-dinitrobenzyl)pyridine have the most unusual property of turning a very deep blue color when exposed to sunlight. During storage in the dark, the crystals revert to their original sandy color. Color formation takes only a few minutes in bright sunlight, but the loss of color takes about a day. The interconversion appears to be completely reversible any number of times.

One explanation of the phenomenon is the formation of a tautomeric form by action of the light.

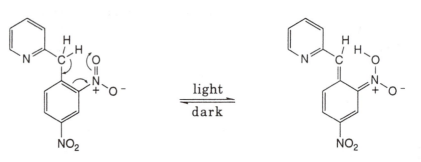

brown-sugar brown   **blue-jeans blue**

2-(2,4-Dinitrobenzyl)pyridine can be prepared from 2-benzylpyridine by treatment with a mixture of nitric acid and sulfuric acid.

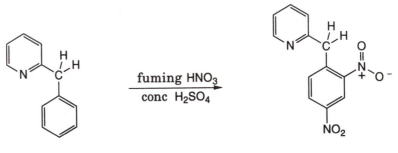

2-benzylpyridine   **2-(2,4-dinitrobenzyl)pyridine**

Scale: 2.5 mL (0.016 mole) of 2-benzylpyridine.

Procedure

Place 12 mL of concentrated sulfuric acid (0.22 mole) in a 100-mL boiling flask. Cool the acid to 5°C or below in an ice bath. Support the flask so that it can be stirred or swirled in an ice bath, and, while mixing well, gradually add 2.5 mL (2.6 grams; 0.016 mole) of 2-benzylpyridine.

To this well-stirred mixture, add dropwise over a period of about 3 minutes, 1.5 mL of concentrated nitric acid (density, 1.42 grams/mL; 2.1 grams; 0.024 moles; Note 1). The addition of the first few drops of the nitric acid will cause a color change to deep brown, but the mixture will become lighter in color as the remainder of the acid is added. After all the nitric acid has been added, heat the mixture for about 20 minutes on the steam bath.

At the end of the heating period, pour the mixture onto about 200 grams of ice in a 1-liter Erlenmeyer flask (Note 2). Basify the solution to a pH of about 11 by adding almost all of a solution of 20 grams (0.5 mole) of sodium hydroxide in about 250 mL of water (Note 3). Toward the end of the addition of base, the product separates to give a milky yellow suspension.

Add about 200 mL of ether, and stir the mixture for 10–15 minutes to extract the product into the ether layer.

Separate the ethereal solution (Note 4), dry it over anhydrous magnesium sulfate for a few minutes, filter it, and distill it to reduce its volume to about 50 mL (Note 5). The product will sometimes start to crystallize during the concentration of the solution.

Complete the crystallization by cooling the mixture in a ice bath. Collect the large sandy prisms by suction filtration, and wash them with a small amount of cold 95% ethanol. Yield: about 2 grams (about 50%). The product can be recrystallized from 95% ethanol with 90% recovery using about 10 mL of ethanol per gram; decolorizing carbon helps only a little.

Notes

1. The procedure originally called for 1.5 mL of red fuming nitric acid (density, 1.5 g/mL; 0.035 mole), but we discovered by accident that ordinary concentrated nitric acid works almost as well.

2. A 1-liter beaker is almost as satisfactory, but it is easier to spill from a beaker than from an Erlenmeyer flask.

3. If you can make the sodium hydroxide solution from "50% NaOH," add 26 mL of "50% NaOH" (40 grams of 50% by weight NaOH; 20 grams of NaOH; 0.5 mole of NaOH) to 230 mL of water.

   If you must make this solution from solid NaOH, put 20 grams of solid NaOH (0.5 mole NaOH) into a 500-mL Erlenmeyer flask or beaker and add at first only about 20 mL of water. The mixture will get very hot, and the sodium hydroxide will dissolve quite rapidly. After almost all the solid has dissolved, add the remaining 225 mL of water. If you add all the water at once, dissolution takes much longer!

4. This separation can be messy. First, the total volume will be about 500 mL, and separatory funnels large enough to hold the entire mixture are not so abundant as the smaller sizes. Second, the layers sometimes do not separate cleanly, and there is always some emulsion present at the interface.

   One approach to the volume problem is to use a pipette to remove most of the bottom layer, and to decant some of the top layer until the volume of the remaining mixture will fit into an available separatory funnel.

5. Collect your ether and give it to others to use for their extractions.

Time: About 3 hours.

Disposal

> **Sink:** concentrated sulfuric acid (after dilution); concentrated nitric acid (after dilution); aqueous washes
>
> **Solid waste:** filter paper
>
> **Nonhalogenated liquid organic waste:** 2-benzylpyridine; ether; ethanol
>
> **Solid inorganic waste:** sodium hydroxide; magnesium sulfate
>
> **Solid organic waste:** 2-(2,4-dinitrobenzyl)pyridine

References

1. J. A. Sousa and J. Weinstein, *J. Org. Chem.* **27,** 3155 (1962); and A. L. Bluhm, J. Weinstein, and J. A. Sousa, *J. Org. Chem.* **28,** 1989 (1963).

2. A. Ault and C. Kouba, *J. Chem. Educ.* **51,** 395 (1974).

3. K. Schofield, *J. Chem. Soc.* **1949,** 2411; and A. J. Nunn and K. Schofield, *J. Chem. Soc.* **1952,** 586.

4. A. E. Tschitschibabin, B. M. Kuindshi, and S. W. Benewolenskaja, *Chem. Ber.* **59,** 1580 (1925).

E87. A Chemiluminescent Compound: Luminol

The production of "cold" light by the American firefly is a familiar example of light emission that is nonthermal in origin; it is luminescence rather than incandescence. The light of the firefly is produced during oxidation of luciferin to oxyluciferin by oxygen in the presence of adenosine triphosphate (ATP), magnesium ion, and the enzyme luciferase.

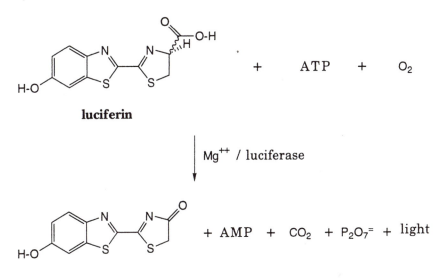

Apparently, oxyluciferin is produced in an electronically excited state, and the energy released when the corresponding ground state is formed is given off as light.

A number of other substances have been shown to undergo chemical reaction with simultaneous production of light. A good example that is easy to observe is that of luminol reacting with oxygen and base in dimethylsulfoxide solution.

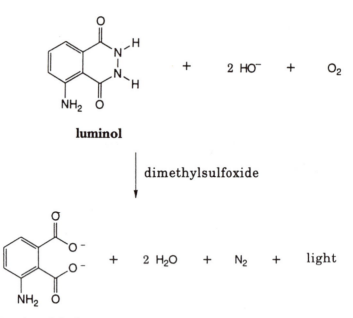

luminol   +   2 HO⁻   +   O₂

dimethylsulfoxide

3-aminophthalate   +   2 H₂O   +   N₂   +   light

Luminol can be prepared from 3-nitrophthalic acid by treatment with hydrazine followed by reduction of the first product by sodium dithionite.

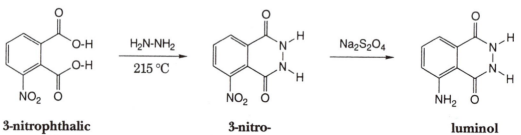

3-nitrophthalic acid   **3-nitro-phthalhydrazide**   **luminol**

Preparation of Luminol from 3-Nitrophthalic Acid via 3-Nitrophthalhydrazide

Scale: 1.0 gram (0.005 mole) of 3-nitrophthalic acid.

Procedure

First put a flask containing 15 mL of water on the steam bath to get hot. Then add to a 20 mm × 150 mm test tube 1.0 gram (0.005 mole) of 3-nitrophthalic acid and 2 mL of a 5% aqueous solution of hydrazine (0.005 mole of hydrazine; Note 1). Heat the mixture over a small burner flame until all the solid has gone into solution.

Now add 3 mL of triethylene glycol, clamp the tube in a vertical position over the burner, and insert both a thermometer and a tube connected to the water aspirator. Boil the solution vigorously to distill the excess water, removing the vapors by means of the aspirator. Let the temperature rise rapidly until (in 3–4 minutes) it reaches 215°C. Remove the burner, notice the time, and by intermittent gentle heating, maintain the temperature at 215–220°C for 2 minutes.

At this time, remove the burner, allow the temperature of the mixture to fall to 100°C (crystals of product may appear), add the 15 mL of hot water (Note 2), and collect the precipitated 3-nitrophthalhydrazide by suction filtration.

Transfer the damp nitro compound to the test tube in which it was prepared; do not clean the tube before making the addition. Add 5 mL of 10% sodium hydroxide solution (0.014 mole NaOH, stir the mixture to dissolve the solid, and to the resulting deep brown-red solution add 3.5 grams of "85% technical" sodium dithionite (17 mmole). Wash the solid down from the walls of the test tube with a little water. Heat the mixture to the boiling point, stir it, and keep it hot for 5 minutes; during this time, some of the reduction product may separate from the yellow-brown solution.

Now add 2 mL of acetic acid (35 mmole), cool the mixture thoroughly by swirling the test tube in a beaker of cold water, and collect the resulting precipitate of light yellow luminol by suction filtration. The damp crude product is suitable for demonstration of chemiluminescence.

Notes

1. The 8% hydrazine solution can be prepared either by diluting 30.0 mL (31.2 grams) of 64% hydrazine to 250 mL or by diluting 22.3 mL (23.5 grams) of 85% hydrazine to 250 mL.
2. The reason for using hot water rather than cold is that the solid is then produced in a form that is more easily filtered.

Time: 2 hours.

Demonstration of the Chemiluminescence of Luminol

Procedure

Blue-green light!

Put 100 mL of dimethylsulfoxide in a 250-mL bottle or flask, add 1 mL of 50% aqueous sodium hydroxide, and swirl to mix. Then add 10–20 mg of luminol, stopper the container, and shake the contents vigorously. Within 60 seconds, the contents of the bottle or flask will emit a blue-green light, which will be sufficiently bright to allow you to easily read this book in the dark. When the light fades, it can be restored by shaking the bottle or flask again. When all the luminol has reacted, further portions can be added, even after several weeks.

Disposal

Sink: aqueous sodium hydroxide; acetic acid

Solid waste: filter paper

Nonhalogenated liquid organic waste: hydrazine hydrate; triethylene glycol; dimethylsulfoxide

Solid inorganic waste: sodium dithionite

Solid organic waste: 3-nitrophthalic acid; luminol

Questions

1. Why is 3-nitrophthalhydrazide soluble in base?
2. Why does the addition of acetic acid cause the luminol to precipitate?

References

1. L. F. Fieser, *Organic Experiments,* 2nd edition, Heath, Boston, 1968, p. 239.
2. E. H. White and D. F. Roswell, *Accounts Chem. Res.* **3,** 54 (1970); and E. H. White, *J. Chem. Educ.* **34,** 275 (1957).
3. H. W. Schneider, *J. Chem. Educ.* **47,** 519 (1970); and M. T. Beck and F. Joo, *J. Chem. Educ.* **48,** A559 (1971).
4. W. D. McElroy and M. DeLuca, in *Chemiluminescence and Bioluminescence,* Cormier, Hercules, and Lee, editors, Plenum Press, New York, 1973, p. 285.

Synthetic Sequences: Experiments that Use a Sequence of Reactions

High yield is most important when each reaction uses as starting material the product of a previous reaction. For example, if three successive reactions can be carried out at 90% yield each, the overall yield at the end of the third reaction, is $(0.9)3 \times 100 = 73\%$. If each reaction is carried out at half that yield, the overall yield at the end of the third reaction is $(0.45)3 \times 100 = 9\%$, or one-eighth as much. For this reason, synthetic sequences are a good test of the experimenter's laboratory skill.

Six synthetic sequences are outlined in the next section. Many more can be made up from reactions in the preceding sections.

The syntheses presented in this section are:

- Δ^4-Cholestene-3-one from Cholesterol

- Tetraphenylcyclopentadienone from Benzaldehyde and Phenylacetic Acid

- Sulfanilamide

- *p*-Phenetidine from *p*-Phenetidine; a Bootstrap Synthesis

- 1-Bromo-3-chloro-5-iodobenzene

- MOED: a Merocyanine Dye

STEROID TRANSFORMATIONS

Δ^4-Cholestene-3-one from Cholesterol via Cholesterol Dibromide, 5α,6β-Dibromocholestane-3-one, and Δ^5-Cholestene-3-one

As an initial step in exploring the chemistry of a rare and expensive substance, preliminary experiments can be done in which a less expensive substance of similar structure is used as a model compound. Cholesterol, which can be isolated in quantity from beef spinal cord and brains (and from human gallstones; see Section E1), has often served as a model compound for the exploration of the chemistry of a class of polycyclic aliphatic compounds known as *steroids,* the most interesting members of which are the sex hormones.

THE SEX HORMONES

The sex hormones are the substances directly responsible for the development of sex characteristics and for the sexual functioning of the human organism. These substances are produced in the ovaries and the testes in response to stimulation by the gonadotropic hormones secreted in the anterior lobe of the pituitary gland. In contrast to the proteinaceous gonadotropic hormones, the sex hormones are polycyclic aliphatic substances. The three primary human sex hormones are estradiol (the primary female sex hormone, or estrogen), testosterone (the primary male sex hormone, or androgen), and progesterone, a substance secreted by the corpus luteum of the female, which prepares the bed of the uterus for implantation of the fertilized ovum.

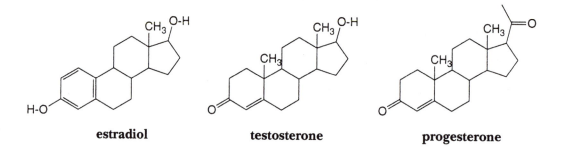

estradiol   testosterone   progesterone

These substances were originally obtained by isolation from natural sources. Estradiol was first isolated in 1934 from sow ovaries at a yield of about 12 mg from some 3,600 kg (four tons) of ovaries, testosterone in 1935 from steer testes at a yield of 10 mg from 100 kg of steer testis tissue, and progesterone in 1934 from sow ovaries at a yield of 20 mg from 625 kg of ovaries, from 50,000 sows. Further development of isolation methods later gave somewhat larger yields, but there was an obvious need for vastly more efficient procedures for the production of these hormones.

In 1935, several research groups determined that cholesterol (see Experiment E1) as the acetate dibromide could be degraded by vigorous chromic acid oxidation to give androstenolone as a minor product.

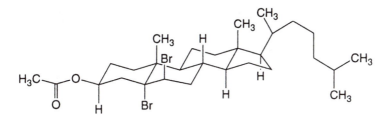

cholesterol acetate dibromide

Androstenolone, which had originally been isolated from male urine the year before, then served as a starting material for syntheses of the various sex hormones.

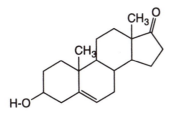

androstenolone

The yield of androstenolone, isolated as the semicarbazone acetate, was originally less than 1%, but by various improvements in the method it was raised over the years to about 8%.

A far more satisfactory alternative for hormone syntheses was provided in 1939 by Russell Marker's discovery that certain substances of vegetable origin, such as diosgenin, could be easily degraded in high yield to give suitable starting materials.

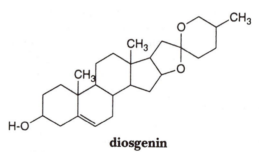

diosgenin

For example, having established the spiroketal structure of the side chain of diosgenin, Marker showed that it could be converted in three steps to 16-dehydropregnenolone acetate in more than 60% yield.

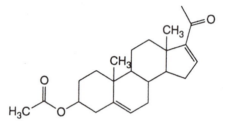

16-dehydropregnenolone acetate

The double bond in the five-membered ring could then be selectively hydrogenated to give pregnenolone acetate.

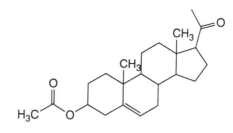

pregnenolone acetate

In 1943, Marker demonstrated the potential of his discovery by preparing 3 kg of progesterone from diosgenin. At that time, this amount was worth a quarter of a million dollars and was equivalent to the yield from the corpora lutea of 30 million sows. The source of the diosgenin was the roots of the wild Mexican yam. Marker worked directly in Mexico, achieving his feat under incredibly primitive conditions with the help of untrained assistants. Marker's unusual and generally unappreciated professional life is reviewed in an article in the *Journal of Chemical Education* (Reference 1). The article ends with the following quotation:

> "When I retired from Chemistry in 1949, after 5 years of production and research in Mexico, I felt I had accomplished what I had set out to do. I had found sources for the production of steroidal hormones in quantity at low prices, developed the process of manufacture, and put them into production. I assisted in establishing many competitive companies in order to insure a fair price to the public and without patent protection or royalties from the producers.
>
> "Since retiring from the laboratory 20 years ago, I have never returned to chemistry or consulting, and have no shares of stocks in any hormone or related companies. My only appearance in public was recently on April 23, 1969 to accept an award by the Mexican Chemical Society showing their appreciation for the work I had accomplished."

*Russell E. Marker, May 15, 1969**

The book *Steroids* by Fieser and Fieser gives many interesting details about the isolation, structure determination, preparation, and total synthesis of sex hormones and other steroids (Reference 2).

SYNTHESIS OF SEX HORMONES

Androstenolone, obtained from cholesterol, was used for the preparation of testosterone, as indicated here.

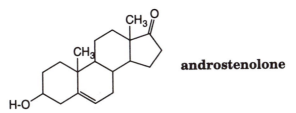

androstenolone

*Quoted with permission from the *Journal of Chemical Education*, **50**, 199 (1973).

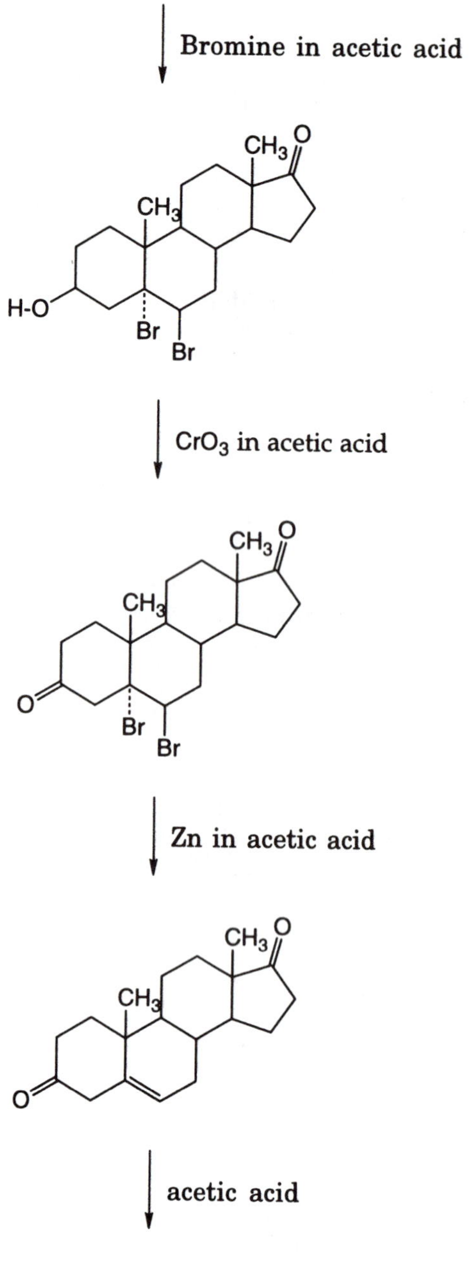
Bromine in acetic acid
CH3
O
CH3
H-O
Br
Br
CrO3 in acetic acid
CH3
O
CH3
O
Br
Br
Zn in acetic acid
CH3
O
CH3
O
acetic acid

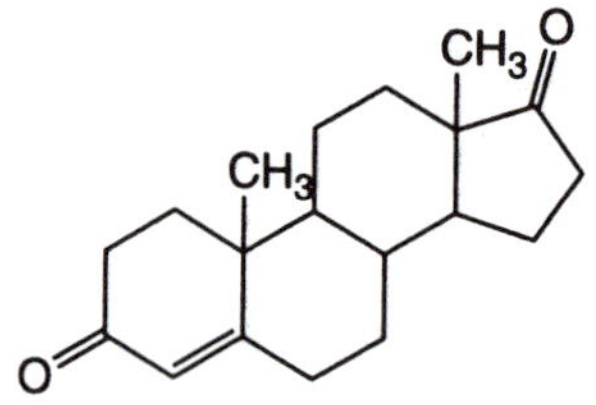
CH3
O
CH3
O

androstenedione

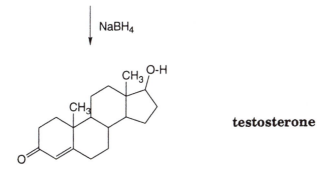

testosterone

In a similar way, pregnenolone, obtained from diosgenin via pregnenolone acetate, was an important starting material for the synthesis of progesterone.

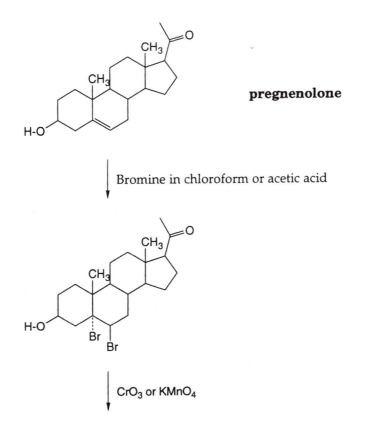

pregnenolone

Bromine in chloroform or acetic acid

CrO_3 or $KMnO_4$

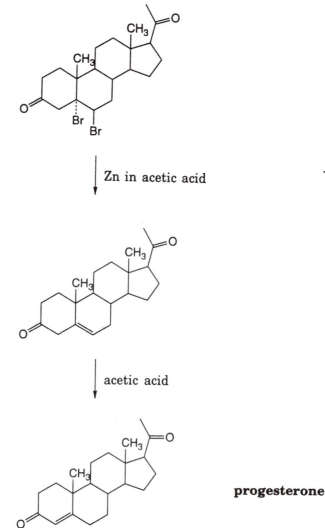

Both these syntheses require the transformation of a β,γ-unsaturated alcohol into the corresponding α,β-unsaturated ketone.

STEROID TRANSFORMATIONS

In the sequence of reactions described in this section, exactly the same transformations are carried out, except that cholesterol is used as an inexpensive model compound. The reactions, which follow, consist of addition of bromine to the double bond (*anti*-mechanism of addition) to protect it from the oxidizing agent, chromate oxidation of the secondary alcohol to the corresponding ketone, reductive elimination by zinc of bromine from the vicinal dibromide to regenerate the carbon–carbon double bond, and finally acid-catalyzed isomerization of the double bond from the unconjugated β,γ-position to the more stable conjugated α,β-position.

There are no good stopping points in this experiment until the Δ5-cholestene-3-one solution has been set aside for crystallization (Experiment E90). If the bromine solution (Experiment E88) and the sodium dichromate solution (Experiment E89) are prepared ahead of time, it will take an inexperienced person about 4 hours to reach this point.

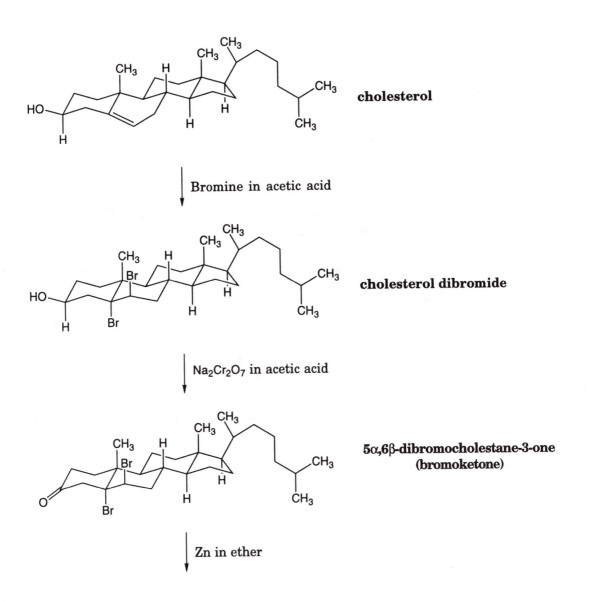

cholesterol

Bromine in acetic acid

cholesterol dibromide

$Na_2Cr_2O_7$ in acetic acid

5α,6β-dibromocholestane-3-one (bromoketone)

Zn in ether

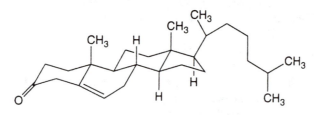

Δ⁵-cholestene-3-one

oxalic acid in ethanol

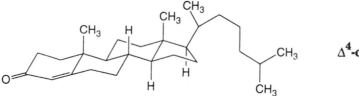

Δ⁴-cholestene-3-one

E88. Cholesterol Dibromide from Cholesterol

The first step in the conversion of cholesterol to Δ^4-cholestene-3-one is addition of bromine to the carbon–carbon double bond of cholesterol to form cholesterol dibromide.

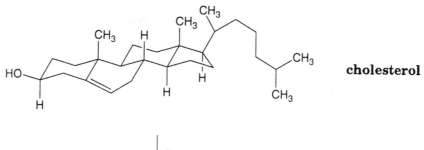

cholesterol

Bromine in acetic acid

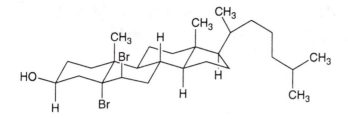

cholesterol dibromide

Scale: 3.0 grams (7.8 mmole) of cholesterol.

Procedure

Place 3.0 grams (7.8 mmole) of cholesterol in a 50-mL Erlenmeyer flask and add 20 mL of anhydrous ether. Dissolve the cholesterol by swirling the flask in a beaker of warm water until the ether just starts to boil. Cool the solution to about room temperature by swirling the flask in a beaker of cool water. Add 12 mL of a solution of bromine and sodium acetate in glacial acetic acid (Note 1; 8.5 mmole of bromine) and swirl to mix. The contents of the flask should promptly solidify to a stiff paste of cholesterol dibromide. Now place the flask in a beaker of cold water to cool.

   While the reaction mixture cools, prepare a mixture of 9 mL of anhydrous ether and 21 mL of glacial acetic acid. Cool this mixture in an ice bath while you collect the cholesterol dibromide by suction filtration. Use the cooled mixture of ether and acetic acid to rinse the product from the Erlenmeyer flask into the funnel and to wash the product in the funnel free of the yellow mother liquor. Continue to apply suction to the product until the wash liquid has almost stopped dripping from the stem of the funnel. The damp dibromide is suitable for use in the next reaction.

Note

1. The solution can be prepared by dissolving 1.0 gram of anhydrous sodium acetate (0.012 mole) in 120 mL of glacial acetic acid and then adding 4.10 mL (13.6 grams; 0.085 mole) of bromine. This is enough reagent for 10 3-gram runs.

Time: about 1 hour.

E89. 5α,6β-Dibromocholestane-3-one from Cholesterol Dibromide

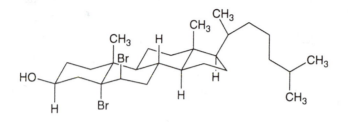

cholesterol dibromide

Na₂Cr₂O₇ in acetic acid

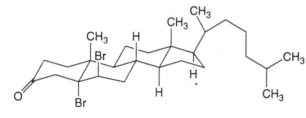

5α,6β-dibromocholestane-3-one
(bromoketone)

Procedure

Add the moist dibromide obtained from the procedure of Section E88 to 40 mL of glacial acetic acid in a 125-mL Erlenmeyer flask. To this, add, all at once, 40 mL of a solution of sodium dichromate in acetic acid that has been preheated to 105°C. (Note 1; sufficient dichromate to oxidize 16 mmoles of secondary alcohol to the corresponding ketone). After the mixture has been swirled briefly, the temperature should rise to between 55 and 60°C. The temperature of the mixture must be maintained between 55 and 58°C for as long as it takes the solids to dissolve (3–5 minutes) and then for 2 minutes more (Note 2).

After allowing the solution to stand at room temperature for about 20 minutes, add 8 mL of cold water, swirl to mix, and then cool the suspension of dibromoketone to about 15°C by means of a cold water or ice bath.

Collect the product by suction filtration, using 10–12 mL of cold methanol for rinsing and washing. Transfer the crude product to a 100-mL beaker containing 25 mL of ice-cold methanol. Stir the mixture to thoroughly wash the crystals and then collect the product again by suction filtration. The damp dibromoketone is suitable for use in the next reaction.

Notes

1. The solution can be prepared by dissolving 16 grams of sodium dichromate dihydrate (0.054 mole) in 400 mL of glacial acetic acid. This is enough reagent for 10 runs of the scale described.

2. If the temperature fails to reach 55°C or shows signs of falling below 55°C, the mixture should be heated briefly to reach or maintain the specified temperature.

Time: 1 hour.

E90. Δ⁵-Cholestene-3-one from 5α,6β-Dibromocholestane-3-one

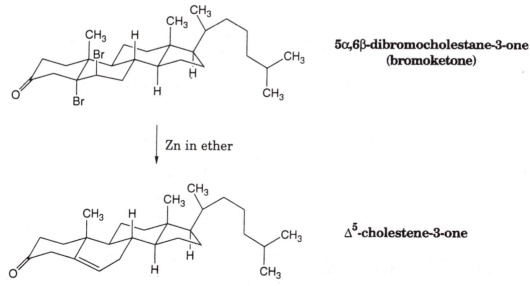

5α,6β-dibromocholestane-3-one
(bromoketone)

Zn in ether

Δ⁵-cholestene-3-one

If the reaction described in the following procedure is successful, the product will have an infrared spectrum like that of Figure E90-1. If the reaction does not go as expected, the dibromoketone will be recovered unchanged; see Note 4.

Procedure

Place the damp product obtained from the preceding oxidation in a 125-mL Erlenmeyer flask, and add to the flask 40 mL of anhydrous ether and 0.5 mL of glacial acetic acid. Cool the contents of the flask to between 15 and 20°C by swirling it in a beaker of cold water. Then add, in several portions over about 3–4 minutes, 0.8 gram (12 mmole) of zinc dust. During this time, the flask must be swirled vigorously to suspend the zinc; the temperature of the contents of the flask should show a definite tendency to rise but should be held below 20°C by very brief swirling of the flask in the beaker of cold water. Most but not all of the zinc will dissolve. After the zinc has been added and the mildly exothermic reaction is over, allow the mixture to stand at room temperature for 10 minutes.

Then add 1.4 mL (17 mmole) of pyridine (to precipitate ionic zinc by complex formation), swirl well to mix, and remove the white precipitate by suction filtration (Note 1). Wash the filter cake well with several small portions of ether, collecting these washings along with the original filtrate.

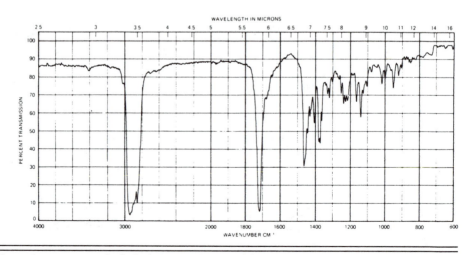

Figure E90-1. Infrared spectrum of Δ^5-cholestene-3-one; CCl_4 solution.

Transfer the filtrate to a separatory funnel and wash it with three 15-mL portions of water and one 15-mL portion of 5% sodium bicarbonate solution (Note 2).

Dry the ethereal solution over anhydrous magnesium sulfate, and then filter it by gravity into a 50-mL Erlenmeyer flask that has been marked to show levels at which it contains approximately 20 and 15 mL (Note 3).

Add a boiling stone and evaporate or distill the ether until the volume has been reduced to 20 mL. The flask should be warmed either on the steam bath or in a beaker of warm water; no burners should be used because of the risk of setting ether vapors on fire. Add 10 mL of methanol to the flask and continue to concentrate the solution until the volume has been reduced to 15 mL. Stopper the flask and set it aside for crystallization.

After crystallization has proceeded at room temperature for at least one-half hour, cool the flask in an ice bath and then collect the unsaturated ketone by suction filtration (Note 4). A small amount of cold methanol can be used for rinsing and washing. Δ^5-Cholestene-3-one is obtained as white granular crystals in a yield of about 1.5 grams (about 50% from cholesterol) with a melting point of 126–128°C. Reported literature values (Reference 3) include m.p., 126–129°C; [α], -2.5° (2.03 grams/100 mL of chloroform).

Notes

1. It is the filtrate you want; be sure your filter flask is clean.

2. The purpose of the sodium bicarbonate wash is to remove any trace of acid that could prematurely catalyze the isomerization of the double bond. Dip a piece of moist blue litmus paper into the ethereal solution to make sure it is not acidic.

3. This can be done by adding the appropriate volume of water to a second, similar flask and then making a mark with a wax pencil on the flask you will use at about the same height as the level of liquid in the second flask.

4. If the debromination with zinc was unsuccessful, the material obtained at this point will be the unchanged 5α,6β-dibromocholestane-3-one. It will melt at about 70°C with decomposition to give a brown-orange liquid and will turn pink to purple upon storage. If the debromination with zinc was unsuccessful *and* the solution was set aside for crystallization for a week, the product obtained may be 6β-bromocholest-4-ene-3-one.

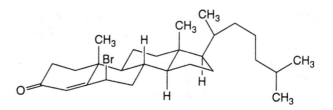

6β-bromocholest-4-ene-3-one

This substance is reported to have a melting point of 122–124°C and can be prepared as follows. Add 1 gram of 5a,6b-dibromocholestane-3-one to 10 mL of pyridine at 5°C, stir the mixture at room temperature for 2 hours, pour the resulting solution into water, collect the precipitate by suction filtration, and recrystallize using 8 mL of hexane [see *Chemical Abstracts* **60,** 590d (1964)]. The infrared spectrum of 6b-bromocholest-4-ene-3-one is shown in Figure E90-2. This compound is recovered unchanged when treated according to the procedure of Experiment E91.

Time: 2 hours.

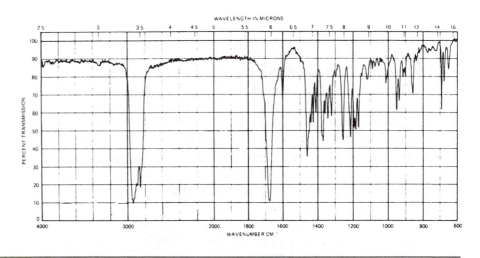

Figure E90-2. Infrared spectrum of 6β-bromocholest-4-ene-3-one; CCl₄ solution; see Note 4.

E91. Δ⁴-Cholestene-3-one from Δ⁵-Cholestene-3-one

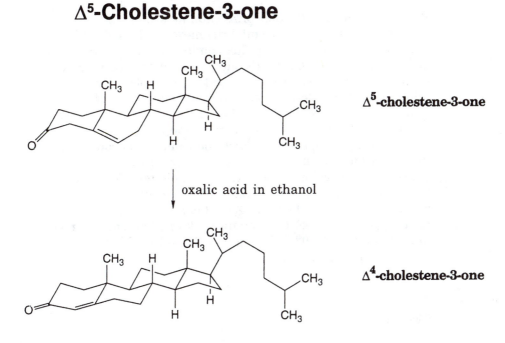

Δ⁵-cholestene-3-one

oxalic acid in ethanol

Δ⁴-cholestene-3-one

The infrared spectrum of Δ⁴-cholestene-3-one is shown in Figure E91-1.

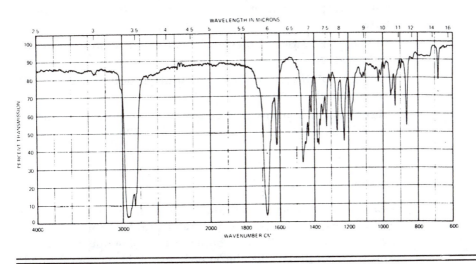

Figure E91-1. Infrared spectrum of Δ⁴-cholestene-3-one; CCl₄ solution.

Procedure

Place 1.0 gram (2.6 mmole) of Δ^5-cholestene-3-one and 100 mg of anhydrous oxalic acid in a 25-mL Erlenmeyer flask. Add 8 mL of 95% ethanol, fit the flask with a reflux condenser, and boil the mixture gently until all the solids have gone into solution (about 15 minutes of boiling is required). Continue to heat the solution for 10 more minutes and then allow it to cool to room temperature. If necessary, induce crystallization by scratching the inner surface of the flask at the liquid level with a stirring rod or by adding seed crystals.

After crystallization appears to be complete at room temperature, cool the flask in an ice bath for about 15 minutes and then collect the large colorless needles of Δ^4-cholestene-3-one by suction filtration. A small amount of cold methanol may be used for rinsing and washing. Yield: about 0.9 gram (about 90%) of crystals with a melting point of 79–81°C. Literature values (Reference 3) include: m.p., 81–82°C; $[\alpha]$, +92.0° (2.01 grams/100 mL of chloroform); λ_{max}, 242 nm; ε, 17,000 (ethanol).

Time: 2 hours.

Disposal

> **Sink:** acetic acid; aqueous washes
>
> **Solid waste:** filter paper
>
> **Nonhalogenated liquid organic waste:** ether; methanol; pyridine
>
> **Solid inorganic waste:** sodium acetate; zinc; magnesium sulfate
>
> **Solid organic waste:** cholesterol; oxalic acid; all intermediates and products
>
> **Bromine solution:** place waste or spills in bottle reserved for this material
>
> **Chromic acid solution:** place waste or spills in bottle reserved for this material

Questions

1. How many stereoisomeric 5,6-dibromocholestane-3-ones can there be? How many of these stereoisomers could be formed by an *anti*-mechanism of addition of bromine to cholesterol? Suggest an explanation for the fact that the 5α,6β-isomer is the only one formed. Which isomer would you expect to be the most stable?

2. Write a reasonable mechanism for the debromination of 5α,6β-dibromocholestane-3-one by zinc.

3. Comment on the mechanism and circumstances of formation of 6β-bromocholest-4-ene-3-one from 5α,6β-dibromocholestane-3-one as described in Note 4 of Experiment E90.

4. The acid-catalyzed isomerization of the α,β-unsaturated ketone to the α,β-unsaturated isomer probably goes via the intermediate formation of the conjugated enol. Write out the details of the mechanism, and be sure to indicate the possibilities for resonance stabilization of each intermediate cation.

References

1. P. A. Lehmann, F. A. Bolivar, and R. Quintero, *J. Chem. Educ.* **50,** 195 (1973).

2. L. F. Fieser and M. Fieser, *Steriods,* Reinhold, New York, 1959

3. L. F. Fieser, *J. Am. Chem. Soc.* **75,** 5421 (1953).

4. L. F. Fieser, *Organic Syntheses,* Coll. Vol. 4, Wiley, New York, 1963, p. 195.

TETRAPHENYLCYCLOPENTADIENONE: A PURPLE COMPOUND

In the synthesis of Δ^4-cholestene-3-one presented in Experiments E88 through E91, a *linear strategy* was used. That is, if we think of the desired product to be D and the starting material to be A, a linear strategy would be the conversion of A to B, B to C, C to D, a total of three reactions, run one after the other. If each reaction were to take place with a yield of 90%, the overall yield of D from A would be $(0.90)^3$, or 73%.

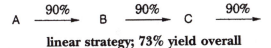

linear strategy; 73% yield overall

If, however, we could prepare intermediates 1 and 2 from different starting materials and then combine 1 and 2 to form product 3, we would again be running a total of three reactions but, again assuming a yield of 90% for each, the overall yield of the final product by this *convergent strategy* would be $(0.90)^2 = 81\%$. Assuming similar yields, convergent strategies generally give higher overall yields and are thus more efficient and preferred over linear strategies.

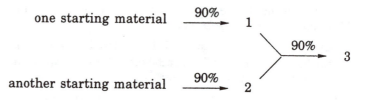

one starting material → 90% → 1

another starting material → 90% → 2

90% → 3

convergent strategy; 81% yield overall

The next four sections present the synthesis of the dark purple compound tetraphenylcyclopentadienone by the following convergent strategy.

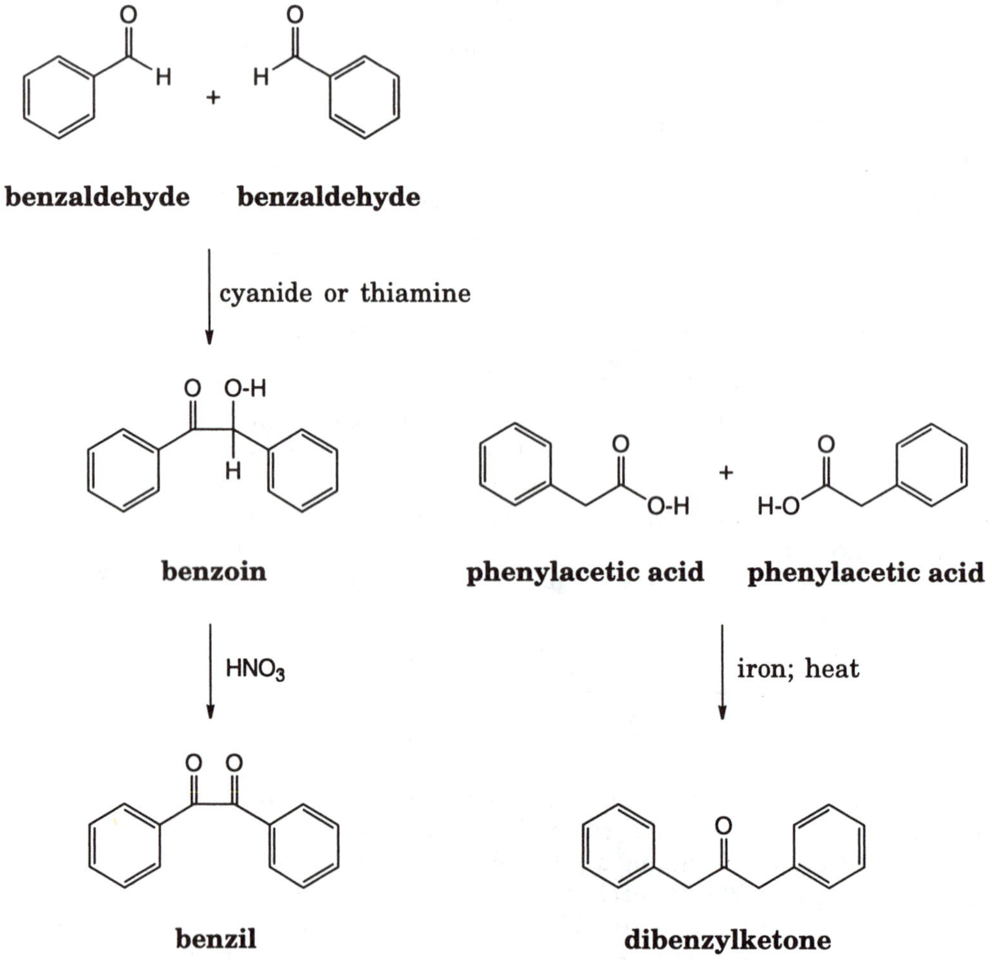

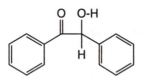

benzaldehyde   **benzaldehyde**

cyanide or thiamine

benzoin

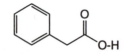

phenylacetic acid   **phenylacetic acid**

HNO_3

iron; heat

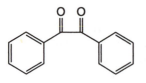

benzil   **dibenzylketone**

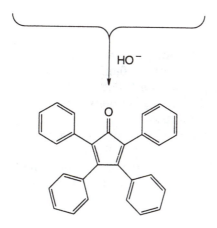

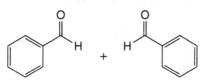

**tetraphenylcyclopentadienone
dark purple**

E92. Benzoin from Benzaldehyde

In this procedure, benzoin is prepared from benzaldehyde by use of
cyanide as the catalyst. It can also be prepared by catalysis with thiamine as described in Experiment E80. The mechanisms of catalysis
are shown in Experiment E80.

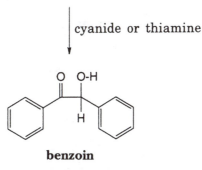

benzaldehyde   **benzaldehyde**

cyanide or thiamine

benzoin

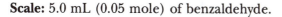

Scale: 5.0 mL (0.05 mole) of benzaldehyde.

Procedure

Dissolve 0.5 g (0.008 mole) of potassium cyanide (poison, see Section 1.7) in 5 mL of water in a 50-mL Erlenmeyer flask. Add 10 mL of 95% ethanol followed by 5 mL (5.2 g; 0.049 mole) of benzaldehyde (Note 1). Fit the flask with a condenser, and boil the mixture on the steam bath for 30 minutes.

Allow the mixture to cool to room temperature, and swirl the flask occasionally to speed crystallization.

Cool the mixture in an ice bath, collect the benzoin by suction filtration, and wash it, first with two 5-mL portions of cold 50% ethanol and finally with water. Benzoin can be recrystallized from methanol using 12 mL per gram.

Note

1. The benzaldehyde should be free of benzoic acid to avoid liberating hydrogen cyanide gas.

Time: less than 2 hours.

E93. Benzil from Benzoin

In this procedure, benzoin is oxidized to the corresponding diketone benzil by nitric acid.

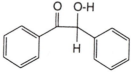

benzoin

HNO$_3$

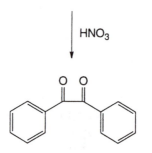

benzil

Scale: 4.2 grams (0.020 mole) of benzoin.

Procedure

Place 4.2 g (0.02 mole) of benzoin in a 125-mL ground-glass-stoppered Erlenmeyer flask and then add 11 mL (20 g; 0.22 mole) of concentrated nitric acid. Heat this mixture in the hood (Note 1) on a steam bath for 10 to 12 minutes. During the heating period, swirl the mixture occasionally (Note 2).

At the end of the heating period, add to the flask 75 mL of cold water, and swirl the flask to mix the contents. Then add a seed crystal of benzil, stopper the flask, and shake it to cause the oily product to solidify in small lumps.

Collect the bright yellow solidified benzil by suction filtration, and wash it thoroughly with water to remove the nitric acid. Yield: about 4 g (about 95%). Benzil can be recrystallized from 95% ethanol, using 5 7 mL per gram.

Notes

1. Nitrous oxide fumes are evolved.
2. The solid will gradually dissolve, and an oil (molten benzil) will form.

Time: 1 hour.

E94. Dibenzylketone from Phenylacetic Acid

This transformation takes place in two stages. The first is the reaction of the acid with elemental iron to form the ferrous salt of the acid.

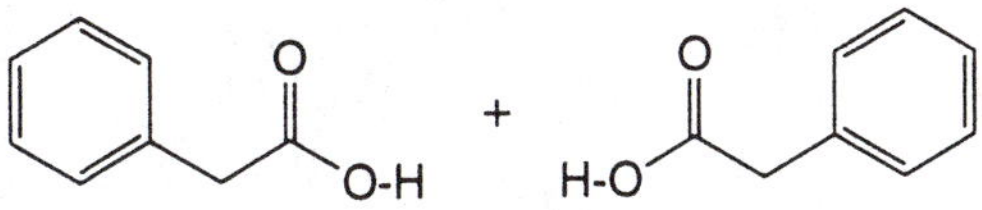

phenylacetic acid   **phenylacetic acid**

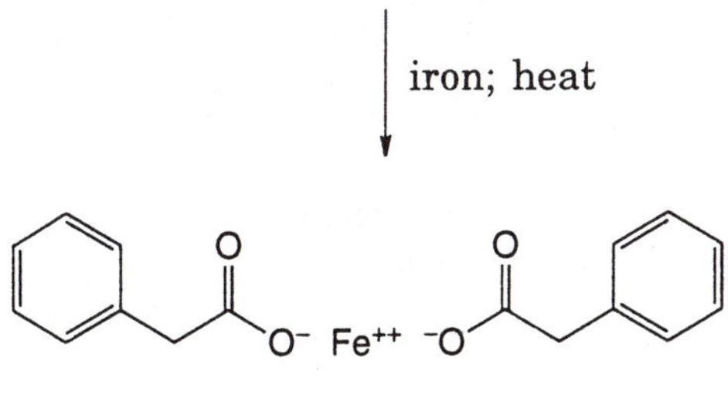

iron; heat

The second stage is the loss of the elements of carbon dioxide and ferrous oxide by pyrolysis of the ferrous salt.

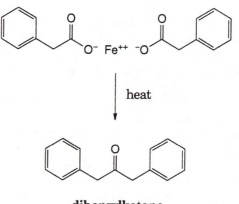

dibenzylketone

The dibenzylketone produced in this experiment should have an infrared spectrum like that shown in Figure E94-1.

Scale: 6.8 grams (0.050 mole) of phenylacetic acid.

Procedure (Reference 1)

Place 6.8 g (0.050 mole) phenylacetic acid and 1.5 g (0.027 mole) iron powder in a 50-mL distilling flask (Note 1). Fit the flask with a

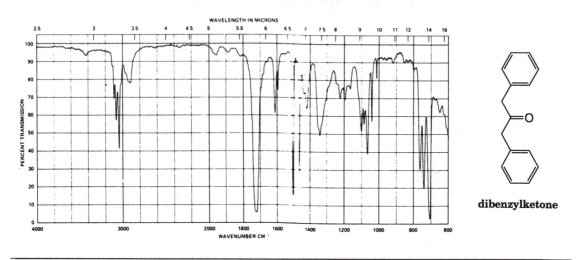

Figure E94-1. Infrared spectrum of dibenzylketone; thin film.

side-arm test tube as a receiver, and connect a length of rubber tubing to the side arm; lead the tubing to an area free of flames (Note 2). Arrange a thermometer in the top of the distilling flask so that the bulb almost touches the bottom of the flask. Heat the mixture of solids until it melts, and gradually raise the temperature to 320°C (Note 3). When the temperature of the mixture reaches 320°, stop heating and raise the thermometer into position for a normal distillation. Then distill the contents of the flask, collecting the product at about 324°C (Note 4). Yield: about 60%.

Notes

1. Use an inexpensive flask, as this reaction mixture etches the glass badly.
2. Hydrogen gas (flammable, explosive) is evolved.
3. During the heating period, which should be about 15 minutes, the contents of the flask will liquefy, gas will be evolved (H_2), and the contents of the flask will darken and solidify and then will remelt with further gas evolution (CO_2). Do not heat so quickly that anything distills over.
4. The product will be colored yellow to brown. Do not attempt to distill to dryness. To clean the distilling flask, use acetone and soapy water until they seem to have no more effect. Then cover the residue in the flask with concentrated hydrochloric acid and allow it to stand in the hood.

Time: 3 hours.

Reference

1. R. Davis and H. P. Schultz, *J. Org. Chem.* **27,** 854 (1962).

E95. Tetraphenylcyclopentadienone from Benzil and Dibenzyl Ketone

In this reaction, the 5-membered ring is formed by a pair of aldol condensations between the carbonyl groups of the diketone, benzil, and the methylene groups of dibenzylketone. Water is spontaneously eliminated from the intermediate to form the final product, which separates from the solution as deep purple flakes.

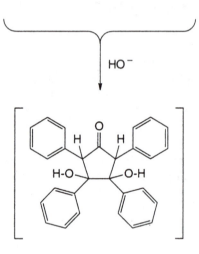

benzil   **dibenzylketone**

intermediate

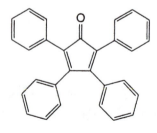

tetraphenylcyclopentadienone
dark purple

Scale: 2.1 grams (0.010 mole each) of benzil and dibenzylketone.

Procedure (Reference, 1)

Dissolve 2.1 g (0.010 mole) of benzil (Section E93) and 2.1 g (0.010 mole) of dibenzylketone (Section E94) in 10 mL of triethylene glycol in a 25 mm × 150 mm test tube. While supporting the test tube with a clamp and stirring the contents with a thermometer, heat the mixture with a small flame until the benzil is dissolved. Adjust the temperature to exactly 100°C, add 1 mL of 40% benzyl-trimethylammonium hydroxide in methanol, and stir the mixture well with the thermometer.

When the temperature has dropped to 80°C, cool the mixture almost to room temperature, thin it with 10 mL of methanol, and collect the product by suction filtration. Wash it with methanol until the filtrate is purple-pink rather than brown. The very dark purple tetraphenylcyclopentadienone can be recrystallized from triethylene glycol, using 10 mL per gram.

Time: less than 2 hours.

Disposal

> **Sink:** nitric acid (after dilution); aqueous washes
>
> **Solid waste:** filter paper
>
> **Nonhalogenated liquid organic waste:** benzaldehyde; methanol; ethanol; triethylene glycol; benzyltrimethyl-ammonium hydroxide
>
> **Solid organic waste:** benzoin; benzil; phenylacetic acid; dibenzylketone
>
> **Potassium cyanide:** place waste or spills in the container reserved for this material

Reference

1. L. F. Fieser, *Organic Experiments,* Heath, Boston, 1964, p. 303.

SULFANILAMIDE

Sulfanilamide was the first substance to be used systematically and effectively as a chemotherapeutic agent for the prevention and cure of bacterial infections in humans.

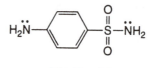

sulfanilamide

Although it was first prepared in 1908, more than 25 years had to pass before its therapeutic value was discovered. The discovery was, typically, indirect. In 1909, German dye chemists of the I. G. Farbenindustrie found that azo dyes containing sulfonamide groups were superior in color fastness, especially toward the proteins of wool and silk. This finding led others to an investigation of the use of these dyes as selective staining agents for bacterial protoplasm. In the course of this work, it was gradually recognized that certain dyes had the ability to kill some bacteria in vitro (in a bacterial culture outside a living host). This, reasonably enough, inspired the hope that a similar antibacterial action could occur in vivo (within a living host), but research along these lines was not pursued very vigorously. Then in the early 1930s, further work at the I. G. Farbenindustrie with a new sulfonamide-containing dye, Prontosil, showed that this new substance could protect mice with streptococcal and other bacterial infections despite the fact that it had no in vitro antibacterial action. This discovery was recognized by the awarding of the Nobel prize in medicine for 1939 to Gerhard Domagk, one of the directors of research of the I. G. Farbenindustrie.

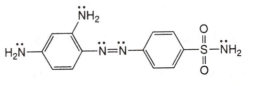

Prontosil

In 1935, research workers at the Pasteur Institute in Paris, who had been intrigued by the fact that Prontosil was effective in vivo but not in vitro, reported a fascinating and far-reaching explanation: the reason was that, in a reaction that could take place only in vivo, Prontosil could be reduced by the host to sulfanilamide, which was the actual effective agent. Further research in England and the United States confirmed and extended this finding, with the result that within the next 10 years more than 5,400 related substances had been prepared and tested for antibacterial activity. Fewer than 20 ever attained therapeutic importance.

As with all drugs, the next question was "How does it work?" How is it that sulfanilamide attacks bacteria specifically, with no apparent effect on the host? First, sulfanilamide does not usually kill the bacteria, but it greatly reduces their rate of growth and reproduction. On the molecular level, sulfanilamide competes with *p*-aminobenzoic acid in the bacterial synthesis of folic acid, a substance required for bacterial growth.

sulfanilamide   ***p*-aminobenzoic acid**

Sulfanilamide rather than *p*-aminobenzoic acid enters into the reaction that normally would produce folic acid, and the product, containing an O=S=O group in place of the C=O group of the amide function, is unable to carry out the function of folic acid. Because the bacteria are able under normal conditions to synthesize their own folic acid, they have no mechanism for obtaining folic acid from their environment at the low concentrations at which it is present.

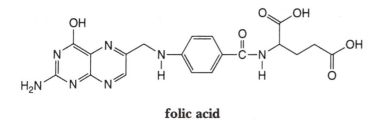

folic acid

Inhibition or diversion of folic acid synthesis, then, accounts for the bacteriostatic action of sulfanilamide. In contrast, the host, a "higher organism," does not have the ability to synthesize folic acid but has instead the capacity to make use of the folic acid present at low concentrations in its diet. Obviously, the presence or absence of sulfanilamide is irrelevant to this capacity to acquire folic acid from the environment, and this accounts for the lack of effect of sulfanilamide on the host organism.

Sulfanilamide can be prepared from benzene in six steps, as indicated by the scheme shown following.

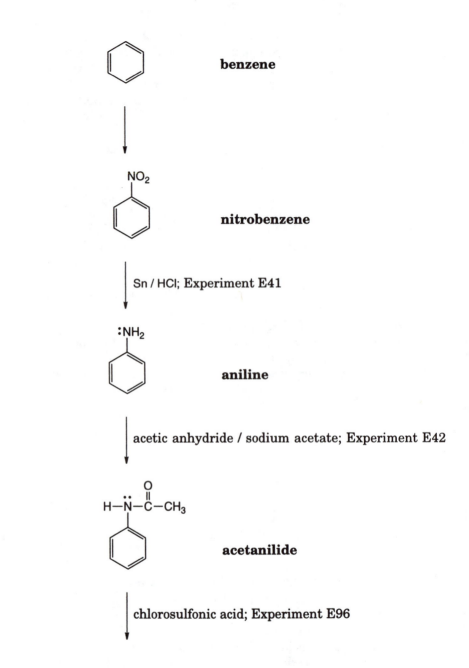

benzene

nitrobenzene

Sn / HCl; Experiment E41

aniline

acetic anhydride / sodium acetate; Experiment E42

acetanilide

chlorosulfonic acid; Experiment E96

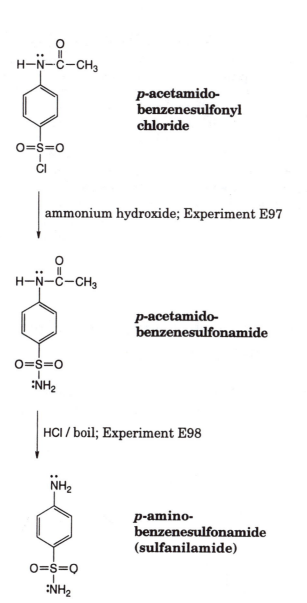

p-acetamido-
benzenesulfonyl
chloride

ammonium hydroxide; Experiment E97

p-acetamido-
benzenesulfonamide

HCl / boil; Experiment E98

p-amino-
benzenesulfonamide
(sulfanilamide)

The last three steps of this synthesis are presented in this chapter; the procedures for the second and third steps can be found in earlier sections of the book. The reaction by which benzene is converted to nitrobenzene has been omitted from this edition of this book; many chemists believe that the toxicity of benzene is such that it should not be used in the introductory organic laboratory.

According to this approach, the aromatic amino group is first acetylated and then later deacetylated. There are at least two reasons for this. First, if one were to try to chlorosulfonate aniline itself, chlorosulfonic acid, an acid chloride, could be expected to react with the amino group of aniline. Then, even if it were possible to prepare *p*-aminobenzenesulfonyl chloride, the amino group of one molecule might be expected to react with the sulfonyl chloride group of another.

E96. *p*-Acetamidobenzenesulfonyl Chloride from Acetanilide

In this reaction, chlorosulfonic acid effects an electrophilic aromatic substitution on the benzene ring *para* to the activating group.

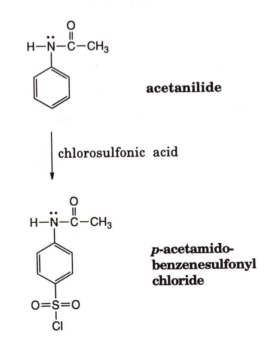

acetanilide

chlorosulfonic acid

p-acetamido-
benzenesulfonyl
chloride

Scale: 3.38 grams (0.025 mole) of acetanilide.

Procedure

Place 3.38 g (0.025 mole) of completely dry (Note 1) acetanilide in a 125-mL Erlenmeyer flask. Gently heat the flask with a flame to melt the acetanilide, and swirl the flask containing the molten acetanilide as it cools, so that it will solidify in a thin layer on the bottom and lower walls of the flask. Prepare a trap to absorb hydrogen chloride, and arrange for it to be connected to the Erlenmeyer flask by a length of rubber tubing (Section 36). Make certain that there is no possibility for water to be sucked back into the flask (caution: Note 1).

Cool the flask in an ice bath. Then remove the flask from the ice bath and add to the flask all at once 8.5 mL (15 g; 0.129 mole) of chlorosulfonic acid (caution: Note 2) and connect the flask to the gas trap. Swirl the mixture until part of the solid has dissolved and hydrogen chloride is being rapidly evolved. Cool the flask in the ice bath if necessary to moderate the reaction.

When all but a few lumps of acetanilide have dissolved (5–10 minutes), heat the mixture on the steam bath for 10 minutes and then cool the flask to room temperature. Slowly and cautiously (Note 1) pour the mixture onto 50 grams of ice in a beaker in the ice bath in the hood. Stir the precipitated product for a few minutes to obtain an even, granular suspension, and then collect it by suction filtration. The crude product should be used immediately.

Notes

1. Chlorosulfonic acid reacts violently with water.
2. Measure the chlorosulfonic acid in a dry graduate, as it will react violently with water.

Time: less than 2 hours.

Questions

1. What do you suppose is the mechanism of this electrophilic aromatic substitution reaction? Why is an excess of chlorosulfonic acid used?
2. Write out the equation for the reaction that takes place between chlorosulfonic acid and water.
3. What might be the product of reaction of chlorosulfonic acid and aniline?

4. What might be the product of reaction of *p*-aminobenzenesulfonyl chloride with another molecule of the same compound?

5. Why would we not expect a similar reaction between two molecules of *p*-acetamidobenzenesulfonyl chloride?

6. Although aminobenzenesulfonic acid is readily available, it can be acetylated only in the form of its sodium salt. Why should this be so? As a matter of fact, the sodium salt of *p*-aminobenzenesulfonic acid is not very soluble in acetic anhydride, and the reaction proceeds so poorly that this is not a practical way to make *p*-acetamido-benzenesulfonic acid, which could then be converted to *p*-acet-amidobenzenesulfonyl chloride.

E97. *p*-Acetamidobenzenesulfonamide from *p*-Acetamidobenzenesulfonyl Chloride

In this reaction, the sulfonic acid chloride is converted to the corresponding amide by sulfonyl transfer to the basic ammonia molecule.

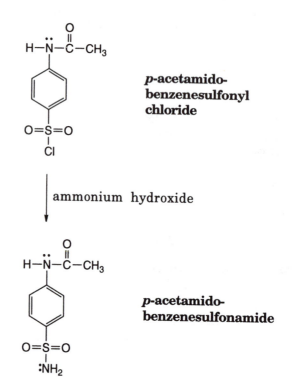

p-acetamido-benzenesulfonyl chloride

ammonium hydroxide

p-acetamido-benzenesulfonamide

Procedure

Transfer the crude product of the procedure of Experiment E96 to a
50-mL Erlenmeyer flask, and add 10 mL (9.0 g; 0.150 mole) of concentrated
ammonium hydroxide and 10 mL of water. By means of a
flame and in the hood, heat the mixture almost to boiling; in about
5 minutes, with occasional swirling, the granular sulfonyl chloride
will be converted to the more pasty sulfonamide. Cool the mixture
well in an ice bath, collect the product by suction filtration, and suck
it as dry as possible.

Time: 1 hour.

Question

1. Estimate the relative nucleophilicity of water, ammonia, and hydroxide
   ion toward the sulfonyl chloride group.

E98. Sulfanilamide from *p*-Acetamidobenzenesulfonamide

In this reaction, the acetyl group is removed from the aromatic amino
group, converting the amide to the free amine.

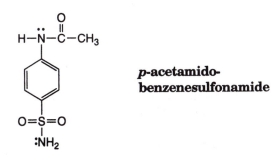

***p*-acetamido-
benzenesulfonamide**

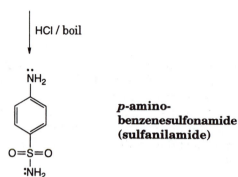

**p-amino-
benzenesulfonamide
(sulfanilamide)**

Procedure

Transfer the moist product of the procedure of Experiment E97 to a 50-mL boiling flask, and add 3.5 mL (4.1 g; 0.042 mole) of concentrated hydrochloric acid and 7 mL of water. Fit the flask with a condenser, and gently boil the mixture under reflux until all the solid has dissolved (about 10 minutes) and then for 10 minutes more.

Cool the solution to room temperature (Note 1), add a little decolorizing carbon, and filter with suction.

Transfer the filtrate to a beaker, and add sodium bicarbonate as a saturated aqueous solution until the mixture is no longer acid to litmus (Note 2). Cool the suspension of precipitated sulfanilamide in the ice bath, and then collect it by suction filtration.

Recrystallize the sulfanilamide from water, using 12 mL per gram (add decolorizing carbon, if necessary).

Notes

1. If solid separates upon cooling, reheat and boil the mixture for a while longer.
2. About 3.8 g (0.045 mole) will be needed.

Time: less than 2 hours.

Disposal

Sink: aqueous washes; ammonium hydroxide (after dilution); concentrated HCl (after dilution); aqueous sodium bicarbonate.
Solid waste: filter paper
Solid organic waste: acetanilide; intermediates and product
Chlorosulfonic acid: add *very cautiously* to dilute aqueous base

Questions

1. Estimate the relative rates of hydrolysis of carboxylic acid amides and sulfonic acid amides.

2. We are often told that the amide is the most stable of all of the acyl derivatives. Why, then, is the amide converted to the amine plus carboxylic acid in this experiment?

3. What is role of the sodium bicarbonate?

A BOOTSTRAP SYNTHESIS:
p-PHENETIDINE FROM *p*-PHENETIDINE

This synthesis, originally presented as a project in the first edition of this book, was reported in the German patent literature as a method by which *p*-phenetidine could be prepared for the synthesis of phenacetin, the analgesic and antipyretic described in Section E74.

The feature of greatest interest is the fact that the product from one molecule of the starting material is two molecules of that same compound.

Briefly, *p*-ethoxyaniline is diazotized and coupled at the *para* position of phenol, the phenolic group is ethylated, and the resulting symmetrical azo compound is reductively cleaved to give two molecules of *p*-ethoxyaniline.

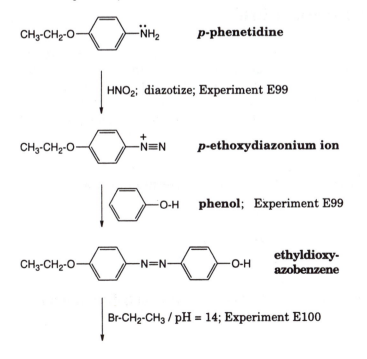

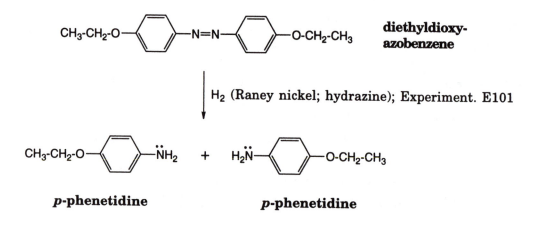

diethyldioxy-
azobenzene

H_2 (Raney nickel; hydrazine); Experiment. E101

p-phenetidine   *p*-phenetidine

The procedures described here are essentially those of the patent (Reference 1, Section 10), except that we carry out the reduction of diethyldioxyazobenzene to 2 moles of phenetidine with hydrazine in the presence of Raney nickel (Reference 2) rather than with zinc metal and hydrochloric acid.

E99. Ethyldioxyazobenzene from *p*-Phenetidine

In this experiment, *p*-phenetidine (*p*-ethoxyaniline) is converted to the corresponding diazonium ion by treatment with nitrous acid (hydrochloric acid plus sodium nitrite). When the diazonium ion is added to the solution of phenol in aqueous sodium carbonate, an electrophilic aromatic substitution takes place at the *para* position of the phenol.

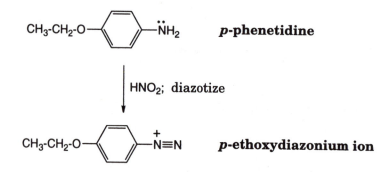

p-phenetidine

HNO_2; diazotize

p-ethoxydiazonium ion

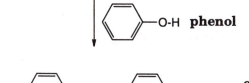 **phenol**

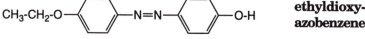 **ethyldioxy-
azobenzene**

Scale: 3.2 mL (3.43 grams; 0.025 mole) of *p*-phenetidine.

Procedure

Place 3.2 mL (3.43 grams; 0.025 mole) of *p*-phenetidine in a 125-mL Erlenmeyer flask. To this, add 5 mL of water and then 4.1 mL (0.049 mole) of concentrated HCl. Swirl this mixture until all the amine has dissolved (Note 1).

Next, prepare a solution of 1.73 grams (0.025 mole) of sodium nitrite in 12 mL of water. Finally, in a 250-mL Erlenmeyer flask, prepare a solution of 2.35 grams (0.025 mole) of phenol and 5.3 grams (0.05 mole) of sodium carbonate in 90 mL of water.

Now, add the solution of sodium nitrite to the solution of the amine; swirl to mix (Note 2). After allowing this mixture to stand for about 60 seconds, pour it slowly while mixing well into the solution of phenol and sodium carbonate. A voluminous yellow precipitate of ethyldioxyazobenzene forms immediately.

After allowing the resulting suspension to stand for 10 minutes or so, collect the solid by suction filtration, using some cold water to rinse the material into the suction funnel and to wash the product on the funnel. The filter cake should be sucked free of as much water as possible so that it can be used in the next reaction without further drying. Yield of dried material: 5.95 grams (98%).

Notes

1. If the solution is highly colored, it can be treated with decolorizing carbon, which can be removed by gravity filtration; most of the color can be removed in this way.
2. The diazotization of this amine can be done at room temperature.

Time: About 1 hour.

E100. Diethyldioxyazobenzene from Ethyldioxyazobenzene

In this experiment, the conjugate base of ethyldioxyazobenzene receives "+CH$_2$–CH$_3$" from ethyl bromide.

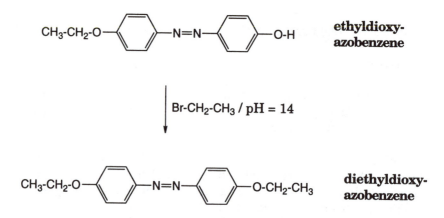

Scale: 6.0 grams (0.025 mole) of ethyldioxyazobenzene.

Procedure

Place 6.0 grams (0.025 mole) of dry ethyldioxyazobenzene in a 100-mL boiling flask, and then add to the flask 25 mL of 1-propanol, followed by 1.5 mL (0.029 mole) of 50% aqueous sodium hydroxide solution. Fit the flask with a reflux condenser, and support the apparatus so that the contents can be heated at about 80°C for 4 hours (Note 1). Now add 8 mL (11.7 grams; 0.10 mole) (Note 2) of ethyl bromide down the condenser, and heat the deep-red solution under very gentle reflux for about 4 hours. During the heating period, a quantity of solid will crystallize from the solution.

At the end of the heating period, cool the flask by immersing it in an ice bath for a few minutes, and then collect the crystalline diethyldioxyazobenzene by suction filtration, using about 25 mL of methanol to assist in transferring the solid to the suction funnel and to wash the product. The product should be sucked as free from solvent as possible. Yield of dried material: 6.6 grams (98%).

Notes

1. An electrically heated oil bath is ideal for this purpose because it can be set and left unattended. A steam bath can

also be used, but maintaining a constant rate of heating is more difficult with it.

2. Only 1 equivalent of ethyl bromide is required. We use 4 equivalents so as to decrease the time needed for the reaction.

Time: About 1 hour working time plus 4 hours reaction time.

Questions

1. What is the role of the sodium hydroxide?
2. How long would the reaction take if only half as much ethyl bromide had been used?

E101. *p*-Phenetidine from diethyl-dioxyazobenzene

In this experiment, *p*-phenetidine is formed in the reduction of diethyldioxyazobenzene by hydrazine in the presence of Raney nickel.

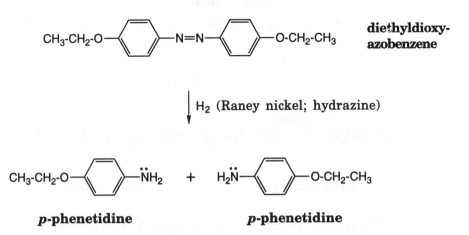

H_2 (Raney nickel; hydrazine)

p-phenetidine   *p*-phenetidine

Raney nickel, a finely divided form of nickel metal, catalyzes the decomposition of hydrazine into hydrogen gas and diimide, the actual reducing agent.

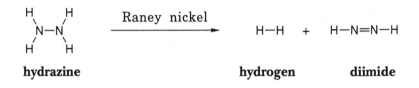

hydrazine   **hydrogen**   **diimide**

Then, diimide first reduces the azo compound to the hydrazo compound,

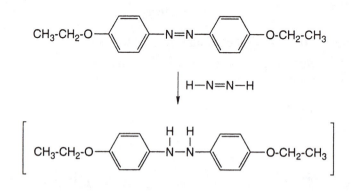

and finally reduces the hydrazo compound to 2 moles of the aromatic amine.

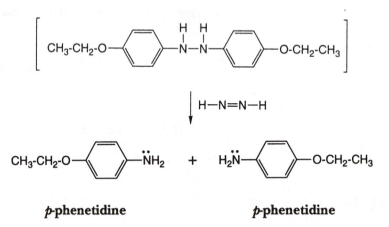

p-phenetidine   *p*-phenetidine

Because the aromatic amine is a liquid in this experiment, it is more convenient to isolate it as the solid acetyl derivative, phenacetin, a substance used as a mild analgesic and antipyretic (agent for the relief of pain and reduction of fever; Experiment E74).

Scale: 2.7 grams (0.010 mole) of diethyldioxyazobenzene.

Procedure

Place 2.7 grams (0.010 mole) of diethyldioxyazobenzene in a 125-mL Erlenmeyer flask. Add to the flask 20 mL of methanol, 1.5 mL of hydrazine (1.5 grams; 0.05 mole), and about 100 milligrams of Raney nickel (Note 1). Fit the flask with a reflux condenser, and support the flask so that it can be heated on the steam bath. Cautiously warm the flask on the steam bath until the contents start to foam. Heat the flask as long as the contents foam and until both the yellow solid and the yellow color, which indicate the presence of unchanged diethyldioxyazobenzene, are gone; about 1 hour of heating will be required (Note 2).

After the starting material and excess hydrazine are both gone, remove the flask from the steam bath and disconnect the condenser from the flask. Add to the flask 20 mL of cold water and then 1.7 mL (0.02 mole) of concentrated HCl. Add a modest portion of decolorizing carbon to the solution, and, after swirling the flask and then allowing it to stand for a few minutes, remove the carbon by gravity filtration. Collect the filtrate in a 125-mL Erlenmeyer flask.

While you are waiting for the solution to flow through the filter, prepare a solution of 3.0 grams (0.022 mole) of sodium acetate trihydrate in 10 mL of water. After filtration is complete, add 2.1 mL (0.022 mole) of acetic anhydride to the filtrate, and swirl the flask to mix the contents. Then add, all at once and while swirling, the solution of sodium acetate; phenacetin will immediately separate from solution. Warm the suspension of phenacetin on the steam bath until all of the solid has dissolved. Then add 50 mL of hot water, swirl to mix, and set the solution aside to cool slowly for crystallization (Note 3).

After crystal formation is well underway, the flask can be placed in a beaker of cold water and then in an ice bath to speed up the process. When crystallization is judged to be complete, collect the product by suction filtration, using water to assist the transfer of the solid to the filter funnel and to wash the crystals. Yield: about 80% (Note 4).

Notes

1. Raney nickel can be purchased, or it can be prepared from the Raney alloy of nickel and aluminum. To prepare Raney nickel from the alloy, add about a gram of the alloy, divided

into several portions, to a mixture of 5 mL of 50% sodium hydroxide and 20 mL of water. The mixture will foam strongly after each addition. When the foaming has subsided, heat the mixture on the steam bath until the bubbling ceases. Decant the basic aqueous supernatant liquid, add about 10 mL of water, swirl, and decant. Repeat this washing operation with a second 10-mL portion of water and then with a 10-mL portion of methanol. Transfer portions of the moist suspension to the reaction flask by means of a medicine dropper.

2. The flask must be heated until the starting material is gone (as indicated by the disappearance of the yellow solid and the yellow color) and until the excess hydrazine has been decomposed (as indicated by the cessation of foaming). If the foaming stops before the color has been discharged, add a few more drops of hydrazine down the condenser to the reaction mixture. The reaction mixture must be heated until both the starting material and the excess hydrazine are gone. The time required for heating will depend on the activity of the Raney nickel, the amount used of this catalyst, and the vigor of heating.

3. If the solution is allowed to cool slowly, the product will separate in the form of large spars.

4. If a solution of 0.020 mole of phenetidine is dissolved in 20 mL of methanol and this solution is carried through the process of acetylation, phenacetin is isolated in a yield of about 80%. This implies that the reduction of the azo compound to the amine gives the amine in a yield of about 100%.

Time: less than 3 hours.

Disposal

Sink: concentrated HCl (after dilution); aqueous washings; aqueous NaOH

Solid waste: filter paper

Nonhalogenated liquid organic waste: *p*-phenetidine; 1-propanol; methanol; hydrazine acetic anhydride

Halogenated liquid organic waste: ethyl bromide

Solid inorganic waste: sodium nitrite; sodium acetate

Solid organic waste: phenol; intermediates and product

Raney nickel: place spills or used material in the container reserved for this material

References

1. J. D. Riedel, *D. R. P.* **48543** (1889).
2. D. Balcom and A. Furst, *J. Am. Chem. Soc.* **75,** 4334 (1953).

1-BROMO-3-CHLORO-5-IODOBENZENE

1-Bromo-3-chloro-5-iodobenzene can be prepared from benzene in eight steps, as indicated in the following scheme.

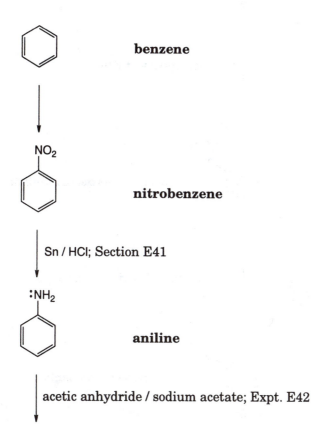

benzene

nitrobenzene

Sn / HCl; Section E41

aniline

acetic anhydride / sodium acetate; Expt. E42

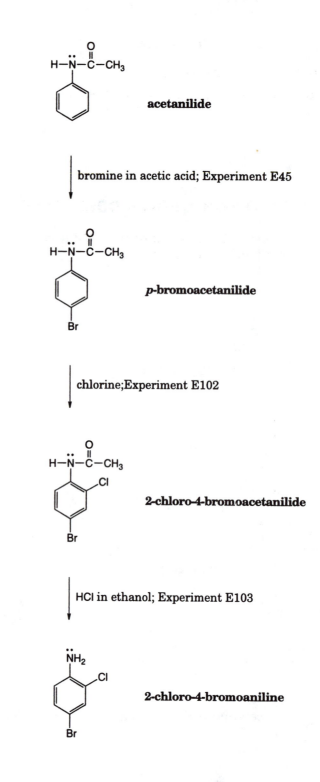

acetanilide

bromine in acetic acid; Experiment E45

p-bromoacetanilide

chlorine;Experiment E102

2-chloro-4-bromoacetanilide

HCl in ethanol; Experiment E103

2-chloro-4-bromoaniline

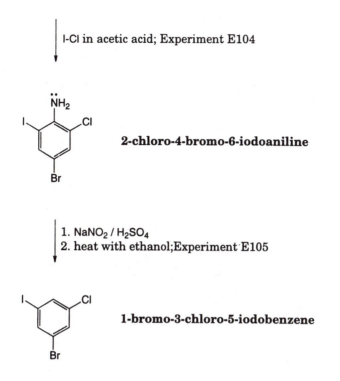

I-Cl in acetic acid; Experiment E104

2-chloro-4-bromo-6-iodoaniline

1. NaNO$_2$ / H$_2$SO$_4$
2. heat with ethanol;Experiment E105

1-bromo-3-chloro-5-iodobenzene

The last four steps of the synthesis are presented in this chapter; procedures for steps two, three, and four can be found in other sections of the book. The reaction by which benzene is converted to nitrobenzene has been omitted from this edition of this book; many chemists believe that the toxicity of benzene is such that it should not be used in the introductory organic laboratory.

E102. 2-Chloro-4-Bromoacetanilide from 4-Bromoacetanilide

Two procedures are given here for the chlorination of *p*-bromoacetanilide. In Procedure A, elemental chlorine is passed into the reaction mixture as a gas, but in Procedure B, chlorine is formed within the reaction mixture itself by a reaction between chloride ion and chlorate ion.

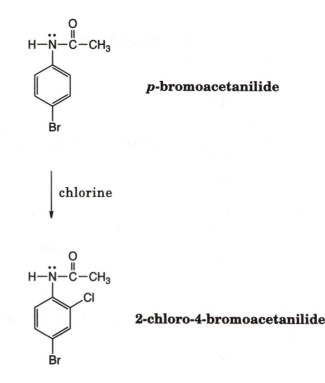

p-bromoacetanilide

2-chloro-4-bromoacetanilide

Scale: 5.35 grams (0.025 mole) of 4-bromoacetanilide.

Procedure A

Suspend 5.35 g (0.025 mole) 4-bromoacetanilide and 2.05 g (0.025 mole) of anhydrous sodium acetate in 20 mL of glacial acetic acid in a 125-mL Erlenmeyer flask. The crude 4-bromoacetanilide can be wet with as much as an equal weight of water without affecting the reaction.

While stirring the solution rapidly, pass in 0.027 mole chlorine gas (1.9 g; equivalent to 1.2 mL liquid chlorine) through a 7-mm tube that leads below the surface of the liquid in the flask (Note 1).

After the addition of chlorine is complete (in about 20 minutes), continue to stir the mixture for 2 or 3 minutes, and then slowly add 80 mL of cold water. During the addition of water, a homogeneous solution will be obtained, and then the product will begin to crystallize.

Treat the resulting suspension with enough concentrated sodium bisulfite solution to remove the yellow–green color, and then collect the crude 2-chloro-4-bromoacetanilide by suction filtration, washing it with water and sucking it as dry as possible; yield: 5.9 g (95%). 2-Chloro-4-bromoacetanilide can be recrystallized from methanol, using 7 mL per gram, with 73% recovery; m.p., 154–156°C (literature value, 151°C).

Note

1. The chlorine is conveniently handled in the following way. A graduated receiver is fitted with a Claisen adapter, and a glass tube is positioned in the center neck of the Claisen adapter by means of a Teflon tubing adapter in such a way that the end of the tube extends to the top of the graduations of the receiver. A drying tube is placed in the outer neck of the Claisen adapter, and the glass tube is connected to a lecture bottle of chlorine gas by a length of Tygon tubing. While the receiver is cooled in a Dry Ice/acetone bath, chlorine gas is passed into it until the required volume of chlorine has been condensed. The end of the Tygon tubing attached to the lecture bottle is then connected to the 7-mm inlet tube of the reaction flask, the drying tube is replaced by a stopper, and the Dry Ice/acetone bath is removed. As the chlorine evaporates, it will pass into the reaction mixture at a convenient rate. If the chlorine starts to boil, the graduated receiver can be cooled momentarily with the cold bath.

Time: 2 hours.

Procedure B (Reference 2)

Suspend 5.35 g (0.025 mole) of 4-bromoacetanilide in a mixture of 12 mL of concentrated hydrochloric acid and 14 mL of glacial acetic acid in a 125-mL flask. Heat the mixture on the steam bath until all the solid has gone into solution. Cool the solution to 0°C in an ice bath. To the cold mixture, gradually add a solution of 1.39 g (0.013 mole) of sodium chlorate dissolved in 3.5 mL of water. During the addition of the sodium chlorate solution, some chlorine gas will be evolved (Note 1). As the addition is made, a yellow precipitate will form and the solution will turn yellow. After the addition is complete, allow the reaction mixture to stand at room temperature for 1 hour, and then collect the solid by suction filtration.

Note

1. You must either work in the hood or make some other provision for removal of the chlorine gas produced (see Section 36).

Time: 2 hours.

Question

1. An attempt was made to prepare 2-chloro-4-bromoacetanilide from acetanilide without isolation of 4-bromoacetanilide, by treating 5 g of acetanilide and two equivalents of sodium acetate dissolved in 40 mL of glacial acetic acid with one equivalent of bromine followed by one equivalent of chlorine. Instead of the anticipated product, 2,4-dibromoacetanilide was isolated in 92% yield. Consideration of this result led us to believe that it might be possible to prepare 2-chloro-4-bromoacetanilide by treating acetanilide, two equivalents of sodium acetate, and one equivalent of sodium bromide in 80% acetic acid with two equivalents of chlorine. The product, however, appears to be a mixture of 2,4-dihalogenated acetanilides. Interpret the observations.

E103. 2-Chloro-4-bromoaniline from 2-Chloro-4-bromoacetanilide

2-Chloro-4-bromoacetanilide is hydrolyzed to the corresponding amine by boiling with HCl in ethanol.

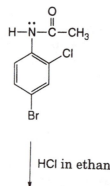

2-chloro-4-bromoacetanilide

HCl in ethanol

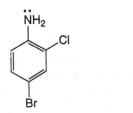

2-chloro-4-bromoaniline

The infrared and proton NMR spectra of 2-chloro-4-bromoaniline are shown in Figures E103-1 and E103-2.

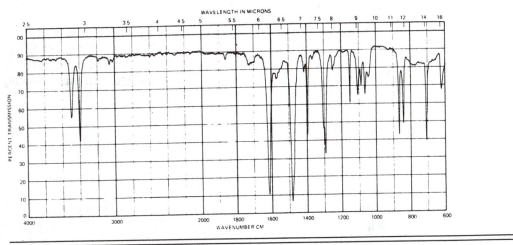

Figure E103-1. Infrared spectrum of 2-chloro-4-bromoaniline; CCl_4 solution.

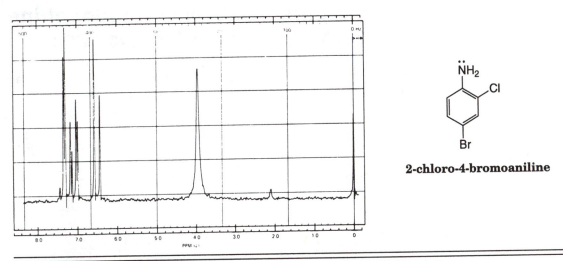

2-chloro-4-bromoaniline

Figure E103-2. NMR spectrum of 2-chloro-4-bromoaniline; $CDCl_3$ solution.

Procedure

Place 6.2 g (0.025 mole) 2-chloro-4-bromoacetanilide in a 100-mL boiling flask, and then add 10 mL of 95% ethanol and 6.5 mL (0.078 moles) of concentrated hydrochloric acid. Fit the flask with a reflux condenser and heat the mixture on the steam bath for 30 minutes. During this time, a clear solution will be obtained, and then a white precipitate will form.

At the end of the heating period, add 40 mL of hot water, and swirl the flask to dissolve the solid. Pour the solution onto about 75 g of ice, and then add to the well-stirred mixture 6 mL (0.115 moles) of 50% sodium hydroxide solution.

Collect the crude product by suction filtration, washing it well with water and sucking it as dry as possible. Yield: 4.7 g (91%). 2-Chloro-4-bromoaniline can be recrystallized from hexane, using 2 mL per gram, with 91% recovery; m.p., 69–71°C (literature value, 70–71°C).

Time: less than 2 hours.

Questions

1. Why is the initial product of the reaction soluble in water?
2. What is the role of the sodium hydroxide?

E104. 2-Chloro-4-bromo-6-iodoaniline from 2-Chloro-4-bromoaniline

In this experiment, 2-chloro-4-bromoaniline is iodinated by iodine monochloride to form 2-chloro-4-bromo-6-iodoaniline.

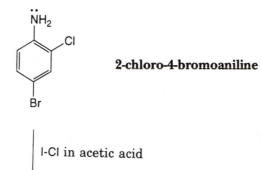

2-chloro-4-bromoaniline

I-Cl in acetic acid

2-chloro-4-bromo-6-iodoaniline

The infrared spectrum of 2-chloro-4-bromo-6-iodoaniline is shown in Figure E104-1.

Scale: 2.5 grams (0.012 mole) of 2-chloro-4-bromoaniline.

Procedure

Dissolve 2.5 g (0.012 mole) 2-chloro-4-bromoaniline in 40 mL of glacial acetic acid, and add to the solution 35 mL of water. (If the crude 2-chloro-4-bromoaniline is wet with water, reduce accordingly the amount of water added to the reaction mixture.)

Then add, over a period of 5 minutes, a solution of 2.5 g (0.8 mL; 0.015 mole) of technical iodine monochloride in 10 mL of glacial acetic acid. Heat the resulting mixture to 90°C on the steam bath, and add enough concentrated aqueous sodium bisulfite solution to lighten the color of the solution to bright yellow; about 10 mL will be needed. Then add an amount of water such that the volume of sodium bisulfite solution plus water added will equal 12.5 mL; about 2.5 mL will be required.

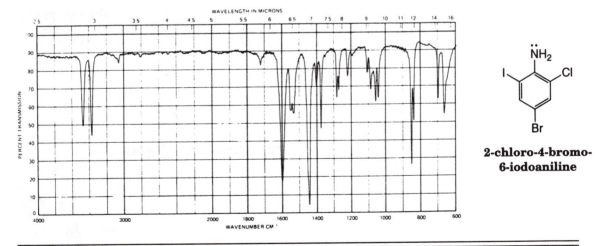

2-chloro-4-bromo-6-iodoaniline

Figure E104-1. Infrared spectrum of 2-chloro-4-bromo-6-iodoaniline; CCl_4 solution.

Allow the solution to cool slowly to room temperature, and finally cool it in an ice bath.

Collect the 2-chloro-4-bromo-6-iodoaniline, which will have separated in a mass of long, almost colorless crystals, by suction filtration. Wash them with a little 33% acetic acid solution and then with water. 2-Chloro-4-bromo-6-iodoaniline can be recrystallized with 80% recovery by dissolving 1 gm 20 mL of glacial acetic acid, slowly adding 5 mL of water to the solution as it is heated on the steam bath, and allowing the resulting solution to cool slowly; m.p., 98–99°C (literature value, 97–97.5°C).

Time: less than 2 hours, plus the time required for the solution to cool.

Question

1. Why does iodine monochloride effect an electrophilic iodination rather than a chlorination?

E105. 1-Bromo-3-Chloro-5-Iodobenzene from 2-Chloro-4-Bromo-6-Iodoaniline

In this procedure, the amine is first diazotized, and the diazonium salt is then reduced by ethanol. The overall result is that the amino group is replaced by a hydrogen atom.

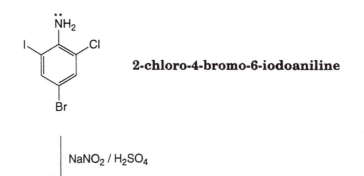

2-chloro-4-bromo-6-iodoaniline

$NaNO_2 / H_2SO_4$

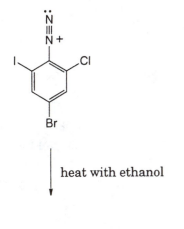

heat with ethanol

1-bromo-3-chloro-5-iodobenzene

The infrared spectrum and the proton NMR spectrum of 1-
bromo-3-chloro-5-iodobenzene are shown in Figures E105-1 and
E105-2.

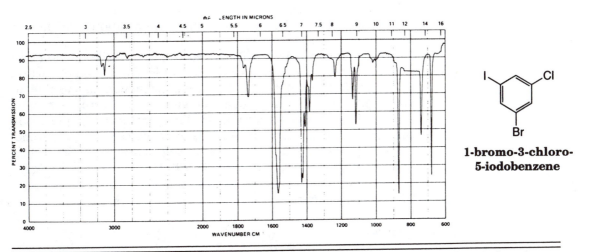

Figure E105-1. Infrared spectrum of 1-bromo-3-chloro-5-iodobenzene; CCl_4 solution.

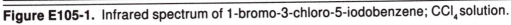

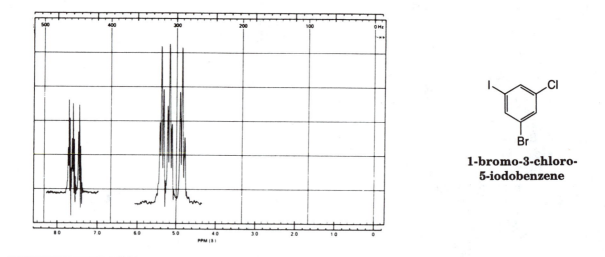

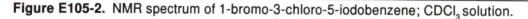

1-bromo-3-chloro-
5-iodobenzene

Figure E105-2. NMR spectrum of 1-bromo-3-chloro-5-iodobenzene; CDCl$_3$ solution.

Procedure

Suspend 2.0 g (0.006 mole) 2-chloro-4-bromo-6-iodoaniline in 10 mL absolute ethanol in a 250-mL boiling flask, and add 1.0 mL of concentrated sulfuric acid dropwise while stirring the mixture. Fit the flask with a reflux condenser. Add 0.70 g (0.01 mole) of powdered sodium nitrite down the condenser, in several portions, to the stirred solution. After the addition of sodium nitrite is complete, heat the mixture on the steam bath for 10 minutes, and then add 50 mL of hot water down the condenser.

Steam distill the mixture, and collect about 100 mL of distillate.

The product, which will have solidified in the condenser and receiver, is most conveniently isolated by dissolution and extraction with ether. Dry the extract with anhydrous magnesium sulfate, filter it, and distill off the ether. A residue of about 1.5 g (80%) crude 1-bromo-3-chloro-6-iodobenzene will remain. Recrystallization of the crude product from 10 mL of methanol will give about 0.9 g (47%) 1-bromo-3-chloro-5-iodobenzene in the form of long, almost colorless needles; m.p., 87.5–89°C (literature value, 85.5-86°C).

Time: about 3 hours.

Disposal

Sink: acetic acid (after dilution); HCl (after dilution); NaOH (after dilution); sulfuric acid (after dilution)

Solid waste: filter paper

Nonhalogenated liquid organic waste: methanol; ethanol; hexane; ether

Solid inorganic waste: sodium bisulfite; sodium nitrite; magnesium sulfate

Halogenated solid organic waste: *p*-bromoacetanilide; 2-chloro-4-bromoacetanilide; 2-chloro-4-bromoaniline; 2-chloro-4-bromo-6-iodoaniline; 1-bromo-3-chloro-5-iodobenzene

Chlorine: allow any excess chlorine to evaporate into water in the hood; add sodium bisulfite solution to reduce the chlorine to chloride. Allow the water to evaporate and discard the residue with solid inorganic waste.

Iodine monochloride: add to water; add sodium bisulfite to reduce to chloride and iodide. Allow the water to evaporate and discard the residue with solid inorganic waste.

Sodium chlorate: put any spilled or extra material in the container reserved for this compound.

Questions

1. How does ethanol reduce the diazonium ion? What becomes of the ethanol?

2. What do you suppose might happen if the mixture of 2,4-dihalogenated acetanilides described in the question in Section E102 were carried on through the rest of the procedures?

References

1. A. Ault and R. Kraig, *J. Chem. Educ.* **43,** 213 (1966).

2. R. M. Roberts, J. C. Gilbert, L. B. Rodewald, and A. S. Wingrove, *An Introduction to Modern Experimental Organic Chemistry,* 2nd edition, Holt, Rinehart and Winston, New York, 1974, p. 350.

MOED: A MEROCYANINE DYE

These final sections describe the preparation of a substance whose color depends on the solvent in which it is dissolved. As the polarity of the solvent is decreased, the wavelength of maximum light absorption becomes longer, and the color changes from yellow through orange to red and finally to violet. The colors are quite intense, and the changes are remarkable.

The first step in the synthesis is the preparation of the quaternary methiodide of 4-methylpyridine from the amine and methyl iodide.

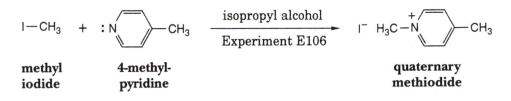

methyl iodide   **4-methyl-pyridine**   **quaternary methiodide**

In the second step, the conjugate base of the methiodide attacks the carbonyl group of *p*-hydroxybenzaldehyde to give an addition product.

conjugate base of the methiodide; present in low concentration

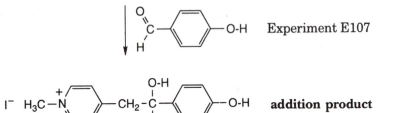

The addition product then undergoes elimination of water to give the bright red 4-(*p*-hydroxystyryl)-1-methylpyridinium iodide.

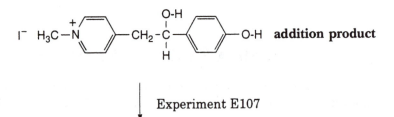

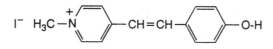

4-(*p*-hydroxystyryl)-1-methylpyridinium iodide; red

The maroon-colored merocyanine dye MOED, the conjugate base of the red substance, is then produced by treatment of the red substance with aqueous potassium hydroxide.

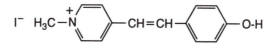

4-(*p*-hydroxystyryl)-1-methylpyridinium iodide; red

KOH   Experiment E108

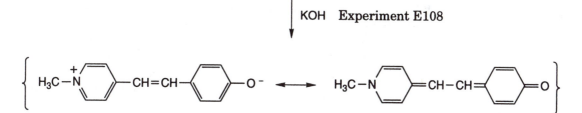

1-methyl-4-[(oxocyclohexadienylidene)ethylidene]-1,4-dihydropyridine; MOED; a maroon solid

E106. 1,4-Dimethylpyridinium Iodide from 4-Methylpyridine and Methyl Iodide

This reaction involves transfer of "+CH₃" from iodide ion to an aromatic amine to form the corresponding quaternary ammonium salt. Even though the temperature of the reaction mixture rises as the exothermic reaction takes place, the product crystallizes from the hot solution.

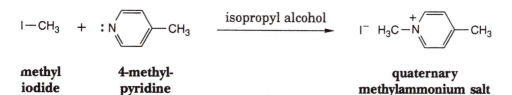

| methyl iodide | 4-methyl-pyridine | | quaternary methylammonium salt |

Scale: 4.9 mL (0.050 mole) of 4-methylpyridine.

Procedure

Place 10 mL of isopropyl alcohol and 4.9 mL (4.7 g; 0.050 mole) of 4-methylpyridine (4-picoline) in a 25-mL Erlenmeyer flask. Fit the flask with a reflux condenser and add to the flask, down the condenser, 3.1 mL (7.1 grams; 0.050 mole) of methyl iodide. Swirl the flask to mix the contents.

Lower a thermometer down the condenser so that the bulb rests on the bottom of the Erlenmeyer flask and indicates the temperature of the liquid.

Heat the flask on the steam bath until the temperature of the liquid has reached 50°C. At this time, remove the flask from the heat and allow the exothermic reaction to take place spontaneously. During the course of the reaction, the temperature will rise to a maximum of about 70–75°C, and the solid product will crystallize from the reaction mixture.

Allow the mixture to cool spontaneously for about 15 minutes after the maximum temperature has been reached, and then collect the product by suction filtration. Portions of isopropyl alcohol can be used to rinse the reaction flask and to wash the product. Yield: about 10 grams (about 85%).

Time: about 1 hour.

E107. Preparation of 4-(*p*-Hydroxystyryl)- 1-Methylpyridinium Iodide from 1,4-Dimethylpyridinium Iodide and *p*-Hydroxybenzaldehyde

In this reaction, the weak base piperidine converts a small fraction of the quaternary ammonium salt to its conjugate base that is then added as a nucleophile to the electrophilic carbon atom of the aldehyde.

conjugate base of the methiodide; present in low concentration

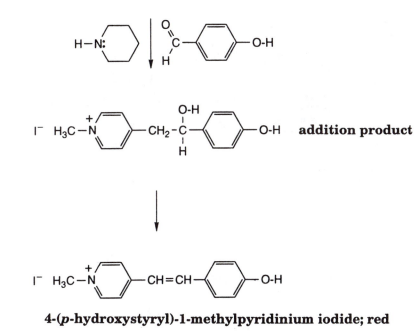

4-(p-hydroxystyryl)-1-methylpyridinium iodide; red

Scale: 2.35 grams (0.010 mole) of 1,4-dimethylpyridinium iodide.

Procedure

Add to a 50-mL Erlenmeyer flask 2.35 g (0.010 mole) of 1,4-dimethyl-pyridinium iodide, 1.22 g (0.010 mole) of *p*-hydroxybenzal-dehyde, 15 mL of 1-propanol, and 1 mL of piperidine. Fit the flask with a reflux condenser, swirl the flask to mix the contents, and then heat the flask on the steam bath for 45 minutes. During the heating time, the reaction mixture will turn red, and the product will separate from solution as red crystals.

At the end of the heating period, cool the flask by immersing it in cold water for a few minutes, and then collect the product by suction filtration. Methyl alcohol can be used to rinse the flask and to wash the crystals. Yield: about 3 grams (about 90%). The product can be recrystallized from methanol, using 75 mL per gram. Recrystallization occurs slowly and therefore the solution should be allowed to stand for several hours before the product is collected by suction filtration.

Time: about 2 hours.

E108. 1-Methyl-4-[(Oxocyclohexadienylidene)-Ethylidene]-1,4-Dihydropyridine (MOED) from 4-(*p*-Hydroxystyryl)-1-Methylpyridinium Iodide

In this procedure, the red 4-(*p*-hydroxystyryl)-1-methylpyridinium iodide is treated with potassium hydroxide, which converts it to MOED.

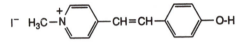

4-(*p*-hydroxystyryl)-1-methylpyridinium iodide; red

KOH

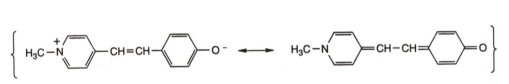

1-methyl-4-[(oxocyclohexadienylidene)ethylidene]-1,4-dihydropyridine; MOED; a maroon solid

Scale: 1.0 grams (0.003 mole) 4-(*p*-hydroxystyryl)-1-methylpyridinium iodide.

Procedure

Add 1.0 g (0.003 mole) of 4-(*p*-hydroxystyryl)-1-methylpyridinium iodide to a solution of 0.25 mL of 45% aqueous potassium hydroxide (0.003 mole KOH) in 50 mL of water in a 125-mL Erlenmeyer flask. Heat the mixture on the steam bath until all the solid has dissolved. Allow the solution to cool to room temperature (Note 1) and then collect the maroon crystals of MOED by suction filtration. Water can be used to rinse the flask and to wash the crystals. Yield: about 0.75 gram (about 75%).

Note

1. The product crystallizes rather slowly, so allow the mixture to stand overnight for the best yield.

Time: about 1 hour.

Demonstration of the Dependence of the Color of MOED on Solvent

Portions of the solution can be used as described below to demonstrate the dependence of the color of MOED on the solvent.

Procedure

Add 20 mg of MOED to 1 mL of water in a small test tube, and heat the test tube in the steam bath until the solid has dissolved. Place two drops of the resulting red solution in a 25-mL Erlenmeyer flask and add 15 mL of water, methanol, isopropanol, or acetone. The resulting color of the solution will be yellow, orange, red, or violet, respectively.

Disposal

> **Sink:** aqueous sodium hydroxide (after dilution); aqueous washes
>
> **Solid waste:** filter paper
>
> **Nonhalogenated liquid organic waste:** isopropyl alcohol; 4-methylpyridine; piperidine; methanol
>
> **Halogenated liquid organic waste:** methyl iodide
>
> **Solid organic waste:** *p*-hydroxybenzaldehyde; intermediates; MOED

Reference

1. M. J. Minch and S. Sadiq Shah, *J. Chem. Educ.* **54,** 709 (1977).

Appendix

All boiling points and melting points are given in °C.

Table A.1. Derivatives of alcohols

Alcohol	b.p.	m.p.	Derivative pNB	DNB	HNP	PU	NU
Methyl alcohol	65		96	108	153	47	124
Ethyl alcohol	78		57	93	158	52	79
2-Propanol	82		111	123	152	76	106
tert-Butyl alcohol	83	26	116	142		136	101
3-Buten-2-ol	96			44			
Allyl alcohol	97		28	50	124	70	108
1-Propanol	97		35	74	146	57	80
2-Butanol	99		26	76	131	65	97
2-Methyl-2-butanol	102		85	116		42	72
2-Methyl-1-propanol	108		69	87	181	86	104
3-Methyl-2-butanol	114			76	127	68	109
3-Pentanol	116		17	101	121	49	95
1-Butanol	118		70	64	147	61	71
3,3-Dimethyl-2-butanol	121			111		66	101
3-Methyl-3-pentanol	123			97		44	84
2-Methyl-2-pentanol	123			72			
2-Methyl-1-butanol	129			70	158	31	82
4-Methyl-2-pentanol	132		26	65		143	88
3-Methyl-1-butanol	132		21	61	166	57	68
2,2-Dimethyl-1-butanol	137				51	66	80
1-Pentanol	138		11	46	136	46	68
2-Hexanol	139		40	39			61
2,4-Dimethyl-3-pentanol	140		155		151	95	99
Cyclopentanol	141			115		133	118
4-Heptanol	156		35	64			80
1-Hexanol	158		5	58	124	42	62
Cyclohexanol	161	25	50	113	160	82	129
1-Heptanol	177		10	47	127	65	62
2-Octanol	179			32		114	63
1-Octanol	195		12	62	128	74	67
Benzyl alcohol	205		85	113	176	77	134
2-Phenylethyl alcohol	220		62	108	123	80	119

pNB = p-Nitrobenzoate ester; Section 30.1
DNB = 3,5-Dinitrobenzoate ester; Section 30.1
HNP = Hydrogen 3-nitrophthalate ester; Section 30.2
PU = Phenylurethane; Section 30.3
NU = α-Naphthylurethan; Section 30.3

Table A.2. Derivatives of aldehydes

Aldehyde	b.p.	m.p.	Derivative			
			Semi	DNP	Ox	Meth
Acetaldehyde	20		162	168	47	139
Propionaldehyde	49		154	150	40	156
Isobutyraldehyde	64		126	187		154
Butyraldehyde	75		104	123		142
Trimethylacetaldehyde	75		190	210	41	
3-Methylbutanal	93		107	123	49	155
2-Methylbutanal	93		105	120		
Pentanal	103		108	107	52	105
2,2-Dimethylpentanal	103			147		
2-Butenal	104		199	190	119	183
Hexanal	131		106	104	51	109
Heptanal	155		109	108	57	135
Octanal	171		101	106	60	90
Benzaldehyde	179		233	237	35	193
Phenylacetaldehyde	194	33	153	121	103	165
Cinnamaldehyde	252		216	255	139	213

Semi = Semicarbazone; Section 30.6
DNP = 2,4-Dinitrophenylhydrazone; Section 30.5
Ox = Oxime; Section 30.7
Meth = Methone derivative; Section 30.4

Table A.3. Derivatives of amides

Amide	m.p.	Components		Derivative
		Acid; m.p.	Amine; m.p.	Xan
Acetanilide	114	Acetic	Aniline	
Phenylacetanilide	117	Phenylacetic; 77	Aniline	
Benzamide	128	Benzoic; 122	Ammonia	124
Phenacetin	134	Acetic	p-Ethoxyaniline	
Phenylacetamide	154	Phenylacetic; 77	Ammonia	196
p-Toluamide	160	p-Toluic; 182	Ammonia	225
Benzanilide	161	Benzoic; 122	Aniline	
p-Bromoacetanilide	167	Acetic	p-Bromoaniline; 66	
p-Chloroacetanilide	179	Acetic	p-Chloroaniline; 72	

Xan = 9-Acylamidoxanthene; Section 30.9

Table A.4. Derivatives of primary and secondary amines

Amine	b.p.	m.p.	Ac	Benz	pTos	PTU	NTU
tert-Butylamine	46		102	134		120	143
n-Propylamine	49			84	52	63	
Diethylamine	56			42	60	34	108
Allylamine	58				64	98	
Isobutylamine	69			57	78	82	
n-Butylamine	77			42		65	109
Isoamylamine	96					102	
n-Amylamine	104				65	69	114
Piperidine	106			48	96	101	
Di-*n*-Propylamine	110					69	
n-Hexylamine	130			40		77	
Morpholine	130			75	147	136	
Cyclohexylamine	134		101	149		148	
n-Heptylamine	155					75	121
Di-*n*-Butylamine	159					86	123
Aniline	184		114	160	103	154	
Benzylamine	185		65	105	116	157	
α-Phenylethylamine	187		57	120			
N-Methylaniline	196		102	63	94	87	
β-Phenylethylamine	198		51	116		135	
2-Methylaniline	200		110	146	186	136	
3-Methylaniline	203		65	125	172	94	
2-Chloroaniline	209		87	99	193	156	
3-Chloroaniline	236		78	120	210	124	
n-Butylaniline	241			56	56		
p-Ethoxyaniline	248	3	137	173	106	136	
N-Benzylaniline	298	37	58	107	149	103	
4-Methylaniline	200	45	147	158	118	141	
4-Bromoaniline	245	66	167	204		148	
4-Chloroaniline	232	72	179	192	119	152	

Ac = N-Substituted acetamide; Section 30.11
Benz = N-Substituted benzamide; Section 30.12
pTos = N-Substituted p-toluenesulfonamide; Section 30.13
PTU = Phenylthiourea; Section 30.14
NTU = Naphthylthiourea; Section 30.14

Table A.5. Derivatives of tertiary amines

Amine	b.p.	m.p.	n_D	den	Pic	MeI	Tos
Triethylamine	89		1.400	0.726	173		
Pyridine	116		1.509	0.978	167	117	139
4-Methylpyridine	143		1.506	0.957	167		
N,N-Dimethylaniline	193	3	1.558	0.956	163	228	161
Tri-n-butylamine	211			0.778	106	180	
N,N-Diethylaniline	218			0.935	142	102	
Quinoline	239		1.627	1.023	203	133	126
Isoquinoline	243	26	1.615	1.097	222	159	163
Tribenzylamine	380	91			190	184	

n_D = Index of refraction; Section 20
den = Density; Section 19
Pic = Picrate; Section 30.15
MeI = Methiodide; Section 30.16
Tos = Methyl p-toluenesulfonate; Section 30.16

Table A.6. Derivatives of carboxylic acids

Carboxylic Acid	b.p.	m.p.	Am	An	pTol	pBr	pNB
			Derivative				
Acetic	118	17	82	114	149	86	78
Propionic	141		81	106	126	63	31
Isobutyric	155		129	105	109	77	
Butyric	164		116	97	75	63	35
3-Methylbutanoic	176		135	110	107	68	
Pentanoic	186		106	63	74	75	
Hexanoic	205		101	95	75	72	
Heptanoic	223		97	71	81	72	
Cyclohexanecarboxylic	233	31	186	146			
Dodecanoic	299	44	100	78	87	76	
3-Phenylpropionic	280	49	105	98	135	104	36
Chloroacetic	189	61	121	137	162	104	
Trichloroacetic	198	58	141	97	113		80
Stearic		70	109	96	102	92	
Phenylacetic	256	76	156	118	136	89	65
Phenoxyacetic	285	100	102	99		149	
2-Methylbenzoic	257	107	143	125	144	57	91
3-Methylbenzoic	263	113	97	126	118	108	87
Benzoic		122	130	160	158	119	89
trans-Cinnamic		133	148	153	168	146	117
2-Chlorobenzoic		142	141	118	131	106	106
Diphenylacetic		148	168	180	173		
3-Chlorobenzoic		158	134	125		116	107
4-Methylbenzoic		180	160	145	165	153	105
4-Methoxybenzoic		186	167	171	186	152	132
3,5-Dinitrobenzoic		205	183	234		159	157
4-Nitrobenzoic		241	201	211	204	137	168
4-Chlorobenzoic		243	179	194		126	130

Am = Amide; Section 30.17
An = Anilide; Section 30.18
pTol = p-Toluidide; Section 30.18
pBr = p-Bromophenacyl ester; Section 30.19
pNB = p-Nitrobenzyl ester; Section 30.20

Table A.7. Derivatives of esters

Ester	b.p.	m.p.	n_D	den	NBz	DNB	Acid
Methyl formate	32		1.346	0.974	60	108	
Ethyl formate	54		1.360	0.922	60	93	
Methyl acetate	57		1.362	0.927	61	108	
Ethyl acetate	77		1.372	0.901	61	93	
Methyl propionate	80		1.378	0.915	47	108	
Isopropyl acetate	88		1.374	0.842	61	123	
Methyl isobutyrate	93		1.384	0.891	88	108	
Ethyl propionate	99		1.385	0.889	47	93	
n-Propyl acetate	102		1.385	0.883	61	74	
Methyl n-butyrate	102		1.388	0.892	38	108	
Ethyl isobutyrate	110		1.390	0.869	88	93	
sec-Butyl acetate	112		1.386	0.872	61	76	
Isobutyl acetate	118		1.390	0.875	61	87	
Ethyl n-butyrate	122		1.400	0.879	38	93	
n-Propyl propionate	123		1.392	0.874	47	74	
n-Butyl acetate	126		1.396	0.881	61	64	
Methyl chloroacetate	132		1.422	1.238		108	61
Isoamyl acetate	142		1.400	0.867	61	61	
Ethyl chloroacetate	145		1.423	1.156		93	61
n-Amyl acetate	149		1.403	0.876	61	46	
n-Hexyl acetate	169		1.411	0.873	61	58	
n-Heptyl acetate	192		1.417	0.871	61	47	
Phenyl acetate	197		1.503	1.078	61	146	
Methyl benzoate	199		1.516	1.089	106	108	122
2-Tolyl acetate	208			1.048	61	135	
3-Tolyl acetate	212	12	1.498	1.049	61	165	
Ethyl benzoate	213		1.506	1.047	106	93	122
4-Tolyl acetate	213		1.500	1.051	61	189	
Methyl 2-toluate	215			1.068		108	108
Methyl 3-toluate	215			1.061	76	108	113
Benzyl acetate	217		1.520	1.055	61	113	
Methyl phenylacetate	220		1.507	1.068	122	108	77
Methyl salicylate	224		1.537	1.184		108	158
Ethyl phenylacetate	228		1.499	1.033	122	93	77
n-Propyl benzoate	231		1.500	1.023	106	74	122
Ethyl 4-toluate	234		1.509	1.027	133	93	180
Methyl phenoxyacetate	245			1.150	86	108	99
n-Butyl benzoate	250		1.497	1.000	106	64	122
Ethyl phenoxyacetate	251			1.104	86	93	99
Ethyl cinnamate	271	7	1.560	1.049	226	93	133
Methyl anthranilate	300	24				108	146
Methyl 4-toluate	223	33			133	108	180
Methyl cinnamate	261	36			226	108	133

n_D = Index of refraction; Section 20
den = Density; Section 19
NBz = N-Benzylamide of acidic component; Section 30.21
DNB = 3,5-Dinitrobenzyl ester of alcohol component; Section 30.22
Acid = Carboxylic acid produced upon hydrolysis; Section 30.23

Table A.8. Derivatives of aromatic ethers

Aromatic Ether	b.p.	m.p.	n_D	den	Derivative	
					Pic	Br
Methoxybenzene	154		1.5221	0.993	81	61
Benzyl methyl ether	171		1.5008	0.965	116	
2-Methoxytoluene	171		1.505	0.985	116	64
Phenyl ethyl ether	172		1.508	0.967	92	
4-Methoxytoluene	176		1.512	0.970	89	
3-Methoxytoluene	177		1.513	0.972	114	55
Benzyl ethyl ether	186		1.496	0.948		
n-Butyl phenyl ether	206		1.505	0.950	112	
1,3-Dimethoxybenzene	217		1.423	1.055	58	
n-Butyl benzyl ether	221		1.483	0.923		
1-Methoxynaphthalene	271		1.694	1.092	130	
Dibenzyl ether	290	4		1.043	78	108
1,2-Dimethoxybenzene	205	23	1.529	1.080	56	93
Diphenyl ether	259	28	1.583	1.073	110	55
1,4-Dimethoxybenzene	213	56			48	142
2-Methoxynaphthalene	273	73			117	84

n_D = Index of refraction; Section 20
den = Density; Section 19
Pic = Picrate; Section 30.15
Br = Bromo derivative; Section 30.30

Table A.9. Derivatives of aliphatic halides

Aliphatic Halide	b.p.	m.p.	n_D	den	Derivative		
					SRP	An	Nap
2-Chloropropane	37		1.378	0.859	196	103	
Ethyl bromide	38		1.425	1.460	188	104	126
Methyl iodide	43		1.532	2.282	224	104	160
1-Chloropropane	47		1.388	0.899	181	92	121
tert-Butyl chloride	51		1.386	0.846		128	147
2-Bromopropane	60		1.425	1.314	196	103	
sec-Butyl chloride	68		1.397	0.804	190	108	129
Isobutyl chloride	69		1.398	0.881	174	109	125
1-Bromopropane	71		1.434	1.353	181	92	121
Ethyl iodide	72		1.514	1.940	188	104	126
tert-Butyl bromide	73			1.211		128	147
n-Butyl chloride	78		1.402	0.886	180	63	112
2-Iodopropane	90		1.499	1.703	196	103	
Isobutyl bromide	91		1.435	1.253	174	109	125
sec-Butyl bromide	91		1.437	1.256	190	108	129
n-Butyl bromide	101		1.440	1.274	180	63	112
1-Iodopropane	102		1.505	1.743	181	92	121
tert-Butyl iodide	103					128	147
1-Chloropentane	106		1.412	0.882	154	96	112
Chlorocyclopentane	115		1.451	1.005			
sec-Butyl iodide	120		1.499	1.592	190	108	129
Isobutyl iodide	120		1.496	1.602	174	109	125
1-Bromopentane	129		1.445	1.219	154	96	112
n-Butyl iodide	130		1.499	1.616	180	63	112
1-Chlorohexane	133		1.420	0.878	157	69	106
Bromocyclopentane	137		1.489	1.387			
Chlorocyclohexane	143		1.462	0.989		146	188
1-Iodopentane	155		1.496	1.512	154	96	112
1-Bromohexane	157		1.448	1.175	157	69	106
Bromocyclohexane	165		1.495	1.336		146	188
Iodocyclopentane	167		1.545	1.710			
Iodocyclohexane	179			1.626		146	188

n_D = Index of refraction; Section 20
den = Density; Section 19
SRP = S-Alkylthiuronium Picrate; Section 30.25
An = Anilide; Section 30.28
Nap = α-Naphthalide; Section 30.28

Table A.10. Derivatives of aromatic halides

Aromatic Halide	b.p.	m.p.	n_D	den	Derivative Acid
Chlorobenzene	132		1.525	1.107	
Bromobenzene	156		1.560	1.494	
2-Chlorotoluene	159		1.524	1.082	141
3-Chlorotoluene	162		1.521	1.082	158
4-Chlorotoluene	162	7	1.521	1.071	240
1,3-Dichlorobenzene	173		1.546	1.288	
1,2-Dichlorobenzene	179		1.522	1.305	
1,4-Dichlorobenzene	173	53			
2-Bromotoluene	182		1.425		150
3-Bromotoluene	184		1.410		155
Iodobenzene	188		1.620	1.831	
2,4-Dichlorotoluene	199		1.249	1.545	164
2,6-Dichlorotoluene	199		1.269	1.551	139
3-Iodotoluene	204			1.698	187
2-Iodotoluene	211			1.698	162
1,3-Dibromobenzene	219		1.606	1.952	
1,2-Dibromobenzene	224	7	1.609	1.956	
4-Bromotoluene	184	28			251
4-Iodotoluene	211	35			270
1-Bromo-4-chlorobenzene	195	67			
1,4-Dibromobenzene	219	89			

n_D = Index of refraction; Section 20
den = Density; Section 19
Acid = Acid formed upon oxidation with permanganate; Section 30.27

Table A.11. Derivatives of aromatic hydrocarbons

Aromatic Hydrocarbon	b.p.	m.p.	n_D	den	Derivative			
					oAr	Acid	Pic	TNF
Toluene	110		1.496	0.867	137	122	88	
Ethylbenzene	136		1.496	0.867	128	122	97	
p-Xylene	138	13	1.496	0.861	148	subl	90	
m-Xylene	139		1.497	0.864	142	348	91	
o-Xylene	144		1.505	0.880	178	208	88	
Isopropylbenzene	152		1.491	0.861	133	122		
1,3,5-Trimethylbenzene	165		1.499	0.865	212	380	97	
tert-Butylbenzene	169		1.493	0.867		122		
p-Isopropyltoluene	177		1.491	0.834	124	subl		
n-Butylbenzene	183		1.490	0.860	97	122		
Bibenzyl	284	53						
Biphenyl	255	71			225			
Naphthalene	218	80			172		149	154
Fluorene	295	114			228		87	179
trans-Stilbene	305	124						
Anthracene	340	216					138	194

n_D = Index of refraction; Section 20
den = Density; Section 19
oAr = o-Aroylbenzoic acid; Section 30.26
Acid = Acid formed upon oxidation with permanganate; Section 30.27
Pic = Picrate; Section 30.15
TNF = 2,4,7-Trinitrofluorenone adduct; Section 30.29

Table A.12. Derivatives of ketones

Ketone	b.p.	m.p.	Derivative Semi	DNP	Ox
Acetone	56		190	128	59
2-Butanone	82		136	117	129
2-Methyl-3-butanone	94		114	120	
3-Pentanone	102		139	156	69
2-Pentanone	102		112	144	58
4-Methyl-2-pentanone	117		135	95	58
3-Methyl-2-pentanone	118		95	71	
2,4-Dimethyl-3-pentanone	124		160	98	
3-Hexanone	125		113	130	
2-Hexanone	128		125	110	49
Cyclopentanone	131		216	146	57
2-Heptanone	151		127	89	
Cyclohexanone	156		167	162	91
2,6-Dimethyl-4-heptanone	168		126	92	210
2-Octanone	173		123	58	
Cyclohexyl methyl ketone	180		177		60
Acetophenone	205	20	199	250	60
Propiophenone	220	20	174	191	54
4-Methylacetophenone	226	28	205	258	88
Butyrophenone	230		191	190	50
4-Chloroacetophenone	232		204	231	95
Valerophenone	249		160	166	52

Semi = Semicarbazone; Section 30.6
DNP = 2,4-Dinitrophenylhydrazone; Section 30.5
Ox = Oxime; Section 30.7

Table A.13. Derivatives of nitriles

Nitrile	b.p.	m.p.	Derivative n_D	den	Acid
Benzonitrile	190		1.529	1.010	122
2-Toluonitrile	205		1.527	0.991	104
3-Toluonitrile	212			1.032	113
Phenylacetonitrile	234		1.521	1.021	77
Phenoxyacetonitrile	240			1.09	99
Cinnamonitrile	256	20			133
4-Toluonitrile	217	29			180
3-Chlorobenzonitrile		41			158
2-Chlorobenzonitrile	232	43			141
4-Chlorobenzonitrile	223	96			240

n_D = Index of refraction; Section 20
den = Density; Section 19
Acid = Acid formed upon hydrolysis; Section 30.8

Table A.14. Derivatives of phenols

Phenol	b.p.	m.p.	NU	Br	ArO
				Derivative	
2-Chlorophenol	176	7	120	76	145
2-Bromophenol	195	5	129	95	
3-Methylphenol	203	12	128	84	103
2-Methylphenol	192	31	142	56	152
2-Methoxyphenol	205	32	118	116	116
3-Bromophenol	236	33	108		108
3-Chlorophenol	214	33	158		110
4-Methylphenol	202	16	146	199	
4-Chlorophenol	217	43	166	90	156
Phenol	183	42	133	95	99
4-Ethylphenol	219	47	128		97
Thymol	234	52	160	55	149
3,4-Dimethylphenol	225	63	142	171	163
4-Bromophenol	238	66	169	171	157
3,5-Dimethylphenol	220	68	109	166	111
2,4,6-Trimethylphenol	220	70		158	142
2,5-Dimethylphenol	212	75	173	178	118
1-Naphthol	280	94	152	105	194
3-Nitrophenol		97	167	91	156
4-*tert*-Butylphenol	237	100	110	67	87
4-Nitrophenol		114	151	142	187
2-Naphthol	286	123	157	84	95

NU = α-Naphthylurethan; Section 30.3
Br = Bromination product; Section 30.30
ArO = Aryloxyacetic acid; Section 30.31

Chemical Substance Index

General Subject Index

655